People in Nature

People in Nature

WILDLIFE CONSERVATION IN
SOUTH AND CENTRAL AMERICA

KIRSTEN M. SILVIUS, RICHARD E. BODMER,
AND JOSÉ M. V. FRAGOSO, EDITORS

COLUMBIA UNIVERSITY PRESS / NEW YORK

COLUMBIA UNIVERSITY PRESS
Publishers Since 1893
New York   Chichester, West Sussex
Copyright © 2004 Columbia University Press

Library of Congress Cataloging-in-Publication Data
People in nature : wildlife conservation in South and Central America / Kirsten M. Silvius,
Richard E. Bodmer, and José M. V. Fragoso, editors.
p. cm.
Includes bibliographical references (p.).
ISBN 0–231–12782–0 (cloth : alk. paper) —ISBN 0–231–12783–9 (paper : alk. paper)
1. Wildlife management—South America.   2. Wildlife management—Central America.   I.
Silvius, Kirsten M.   II. Bodmer, Richard E.   III. Fragoso, José M. V.

SK479.P46 2004
333.954′16′098—dc22

2003069649

Columbia University Press books are printed on permanent
and durable acid-free paper.

Printed in the United States of America

c 10 9 8 7 6 5 4 3 2 1
p 10 9 8 7 6 5 4 3 2 1

To José Márcio Ayres, 1954–2003

That this book is dedicated to Marcio Ayres is powerfully appropriate, in that Marcio touched the lives and the intellects of so many of the authors. Marcio influenced the course of conservation in the Amazon probably more than any other single person in recent memory, and community-based management was at the heart of everything that he did.

Marcio will be forever associated with the creation of the Mamirauá and Amanã Reserves, two huge protected areas in central Amazonia that involve local communities in their management and development. In 1996, when the first was gazetted, Marcio helped introduce a new concept—the "sustainable development reserve." As opposed to a national park, which in Brazil called for the removal of local people from the reserve, the sustainable development reserve actively involved local inhabitants in management. Brazil's President Fernando Henrique Cardoso would later call Mamirauá "a living example of how it is possible to create positive coexistence between the inhabitants of a region and the preservation of that region." This was not empty rhetoric. Marcio had realized early on that in the absence of strong governmental institutions in the Amazon, local people driven by their own self-interest could become the guardians of nature and natural resources. Mamirauá, situated in the flooded forests, contains important wildlife, timber, and especially fish resources. The management plan granted usufruct rights to the local people, allowing them with the help of government agencies to exclude nonresidents from fishing in the reserve. The result was one of those rare "win-win" situations: the average income of local fishermen rose from R$320 in 1999 to R$845 in 2001, based largely on an increase in fish production from management lakes from 6.2 to 15 tons, while at the same time populations of pirarucu (*Arapaima*), the most important fisheries species, tripled in density. And local people have seen a dramatic rise in their educational achievement and health.

Marcio was broadly recognized for his accomplishments. He moved from the national to the international stage (serving for example as the Deputy Chair of the Species Survival Commission). He moved in and out of the Brazilian government. He was the Carter Chair in Rain Forest Ecology with the Wildlife Conservation Society. He received the Conservation Award from the American Society of Primatology in 1987, the World Wildlife Fund Gold Medal in 1992, the Augusto Ruschi Award Medal from the Brazilian Academy of Sciences in 1995, and the Rolex Award in 2002.

But what underlay his accomplishments was a deep trust in the power of scientific knowledge. Marcio was by training a forest ecologist with an interest in primates. It was the white uakari monkey, shy denizen of the flooded forest, that led him as a doctoral student to Mamirauá in the first place. Starting in 1987, he assembled a scientific team that divided the proposed reserve into different management zones, some to protect spawning areas for fish, others to allow commercial harvests, and others for subsistence only. Biological and socioeconomic conditions continue to be monitored, and allow for adaptive changes in the management regimes. The success of Mamirauá and Amanã are a testament to the importance of knowledge in conservation.

Amazonian conservation lost a champion when Marcio passed away. Conservation lost a leader. But the world is a better place because of what he did.

John G. Robinson
July 15, 2003

Contents

Part II. Economic Considerations

Part III. Fragmentation and Other Nonharvest Human Impacts

Part IV. Hunting Impacts — Biological Basis and Rationale for Sustainability

List of Contributors

JONATHAN D. BALLOU Smithsonian National Zoological Park, Smithsonian Institution, Washington, D.C., 20008

RICHARD E. BODMER Durrell Institute of Conservation and Ecology, Eliot College, University of Kent, Canterbury, England CT2 7NZ

JORGE E. BOLAÑOS Departamento de Ecología y Sistemática Terrestres, El Colegio de la Frontera Sur, Ap. 63, San Cristóbal de Las Casas, Chiapas 29290, México.

CLAUDIA CAMPOS-ROZO Fundación Natura, Calle 61 No. 4–26, Bogotá, Colombia

LEANDRO CASTELLO Instituto de Desenvolvimento Sustentável Mamirauá, Cx. Postal 38, 69470-000, Tefé, Amazonas, Brazil (*current address:* SUNY-ESF, 1 Forestry Drive, Syracuse, N.Y. 13210)

SANDRA M. C. CAVALCANTI Wildlife Conservation Society, Pantanal Program-(Centro de Conservação do Pantanal), P.O. Box 132, Miranda, MS 79380-000, Brazil

WILLIAM G. R. CRAMPTON Florida Museum of Natural History, University of Florida, Gainesville, FL, 32611-7800

PETER G. CRAWSHAW, JR. Floresta Nacional de São Francisco de Paula, IBAMA/CENAP, Caixa Postal 79, São Francisco de Paula, RS-954000-000, Brazil

LAURY CULLEN, JR. IPÊ (Instituto de Pesquisas Ecológicas), Parque Estadual do Morro do Diabo, C.P. 91, Teodoro Sampaio, São Paulo, CEP 19280-000 Brazil.

SELENE SIQUEIRA DA CUNHA NOGUEIRA Departamento de Ciências Biológicas, Universidade Estadual de Santa Cruz, UESC, Rodovia Ilhéus-Itabuna, km 16, Ilhéus, Bahia, Brazil 45650-000

JOSÉ MARIA B. DAMASCENO Instituto de Desenvolvimento Sustentável Mamirauá, Cx. Postal 38, 69470-000 Tefé, Amazonas, Brazil

AUGUSTO FACHÍN-TERÁN Universidade do Estado do Amazonas, Escola Normal Superior, Avenida Djalma Batista 3578, Bairro Flores, CEP: 69050-030, Manaus, AM, Brasil

TULA G. FANG Durrell Institute of Conservation and Ecology, University of Kent, Canterbury, Kent, UK CT2 7NS

JOSÉ M. V. FRAGOSO Faculty of Environmental and Forest Biology, SUNY-ESF, 1 Forestry Drive, Syracuse, NY, 13210 (*current address:* Department of Botany, University of Hawaii at Mānoa, Honolulu, HI 96822)

JOSÉ A. GONZÁLEZ Área de Fauna Silvestre y Parques Nacionales, Universidad Nacional Agraria La Molina, Apdo. Postal 456, Lima, Perú (*current address:* Agencia Española de Co-operación Internacional, Proyecto Integral Araucaria-Galápagos, Vancouver E5-32 e Italia, Quito, Ecuador)

MICHELLE M. GUERRA Departamento de Ecología y Sistemática Terrestres, El Colegio de la Frontera Sur, Ap. 63, San Cristóbal de Las Casas, Chiapas 29290, México

ERIC HANSEN Office National de la Chasse et de la Faune Sauvage, Direction Etudes et Recherche, Silvolab, BP 316 97379, Kourou, French Guiana/France

CIBELE INDRUSIAK Instituto Brasileiro do Meio Ambiente e dos Recursos Naturais Ren-ováveis-Ibama/Associação Pró-Carnívoros, Rua Augusto de Lourenço Martins, 70, Porto Ale-gre, RS, Brazil, CEP 91.740.510

MARIA RENATA P. LEITE-PITMAN Center for Tropical Conservation, Duke University, P.O. Box 90381, Durham, N.C. 27708-0381

ROSA M. LEMOS DE SÁ WWF-Brazil, SHLS EQ QL 618, Conjunto E 2° Andar 71620-430, Brasília-DF, Brazil

JAN K. F. MÄHLER, JR. Instituto Brasileiro do Meio Ambiente e dos Recursos Naturais Ren-ováveis-Ibama/Associação Pró-Carnívoros, Rua Augusto de Lourenço Martins, 70, Porto Ale-gre, RS, Brazil, CEP 91.740.510

EDUARDO J. NARANJO Departamento de Ecología y Sistemática Terrestres, El Colegio de la Frontera Sur, Ap. 63, San Cristóbal de Las Casas, Chiapas 29290, México

SÉRGIO LUIZ GAMA NOGUEIRA-FILHO Departamento de Ciências Agrárias e Ambientais, Universidade Estadual de Santa Cruz, UESC, Rodovia Ilhéus-Itabuna, km 16, Ilhéus, Bahia, Brazil 45650-000

ANDREW J. NOSS WCS-Bolivia, Casilla 6272, Santa Cruz, Bolivia

ANDRÉS J. NOVARO Wildlife Conservation Society and Centro de Ecología Aplicada del Neuquén, Junín de los Andes, Neuquén, Argentina

MICHAEL D. PAINTER WCS-Bolivia, Casilla 6272, Santa Cruz, Bolivia

ETERSIT PEZO LOZANO Facultad de Biología, Universidad Nacional de la Amazonia Perua-na, Plaza Serifin Filomeno, Iquitos, Loreto, Peru

LAURENZ PINDER Cerrado Biome Manager, The Nature Conservancy do Brasil, SHIN Conj. J., Bl. B, salas 301-309, Brasilia DF 71-503-505

PABLO E. PUERTAS Wildlife Conservation Society, Calle Malecón, Tarapacá N⁻ 332, Altos, Iquitos, Perú

CÉCILE RICHARD-HANSEN Office National de la Chasse et de la Faune Sauvage, Direction Etudes et Recherche, Silvolab, BP 316 97379, Kourou, French Guiana/France

JOHN G. ROBINSON Wildlife Conservation Society, 2300 Southern Boulevard, Bronx, New York 10460

HEIDI RUBIO-TORGLER Fundación Natura, Calle 61 No. 4-26, Bogotá, Colombia

CATHERINE T. SAHLEY Asociación para la Investigación y Conservación de la Naturaleza CONATURA, Aptdo. 688, Arequipa, Peru

JESÚS SÁNCHEZ VALDIVIA Asociación para la Investigación y Conservación de la Naturaleza CONATURA, Aptdo. 688, Arequipa, Peru

ANDRÉS E. SEIJAS Universidad Nacional Experimental de los Llanos Occidentales "Ezequiel Zamora" (UNELLEZ), Guanare, Portuguesa, Venezuela

KIRSTEN M. SILVIUS Faculty of Environmental and Forest Biology, SUNY-ESF, 1 Forestry Drive, Syracuse, NY 13210 (*current address:* Environmental Center, University of Hawaii at Mānoa, Honolulu, HI 96822)

SALVADOR TELLO Programa de Ecosistemas Acuáticos, Instituto de Investigaciones de la Amazonía Peruana, Av. Abelardo Quiñones, Apartado Postal 784, Iquitos, Peru

JOHN B. THORBJARNARSON Wildlife Conservation Society, 2300 Southern Boulevard, Bronx, New York 10460

JORGE TORRES VARGAS Asociación para la Investigación y Conservación de la Naturaleza CONATURA, Aptdo. 688, Arequipa, Peru

WENDY R. TOWNSEND Museo de Historia Natural Noel Kempff Mercado, Casilla 6266, Santa Cruz, Bolivia, and The Field Museum, Chicago, IL

ASTRID ULLOA Instituto Colombiano de Antropología e Historia-ICANH, Calle 12, No. 2-41, Bogotá, Colombia

CLAUDIO VALLADARES PADUA IPÊ (Instituto de Pesquisas Ecológicas), Parque Estadual do Morro do Diabo, C.P. 91, Teodoro Sampaio, São Paulo, CEP 19280-000, Brazil

JOÃO PAULO VIANA Secretaria de Coordinação da Amazônia, Ministério do Meio Ambiente Esplanada dos Ministérios, Bloco B, 9° Andar, Brasília, DF 70.068–900.

RICHARD C. VOGT Instituto Nacional de Pesquisas da Amazônia (INPA), Aquatic Biology-CPBA, Alameda Cosme Ferreira 1756, Aleixo, C.P. 478, CEP 69.011-970, Manaus, AM. Brazil

People in Nature

1

Introduction — Wildlife Conservation and Management in South and Central America

MULTIPLE PRESSURES AND INNOVATIVE SOLUTIONS

JOSÉ M. V. FRAGOSO, RICHARD E. BODMER,
AND KIRSTEN M. SILVIUS

THE SOUTH AND CENTRAL AMERICAN CONTEXT

South and Central American (including Mexico) approaches to wildlife conservation are rooted in traditions of resource use derived from interactions between complex biological, cultural, and socioeconomic systems. South and Central American peoples inhabit a land rich in biological diversity and complexity, with several nations considered megadiversity countries (e.g., Brazil, Colombia, and Ecuador) (see Mittermeier, Robles-Gil, and Mittermeier 1997). The most extensive tropical forests and wetlands of our planet occur in South and Central America. Unlike the situation in many parts of the world, most of these ecosystems still function as intact ecological entities little disturbed by human activities (Mittermeier et al. 1998). The Amazon rain forest, for example, extends over 2500 km from east to west and about 2000 km from north to south. It is the largest continuous tropical forest on earth and the second largest forested ecosystem after the Eurasian Boreal forest. The world's largest wetland, the Pantanal, is located in south central Brazil and northern Paraguay, and the Andean Mountain range supports some of the most extensive montane forests and grasslands in existence. With the exception of high altitude Andean habitats and Atlantic forests, these "natural areas" are relatively unfragmented and continue functioning as continental level "natural" ecosystems. Many are considered as some of our planet's last great wilderness areas (Dinerstein et al. 1995; Mittermeier et al. 1998). The "intact" condition of South American biomes is unusual, given the high levels of species extirpations and ecosystem fragmentation that have occurred in North America, Europe, Africa, and much of the rest of the world.

The persistence of intact ecosystems in South America, and to a lesser degree in Central America, is to a large extent due to the region's unique mixture of peoples, cultures, and history. Before the arrival of Portuguese and Spanish colonists, over a thousand distinct indigenous nations and cultures inhabited South and Central America (Steward and Faron 1959; Ramos 1998). Although many of these peoples disappeared after the European invasion, many others, including over 200 groups of "first peoples" in the Amazon region, still inhabit their traditional lands (Ricardo 1995; Ramos 1998). From these cultures South and Central America inherited the view of nature characteristic of peoples whose lives depended on understanding and integrating nuances of nonhuman creatures and ecological rhythms. These cultures maintain a world view in which nature is not "red in tooth and claw," but is instead a society where all creatures are considered close relatives. Surviving indigenous peoples like the Embera (Ulloa, Rubio-Torgler, and Campos-Rozo this volume), the Yanomami (Fragoso this volume), Xavante (Silvius this volume), and others continue reminding the larger society that nonhuman nature is an integral part of human lives and spirituality.

These indigenous cultures live alongside the new Americans, descendants of African, Italian, German, Polish, East Indian, and other immigrants who followed or were brought over by the original Portuguese and Spanish colonizers to labor on the land. During 500 years of human intermingling, members of these groups fused and created a dynamic and vital "Latin American" ethnicity, each country exhibiting a unique strain that, despite linkages with European and Christian world views, is also deeply rooted in the local environmental conditions and landscapes (Pratt 1992). Thus the Llaneros of Venezuela are intimately tied to the llanos, for example, as are the Pantaneros of Brazil to the Pantanal. The Amazonian rural groups, variously classified as *caboclos*, detribalized indigenous peoples, or Amazonian peasants, have evolved their own distinctive, subsistence-influenced societies (Nugent 1993).

A new conservation philosophy or attitude has developed along with the new people. This philosophy differs significantly from Northern perspectives in that it is more resistant to converting nature into human-dominated landscapes and to completely replacing wildlife with domesticated animals. Just as North America, with its own blend of peoples and world views (which early on excluded and resisted most of the potential contributions of indigenous and African cultures) developed its own unique philosophy of conservation, so did the Latin American regions, with their blending of American Indian, African, European, and, to a lesser extent, Asian world views. This Latin American philosophy of conservation was first characterized during a special panel discussion at the 1997 International Conferences on Wildlife Management and Conservation in Latin America and the Amazon, held in Iquitos, Peru.

Cultural diversity goes hand in hand with diversity of socioeconomic systems, in the broad sense of the word. In the southern continent "highly advanced" (consider the international nature of the stock market in São Paulo, Brazil) and "highly traditional" (consider the kinship-based economic systems of the Yanomami and other indigenous peoples of South America) socio-ecological-economic systems coexist and cofunction. Between these systems lie others that incorporate different amounts of the "advanced" or "traditional" patterns. For example, the socio-ecological-economic systems of rubber-tappers and *ribereños* (river peoples) are similar to those of the Yanomami, while those of ranchers, farmers, and city slum dwellers are probably more similar to those of the inhabitants of São Paulo. To the outsider the coexistence of such divergent systems may seem discordant. Most South Americans, however, know and value the way in which all these systems continue functioning in their countries. It is in the context of this rich inter- and intraethnical milieu that wildlife and conservation biologists strive to influence local, national, and international policies regarding the use and abuse of "wild" species and "wild" spaces.

Although researchers trained in North American and European management strategies are clearly influencing emerging policies of the South, the ecological, cultural, and economic setting of South and Central America make it both inevitable and imperative that effective wildlife conservation strategies will differ greatly from those that evolved in North America or Europe. The International Conferences on Wildlife Management and Conservation in Latin America and the Amazon (henceforth the Conferences), held biannually since 1992, have been a nucleus for the development and presentation of innovative management solutions applied by national academics, students, practioners, businesspeople, indigenous Americans, and other local peoples.

THE CONFERENCES

J. G. Robinson and K. H. Redford's 1991 "Neotropical Wildlife Use and Conservation" helped define the field of South and Central American wildlife management by describing issues of subsistence hunting, market hunting, and captive breeding. The five Conferences held since then have essentially charted the development of the field. The first conference was held in Belém, Brazil, in 1992; the second in Iquitos, Peru, in 1995; the third in Santa Cruz, Bolivia, in 1997; the fourth in Asunción, Paraguay, in 1999; and the fifth in Cartagena, Colombia, in 2001. The sixth conference will be held again in Iquitos in 2004. The meetings were hosted by local nongovernmental organizations (NGOs) and academic institutions (Museu Paraense Emílio Goeldi, Universidad Nacional de la Amazonia Peruana, Museo de Historia Natural Noel Kempff Mercado, Fundación Moises Bertoni, CITES-Paraguay, Fundación Natura, Ministerio del Medio Ambiente-Colombia, and Instituto Amazónico de Investigaciones Cientificas-Sinchi). They were funded and supported by a diversity of national and international organizations (MacArthur Foundation, Wildlife Conservation Society, Sociedade Civil Mamirauá, Liz Claiborne Art Ortenberg Foundation, World Wildlife Fund, Instituto de Pesquisas Ecológicas, CNPq-Brasil, Tropical Conservation and Development Program-University of Florida, U.S. Fish and Wildlife Service, Universidad Nacional de Colombia, Instituto de Ecología de la Universidad Mayor de San Andrés, and UNDP/

GEF). Proceedings have been published in Latin America for all the conferences in Spanish and Portuguese (Fang et al. 1997; Valladares-Padua and Bodmer 1997; Fang, Montenegro, and Bodmer 1999; Cabrera, Mercolli, and Resquin 2000; Polanco-Ochoa 2003).

International researchers were key voices in the early meetings, as were national professionals in ecology and anthropology and representatives of indigenous peoples (e.g., Xavante leaders participated in the first meeting in Belém, Cocama-Cocamilla representatives in the second meeting in Iquitos, Siriono and Izoceño communities in the third meeting in Santa Cruz, and Aché representatives in the fourth meeting in Asunción). Although international researchers are key participants at the meetings, the majority of those in attendance have always been South and Central American professionals, academics, indigenous peoples, and graduate and undergraduate students. Over the last ten years, all of these people have been strongly influenced by the experiences of the meetings. Indigenous and other local peoples attended the meetings both to learn what Western science has to offer them about wildlife management, and to present their own perspectives. In many cases, indigenous representatives are conducting their own projects. This level of inclusiveness at a professional meeting contrasts greatly with similar meetings held in North America but mirrors the blended nature of South and Central American societies.

Our purpose with this book is to highlight South and Central American approaches to wildlife management and to make the information available to the English-speaking public. By collating a selection of Conference presentations, we are documenting both the current state and the historical development of a Latin American conservation and management strategy by people whose perspectives acknowledge the realities of South and Central America, both from biological and socioeconomic points of view. Through our selection of papers we ask, and answer: How can a South and Central America perspective of sustainability and wildlife conservation be incorporated into research and action? What are the questions people are asking in the "south," and what are the solutions being pursued?

As editors we have chosen to emphasize a broad range of topics not completely covered in texts that focus on either hunting, protected areas, or resource use by local peoples. The papers presented here do not analyze the social and cultural factors that result from a subsistence-based economy, rather they link wildlife ecology with the livelihoods of rural people. Most of the researchers featured in our book are either South or Central American or people who have lived much of their lives in the region. Many of these researchers received their academic training in wildlife ecology at universities in the United States, Canada, or Europe. Their approaches therefore reflect the tension between temperate models and tropical realities that currently characterize the field of South and Central American wildlife management and conservation. This tension is another factor contributing to the unique cultural/philosophical perspective of the region.

CONTINENTAL-SCALE DIFFERENCES

In the tropical forests of Asia and Africa, there is much concern about the "bush meat crisis." In these regions wildlife hunting for meat is driving many species to the verge of extinction (Martin 1983; Srikosamatara, Siripholdej, and Suteethorn 1992; Robinson and Bennett 2000.) In the tropics of South America, as in Africa and Asia, the pressure on animal communities also comes primarily from subsistence hunting. The commercial uses and sport hunting that are important in Africa (Hasler 1996; Hurt and Ravn 2000) and the commercial use of animals for the medicinal trade that are important in Asia (Martin and Martin 1991; Srikosamatara, Siripholdej, and Suteethorn 1992) are less important in South and Central America (Robinson and Redford 1991). These differences are largely due to a consistent and dedicated group of people who have promoted wildlife management and conservation throughout the Neotropics during the past three decades. This group of people, all participants in the Conferences, has helped avert a crisis. Thus, even though subsistence hunting is a key impact on wildlife in all tropical regions, the main difference between the continents is in the implementation of management, which has a much longer history in South America and has in the last decade been stimulated and coordinated by the Conferences.

Managing subsistence hunting and fishing remains a key issue for wildlife conservation in South and Central America. Subsistence peoples in the Neotropics usually live in rural communities in isolated areas. Extraction of animals for subsistance uses is often much greater than for commercial uses (Tello this volume; Crampton et al. this volume; Bodmer, Pezo, and Fang this volume). Community-based approaches to wildlife management have therefore been a focus for wildlife conservation in Latin America. In this volume we see how community-based approaches are vital to wildlife conservation. In South and Central America local peoples demonstrate a sincere willingness to manage their own wildlife resources, despite an "economic underdevelopment" and a lack of basic necessities. In the "south" it is not only the scientific Western world view that matters—the traditions of indigenous groups and rural communities hold equal sway with the precepts of science. This occurs not only because indigenous and rural people control large areas of undeveloped lands in South America, but because the society at large has incorporated aspects of the other world views into the mainstream.

In this volume, several authors explore the benefits and complications that arise from developing wildlife management plans that explicitly incorporate distinct world views. Ulloa, Rubio-Torgler, and Campos-Rozo explore the complex social and cultural processes required to develop fully participatory management alternatives for the overlap zone between a national park and an indigenous reserve in Colombia. Silvius explores the congruencies and divergences between the traditional management techniques of the Xavante people in Central Brazil and the management approaches of biologists trained in the Western tradition. Townsend, one of the most successful promoters of the participatory method of management with indigenous peoples in South America, encapsulates in a pithy, and characteristically to-the-point manner the true definition of participatory management. The willingness of several countries to establish and find ways to manage such overlap areas is a key theme in Latin American conservation and perhaps one of the key lessons to emerge from the "south." Crampton and colleagues contribute two articles that trace the development of community management by local, nontribal riberinho peoples in the Mamirauá Sustainable Development Reserve, from the historical overexploitation of turtles, manatees, and fish to the current system of lake rotations and internally set quotas.

Unlike the situation in developed countries, governments in the South often lack financial resources to adequately implement wildlife conservation and management, and rural areas of South and Central America are left to find their own solutions. There are not enough trained biologists to collect the required data to develop biologically sound management plans. Management plans, however, are often required for communities to retain legal rights to the resources on which they depend. Therefore communities take the initiative to develop the management plans and, with the often intermittent help of biologists and NGOs, set out themselves to collect the data they need to set realistic harvesting levels for wildlife and other resources.

Management plans are often based on analysis of sustainability. One of the first questions that a community will ask is "How many animals can we hunt?" Many studies conducted with local communities in Latin America are looking for ways to evaluate the sustainability of hunting. In this volume several papers deal directly with this question. Bodmer and Robinson review simple population models that are used by many projects throughout Latin America to evaluate sustainability of hunting. Naranjo and colleagues apply these models in Chiapas, Mexico, to evaluate the sustainability of hunting for rural and indigenous hunters. Novaro explores in more detail potential applications and theoretical predictions of the source-sink model for managing hunting in both disturbed and undisturbed areas. Puertas and Bodmer show how catch per unit effort can be used to link community participation in wildlife management plans with an analysis of hunting sustainability. González examines how subsistence and commercial uses affect the viability of bird populations in Amazonian flooded forests. Fachín-Terán, Vogt, and Thorbjarnarson look at the sustainability of the Amazonian turtle fishery, while Tello examines the sustainability of subsistence and commercial fishing in Peru's Pacaya Samiria National Reserve.

Economics is an important part of wildlife use and conservation in the Neotropics. Rural economies depend on wildlife products, many of which are sold in urban centers. Viana et al. follow up on the overview essays by Crampton et al. to describe in detail the economic importance of one Amazonian fishery and the economic balance sheet of local involvement in fisheries management. Sahley, Vargas, and Valdivia describe the clash that occurs when a traditional use system, vicuña-shearing for commercial wool production in Peru, is altered by political and other demands, resulting in an ongoing conflict between profit, culture, and ecology. Contrasting with the vicuña experience, and with the rejection of captive breeding by the Embera documented by Ulloa and colleagues, Nogueira-Filho and Nogueira summarize the intensive research that has made captive breeding of two native species, the collared peccary and capybara, economically viable and culturally acceptable in Brazil. Bodmer, Pezo, and Fang look at the relative importance of wildlife products in rural and urban areas and show the relative insignificance of the urban and international market with respect to the local rural market.

But community-based approaches are not the only focus of wildlife management and conservation in Latin America. Fragmentation and other forms of human encroachment are major concerns in many regions. In this volume Cullen et al. describe the synergy between hunting and fragmentation in the Atlantic Forest of São Paulo state, Brazil, and propose innovative ways in which land users, many of them illegal land invaders, can contribute to the reconstruction of an area whose environmental deterioration started long before they arrived. Working further south in the Atlantic Forest, Crawshaw et al. document the unexpected but potentially ephemeral survival of a jaguar population and highlight the importance of connecting existing large forest fragments that will allow metapopulation-level connectivity of large predators in island parks. Seijas uses GIS techniques to document the spatial patterns of human pressures on the Orinoco crocodile in one river basin in Venezuela, finding unexpected relationships between the presence of humans and crocodiles. Lemos records the wavelike pattern of change sweeping through a primate community following the flooding of a 500-km² area in the southwestern Brazilian Amazon. Pinder explores niche partitioning and coexistence for native ungulates and introduced cattle in the Brazilian Pantanal, while Fragoso discusses how the western penetration and continuing colonization of remote areas of the Amazon may be having severe impacts on ungulate populations through the introduction of exotic diseases.

The high levels of biodiversity and complex ecological communities that characterize many South and Central American ecosystems demand their own detailed ecological studies and management approaches. The single-species models that are suited to altered ecosystems in temperate zones are not feasible in South America if a management goal is to protect biodiversity and maintain ecosystem function. At the same time the large extent and availability of intact habitats make possible management based on the concepts of metapopulations and source-sink models, and several authors in this book discuss the implications of these models for wildlife conservation in South and Central America.

Local, nonmarket economies, as well as local, national, and global economies, are all involved with wildlife use and must be considered in conservation and management. Pressures on individual species occur at a multiplicity of socioeconomic scales, and therefore management recommendations that consider all these scales must be implemented. It is just as important to understand the decision-making process of an Amazonian fisherman who, faced with a nesting turtle on a river beach, chooses either to kill it or to let it complete the reproductive cycle, as it is to understand the pressures on national governments that grant concessions to international corporations or become signatories to international conservation treaties. Noss and Painter describe this multiscale approach to conservation on several million hectares in the Bolivian Chaco, the result of an ambitious collaboration between the Izoceño-Guaraní people and the Wildlife Conservation Society.

International conservation pressure is influential in many regions, especially in the Amazon, claiming equal voice with local management goals in wildlife conservation and management. Where the United States and Canada achieved most of their development free of the constraints of international supervision and influence, South and Central American countries must often make decisions, both for and against protecting the environment, that are not influenced solely by their internal practices and traditions. (e.g., debt for nature swaps, U.S. aid agency projects, Global Environmental Facility of the United Nations Development Program projects, World Bank projects, Inter American Development Bank projects, International NGO projects, and World Conservation Strategy projects such as Biosphere Reserves). This influence is clearly seen in the case of French Guiana, one of the last nonindependent states in South America. Richard-Hansen and Hansen describe the intriguing process through which an overseas French national agency is relying on the outcomes of the Conferences to institute a territorial system of wildlife management in a place that almost completely lacks preexisting, locally adapted management strategies.

Finally, unlike the situation in North America and Europe, wildlife managers in the "south" play a role not only in natural resource management but also in the political, social, and economic development of their countries. Biologists and managers with Bachelor's, Master's, or Ph.D. degrees are among the educated elite in these countries. In the Brazilian state of Acre, for example, on the border with Peru and Bolivia, the government bills itself the "government of the forest." Several government functionaries, including the governor, are foresters or biologists. The concept of sustainability is thus permeating society from several sources, including the ideals of trained environmental scientists as well as the needs of indigenous and other rural peoples.

PART I

———

Local Peoples and Community Management

2

Conceptual Basis for the Selection of Wildlife Management Strategies by the Embera People in Utría National Park, Chocó, Colombia

ASTRID ULLOA, HEIDI RUBIO-TORGLER,
AND CLAUDIA CAMPOS-ROZO

The depletion of natural resources and the consequent deterioration in quality of life for humans have in recent decades generated the urgent need to rethink the relationship between human groups and nature. Conservation strategies and actions directed toward natural resource management—especially game animals—are now subjects of interest to governments, to nongovernmental organizations (NGOs), to biologists and anthropologists, and of course, to local peoples.

One frequently used tool for conservation is the creation of protected areas. These areas, however, can be a source of conflict in cases when they are superimposed on the lands of local peoples because they bring a normative structure that regulates local peoples' use of their principal economic base—natural resources such as game animals. Different game management strategies have been attempted, many of them developed by NGOs and academic scientists working jointly with local peoples. Governments have also initiated efforts to support resource management strategies, especially those forms of management that include a wide range of options and rely on methodologies that stimulate participation.

Despite these efforts, it remains a priority to understand the relationship of local people to their land and to conservation areas in order to generate long-term management strategies that are guided by an interdisciplinary and intercultural vision that in turn facilitates their implementation with local communities. Wildlife management strategies aimed at sites where protected and human-use areas overlap will only be effective if they harmonize the use and management of resources with the local inhabitants and if they include plans to recover local resources and encourage the sustainable use of species that are of cultural and ecological interest.

In Colombia there are eighteen protected areas that overlap with indigenous territories, in particular with the legal figure of indigenous reserves, or *resguardo* (in Colombia, resguardos are lands to which indigenous communities hold legal collective title). Just as in other countries, many of these conservation areas were created without taking into account the social and cultural characteristics of the people and without seeking their participation. Furthermore, in many cases local peoples have been marginalized by the areas' management system. However, the Colombian Ministry of Environment is investing considerable effort to change the situation.

Article 7 of Law 622 recognizes as a legal land category the overlap zones between national parks and indigenous reserves in Colombia, implying a mandate for joint, participatory management of these areas. This study describes a long-term effort to develop a management strategy for one such overlap zone. The project relied on the historical and cultural relationships of Embera indigenous people with their natural resources to create a strategy that is consonant with recent state level conservation goals and culturally as well as ecologically viable. Following a brief summary of the overall scope of the project, this paper focuses on the process of selecting and reaching an agreement on wildlife management strategies in the national park-indigenous reserve overlap zone. Detailed explanations of the various phases of the project and the participatory methodologies used have been published in Campos, Ulloa, and Rubio 1996; Ulloa et al. 1996; Rubio-Torgler et al. 1998; Rubio-Torgler, Ulloa, and Campos-Rozo 2000; and Campos-Rozo, Rubio-Torgler, and Ulloa 2001.

ISSUES IN RESOURCE USE AND CONSERVATION BY THE EMBERA

There are eight conservation areas in the Chocó Biogeographic Province: seven National Natural Parks and one Flora and Fauna Sanctuary. The 53,200-ha Utría National Natural Park (UNNP) was created in 1987. Eighty-five percent of its land surface area overlaps with three Embera Indigenous Reserves, and this overlap zone supports four communities with a joint population of 600 people (fig. 2.1).

This project arose out of certain conditions existing in the overlap zone: (a) the interaction between two conceptualizations of wildlife management, that of the Embera and that of the national society, which are based on different logics and ways of thinking; (b) the implications of state and local politics and projects related to land and resource management; (c) the process of interaction with other societies in which the Embera communities are immersed, and (d) a reduction in game populations in general and of large primates in particular (howler monkeys, *Alouatta palliata*, and spider monkeys, *Ateles fuscipes*); the extinction of the tapir (*Tapirus bairdii*); and the near-disappearance of the white-lipped peccary (*Tayassu pecari*). The latter two species are of great symbolic, dietary, and ecological importance for the Embera.

These conditions stimulated the search for commonalities to be used in the joint management of the overlap area, generating an intercultural, consensus-building process that combined Embera and Western approaches to wildlife management,

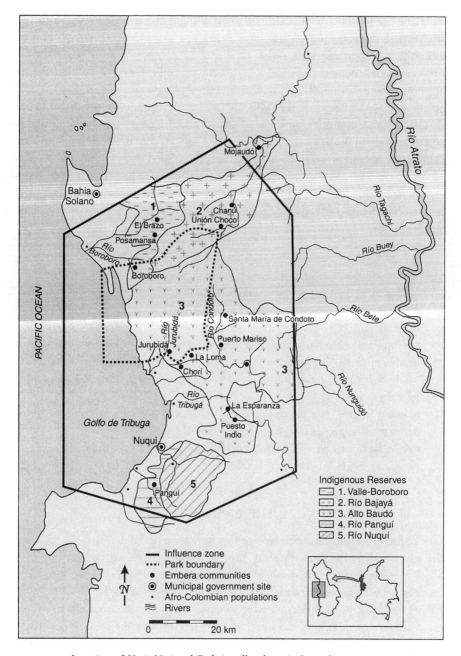

FIGURE 2.1 Location of Utría National Park (small polygon), the indigenous reserves (resguardos), numbered 1 through 5, and the Embera communities falling within the park's influence zone (large polygon).

and their visions of the relationship between humans and nature, in order to arrive at a site-specific intercultural and interdisciplinary management strategy. This effort arose as a social decision rather than as a state imposition and relied on the participation of members of sixteen communities, indigenous researchers, the indigenous organization OREWA (Embera Wounan Regional Organization), a state institution (UAESPNN, in the Ministry of the Environment), an NGO (Fundación Natura-Colombia), and several organizations that provided technical support (Colombian Institute for Anthropology and History-ICANH) and funding (The Wildlife Conservation Society, Conservation Food and Health, and the Organization of Ibero-American States). This was one of the first formal attempts in Colombia to arrive at joint wildlife management between an indigenous organization, an NGO, and a governmental institution for a park-reserve overlap area. The objectives were

1. to improve the relationship between the state and local peoples,
2. to increase the level of participation of local peoples and the OREWA in the management of the area, taking into account their social processes and cultural practices,
3. to counteract the scarcity of game caused by the impact of recent hunting practices and by the hunting of species with small, at-risk populations or of species vulnerable to anthropic processes, given that a severe impact on wildlife in the area would both reduce the quality of life of the indigenous population and alter ecological processes,
4. to rescue or give voice to the Embera's interests and proposals for wildlife management by linking them to those of the national society,
5. to have an impact on wildlife management policies in national parks and indigenous reserves in Colombia in general.

THE EMBERA

The Embera live in rain forest territories where they maintain symbolic, productive and social exchanges with other Embera communities, with other worlds for which they conceive the existence of different beings, and with other cultures. The Embera have traditionally settled in river headwaters in accordance with family linkages and today are concentrated in villages. Currently 70,000 Embera people are distributed primarily in the upper and middle headwaters of the many rivers that drain to the Colombian Pacific or along the Atrato River in the Chocó, Cauca, and Valle Departments. Some Embera also live in Córdoba and in the mountainous and foothill regions of the Cordillera Occidental in Antioquia, Caldas, Risaralda, and Caquetá. Embera economy is currently based on hunting, fishing, gathering, diversified agricultural production, and husbandry of small domestic species. The Embera also market a small agricultural surplus.

For the Embera the universe is structured into three worlds, inhabited by different beings with which humans interact by means of concepts, representations, and practices (fig. 2.2). In the upper world live the creator (*dachizeze*), the spirits of the dead, and the primordial beings. In the lower world live some *jai* (vital principles of all beings), *wuandras* (the mothers of the species), and other entities. The middle world is inhabited by humans, animals, plants, and diverse entities with human and/or animal appearance. Natural resources and their use are underpinned by the concept of wuandra or mothers of the plants and animals. The most important mother is that of the white-lipped peccary, because it determines the abundance or scarcity of species and allows humans to maintain access to and exchange relationships with nature by means of individual practices and the practices of the *jaibaná* or shaman. The jaibaná is a man or woman who after a long learning process ac-

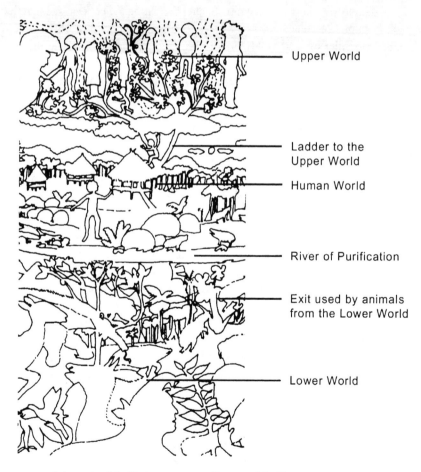

Upper World

Ladder to the
Upper World

Human World

River of Purification

Exit used by animals
from the Lower World

Lower World

FIGURE 2.2 Schematic of the Embera conceptualization of the Universe.

quires knowledge vital to Embera culture and mediates the interactions between humans and nature.

The Embera's complex body of knowledge of the environment arises from their long historic-cultural process and their relationship with the territory. This knowledge is expressed in their strategies for the management of nature and of resources and in their symbolic, productive, and social activities. Embera wildlife management integrates the human and the nonhuman in a process of reciprocal relationships. Relationships with animals are established by means of (a) the jaibaná, who regulates hunting by designating territories and species as sacred and/or forbidden; (b) selective hunting; (c) interactions among spaces assigned to different uses (shifting agriculture, semi-nomadic, or rotational hunting); (d) diversified production (relationships between hunting, fishing, agriculture, and gathering); and (e) cycles of production associated with seasonal species that provide varied sources of animal and plant proteins.

Game scarcity (see next section) has affected the Embera diet because meat that previously came from now-scarce species, especially white-lipped peccaries, is now provided by smaller species, which formerly were not preferred. Similarly, the symbolic importance of the white-lipped peccary—on which a large part of Embera culture is based—means that game scarcity affects Embera culture as well as their diet. Today, the Embera have several explanations for game scarcity, including the activities of the jaibaná, human population increase, increased demand for game meat, the introduction of firearms, the more frequent use of nonselective hunting with dogs, and forest fragmentation.

BRIEF PROJECT DESCRIPTION

The first phase of the project, which extended from 1990 to 1992, had as its two main objectives to determine the way in which the Embera use wildlife, including hunting practices and use of space, and to promote the implementation of a management agreement for the overlap zone between the communities, the OREWA, and the state government. From June 1990 to November 1991, participatory methods were used to gather data among the four communities in the overlap zone (Rubio-Torgler 1992). Two of the communities are relatively isolated from contact with the national society, while the other two communities are close to Afro-Colombian populations on the coast. In consequence, the latter two communities interact more with the national society than the former, and their lands are impacted by nonindigenous peoples.

In each community one trained participant recorded data on every hunted animal brought into the village, writing down species, weight, sex, location of kill, and method used to hunt the prey. Hunting locations were recorded using participatory mapping methods. Throughout the study period, during their visits to the communities, researchers used observational methods to record Embera perceptions of hunting and of animal ecology (Rubio-Torgler 1992).

Over an 18-month period, the four communities hunted a total of 1,079 animals of 5 reptilian, 6 avian, and 18 mammalian species, representing a biomass of 9,015 kg. Seventy-eight percent of individuals hunted belonged to 5 species: *Dasyprocta punctata* (269 individuals), *Agouti paca* (230), *Dasypus novemcintus* (135), *Tayassu tajacu* (112), and *Mazama americana* (95). Seventy-five percent of the biomass came from the same species, with *Tayassu tajacu* contributing the greatest amount and *Dasypus novemcinctus* the smallest. No tapirs or white-lipped peccaries were killed, and community members reported that these species had not been seen for a long time. Large primates also were rarely encountered and killed by hunters. The isolated communities hunted more animals overall and more individuals of the larger species than the communities near the coast. This difference is probably due to a combination of game depletion, spurred by economic booms and trade, and cultural change, which alters activity budgets in the communities near the coast (Rubio-Torgler 1992).

As a result of game depletion, the Embera in these communities hunt smaller animals than they did 15 years ago. They want to protect the populations of large species to increase their abundance, but have no interest in protecting "pest" species such as jaguars. Their culture does not contain the concept of biological extinction, and they believe that the jaibaná hides away the animals either as punishment for overhunting by humans or out of malice. However, they are willing to use both traditional and Western scientific methods to lead to the recovery of game animal populations.

The second phase of the project (1994 to 1996) built consensus on alternative strategies for wildlife management among members of sixteen Embera communities located in the influence zone of Utría National Park and representatives of the OREWA, all of whom were part of the project's core team and participated in research, coordination, evaluation, and budget management. During this stage project participants concentrated on exploring with the indigenous communities the strategies used by the Embera and the national society (both government and civil society) to achieve sustainable wildlife management, identifying social, cultural, and biological aspects that needed to be considered in order to assess the feasibility of each strategy. Project participants also continued the effort to understand Embera conceptualizations and practices related to wildlife, and inquired into systems of perception and representation and the social processes involved in decision making about wildlife. Finally, project participants carried out a feasibility analysis of the different wildlife management strategies that were proposed for the five communities that would be directly or indirectly involved with the implemented strategies and that expressed interest in participating in the management plan. These communities were Alto Bojayá, Alto Baudó, and Boroboro-Valle (direct involvement), as well as Nuquí and Paguí (indirect involvement).

In the third phase of the project, from 1997 through the present, the Embera people have been implementing some of the selected wildlife management strategies within their territories. Unfortunately, because of political problems among the OREWA, Fundación Natura, and the Ministry of Environment, as well as social and political problems at both regional and national scales, these actions have not yet been articulated at a regional level.

CONCEPTUAL FOCUS OF THE PROJECT

This project was conceived as a new approach to wildlife management in protected areas inhabited by local peoples and was based on consensus building and the participation of all stakeholders. Its conceptual focus is based on seven key premises:

1. Long-term conservation is feasible if it is taken up as a social decision in which local stakeholders put forth their own solutions, rather than having them imposed by the state.
2. Natural resource management strategies must be planned and implemented in a joint manner, taking into account the wildlife management solutions of the local people, of Western science, and of state policies.
3. The interaction between western scientific and local knowledge should be pursued and explored.
4. Management strategies cannot be exported wholesale from one region to another.
5. The construction of wildlife management strategies must be carried out with an interdisciplinary and intercultural vision.
6. Different management options must be integrated, with special emphasis on cultural, conservationist, and productive elements.
7. The conceptualizations, cultural practices, and social and political organization of the local people must always be taken into account to ensure that decision making is autonomous and that there is full participation in the planning, diagnosis, evaluation, analysis, and implementation of conservation actions.

The above premises emerge from the framework provided by the dialogue between local indigenous knowledge and Western knowledge, by the interaction of the disciplines of biology and anthropology, and by state policies (fig. 2.3).

Local indigenous knowledge was defined as the conceptualization and perception of the world by the Embera, with emphasis on their concept of territory and of the relationships between humans and territory and between humans and animals. This knowledge gives rise to particular social activities and practices involved in decision making and in the use and management of productive, social, and symbolic spaces, all of which were considered in the definition of indigenous knowledge.

The interdisciplinary and intercultural dialogue included contributions from biologists, anthropologists, and individuals specialized in other disciplines who participated at specific points in the project. From the anthropological perspective, we set out to understand the stage upon which the interacting actors develop a way of constructing the relationship between humans and nature. This was done primarily by observing the transformations generated by the development and presentation of conservation solutions, mediated by the definitions and cultural practices partic-

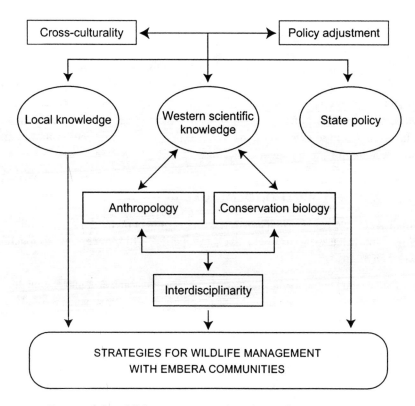

FIGURE 2.3 Framework for wildlife management with Embera indigenous communities.

ular to each actor. This understanding was facilitated by the concept of culture as a permanent process of restructuring meaning, in which society is constantly being reinterpreted, elaborated, and constructed, with structuring categories at the symbolic level that permit the continuity of culture as a dynamic process (García Canclini 1982). In this case one aspect of reality is constructed on the basis of the interests of the local peoples, of an indigenous organization, of a government organization, of an NGO, and of biology and anthropology researchers. In addition, researchers act as mediators that link the relationships proposed by the different actors, because their knowledge of different situations and practices can help the actors to interact under conditions where the political equality of all actors is recognized.

The goals of conservation biology are to research human impacts on biological diversity and to develop practical means to prevent the extinction of species (Soulé and Wilcox 1980; Wilson 1988; Primack 1993). The project strove to accomplish these two main goals through the diagnosis of hunting and its impact on wildlife and through the generation of strategies to protect species, always reconciling the needs of animal species with the needs of the people. In this particular case we strove for the conservation of species that have great symbolic and dietary importance for local indigenous peoples as well as of species of great ecological importance.

The indigenous researchers that participated in the core team were members of the indigenous communities and representatives of the OREWA. Each group carried out research on their own reality, based on their own ways of approaching their world, always maintaining interchange with the biologists and anthropologists in order to find solutions to problems. Diverse peoples from the communities also participated in this exchange of knowledge.

METHODOLOGICAL APPROACH

Initially conceived as a participatory process, the project methodology gradually developed into what we term interactive participation. This concept brings together several elements:

1. *Participation* promotes the presence of, as well as the giving of opinion by and the action on the part of, all actors during the entire process. Participation is based on respect for differences and considers local people as key actors in the process.
2. *Autonomy* proposes decision making and action by the local inhabitants with respect to the use of their territory and wildlife.
3. *Equity* works toward equality of political conditions based on the differences between each of the actors, thus generating a respectful dialogue.
4. *Interculturality* facilitates the exchange of knowledge, logics, and ways of acting between the two cultures.
5. *Interdisciplinarity* seeks a joint vision by the social and natural sciences of the problem and its solutions.
6. *Communication* explores the different systems of perception and representation of the two cultures. Communications were complemented by various materials used to socialize information—pamphlets, tapes, maps, calendars, posters, guides, and others.
7. *Continuity* proposes a long-term process that can be adjusted according to political, social, cultural and environmental conditions.

In order to develop the methodology and include each of the above elements, we took into account cultural aspects of the Embera:

1. social organization (nuclear family and kinship networks);
2. mechanisms of social control and cohesion, such as head of kin, the jaibaná, traditional leaders, the new categories of leaders, and the OREWA;
3. the traditional system for reaching consensus and for decision making (majority);
4. traditional systems of representation (oral, graphical, musical, and others);
5. traditional (oral) and new (schooling) ways of socializing information;
6. perceptions and explanation for the decrease and extinction of game (e.g., the actions of the jaibaná);

7. cultural strategies for game management such as rotational hunting;
8. traditional classification systems for animals (e.g., by habitat or behavior);
9. values and attitudes with respect to wildlife, that is, the absence of the concept of extinction.

Similarly, we also considered the proposals that the communities have for their own future and development, to avoid the external imposition of projects alien to the local cultural or environmental reality, and to allow instead projects to be constructed at the local level in a decentralized fashion. In this way the local inhabitants would make their own decisions about the planning, diagnosis of, implementation of, and follow-up on all aspects of the project that impinge on their territory.

We also took into account parallel aspects in the national society: (a) control mechanisms (e.g., environmental legislation); (b) national policy with respect to other ethnic groups; (c) national policies for participation and the opportunities/mechanisms available for participation; (d) representational systems (oral and written) and mechanisms for socializing these systems (written materials, maps, and data bases); (e) perceptions and explanations for the decrease and extinction of wildlife from the Western perspective (anthropic and environmental factors); (f) legal wildlife management strategies (hunting seasons, wildlife refuges, captive breeding, and others); (g) biological systems of classification; (h) values and attitudes toward wildlife (conservation); and i) wildlife management policies inherent to each of the participating institutions.

CONSENSUS-BUILDING AND SELECTION OF WILDLIFE MANAGEMENT ALTERNATIVES

A basic step required to initiate consensus building among the above mentioned actors, a step that although obvious is often ignored, was to recognize that two different conceptualizations and ways of thinking were interacting with each other. The management strategies to be considered were therefore defined while taking into consideration six elements, that are linked together by means of the project's interactive participation methodology:

1. a study of the Embera population and Embera use of territory and wildlife;
2. Embera wildlife management strategies;
3. national and legal strategies;
4. interests and requirements of the communities that live in the park;
5. environmental policies of the NGO and the government;
6. historical, cultural, and ecological context at the local, regional, national, and global scale.

The three indigenous reserves that overlap with the park participated in the process through representatives from fourteen of their communities. Representatives from two communities from nearby reserves also participated because the proposed solutions would affect not only the overlap zone but also the surrounding populations for political, sociocultural, and environmental reasons. The Embera, through the OREWA, have always proposed solutions that apply not only to one isolated case but that can be extended to all communities. According to their conception, all Embera have access to the territory. Additionally, if the communities from reserves that do not fall within the park were ignored, the project would be overlooking the reality of kinship and social relationships that link these groups directly to the park. Another factor was biological: the movement of animals links spaces and ecosystems. Finally, conservation actions must involve the greatest number of people and institutions possible, as such involvement gives these actions the regional recognition required to make them viable.

The participation of indigenous researchers facilitated not only the interactions among the various actors and institutions but also the construction of a common language. The perceptions of the OREWA and the indigenous researchers were key in guiding the dynamic of a communication process adjusted to Embera parameters. This dynamic was established at two levels: encounters and the preparation of materials.

Encounters are defined as interactive opportunities for reflection and included meetings, workshops, and committees. As the point where collective analysis leading to decision making was initiated, these encounters became the focal element of the dynamic. Two social processes took place during these encounters: participation / consensus building / decision making and exchange of knowledge and feedback.

Workshops were open opportunities for consensus building and participation and were based on the traditional form of agreement, that is, decision by majority. They were complemented by various materials used to socialize information—pamphlets, tapes, maps, calendars, posters, guides, and others. Since encounters are brief, time-limited events, these printed and taped materials gave continuity to the process of reflection. They also served to incorporate into the process other community members who were not present at the workshops.

Elaboration of materials took into account several variables, both thematic and representational in nature, and relied on the communication strategies most appropriate to Embera culture. Graphic and oral materials were best accepted by the communities because they fit into traditional forms of representation. Community members expressed interest in understanding the problems of game scarcity from the Western perspective and the solutions proposed by the national society. Therefore references to these concepts and perspectives were included in the representational materials.

PRESELECTION OF ALTERNATIVE MANAGEMENT STRATEGIES

We carried out a preselection of potentially appropriate management strategies, taking into account our survey of Embera use of territory and wildlife; Embera management strategies; legal strategies; the interests and needs of the communities; the environmental policies of the OREWA, Fundación Natura, and the Ministry of Environment; and the territorial and environmental contexts of the park-reserve overlap zone. We established basic principles to be used as guidelines in this preselection, principles which would enable the long-term continuity of Embera culture as well as that of the ecosystem in which it occurs. Based on an eventual analysis of their social and biological viability, a final set of alternatives would be chosen for implementation by the communities from this pool of preselected alternatives.

PRINCIPLES TO BE CONSIDERED: CULTURAL, CONSERVATIONIST, AND PRODUCTIVE

The cultural principle demands a consideration of the conceptions and knowledge of the human-nature relationship, values and meanings of nature, social practices, productive processes, and processes of interaction with other societies, taking into account the role played by nonhuman nature in all of these factors. The relationship between humans and nature is in most indigenous cultures a continuous process of reciprocal relations, and must be viewed in an integral way, so that one does not consider only the resource. Similarly, one must consider the knowledge of and interest in nonhuman species and the practices, innovations, and cultural strategies that refer to resource management in general and game animals in particular. The cultural principle also implies a recovery and consolidation of local knowledge and management strategies (e.g., culturally restricted territories, agreement on use of animals with the mothers or owners of the animals, and identification of protein sources other than hunted animals). It further implies a need to understand and incorporate Embera explanations for the scarcity of animals, thus adjusting the management alternatives to Embera cultural parameters. Additionally, one must consider the impact of national, regional and local policies and development plans on culture.

The conservationist principle demands a consideration of the environmental conditions, an evaluation of supply and demand of natural resources, biological characteristics of the species, extinction processes, and the environmental carrying capacity. The goal of the conservation element of the strategies is sustainable use of wildlife or at least long-term sustained harvest, attaining the maximum production for human consumption that will not deplete wildlife populations or make them vulnerable to local extinction. Implicit in this principle is management with and for people because long-term conservation is viable only when local practices and knowledge and ethnic rights are considered along with the scientific knowledge of the biological and social sciences.

The productive principle calls for technical improvements in the management of traditional or new resources (introduced species or those that are not regularly used) in order to achieve greater productivity of animal or plant protein, a process that can help reduce pressure on game animals. This principle aims at generating strategies that ensure food quality and security for local inhabitants.

CATEGORIES OF WILDLIFE MANAGEMENT

In accordance with the above principles, four types of management were proposed. They were categorized according to their influence on wildlife by a social group.

Direct management occurs when a human group takes action to control and/or conserve species or groups of species by means of actions that directly affect them or their habitat:

1. Symbolic practices are based on hunting restrictions or prohibitions associated with symbolic criteria and specific ritual practices; they generate actions and allow control of animals by a group or people or an individual. They derive from cultural conceptualizations and generate social practices.

2. Conservationist practices are those that most human groups use to maintain equilibrium between supply and demand of natural resources used for a variety of purposes, such as food, symbolic, aesthetic or spiritual, among others. These practices may allow the increase of animal populations, for example, when they are released from hunting pressure at certain times or places.

Indirect management occurs when a human group carries out productive practices that provide food security and that decrease pressure on wildlife populations. There are several distinct forms of indirect management:

1. Extensive practices apply traditional economic practices to resources that are not traditionally used. In order to be considered extensive, the resource must be congruent with the cosmology of the social group. It must further be easily attained, and techniques for its use must be easily acquired. Also, it must be near to the territory used by the community, and it must be acceptable in the diet.

2. Technical improvement practices increase the quantity or quality of a resource by improving the technical level of traditional productive practices.

3. Cultural change practices require the modification of the relationship between humans and nature both at the symbolic and the daily use level in order to allow access to a resource. Use of this type of practice requires the acquisition of new knowledge about the resource at the ecological and technological level, and this new knowledge implies sociocultural changes.

EMBERA AND NATIONAL SOCIETY WILDLIFE MANAGEMENT STRATEGIES

Once the ground rules for reaching concordance on any suggested wildlife management strategy had been set, the project proceeded to revisit those strategies of the Embera and national societies that ensure the continuity of both Embera culture and species conservation and that provide an optional source of animal protein (table 2.1). By opening up a discussion with the Embera communities about their own management strategies, we were also able to discuss other management op-

TABLE 2.1 Types of Wildlife Management and
Alternatives Preselected for Analysis by the Embera

EMBERA WILDLIFE MANAGEMENT	NATIONAL SOCIETY (LEGAL AND CIVIL) WILDLIFE MANAGEMENT	PRESELECTED ALTERNATIVES
Symbolic reciprocity relationship with animals — Practices of the Jaibaná		Control by the Jaibaná
Symbolic reciprocity relationship with animals — Practices of the Jaibaná — Diversified production Natural history knowledge Interrelated use areas — Rotation of hunting areas — Rotational agriculture — Rotational hunting	Protected areas Areas with use restrictions — Fauna reserve — Hunting reserves Game Source areas Communal reserves Rotation of hunting areas	Game refuges
Selective hunting Diversified production Symbolic reciprocity relationship with animals — Practices of the Jaibaná Natural history knowledge	Regulation of wildlife use Hunting bans	Selective hunting or hunting bans — Ban during reproductive season — Ban by sex — Ban by age — Total bans
Natural history knowledge Diversified production — Broadening the resource base	Substituting protein sources	Use of new resources — Marine fisheries
Diversified production Natural history knowledge	Substituting protein sources	Improved techniques for small animal husbandry — pigs and chickens
Symbolic reciprocity relationship with animals — Practices of the Jaibaná Raising pets	Captive wildlife production — repopulation — commerce — food source	Captive wildlife production — Food source

tions, as viewed from the perspective of their culture. This approach ensured that the proposals would actually be developed and implemented.

Embera wildlife management strategies include the diversified use of wildlife, diversified sources of vegetable and animal protein, symbolic reciprocity relations with animals, diversified production (explicit links between hunting, fishing, agriculture, and gathering activities), productive cycles associated with seasonal species, and interaction of spaces designated for different uses (by means of shifting agriculture, seminomadic or rotational hunting, and selective hunting). Currently, the Embera's ability to use the full range of strategies is limited by social, environmental, and territorial changes. These limitations dictate the need for agreement on new strategies, which although based on traditional Embera practices, have a new connotation due to the Embera's current situation.

We also took into consideration the dynamic nature of Embera society, a society that has experienced a series of interactions that have contributed to the introduction—or rather imposition—of new social structures within the communities and of new ways of organization. These new ways have persisted in some situations because, paradoxically, they serve to retain cultural identity. This means that day-to-day new knowledge and social processes arise that must or will be appropriated, transformed, or given new meaning.

At the same time we also took into account governmental proposals for in situ (protected areas, establishment of restricted use areas, regulations on wildlife use, and reintroduction and repopulation of species) and ex situ (captive breeding) wildlife protection. Parallel to these legally accepted strategies we also included proposals from the civil society with articulate conservation actions that take into account the interests of local peoples, such as community reserves, protection of wildlife sources, protection of water bodies, substitution of protein sources, and rotation of hunting zones, among others.

Multiple indirect strategies for wildlife management exist, such as ecotourism, craft sales, photo safaris, collection of vegetable ivory, agroforestry, pisciculture, and use of such pest species as pigeons. Project participants preferred, however, to search for solutions to wildlife scarcity that were based on traditional Embera strategies, solutions that would directly or indirectly advance wildlife conservation.

PRESELECTED ALTERNATIVES

Taking into account cultural, conservationist, and productive principles, along with the kinds of management strategies available, project participants preselected the following alternatives: control of animals by the jaibaná, wildlife refuges, selective hunting and hunting bans (total or by reproductive season, age, or sex), use of marine fisheries, breeding of smaller domestic species, and captive breeding of native species. The alternatives in the final proposal are expressed in Western terminology. Conceptually, however, with the exception of captive breeding, they are based on Embera ideas and strategies that legitimize the people's knowledge and their relationship with the environment (table 2.1).

Control of Animals by the Jaibaná The jaibaná sustains the relationship of symbolic reciprocity between animals and the Embera. In order for any management alternative to be viable, control of animals by the jaibaná must be proposed as a vital element that articulates Embera cosmology and that is the starting point from which other alternatives can then be proposed. Control of animals by the jaibaná occurs when the jaibaná interacts with the mothers of the species in order to lock up the white-lipped peccary in the underworld or to uncover the underworld to release the white-lipped peccary. This control causes the abundance of game animals to increase or decrease. The Embera believe that the disappearance of species from their territories is caused by the actions of the jaibaná.

Wildlife Refuges The concept of wildlife refuges encompasses practices such as restrictions on a territory by a jaibaná, rotational and seminomadic hunting, diversified production, diversified protein sources, and interaction among spaces designated for different uses. All of these practices remove pressure from a portion of land and also make available protein from different plant and animal sources. These Embera strategies are related to national society strategies, such as protected areas, restricted use areas (faunal refuges and hunting reserves), faunal source areas, communal reserves and rotation of hunting areas, all of which also have the goal of leaving an area free of pressure.

Based on the conceptual congruence between Embera and national society strategies, we reached consensus on the definition of wildlife refuges as portions of land where a human group decides to stop hunting and extracting animals for a predetermined time in order to allow animal populations to maintain themselves or to increase. These refuges also function as sources of animals that disperse into hunting areas. In setting the boundaries of a refuge, one must consider spaces available for the productive, social, and symbolic activities of the human group as well as the occurrence of habitats used by the animals of interest.

Selective Hunting or Hunting Bans This concept encompasses Embera practices such as selective hunting, diversified production, symbolic reciprocity relationship with animals, and practices of the jaibaná associated with restrictions placed on some species. These strategies coincide at the conceptual level with such national society strategies as regulation of the use of wildlife and hunting bans. Therefore we reached consensus on the definition of selective hunting and hunting bans as occurring when a group of people decide to use or hunt in a selective way one or more animal species during a predetermined time with the objective of allowing the populations of these species to increase or be maintained so that they may be sustainably hunted.

Use of New Resources The concept of use of new resources encompasses both Embera practices of diversified production and the national society's strategy of the substitution of protein sources. We therefore define the use of new resources as the use of an animal species (new to the area or new to use) in a sustainable manner within a local productive practice, with the consequent reduction of pressure on wildlife by means of protein substitution.

Improvement of Animal Husbandry This concept encompasses Embera practices of diversified production, which in turn is related to such national society strategies as substitution of protein sources. Given this congruence in the strategies, the project reached consensus on the use of livestock, defining improvement in small animal husbandry as an increase in the quality and quantity of production associated with small domesticated species in order to decrease pressure on wildlife populations. In this project, chickens and pigs were chosen as target species for increased production.

Captive Breeding for Food The concept of captive breeding brings together the Embera concepts of symbolic reciprocity with animals by the jaibaná, the keeping of pets, and of diversified production with the national society concept of captive breeding. We define captive breeding for food as the rearing and reproduction of wild species in captive or semi-captive conditions that achieves stable productivity, which in the long term serves as a protein option for local peoples.

DEFINING SOCIOCULTURAL AND BIOLOGICAL VIABILITY OF THE PRESELECTED ALTERNATIVE STRATEGIES

Once a set of potential alternative management strategies had been preselected, the project needed to analyze the viability of each one. We defined viability as the ability to sustain traditional practices, to increase the amount of meat available to each individual, and to improve health and nutrition levels, thus ensuring food security to the Embera as well as the recovery or maintenance of game populations. Project participants determined the minimum sociocultural and biological concepts and elements to be taken into account to accurately gauge the viability of each of the alternatives. These factors are not always easy to approach or identify because of limited basic information about the area and the short amount of time available for research afforded by the real need to make timely decisions about management. Nevertheless, these concepts and elements serve as guides to infer the feasibility of the strategies.

SOCIOCULTURAL CONCEPTS AND ELEMENTS USED TO DEFINE THE VIABILITY OF WILDLIFE MANAGEMENT STRATEGIES

To be socioculturally viable, it is not enough for an alternative to provide an optional protein source and result in long-term management of resources. It must also (a) derive from the people's own strategies, (b) be accepted by a group that is representative of the community in terms of numbers, social rank, and gender, (c) not obstruct any cultural, political, or economic processes, and (d) not generate pro-

TABLE 2.2 General Factors for Evaluating the Sociocultural Viability
of Wildlife Management Alternatives

Conceptualization and knowledge of the universe
    a. Interacting time scales
    b. Spaces in which these time scales act
Territory
    a. Capacity for cultural reproduction
    b. Ability to sustain the population
    c. Boundaries
    d. Relationship among use spaces
    e. Resource use processes
    f. Regulation and access
    g. History of sociocultural and environmental processes
Relationship between humans and nature
    a. Entities that make up the universe
    b. Relationships with the entities
    c. Clasification of animals
    d. Strategies for use and management
    e. Perceptions and values of abundance and scarcity
Socialization
    a. Systems of perception and presentation
    b. Socialization processes
Organization, cohesion, control, and social regulation
    a. Internal systems for social regulation and authority
Intercultural relationships and transformations at conceptual, social, and cultural levels
    a. Conceptualized
    b. Socialized
    c. In organization, cohesion, and social control
    d. In productive activities

cesses that cannot be culturally assumed due to cultural concepts or daily practices. Viability analyses must therefore take into account forms of social organization, processes of socialization, and cultural interrelations, along with a people's conceptualization and knowledge of the universe and of their territory and their ideas about the relationship between humans and nonhuman nature (table 2.2).

BIOLOGICAL ASPECTS AND ELEMENTS USED TO DEFINE THE VIABILITY OF WILDLIFE MANAGEMENT STRATEGIES

An alternative is considered biologically viable if it helps wildlife populations to increase and if it allows extraction without affecting population viability. The factors that help to define the biological viability of the alternatives can be determined by studies of how the people use their territory and wildlife, by evaluations of population density, and through theoretical analyses. Other biological aspects such as resilience and resistance of ecosystems and communities are also important in determining the biological viability of wildlife management strategies. However, given the immediate need for conservation of wildlife populations of economic importance to the local people, in this respect it was impossible to carry out the complex, long-term, and expensive studies necessary to evaluate these aspects of the biology of the system.

SOCIOCULTURAL AND BIOLOGICAL FEASIBILITY OF EACH OF THE ALTERNATIVE MANAGEMENT STRATEGIES PRESELECTED FOR THE PARK-RESERVE OVERLAP ZONE

Between 1994 and 1996, the period of analysis, eight workshops were carried out, one for each preselected alternative, in conjunction with local meetings and other consensus-building processes headed by the indigenous researchers at the local level. These local meetings gave continuity to the reflection process started in the workshops.

In the workshops the social, political, and environmental processes that could result from specific management strategies were discussed. Discussions were based on factors previously identified by the researchers. The dynamic consisted in exercises that generated reflection, carried out in Embera language with the support of the indigenous researchers (for example, analyzing changes in gender-specific daily practices that would result from captive breeding). The information was socialized following the perceptional and representational systems of the Embera, that is, using graphical (face and body painting), oral and musical traditions, and other traditions of the material culture.

SELECTION OF THE ALTERNATIVES BY THE COMMUNITIES

After two years (1994–1996) of consensus-building exercises among all the stakeholders and based on the preselected alternatives, five final strategies were selected (table 2.3). These strategies blended cultural, conservationist, and productive elements and consolidated and strengthened Embera knowledge.

Captive breeding was considered to have low social feasibility and was eliminated from the list of preselected alternatives. More specifically, captive breeding was not deemed a valid alternative because in Embera conceptualizations animals are cared for by their wuandras, or mothers, and are under the control of the jaibaná. Therefore wild animals do not require any additional care. Implementation of captive breeding would imply a conceptual change in which humans would replace the wuandras and would have to take care of the animals without mediation by the jaibaná. Additionally, captive breeding would imply changes in daily practices due to the maintenance requirements of the captive animals.

Although the final proposal was formulated in Western terminology, its alterna-

TABLE 2.3 Selected Wildlife Management Strategies

MANAGEMENT TYPE	PRACTICES	ALTERNATIVE	VIABILITY
Direct Management	Symbolic practices	Control of animals by the *jaibaná*	High
	Conservationist practices	Refuges	High
		Selective hunting or bans	Medium
Indirect Management	Extensive practices	Marine fisheries	High
	Practices to improve		
	husbandry techniques	Raising pigs and chickens	High

tives respond directly to Embera ideas and practices and sought legitimacy in their knowledge and ways of relating to nature. To illustrate the sociocultural and biological feasibility analysis carried out for each alternative, we describe below the decisions taken with respect to the establishment of wildlife refuges.

THE WILDLIFE REFUGES

/ TATSIRÂ EMBERA EJUA ANIMARA GUACAUATA / LAND FOR TAKING CARE OF THE ANIMALS WITHIN THE TERRITORY /

The Embera from the park-reserve overlap zone communities chose to establish refuges that would allow the recovery of white-lipped peccaries, collared peccaries, red brocket deer, spider monkeys, howler monkeys, white-faced capuchin monkeys, armadillos, pacas, and agoutis, all species valued for their size, taste, and multiple uses. Additionally, it was noted that refuges would help the recovery of the white-lipped peccary population should these animals return to the region. The area of the refuges was chosen by the communities themselves, guaranteeing that these refuges will be viable in the long term because local people rather than planners unfamiliar with local realities will generate, manage, and assume responsibility for refuges.

Of the 42,300 ha of land surface area in the overlap zone, 14,952 (35.34%) were proposed as refuges, using geographical boundaries that make them easy to locate both on maps and by the local people. The community of Santa María de Condoto selected 3,456 ha; Unión Chocó, 8,271 ha; Boroboro, 1,725 ha; and Jurubidá, 1,500 ha. According to Posada (1991), only 4.8% of the overlap zone is anthropogenically disturbed, while an additional 10.9% is in some way influenced by human activities. These numbers probably have not varied much in the last few years, given the stability of demographic and socioeconomic conditions in the area. This means that approximately 84.3%, that is, 35,658.9 ha, of the overlap zone supports little disturbed forests.

In the diagnosis of use of territory, spaces used for living, agriculture, and hunting by the four communities were determined. Based on these data one can see that the refuges do not overlap with areas used for productive activities such as agriculture, where hunting would also be practiced on a sporadic basis. Similarly, they do not overlap with communal productive areas where activities such as gathering take place and that can experience intensive use by hunters. Furthermore, the eventual expansion of the boundaries of productive activities, especially agriculture, will not interfere with the refuges in the long term because topographical boundaries limit such expansion.

Likewise, based on the Embera vision of the territory, conceptual concordances with refuges were also established. In the proposed refuges, the sites through which animals are locked up and the sites where the animals come out into the world (caves, ponds, and ritual sites) coincide with physical sites in the world inhabited by humans. The fact that there are specific places where these two sites coincide makes it possible for the animals to return through these sites and allows them to disperse among refuges. In Santa María there are also places that have already been protected as refuges by the practices of the jaibaná. Such a site is the mouth of the Omando river, where according to oral tradition, more than thirty years ago the jaibaná Ventura placed a monster as a guardian so that nobody would hunt in the area. Also, in Unión Chocó, at the headwaters of the Bojayá River, there are caves that provide access to the underworld, and from which animals can therefore emerge.

The refuges also coincide with the dwelling place of the *nusi*, or giant fish, of the *pakore*, or grandmother of the animals, and with the exit sites of the wuandras, all vital beings in Embera thought. Since these sacred spaces overlap with the refuges, a relationship is established with the animals by means of the symbolic human-animal reciprocity, and the management alternative becomes quite feasible. Refuge viability is also strengthened by the fact that the relations, classifications, perceptions of abundance and scarcity, and expectations of the Embera coincide for the most part with the biological expectation of increase of animal populations, i.e., whether they are vulnerable, threatened, or common species (table 2.4).

Based on Embera knowledge and by means of surveys of the proposed refuge areas, we observed that the forest is in good conservation shape, that it is composed of a mosaic of habitats, and that the vegetation is not undergoing high levels of intervention. These observations indicate that the habitat should provide the necessary resources for the species of interest. Additionally, the refuges are located toward the center of the overlap zone, and between them there is a continuum of forest that experiences sporadic use for hunting and gathering of plant products, providing a habitat corridor between the refuges. Additionally, relatively undisturbed forests are located to the north and east of the overlap zone. These factors increase the potential for protection of species with large home ranges or migratory habits. They also prevent the fragmentation of the populations as a whole into smaller populations that would be more vulnerable to extinction, and they may maintain genetic variability by permitting the exchange of individuals among subpopulations. Finally,

TABLE 2.4 Species of Interest to Conservation from the Biological and Embera Perspectives

SPECIES THAT THE EMBERA WISH TO RECOVER	LOCAL CATEGORY	BIOLOGICAL EXPECTATION OF RECOVERY*
White-lipped peccary (*Tayassu pecari*)	In the process of extinction	Low
Collared peccary (*Tayassu tajacu*)	Threatened/common	Medium
Brocket deer (*Mazama americana*)	Threatened/common	Medium
Spider monkey (*Ateles fuscipces*)	Threatened	Medium
Howler monkey (*Allouatta palliata*)	Threatened	Medium
White-faced capuchin (*Cebus capucinus*)	Threatened	Medium
Armadillo (*Dasypus novemcinctus*)	Common	High
Paca (*Agouti paca*)	Common	High
Agouti (*Dasyprocta punctata*)	Common	High

*Biological expectation of recovery: refers to the feasibility of recovery or maintenance of the populations; it is defined by taking into account only the intrinsic characteristics of the species, excluding anthropic pressure that is absent in a refuge.

the shape of the refuges is delimited by geographic features and tends to be oval or round so that the borders tend to be distant from the center and edge effects are minimized.

Theoretical analyses that incorporate home range, size of the overlap area, and reproductive rates and that were used to predict the number of individuals or groups that could be sustained by each refuge indicate that the protection of spider monkeys, howler monkeys, and capuchins is feasible (H. Rubio-Torgler unpublished data). Of course, their recovery must be linked to hunting bans because these three species are locally threatened. The numbers of collared peccaries and brocket deer will vary in the refuges, given that their abundances already vary over the entire overlap zone: they are more common in the interior than in the coastal zones where they are threatened (H. Rubio-Torgler unpublished data). For the white-lipped peccary, on the other hand, it is unlikely that the refuges will protect several groups, in as much that the species may currently be undergoing local extinction. Therefore, for threatened populations of all species to recover and disperse throughout the hunting zone, the refuges must be left without hunting for an extensive period of time. Although the decision of when to end the refuge strategy is a social one, it has been suggested by the indigenous people that one of the refuges could be temporarily opened if considered necessary or that use of the refuges for hunting could be rotated based on the recovery of the populations in them.

By proposing refuges linked to the control of animals by the jaibaná and by basing management on culturally established strategies, a process is generated that, although requiring input of new information, contributes to Embera cultural continuity and facilitates the development of the strategy in the long term. Another element to be considered is the fact that conceptually the structure of the refuge is present in the mind of the Embera as a space regulated by productive practices and by the control of the jaibaná.

Nevertheless, as a new form of management, refuges require periodic evaluations of their effectiveness. Monitoring allows one to know whether populations are increasing or whether individuals are dispersing toward hunting zones. Monitoring also implies consensus with people from other surrounding communities, Embera as well as Afro-Colombian and mestizo. Therefore specific individuals must be charged with evaluation of the biological and social effectiveness of the refuges.

Since there are no local mechanisms for evaluating the effectiveness of the refuges, the Embera must determine whether they need outsiders to measure changes in population abundance or whether they can do it by means of traditional roles, such as tongueros (persons who can see the spiritual world by drinking hallucinatory plants), yerbateros (individuals who use plants to cure different illness), and expert hunters. Given these alternatives, the Embera have suggested control by means of traditional and nontraditional forms, such as establishment of a system of regulations, including forms of access to the refuges, and of biological monitoring. At this point it is best not to create new hierarchies of individuals that do not fit into traditional hierarchies or roles. Otherwise, a desire for power could be created among the individuals elected to carry out the control and monitoring, since responsibility would be centered on them. Thus, the implementation process will have to be assumed by all members of the community rather than assigned to a few individuals.

The above factors all suggest that wildlife refuges can become sources of game animals for hunters of the four Embera communities and that they represent a viable form of land use. Currently, wildlife refuges exist in the four communities located in the overlap zone. However, in order for the process to be viable, information about the refuges must be socialized. The most important topics that must be socialized and monitored are minimum area requirements for animal reproduction, population sizes, space used by animals, population growth rates, diet, reproductive behavior, social structure of some of the species, population monitoring methods, and long-term monitoring. In order for the refuges to be viable, they must be implemented in parallel with other alternative strategies of protein procurement and strengthened with hunting bans. In other words, the three basic principles initially proposed must be kept in mind: cultural, conservationist, and productive.

FACTORS THAT AFFECT THE CONTINUITY AND SUCCESS OF THE PROJECT

For continuity to be guaranteed, the process of reaching consensus on wildlife management alternatives must count on the political support and will of all the stakeholders during implementation and follow-up. Among the actions that must be implemented and that must be monitored to ensure continuity are the promotion of activities that will improve the procurement of protein from nonhunting sources, analysis of cultural changes generated by the process, strengthening of intercultural relations, and appropriate application of the communication mechanisms that generate better intercultural relationships and that allow the communities to continue with research and dissemination of results.

However, success and management continuity also depend on two other key factors: cultural continuity and strengthening of participatory processes. Wildlife management based on the conceptualizations of the local people necessarily implies guaranteeing the basic conditions that support the cultural continuity of this human group and of their territory, and recognizing the rights that all communities have to make use of their environment in accordance with their traditional practices. These grant the people the autonomy to decide, under the forms and mechanisms of organization established by their own culture, what actions to take toward management. At the same time the participatory process requires that all stakeholders remain engaged. Such engagement is necessary not only for communities, indigenous organizations, and the State that are directly involved and are the ones immediately affected by the decisions taken about their territories, but also for other human groups, NGOs, etc. indirectly involved and that are members of the civil society. Furthermore, the participatory process must encourage decentralization of State actions, so that real management based on local autonomy can take place. This decentralization implies presenting and assuming the issue of wildlife management as the responsibility of both the local inhabitants and of the State.

Diverse other cultural, ecological, economic, political, and participatory factors and basic conditions affect the success of any consensus and implementation process for long-term wildlife management. These factors can and should be taken into account because they can influence the process depending on the context or level at which they occur (local, regional, national, or transnational). Examples of these factors include development programs instituted at the national or transnational level. While it is clearly necessary to articulate local plans with the nation's reality, it must also be recognized that these two visions are often in conflict because the interests of the nation do not always coincide with the interests of indigenous peoples. A contributing problem in Colombia is that the policies of the Ministry of Environment change depending on the identity of the officials in power, and these changes can accelerate, delay, or otherwise affect the wildlife management process. Finally, social conditions, i.e., the presence of paramilitaries and guerillas, means that consensus-building processes are threatened by violence. Environmental priorities are placed on the back burner because the social conditions necessary for the implementation of a program based on participation and consensus are lacking.

It is therefore not surprising that not all the wildlife management strategies selected by the project have been implemented in the overlap zone. This delay and decoupling of the strategies is also due in part to a lack of interinstitutional coordination and to the political problems existing between the indigenous organization, the National Park, and the NGO Fundación Natura. Additionally, at the local level a series of infrastructure development and tourism projects have been established that will eventually affect the area and its inhabitants. Despite these factors, since 1997 members of the four communities in the overlap zone and several of the neighboring communities have implemented at the local level several conservationist wildlife management strategies: refuges, bans, and reliance on marine fisheries.

3

Bridging the Gap Between Western Scientific and Traditional Indigenous Wildlife Management

THE XAVANTE OF RIO DAS MORTES INDIGENOUS RESERVE, MATO GROSSO, BRAZIL

KIRSTEN M. SILVIUS

Several authors have suggested that indigenous lands in the Neotropics function or could function as important conservation units (Redford and Stearman 1993; Peres 1994; Peres and Terborgh 1995; Redford and Mansour 1996). In the Amazon there are approximately 250 indigenous reserves, representing 44% of government-managed land area (Peres and Terborgh 1995). Twenty percent of the Brazilian Amazon alone is indigenous land. On the basis of land area and of documented levels of species diversity in Amazonia, these lands hold within their boundaries a high proportion of the world's biodiversity, most of it as yet unstudied. Hunting, however, can locally reduce or eliminate vertebrate populations on indigenous lands, especially during the transition from subsistence to market economies, from nomadic to sedentary settlement patterns and from traditional hunting technologies to the use of guns (Bodmer, Fang, and Moya 1988a; Robinson and Redford 1991; Vickers 1991; Redford and Stearman 1993; Peres 1994; Bodmer, Eisenberg, and Redford 1997; Auzel and Wilkie 2000; Eaves and Ruggiero 2000; Fragoso, Silvius, and Prada 2000; Hill and Padwe 2000; Leeuwenberg and Robinson 2000; Mena et al. 2000; Robinson and Bennett 2000a; Yost and Kelley 1983). The fact that hunting is an integral part of indigenous cultures in the Amazon thus brings into question the value of indigenous lands for the conservation of large vertebrates. If indigenous reserves are to play a role in national and regional conservation strategies, the causes of game depletion by indigenous people must be understood and remedied through appropriate management. At stake are not only the biodiversity supported by the vast expanses of forest and savanna ecosystems encompassed by indigenous lands, but also the traditional cultures of the region.

In response to game declines on their reserves and the resultant impact on diet quality and traditional livelihoods, several Neotropical indigenous communities have initiated collaborations with biologists to develop sustainable wildlife use practices (Ulloa, Rubio-Torgler, and Campos-Rozo 1996; Leeuwenberg and Robinson 2000; Souza-Mazurek et al. 2000; Townsend 2000b; Townsend et al. 2001; Noss and Painter this volume.) As highlighted by Ulloa, Rubio-Torgler, and Campos-Rozo (this volume), biologists working with indigenous peoples must balance the ecological rules under which wildlife populations operate with the cultural rules under which indigenous populations operate. A similar situation is experienced by health workers seeking to balance indigenous views of disease with the knowledge of Western medicine (Albert and Gomez 1997). Often, the traditional management practices of the indigenous group in question offer the best means of approximating the management prescriptions of Western science, with the added advantage that traditional practices are more likely to be adhered to than alien practices that do not have a basis in the indigenous world view.

In this article I explore ways in which biologists and indigenous peoples can reach consensus on game management plans by reviewing the motivations, dynamics, and management outcomes of the Xavante Wildlife Management project, which has been under way in central Brazil since 1990 (Leeuwenberg 1997a,b; Fragoso, Silvius, and Prada 2000; Graham 2000; Leeuwenberg and Robinson 2000). I focus on aspects of the Xavante culture and of the project itself that may have influenced the final management decisions made by the community. I then review the findings of biologists and anthropologists working on other indigenous hunting studies and in the process develop a set of loose guidelines for biologists working with indigenous peoples in South and Central America. These guidelines summarize some of the common factors found to affect hunting practices and game depletion in several indigenous reserves.

THE XAVANTE PROJECT

The Xavante people have traditionally lived in the savannas and woodlands of central Brazil's cerrado ecosystem. Their once extensive range and population is now reduced to approximately 9,000 people living on five reserves in Mato Grosso state (Graham 2000). Hunting is a key element of Xavante culture (Maybury-Lewis 1967). Despite extensive contact with Brazilian national society, Eteñitepa, the dominant community in the 330,000-ha Rio das Mortes Reserve, maintains a highly traditional life style. When community members noted a decline in their hunting yields in the late 1980s, they sought advice from World Wildlife Fund (WWF)-Brazil. From 1991 to 1993 wildlife biologist Frans Leeuwneberg worked with the Xavante to collect basic data on hunting effort, hunting areas, sex ratios, and age structure of hunted animals (Leeuwenberg 1997a,b; Leeuwenberg and Robinson 2000), all of which are required to assess the status of wildlife populations on the reserve and to determine the sustainability of the communities hunting practices (Bodmer and Robinson this volume).

After three years of data collection, Leeuwenberg concluded that tapirs (*Tapirus terrestris*) were being hunted unsustainably. He based his conclusion on a comparison of actual harvest rates with potential productivity. Age structure analyses for pampas (*Ozotoceros bezoarcticus*) and marsh deer (*Blastoceros dichotomus*) suggested these species too were overhunted, even though very low numbers of pampas deer were being killed. The situation for giant anteaters (*Myrmecophaga tridactyla*) was less obvious, as adequate data on carrying capacity and productivity were not available for the area. However, these animals were being hunted at such a high rate that there was a good probability that they were being overhunted (Leeuwenberg 1994; Leeuwenberg 1997a,b; Leeuwenberg and Robinson 2000). The same indices showed that other regularly hunted species of concern, including brocket deer (*Mazama americana*), collared peccaries (*Tayassu tajacu*), and white-lipped peccaries (*Tayassu pecari*), were being sustainably hunted.

From 1995 to1997 wildlife biologist José M. V. Fragoso designed a continuous monitoring system and an adaptive management plan, based on data already collected by Leeuwenberg and on new, track-based indices of animal population abundances (Fragoso, Silvius, and Prada 2000). Because most sedentary indigenous groups have a limited hunting radius, Fragoso's design used tracks to assess the relative abundances of key game species at three incremental distances from the village. The rationale was that when areas are left undisturbed for long periods of time due to low hunting pressure far from a village, a source-sink dynamic may be created if the reserve is large and continuous. As long as production is high in the distant areas, animals may move into the hunted areas near the village, in effect maintaining a constant, though low, supply of game. The effect will vary with both the biological parameters of the species and the degree of preference the hunters show for the species. The situation may be stable and sustainable, as it appears to be for a fox-hunting system in the Argentine pampas (Novaro this volume). Because the Rio das Mortes Reserve is a relatively large area, it is possible for source-sink dynamics to be operating within it rather than between the reserve and outside areas, as suggested by Townsend (1995a) for the Sirionó in Bolivia.

On the basis of the combined analyses by Fragoso and Leeuwenberg, management recommendations were made to the community. The community then took several months to consider the recommendations and to reach consensus on a management strategy. In 1997 the community signed an agreement with WWF-Brazil to implement the monitoring and management plan they had chosen. Management has been in place for four years now, and the data from the first two years are being analyzed (R. Lemos de Sá pers. comm.). However, there were key differences between the management strategy recommended by biologists and the strategy chosen by the Xavante community, differences which may be at least partially rooted in cultural perceptions of wildlife.

MANAGEMENT RECOMMENDATIONS AND XAVANTE CHOICES

The analyses by Fragoso, Silvius, and Prada (2000) and by Leeuwenberg and Robinson (2000) indicated that five species were threatened by or vulnerable to overhunting: giant anteater (considered to be overhunted everywhere, with equal abundances at all distances from village, and a source probably outside the reserve), pampas deer (overhunted everywhere, no source in the reserve), giant armadillo (*Priodontes giganteus*, overhunted, low abundances everywhere, unknown natural history), marsh deer (vulnerable or threatened, but potential source area within the reserve), and tapir (vulnerable or threatened, populations low, but higher abundances at greater distances from village). All other species, including collared peccary and white-lipped peccary, showed the expected pattern of low track counts near the village and high track counts far away, suggesting that they have source populations within the reserve. Sample sizes for brocket deer (*M. americana* and *M. gouazoubira*) were very low; however, they actually appeared to have higher abundances near the village than far from it, perhaps because of their ability to use disturbed habitats. Pacas (*Agouti paca*) and agoutis (*Dasyprocta agouti*) were not hunted frequently and had equal abundances at all distances from the village, and so there was no reason to assume that their populations were threatened by any other factors (Fragoso, Silvius, and Villa-Lobos 2000).

SUGGESTED MANAGEMENT SCENARIO BASED ON BIOLOGICAL CONSIDERATIONS

On the basis of the assessments described in the previous section, the following management recommendations were made to the Xavante of the Rio das Mortes Reserve (although the Eteñitepa community retained leadership of the project, WWF-Brazil required that the other communities in the reserve participate in the management plan):

1. Do not hunt giant armadillo and giant anteater in the reserve until populations recover or monitoring indicates that population levels are not likely to become higher and until their biology, including reproductive potential, has been studied.
2. Do not hunt pampas deer in the reserve until their populations recover because population levels appear to be unusually low.
3. Only hunt marsh deer, tapir, and white-lipped peccaries at locations distant from the village in order to allow populations to recover in other areas. Once other areas recover, hunting can be shifted there. In this way, a source-sink system will be maintained not by distance from the village but by design in certain areas, irrespective of distance from the village.
4. Hunt collared peccary, brocket deer, and smaller species at current or higher levels in all areas of the reserve.

XAVANTE MANAGEMENT DECISION

Using their traditional process of achieving consensus through long discussions at the men's council, the Xavante developed a very specific management plan that included well-defined refuge areas and specific hunting periods for species in each area. However, the species they prioritized for protection differed from those recommended by the biologists. The Xavante placed 96,000 ha of the reserve into three different wildlife refuges. These areas were chosen not on the basis of distance from village but rather on the basis of geographical boundaries, location of villages, a perceived need to protect reserve boundaries, and interpretation of the areas as "production zones" for species of concern (F. Leeuwenberg pers. comm.). Animal abundances would be monitored using the track-sampling method, and when abundances increased in an area, hunting would intensify there and decrease in areas where monitoring showed tracks were decreasing. A ban on hunting of tapir and marsh deer, however, would continue in some areas even after they were opened to hunting of other species. The decision to eventually hunt these two species would be made on the basis of track monitoring. Thus, the Xavante preferred to manage species on the basis of refuges rather than on the basis of hunting bans, a decision similar to that of the Embera in Colombia (Ulloa, Rubio-Torgler, and Campos-Rozo 1996, this volume). During the first part of the study, Leeuwenberg (1994) also recommended a ban on two species of concern, the pampas and marsh deer. The Xavante men's council decided against this recommendation, preferring to leave an area unhunted rather than to eliminate the hunting of a particular species altogether.

RATIONALE BEHIND THE MANAGEMENT CHOICES BY THE XAVANTE

To test the hypothesis that the difference between the biologically recommended plan and the Xavante choice lies in culture, tradition, and different interpretation of biological facts given by a very different world view/system of explanation, I searched the anthropological literature for references to use of the different game species by the Xavante. None of the studies carried out by anthropologists at the Eteñitepa community adequately quantified hunting returns, but I have deduced overall patterns from their descriptions of hunting. Maybury-Lewis (1967) gives a qualitative description of the importance of different game species, while Flowers (1983a) monitored meat intake by two households during three days on four occasions representing different seasons. Maybury-Lewis states that peccaries, tapir, and deer were the most prized animals in 1958, describing them in that order. Based on common name roots, he indicates that tapir are classed with peccaries. All deer share the same name root except for marsh deer, which is classed separately. He has little to say about the giant anteater, except that it was abundant and was hunted. Today, this species is the third most frequently captured.

The animals noted by Flowers for twenty-four hunts in 1976 and 1977 are twelve white-lipped peccaries, five brocket deer, eight tapir, and three pacas. She comments that men hunt paca at night in the gardens and that by this time this species had become a much more important aspect of community life and food production than it was during the Maybury-Lewis study. Even though her sample size is small, if giant anteaters were being captured at the same frequency as they are to-day, they should have appeared on her list. The high frequency of tapir captured during Flowers's study is also surprising since today tapir are not brought in any more frequently than the larger deer species, which do not appear on Flowers's list. Still, since different hunters tend to specialize in specific animals, it is difficult to interpret Flowers's small sample size.

Leeuwenberg (1994) comments that primates were never eaten during his study, although they were during Maybury-Lewis's survey. This absence may reflect the influence of white prejudice against primate meat or the fact that currently the Xavante do no like to hunt in forested areas. Community members informed Leeuwenberg that during a past time of low-game availability, capybara, boas, and foxes were also eaten, although they are not used today (Leeuwenberg and Robin-son 2000).

These early changes in hunting parallel the change to a sedentary life style and a greater economic reliance on agriculture. The Xavante have been undergoing constant change since first contact with whites in the 1700s, their retreat from white contact in the mid to late 1800s and their subsequent establishment in their current homeland. When they arrived at their current location (1850 to 1940), the ancestors of the current community members were seminomadic. In the late 1960s they still trekked for most of the year (Maybury-Lewis 1967). By the late seventies, however, they trekked only a few weeks on the year (Flowers 1983a,b). By the early 1990s they relied little on communal hunts or large long-distance fire hunts (Leeuwenberg 1994).

The work by Maybury-Lewis (1967) and Flowers (1983a,b) suggests that tapir were either more abundant in the past or were hunted more intensively. The Xavante may be aware of a decline in tapir populations during the last thirty years that cannot be picked up by the short-term study carried out here. It is possible that the population has stabilized at a lower level than at some past time, as has occurred with hunted ungulate populations in the temperate/arctic zones (Caughley and Sinclair 1994). Anteaters, on the other hand, seem never to have been a preferred or culturally important species. The current intensive use of anteaters is probably recent, suggesting that the Xavante are substituting anteater for other preferred species. This use could be a response to an overall decline in availability of other species or a consequence of hunting near the village. In either case there may not have been time for a reciprocity or respect bond to be established with the species or even for accurate knowledge of its natural history, carrying capacity, or popula-tion characteristics to accumulate. The special concern for the marsh deer and the lesser concern for the pampas deer are difficult to explain from either a biological or cultural point of view and may be due to a simple preference for pampas deer meat.

In conclusion, the anthropological literature suggests that the list of preferred species hunted by the Xavante has changed in parallel with (a) changes in game populations, (b) an increasingly sedentary life, (c) an increased emphasis on gardens, and (d) a more recent reclamation of a traditional hunting culture. Thus the reasons for the differences between the management recommendations of the biologists and the final Xavante choice are practical and biological rather than cultural.

Although there is no evidence of a cultural or spiritual basis for the Xavante's management decisions in this study, other studies do point to the importance of spiritual practices in determining how indigenous peoples interact with game (Albert 1985; Anderson 1996; Colding and Folke 1997). These attitudes are based to an unknown degree on knowledge of natural history and an understanding of the explicit need to manage resources. We do not have sufficient information on the relationship of the Xavante and other indigenous groups with animals at the spiritual level so as to understand how their decisions are made and to predict how decisions will be made in the future. Partly this lack of information is due to the unwillingness of some groups to discuss cosmological issues with outsiders, and biologists must respect and work around this desire for cultural privacy. However, Ulloa, Rubio-Torgler, and Campos-Rozo (this volume) show how a participatory process can successfully allow a community to make management decisions consonant with their world view without the need for an in-depth study of the cultural aspects of hunting.

DYNAMICS ASSOCIATED WITH THE IMPACT OF WESTERN CULTURE ON INDIGENOUS COMMUNITIES

Researchers have identified and, in some cases, studied in detail the aspects of Western culture that can alter indigenous cultures and lead to the overexploitation of game animals. These factors will be operating in all but the most intact indigenous cultures and must be addressed during the elaboration of management plans.

TRANSITION FROM A SEMINOMADIC TO A SEDENTARY LIFE

The most effective factor preventing extreme game depletion may be the traditional seminomadic life of most Amazonian indigenous groups at the time of contact (Vickers 1991; Townsend 1995a; Ulloa, Rubio-Torgler, and Campos-Rozo 1996; Fragoso, Silvius, and Villa-Lobos 2000; Robinson and Bennett 2000b; Stearman 2000). This system creates a shifting impact on game populations over the landscape in time and space. Game populations probably always declined locally to the point at which they were not efficiently hunted with bows and arrows, contributing to the decision to move the village site or to undertake an extended trek. The impact, however, was probably never sufficient to cause local extinction of an animal population. Following contact, many national governments pursued a policy of settling indigenous communities and limiting their ability to practice seminomadic hunting. Limits to reserve size, presence of permanent health posts, and agricultural or livestock projects all lead to sedentarianism. With communities remaining in the same location, sometimes for decades, local game depletion becomes chronic. Game populations may be endangered if they are also under pressure from forces external to the reserve.

The transition to a sedentary life includes not only remaining at a fixed village for more than the two to four years typical of precontact times but also the degree of reluctance to temporarily trek away from this village. The importance of the village versus trekking will vary from community to community. Even historically, the Xavante were likely to have a fixed village site, but they were not there very often (Maybury-Lewis 1967). The Yanomami of northern Brazil and southern Venezuela, on the other hand, both shift the village site and trek extensively (Good 1989).

Currently, the Xavante of Eteñitepa maintain a fixed village and do not trek. This way of life implies strong cultural change, but may have a positive impact on game populations: if no hunting occurs away from the village, then there is little re-duction of game populations and areas distant from the village may in fact become source populations that can be included in the management plan (Fragoso, Sil-vius, and Prada 2000; Novaro, Redford, and Bodmer 2000). However, with access to motorized vehicles, some sedentary communities are now able to hunt distant locations without moving the village site or trekking (Fragoso, Silvius, and Prada 2000; Souza-Mazurek et al. 2000). Without an explicit rotational hunting scheme, the overall impact can be heavier than before: the home site is given no chance to recover, and the distant areas are also hunted

INCORPORATION TO MARKET ECONOMIES

Participation in market economies is considered to be one of the primary contribu-tors to the loss of traditional practices by indigenous groups because such participa-tion brings access to new technologies, shifts traditional power hierarchies in a community, favors sedentarianism, and removes both men and women from tradi-tional practices such as hunting and gardening in favor of seasonal or permanent wage labor. There have been several individual studies of the dynamics within in-dividual communities that lead to participation in market economies and promote overhunting or overfishing (e.g., Gross et al. 1979; Yost and Kelley 1983; Stearman 1990; Stearman and Redford 1992; Godoy, Brokaw, and Wilkie 1995; Godoy, Wilkie, and Franks 1997; Santos et al. 1997; Fragoso, Silvius, and Prada 2000; Townsend 2000b; Godoy, Kirby, and Wilkie 2001).

POPULATION GROWTH

When a community becomes sedentary near a health post, human population is likely to increase due to reduced mortality. Management plans must be able to pre-dict growth on the basis of demographic and cultural factors and to determine how such growth can be prevented from affecting wildlife populations. While management can potentially increase game yields by keeping populations at the most productive level through cropping, the potential is limited and dangerous (Caughley and Sinclair 1994; Bodmer and Robinson this volume). Therefore alternate resources will have to be used unless the community controls its own population growth. In the case of the Xavante, population increase at one site may have been traditionally mitigated by village fissioning. Community division can be detrimental to current management plans, however, by spreading the impact on wildlife population into source areas and by making it difficult for agreement to be reached among communities that have split along hostile faction lines.

HABITAT FRAGMENTATION

Because of increasing land conversion to pasture and agriculture, indigenous reserves like national parks are often isolated from contiguous natural habitats that still support healthy wildlife populations. Management plans must assess the vulnerability of fragmented wildlife populations and look for ways of interconnecting reserves with any other territory. In the case of the Xavante, some wildlife species appear to have population sources outside of the reserve (Fragoso, Silvius, and Villa-Lobos 2000). Here the importance that Ulloa, Rubio-Torgler, and Campos Rozo (this volume) assign to the meshing of local, state, national, and international goals becomes a key issue. To the extent that state and national governments are charged with biodiversity protection, indigenous reserves with a good management plan can provide this function and therefore could receive govermental benefits in exchange for this service.

TRADITIONAL FORMS OF MANAGEMENT AND PATTERNS OF RESOURCE USE

Traditional management practices, such as food taboos, protection of sacred sites, explicit management by shamans and elders, the concept of "owners" of the game, and others, persist in many communities and exert bottom-up pressure for resource protection. These practices open up the potential for effective, innovative resource management by indigenous peoples if they are integrated with, and in some cases substituted for, Western scientific precepts for management. Because these practices and cosmological factors are in some cases the same or similar for several indigenous groups, they are worth examining as focal points in the elaboration of indigenous management plans.

WHITE-LIPPED PECCARIES

In terms of biomass the white-lipped peccary is usually one the most important sources of protein for Neotropical indigenous peoples (Good 1989; Stearman 1995; Mena et al. 2000; Townsend 2000b). It also figures prominently in spiritual or reciprocity systems (Ulloa, Rubio-Torgler, and Campos-Rozo; R. dos Santos pers. comm.; J. Fragoso pers. comm.). The status of white-lip populations and of the traditions associated with the species may serve as indicators of the health of both wildlife and culture, and the species may be a good focus for educational campaigns and management plans. In cases where the white-lipped peccary is not a common animal or is not commonly hunted, another species may substitute it— e.g., marsh deer in the case of the Xavante (Fragoso, Silvius, and Prada 2000; Leeuwenberg and Robinson 2000) or rheas in the case of the Izoceño-Guaraní (Noss and Painter this volume).

EXISTENCE OF SACRED AREAS

Many cultures have traditional reserve systems, sacred sites in which hunting is prohibited because of their cosmological significance (Ventocilla 1992; Ulloa, Rubio-Torgler, and Campos-Rozo 1996; Fragoso, Silvius, and Villa-Lobos 2000). This system provides a population source area for animals and in some cases is recognized as such by the communities themselves (Fragoso, Silvius, and Villa-Lobos 2000). With the loss of traditional cosmology and values, such sites are no longer protected.

Sacred areas or other areas with special status should be assessed for their value in protecting biodiversity and be involved in management plans, either directly as actual refuges, as in the case of the Xavante and the Embera, or as a conceptual tool to explain Western systems of protected areas. If an area has spiritual as well as natural value, it is more likely to be protected in the long term, and its protection will also serve to reinforce cultural values.

HUNTING LARGE ANIMALS FOR RITUALS

Many groups use small mammals daily but focus on white-lipped peccaries and other large mammals for rituals, such as when members of other communities are invited as guests or when weddings, funerals, or rites of passage are celebrated. If this pattern can be reinforced, it may be a good management tool, increasing pressure on small mammal species with healthy populations and high reproductive potential (Bodmer 1995b; Bodmer, Eisenberg, and Redford 1997) and reserving large species for special occasions. The availability of shotguns, which make hunters more likely to pursue large game, and of motorized vehicles, which make it easier for large game to be carried back to the village, currently works against this tradition (Hill and Hawkes 1983; Peres 1990; Souza-Mazurek et al. 2000).

HUNTING IN GARDENS

Hunting in gardens occurs to some extent in all communities but seems to be more common in the more acculturated ones. The Xavante currently reject this option (F. Leeuwenberg pers. comm.), but it seems that twenty years ago they may have used it (Flowers 1983b). It needs to be seriously considered, as it will put pressure on smaller animals whose populations can sustain higher harvest rates (Bodmer, Eisenberg, and Redford 1997).

SHAMANS

Among many indigenous peoples certain older hunters and shamans hold the role of "owners of the game" (Fragoso, Silvius, and Prada 2000). These men have extensive knowledge of the natural history of a particular species and are believed to be in spiritual contact with either the animals or the spirits that represent or mediate for the animals. They can decide whether or not a particular species should be hunted at a particular time and how many individuals should be taken (Ulloa, Rubio-Torgler, and Campos-Rozo 1996; Rubio-Torgler 1997; Fragoso, Silvius, and Prada 2000; Leeuwenberg and Robinson 2000; R. dos Santos pers. comm.). Like taboos, these practices are lost when younger men with access to money, education, and market goods and communication with the national society gain authority and power in the community (Stearman 1995; Fragoso, Silvius, and Prada 2000).

Flowers (1983a,b) and Leeuwenberg (per. comm.) both describe the practice of "owner of the game" among the Xavante, where one man is responsible for directing the hunting of a particular species. The amount of control the person has in determining when and how many individuals of a species should be hunted is not known; the practice may not be comparable to hunting decisions made by shamans in other tribes or to the controlling power ascribed to them by the community (Ulloa, Rubio-Torgler, and Campos-Rozo 1996, D. Yanomami pers. comm.). In the case of the Xavante, such decisions are made by the entire men's council, and it is unclear how much influence one person, the "owner" of a particular game species, might have.

Projects aimed at developing management plans for indigenous areas should ascertain the degree of respect accorded to the shaman or to the equivalent person. The shaman can be a focus for management if he/she has been traditionally viewed as a manager. As described by Ulloa, Rubio-Torgler, and Campos-Rozo (this volume), the Embera stated that shamans should be given decision-making control over animal management because that is how their traditional role is now perceived. The translation between the spiritual control originally attributed to the shaman (covering and uncovering the entrance from the underworld from which animals emerge) and the biological management necessary today is unclear. It is best to make use of this strategy by viewing the shaman as a knowledgeable person who can lead a community in coherent game use in accordance with biological requirements.

TABOOS

For many Amazonian indigenous groups, taboos prevent seasonal or complete use of certain animal species (MacDonald 1977; Ross 1978; Colding and Folke 1997) and may thus decrease hunting pressure on animals, including those that are vulnerable to local extinction (Colding and Folke 1997). However, taboos are rapidly lost following contact: the Huaorani of Ecuador shifted their target species over a fifteen-year period once certain species became amenable to hunting with guns or were valued in the market (Mena et al. 2000). Ulloa, Rubio-Torgler, and Campos-Rozo (1996) also indicate that there has been considerable breakdown of taboos in the Embera society, probably due to the extreme reduction in animal populations. Leeuwenberg (1994) mentions that the Xavante have taboos on armadillo, brocket deer, and peccaries, which cannot be eaten by parents for the six months after the birth of a child. The author does not comment on how strictly these partial prohibitions are adhered to. He also indicates that the use of capybara, fox, and large snakes in the past was restricted by a partial taboo to elderly people and that even today the lesser anteater (tamandua) is eaten primarily by elderly people (F. Leeuwenberg pers. comm.).

Because taboos are flexible and situation-specific, they may or may not function as an adequate management tool. Projects should assess the degree to which traditional taboos are followed. If adherence is lax, then game populations are probably in trouble, and commerce in game meat may be important.

ROLE OF FISHING

Fishing is easily incorporated into and accepted by traditional societies. The reasons for this easy acceptance may be because fishing is similar to hunting (killing a large animal that requires stalking and other skills), although not as strenuous, and because fishing has always been used to some degree. Both the Xavante and the Embera chose an increased emphasis on fishing as a preferred management tool (Fragoso, Silvius, and Prada 2000; Ulloa, Rubio-Torgler, and Campos-Rozo this volume).

However, if there is pressure on fish stocks from human populations outside the reserve or by illegal incursion into the reserve, then the alternative may not be viable, or it may lead to conflict between ethnic groups. Management of the fish stocks will also be necessary to prevent overfishing, especially if the group does not have a fishing tradition with culture-specific management practices. This is of special concern given the ease with which fish populations are driven to extinction by the use of traditional fishing methods such as poison and such introduced methods as gill nets, dynamite, and bleach. Fishing is thus best incorporated into a seasonal shift in resource use, as a complementary diet source rather than as an absolute substitue.

Projects should always assess the willingness of the community to use alternate protein sources rather than introducing a source that is not already acceptable, such as domestic animals, because such acceptance is likely to weaken the traditional culture (Ulloa, Rubio-Torgler, and Campos-Rozo this volume). For example, when missionaries provided domestic chickens to the Yanomami of the Catrimani area as an alternative protein source to hunting, community members simply left the animals to starve in their cages when they went on trek because they had no tradition of or philosophical basis for domestic animal husbandry (P. Guillerme pers. comm.)

CONCLUSION

The above factors serve both as indicators of health of game animals in indigenous areas and as management tools for a community. Traditional reserves can work as well or better than reserves chosen for purely ecological reason because the community will already respect their boundaries. Management for white-lipped peccaries will automatically protect large areas or habitat mosaics that are key to smaller species, at least in a forested habitat. Strengthening the traditional role of shamans will also strengthen the importance given to traditional natural history observations and management systems such as hunting or burning seasons (Fragoso, Silvius, and Prada 2000). Managers or management plan designers should survey the status of the above factors to determine the context in which the management plan will take place. They could focus wildlife management and education on key species in which the hunters have a traditional interest, using these animals as conceptual as well as ecological umbrella species. The type of analysis used by Ulloa, Rubio-Torgler, and Campos-Rozo (this volume) if included from the start of the study will increase the study's coherence and facilitate decision making.

4

Increasing Local Stakeholder Participation in Wildlife Management Projects with Rural Communities

LESSONS FROM BOLIVIA

WENDY R. TOWNSEND

Conservation professionals must consider how to improve their ability to promote wildlife management as a viable development alternative in Latin America. Although not all countries have the same socioeconomic situation as Bolivia, we can probably all agree that natural resource management implies more than just extraction. In legal terms (e.g., Forestry Law #1700 of Bolivia), commercial harvest should require an indication of sustainability documented in the form of a management plan. To produce a management plan, it is essential to have a clear view of the social reality within which wildlife management is to be carried out. For many people, especially for indigenous peoples, wildlife resources make survival possible in the marginalized informal economy they experience in the countryside. For this reason a wildlife management plan requires the participation of local game users to achieve success in a way that a timber management plan may not. The challenge for wildlife managers is to promote biodiversity conservation while recognizing the dignity of the local people, decreasing poverty, and promoting self-administration. These needs imply the participation of local people at all stages of planning and implementation.

Subsistence hunting is a daily activity for rural people in Bolivia and other regions of Latin America. Hunters are naturally interested in wildlife issues and are therefore easily involved in wildlife management. It is particularly important that hunters participate because they make the decisions as to what animals to hunt and when, although sometimes this decision is instantaneous. That moment of decision during a (usually) solitary activity is important, and one that needs to be adequately understood in order to reinforce the behavioral changes needed for sustainable management of wildlife resources. With these considerations in mind, we must examine both the social and legal contexts in which rural inhabitants live because increasing the hunters' involvement will help make them responsible for the management of their wildlife resources. This article offers some ideas for achieving this goal by describing some participatory management techniques, along with their limitations.

DEFINING PARTICIPATION AND PARTICIPATORY

Participation is a very popular concept in many projects, especially in Bolivia, where it is also guaranteed and regulated by legislation. Given that wildlife management requires that hunters participate, it is interesting to contemplate what such participation implies. In the Larousse Spanish Dictionary (García-Pelayo and Gross 1979), we find three interpretations of the word:

1. "Act of participating and its result";
2. "Notice, warning";
3. "System by means of which the employees of a firm are linked to its profits and eventually to its administration."

These definitions accurately reflect the different levels of participation that are found in natural resource management. According to the second definition, for example, only the giving of information is required, while the third definition leads to self-motivated management.

Participation in natural resource projects is carried out in many different ways and to different degrees. In Ecuador, The Nature Conservancy (TNC) designed a system of categories to evaluate participation in its PALOMAP project in the Cayembe-Coca Ecological Reserve (B. Ulfelder pers. comm.). I use the ideas here because they are useful in helping wildlife managers develop their approach to community management plans. In figure 4.1, these categories are represented in

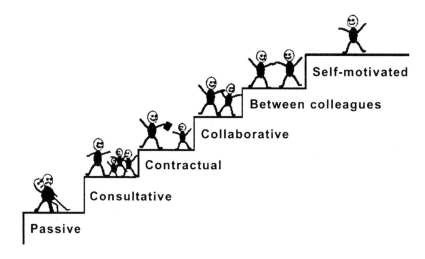

FIGURE 4.1 Stairway to self-motivated participation in resource management.

the form of a ladder, with the steps leading toward increasing participation until the most desired outcome, autonomous and self-motivated management, is reached. The categories that represent the different levels of community participation are described below.

PASSIVE

Passive participation occurs when plans are produced in an office without consultation or intervention by stakeholders. From the point of view of conservation, passive participation is not very useful because it usually does not consider the local reality and thus does not seek, nor does it achieve, consensus with local people. Additionally, this method can lead to serious conflicts because of misunderstandings among local stakeholders, managers, and institutions.

CONTRACTUAL

Contractual participation implies that local people are contracted to carry out a project that was developed by others on behalf of the community. It is akin to paying local people to manage their park but without giving them any power of decision over park management. To a certain extent this participation is analogous to the situation that occurs when logging companies pay for standing wood (in indigenous territories) and take charge of management plans, permits, and administration without any local participation.

CONSULTATIVE

In consultative participation, the institution comes to the community to present its ideas and objectives, but it provides little room for an exchange of ideas. Often, a language of exclusion is used, which relies on technical words and which might even require the use of translators. Consultation may occur unexpectedly without sufficient advance warning for the community to reflect on the topic. Sometimes consultation is carried out independently with different segments of society, typically with men and women as separate focal groups, in an misguided attempt to be "gender sensitive." A consultation meeting may end in an agreement that is not well understood by all the participants. The agreement may easily be forgotten by the time the project reaches the implementation phase. A participatory consultation process can be very useful for gathering initial information, but attempts should be made to ensure that there is information sharing as well as information extraction.

BETWEEN COLLEAGUES

Participation between colleagues requires information to flow horizontally and bidirectionally. It implies the building of a true process of communication with the community during the planning and implementation stages. Achieving this level of communication may require a change in professionals' attitudes and in particular in the technical language they have become accustomed to use. This level of participation also requires mutual respect from the professional and the local people, with each participant listening, attempting to understand, and fulfilling the role they must play in the management plan.

SELF-MOTIVATED

The most complete type of participation is self-motivated, and is attained when the community invests its own effort in participating, exercising its power of decision, seeking information about its own problems, and implementing possible solutions. At this point the community is autonomous in its management decisions. Such autonomy does not mean that community members can or will wish to perform all the tasks necessary for management actions by themselves, but rather that they are prepared and able to make the management decisions that they are convinced are necessary.

COMMUNITY PARTICIPATION AND THE CHALLENGE OF ACHIEVING SELF-MOTIVATION

The real challenge is to attain self-motivated participation by communities in wildlife management plans. We know that wildlife issues are able to attract the attention of stakeholders. For example, in the United States hunters supported levying taxes on themselves from gun and ammunition sales to fund game management (Lacy Act). Although similar initiatives might also be possible in some Latin American countries, most rural subsistence hunters do not have the wherewithal to pay taxes. Perhaps by taxing adventure tourism and other nature-seeking tourists, a fund could be developed, but for most Latin American countries, it would be irresponsible to even dream that taxes could lead to wildlife conservation and the timely protection of critical wildlife populations. For wildlife professionals in rural South America, it should be clear that the people who use the game must be responsible for the sustainability of their resource use. It is our challenge as professionals to provide them with the tools to facilitate the process.

Achieving stakeholder-motivated participation, although an attainable goal, faces many difficulties, especially those based in human limitations. First, there is the diversity of personalities involved—not all biologists have a vocation for working with people. On the contrary, some biologists chose their field of study because they prefer animals to humans! Nevertheless, professionals can educate themselves and become more sensitive to participatory issues, and local people can start to demand mutual respect for their right to participate. Each professional must recognize his or her own strengths, abilities, skills, weaknesses, and limitations. Most importantly, they must honestly evaluate their own commitment as a colleague with community members. This is the only way in which professionals will be able to stimulate community participation through the processes of information gathering and informed decision making.

Another constraint on achieving self-motivated participation in natural resource management is time. Community processes tend to occur slowly, although occasionally communities can be stimulated to take major actions quickly. It can be difficult to obtain funding for long-term programs because funding agencies prefer three-year projects. Therefore community-level processes run the risk of being pressured and thus distorted by professionals in order complete the logical framework and to ensure financing for the following phases. It is extremely important that funding agencies begin to consider the amount of time required for true community-level processes.

Even though most wildlife management concepts are intuitive and logical and have long been observed by rural people with their domestic animals, the language used to express the concepts can be one of the primary obstacles to participation. Not only can Spanish sometimes be the second language of the participants, there is also an exclusionary effect due to the use of technical terminology. Professionals tend to use technical terms as though they were a part of daily language. These words can frighten or confuse rural listeners, and as a consequence, they are rarely inclined to ask for clarification. The result is insufficient informed participation and the inability to attain a good consultation.

Thus it is very important to create dictionaries and glossaries to explain the terminology in each local language so that professionals and technicians can stimulate participation through use of a people's own language to explain technical concepts. The need for translations between two kinds of knowledge—academic and local—becomes even more obvious when we try to achieve self-management because local people will need to integrate all the information from all sources in order to make sound management decisions.

PROMOTING COMMUNITY PARTICIPATION IN WILDLIFE MANAGEMENT

To increase community participation in wildlife management, one must move up the ladder, converting consultation into self-motivated management (fig. 4.1). There are certain attitudes, techniques, and actions that broaden participation:

1. Return all work done with the community back to the community in a permanent way that they will understand. This is the primary complaint of indigenous peoples. They point out that others have studied them for many years, but the majority of this information has not been returned to them. Libraries in Europe and the United States sometimes contain more information about Latin American indigenous peoples than libraries in the region itself. Fortunately, the internet is increasing access to information, but this source of information is not available to most of these communities. The use of workshops to return information is only one step; it is important that communities be left with a written record of the work accomplished.

2. Dialog must be multidirectional: All the actors, at all levels, must be included in the dialog. Professionals must learn to listen and read the message between the lines. Sometimes, with the opening of discussion, conflict arises. This conflict may have existed prior to the dialog, but previously it would have surfaced in ways and at moments that were unfavorable for reaching resolution. By promoting discussion among stakeholders, conservation professionals can guide the search for solutions before wildlife populations reach a critical point. During a recent round table discussion on wildlife management, the question arose of "What do indigenous peoples want the professionals to do?" This is a very difficult question for indigenous people to answer because most of them do not understand what skills or services we have to offer. We need to open the dialog and thus become informed of each other's roles in order to reach a mutual understanding of the necessary steps toward productive community wildlife management. It is our obligation to explain the ways in which our knowledge is useful and to describe the tools we bring with wildlife management. For their part local communities can contribute considerable specific information about the local environment, which can be crucial to the success of management efforts.

Local knowledge is a truly valuable source of secondary information and one that can serve as a baseline for future research. In some localities there still exists local knowledge, accumulated over centuries, with details about distributions, diets, behavior, and other factors necessary for the elaboration of wildlife management plans.

However, not all representatives of a culture know every detail, as there are usually specialists on the subject of wildlife. Hunters are the main group of specialists, but there can be others who transfer their knowledge in the oral history (alternatively called mythology). This has been my own experience with the Murui of the Caquetá river (Townsend, Nuñez, and Macuritofe 1984; Townsend 1995b) with respect to primates in the area of Araracuara, Amazonas, Colombia. When that research was done, there was very little acceptance by the academic community of the utility of indigenous knowledge except with respect to medicinal plants (Posey 1985; Posey and Balée 1989; Clay 1988). The oral history of the Murui is almost an instruction manual of natural history because it includes information about food species, activity patterns, and other details important for wildlife management (Townsend, Nuñez, and Macuritofe 1984). When the information obtained from the Murui expert Vicente Macuritofe Ramirez was compared with that published in the scientific literature, there was a 97% agreement rate (Townsend 1995b). But this information was obtained through interview work with a culturally known specialist of the Murui, not through a participatory meeting.

It is important to understand that there are analytical limitations to the information obtained through participatory meetings and interviews. The first key point is that the information comes from opinion and memory and therefore requires different analytical methods than those used for direct measurements with which we, as biologists, are more familiar. For example, a numerical value agreed upon during a participatory meeting is not an average, as it has no range of values. Therefore it cannot be statistically described unless it is compared to numerical values agreed upon in other meetings, the value at each meeting being the unit of analysis.

Similarly, participatory meetings and interviews should not be used to determine the amount of wildlife harvested for the management plan. Although the people may have a good idea about how much they use, the frailties of human memory and character make it risky to quantify game use in this way for the development of a management plan. Other, more reliable, techniques are available to evaluate wildlife harvest levels to improve management efficiency.

Participatory meetings can facilitate reaching consensus about specific wildlife issues and the information gathered. Although sometimes these meetings are dominated by a few participants, if guided properly, they can provide the opportunity to gather basic information, such as fruiting cycles of wildlife food species. Participatory meetings have produced lists of wildlife food species and of their availability cycles in several places, including the Izozog, Ibiato, TIPNIS, Pilón Lajas, Madidi, and Lomerío where workshop participants identified more than sixty species (Townsend 1995a, 1996; Ino, Kudrenecky, and Townsend 2001; Townsend 2002). Other important points that can be researched during participatory meetings and interviews are the areas and resources critical for wildlife. This information is imperative for any wildlife management plan (Townsend 1996a).

Participatory meetings are also good communication spaces for registering the local names of habitat types in the area. Community maps developed in these meetings facilitate the naming and the distinguishing of these habitats. This work can provide a baseline for communication between professionals and local hunters because it stimulates the interchange of technical descriptions and local knowledge. In this way local people become involved in data collection for management plans for all their resources, including wildlife. Local knowledge is subject to confirmation by observations, and local peoples have shown themselves to be excellent at making these observations. If given the tools, local people can collect data on resource extraction, resource availability, and phenology, as well as other parameters critical to monitoring sustainable natural resource management.

One tool that has been especially useful in promoting participation in data collection for wildlife management has been for hunters to self-monitor their game harvest. In Bolivia this system has had moderate success in promoting community participation but as yet has not led to a self-motivated management system. For example, the Yuracaré of San Pablo, Territorio Indígena Isiboro Secure, registered the use of five classes of resources during one year without much oversight by an NGO (Z. Lehm pers. comm.). Seven families in San Pablo registered their hunting, fishing, fruit collection, garden produce, and the trees that they cut during more than one year. They stated that they believed this information to be important for community planning. There are various experiences with game harvest monitoring by hunters from a variety of indigenous groups: Chiquitano (Lomerio: Guinart 1997), Sirionó: (Ibiato: Townsend 1997), Guaraní (Izozog: Noss and Painter this volume) and Tsimane (Pilon Lajas: Townsend 2002).

The key to achieving self-monitoring success is to create the mechanism in conjunction with the community. For example, one may invent a way in which hunters can easily report their harvest without affecting their cultural habits; for instance, by collecting the skulls of hunted animals. Alternatively, more detailed information can be reported on participatory data sheets. Data sheets have been most successful when they have been designed together with the hunters, thus giving them the opportunity to learn why each piece of information is important. Hunters are more likely to collect the information and use the data sheets if they understand how the information will add to their management plan.

In Bolivia people have responded to this process in a very positive way. The Chiquitanos, for example, insisted on writing the questions on the data sheet in their own language for their children and in Spanish for the hunters (figure 4.2). Linguists have refined the Chiquitano alphabet to the point that some hunters actually have problems reading with the new alphabet, but since the school children use the new way, the hunters wanted it on their form (Townsend 1996a). Some Sirionó have registered their game harvest for more than five years (F. Billon pers. comm.). The 'Tsimane Indians of Pilón Lajas Biosphere Reserve and Indigenous Territory have voluntarily registered their game for two years, and the results have generated community discussions on protecting some areas as reserves and for ecotourism (Townsend 2002). Experiences in Bolivia have also shown that hunters

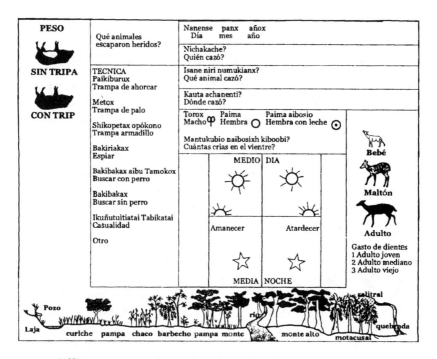

FIGURE 4.2   Self-Monitoring game harvest form designed with Chiquitano hunters of Lomerío.

who know how to write appreciate the opportunity to use this skill. Professionals need to be creative in looking for the ways to explain the utility of the information and in making the data sheets attractive and even fun to use. The information gathered must always be summarized in consultation with the hunters, who then are the first to know what their hunting totals are.

The success of self-monitoring of resource use and availability lies in its twofold utility—first as an impact on the process of community development and second in its ability to obtain the needed data to formulate a management plan. When participatory research tools are given to the resource users, these users also receive the power to inform themselves as a community about the state of their natural resources (Ino, Kudrenecky, and Townsend 2001; Salvatierra et al. 2001). This process may not function in all communities or in all situations where it is attempted. However, the results can be surprising in terms of the assumption of responsibility by the communities (Townsend et al. 2001; Ellis 2002; Townsend 2002). Participatory research allows communities to feel they are the owners of the process and of the information gathered. The sense of ownership of the process is a large step toward self-motivated wildlife management.

5

Community-Based Wildlife Management in the Gran Chaco, Bolivia

ANDREW J. NOSS AND MICHAEL D. PAINTER

Since 1991 the Wildlife Conservation Society (WCS) and the Capitanía de Alto y Bajo Izozog (CABI) have collaborated in the design and implementation of a community wildlife management program in the Bolivian Chaco. WCS is an international conservation organization that works to conserve wild areas and wildlife and carries out research on wildlife species and ecology. CABI is the indigenous organization that represents approximately 9,000 Izoceño-Guaraní inhabitants of twenty-threee communities of the Izozog along the Parapetí river south of the Bañados de Izozog wetlands (declared a RAMSAR site in 2001). This article examines the collaboration between WCS and CABI through the present time, emphasizing activities at the regional and local levels as well as at the institutional and biodiversity levels.

REGIONAL AND INSTITUTIONAL CONCERNS

At the beginning of their collaboration, the interests that governed the respective positions of WCS and CABI for conservation in the Chaco were fundamentally different. On the basis of extensive field research in the Gran Chaco of Argentina, Paraguay, and Bolivia (Taber 1991; Taber et al. 1993, 1994, 1997), WCS became concerned that Bolivia was the only country in the region where large expanses of Chacoan ecosystems and habitats remained relatively intact, and perceived the creation of a protected area as a first step toward conservation of the region. Dry forests are the most threatened ecosystems in lowland Bolivia (Taber, Navarro, and Arribas 1997), while tropical dry forests represent one of the most endangered biomes globally (Janzen 1988; Redford, Taber, and Simonetti 1990).

CABI, as political representative of the Izoceño communities, began its fight for land rights in the 1940s as the Izoceños sought to recover from the devastating Chaco war. Following protest marches to La Paz, the Izoceños received their first land titles in 1948 but promptly saw the titles nullified by the agrarian reform of 1953. Forced to start over, not until 1986 did they once again receive land titles totalling 65,000 hectares. This grant was an important landmark but nevertheless of greater symbolic than real value: the area was insufficient to assure the long-term survival of the Izoceño people because it covered only a part of the land occupied at the time by the Izoceño communities. The land grant also failed to provide for population growth.

CABI therefore began to seek alternative mechanisms to consolidate territory in order to guarantee the long-term survival and security of the Izoceño people and to stop the expanding agricultural frontier. At the same time, CABI sought to identify alternate means of livelihood for the Izoceño people, which did not include the negative environmental, socioeconomic, or cultural impacts associated with the forms of farming and ranching that fueled the city of Santa Cruz's agroindustrial growth since the 1950s. Independently of WCS, CABI leaders reached the conclusion that a protected area would provide the legal basis for halting the expanding agricultural frontier, as well as a focal point for defining new production alternatives.

On the basis of these complementary interests, different but convergent with respect to the future of the Bolivian Chaco, CABI and WCS began to work together in the region in 1991. In the case of the Bolivian Chaco, WCS considered collaboration with and participation by CABI and the Izoceño communities to be the best and only option for long-term biodiversity conservation.

The key success of the collaboration between CABI and WCS was the creation of the Kaa-Iya del Gran Chaco National Park and Integrated Management Area (KINP) in September 1995 (fig. 5.1). CABI presented the proposal for the creation of the park to the government of Bolivia. WCS provided CABI with technical assistance in preparing the proposal and assisted CABI through the review process. Following the creation of the park, CABI was named its coadministrator (together with the National Park Service SERNAP). Covering 3,440,000 hectares, the KINP is Bolivia's largest protected area and the largest tropical dry forest protected area in the world (Taber, Navarro, and Arribas 1997). It is also the only national park in South America created as a result of the initiative of an indigenous people and the only one in which a Native American organization shares fundamental administrative responsibilities with the national government.

In addition to the national park, CABI pursued a second and complementary path to safeguard Izoceño interests as a people. During 1996 CABI played a leading role in the successful effort of lowland indigenous peoples to include the concept of indigenous territory in Bolivia's new agrarian reform law (Ley de Reforma Agraria, Ley 1715, 1996). Called Tierra Comunitaria de Origen (TCO) in Bolivia, the concept refers to territorial rights as defined under International Labor Organization (ILO) Convention 169 on the rights of indigenous people. These rights are based on the historical occupation or use of an area and on the spatial requirements of a people needed to satisfy its subsistence requirements in a manner con-

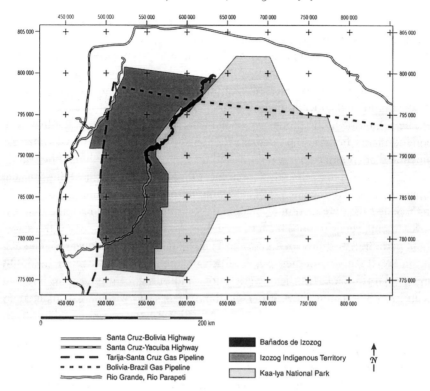

FIGURE 5.1   Map of Kaa-Iya National Park, Izozog Indigenous Terrritory, showing the major highways and gas pipelines.

sistent with its culture and way of life. In early 1997, taking advantage of the stipulations of the new law, CABI presented a demand for a TCO totalling 1,900,000 hectares (fig. 5.1) that adjoins, but does not overlap with, the KINP.

Although many other indigenous peoples also presented territorial demands at about the same time, CABI's case was unique because the territorial demand, like the proposal for the creation and administration of the KINP, derived from a vision of establishing an area in the Chaco that would safeguard the survival of the Izoceños as a people. As a result, contrary to other cases in Bolivia where protected areas and indigenous territories overlap and generate mutually exclusive land use and ownership conflicts, CABI's vision created the opportunity to manage 5,300,000 hectares of Bolivian Chaco—an area nearly the size of Costa Rica—under principles of conservation and sustainable use of wildlife and other natural resources.

The creation of the KINP, together with CABI's role as coadministrator, opened the door for a more extensive community conservation effort. In 1995 USAID/Bolivia joined the CABI-WCS association and began to provide financial assistance that continued through 2003. WCS and USAID technical and financial assistance through the Kaa-Iya Project concentrates on four important areas: (a) institutional strengthening of the technical arm of CABI, the Ivi-Iyambae Foundation, (b) participatory research on wildlife populations and ecology and development of wildlife management practices, (c) planning and environmental monitoring, and (d) environmental education (Painter and Noss 2000; Painter et al. 2003).

The beginning of the project coincided with a rapid expansion of the hydrocarbon industry in Bolivia, in particular with the construction of the Bolivia-Brazil gas pipeline that crosses or borders the KINP for 250 kilometers. Among other things the agreement provided $1,500,000 for titling indigenous territories in the pipeline's area of influence. CABI subsequently proposed a methodology for working together with the Instituto Nacional de Reforma Agraria (INRA), the Bolivian government agency responsible for land titling. Using this methodology, the agreement permitted the titling of the Ayoreo TCO demand for a territory north and east of the KINP. Although other donors, especially the Danish Aid Agency DANIDA and the World Bank, had been working on land titling for several years and had been investing greater sums of money, the Ayoreo TCO was the first indigenous territory to be titled in Bolivia. The Izoceño TCO is being surveyed and third-party claims are being addressed under a process that produced the first title in September 2001 for 160,000 hectares. Delivery of the final title in 2005 will complete the full 1,900,000 ha. Additional funds of approximately $4,000,000 under the agreement with the pipeline consortium included $1,000,000 to start a trust fund that will provide a source of permanent income for the KINP, funds to cover the costs of monitoring the construction of the pipeline within the KINP, and funds for research on flora and fauna that may be affected by the pipeline's construction and use.

As coadministrator of the KINP, CABI has played a crucial role in generating the funds necessary to cover its basic operating budget. During the period 1998 to 2000, the Bolivian government was unable to meet its obligations, and CABI covered over 51% of the operating budget. This allowed the KINP to continue functioning through the government's financial crisis and demonstrated to all the potential importance of the coadministration mechanism for the long-term viability of Bolivia's national parks.

When the TCO titling process is complete, CABI must present a TCO management plan to the government. This plan will include (a) a land use plan, based on a zonification according to the potential of the land and (b) an investment plan defining ways to finance productive activities defined in the plan. CABI's intention is that the zonification of the TCO be an extension of the zonification exercise completed for the KINP, using the same biological and socioeconomic criteria, though obviously with a greater emphasis in the TCO on productive activities as opposed to conservation.

In 2002 CABI was faced again with challenges from further hydrocarbon and other large-scale infrastructure developments. These developments included highway projects connecting Santa Cruz with Brazil and Argentina and gas pipeline projects parallel to both highways (see fig. 5.1). CABI played a leadership role in negotiating the first agreement with the consortium backing the Bolivia-Brazil gas pipeline. This agreement provided the precedent for a subsequent environmental and socioeconomic impact package negotiated by a group of conservation organizations with the consortium building the second Bolivia-Brazil gas pipeline through the Chiquitano forest, and for negotiations by municipalities and indigenous organizations with the donors supporting the bioceanic highway corridor project. These experiences were again the basis for negotiations underway in 2002 between an alliance of seven CABI-led Guaraní TCOs with the consortium building the second pipeline, which will connect the gas fields in southern Bolivia with the principal Bolivia-Brazil pipeline. This pipeline runs along the western edge of the Izoceño TCO for 150 km. Separate negotiations got underway with the consortia building pumping stations and the second gas pipeline to Brazil, which passes through or alongside the KINP for 250 km and the Izoceño TCO for 80 km. One pumping station is within the KINP, and another is on its border. Reflecting the interests of CABI and WCS, and the general objectives of their institutional collaboration, all of these agreements include funds for development and natural resource management activities, for biodiversity research and environmental monitoring, and for land titling processes.

Although these types of activities may appear to lie outside the scope of a conservation project like the Kaa-Iya Project or outside the mandate of a conservation organization like WCS, upon careful consideration it becomes evident that they are critical to long-term conservation in the region. CABI and WCS must address these enabling factors (land titling and institutional strengthening) and regional threats (hydrocarbon development and highway construction) for long-term community conservation efforts to succeed. An unwillingness or inability to address issues at the regional scale doom local efforts to failure. Success in addressing these issues has strengthened CABI as a regional actor in conservation, generated important funds and funding mechanisms for conservation efforts and has provided models—for relations between indigenous groups on the one hand and government, private companies, and NGOs on the other—that strengthen conservation efforts far beyond the Izozog and the Bolivian Chaco.

LOCAL AND WILDLIFE CONCERNS

Parallel to the activities at the regional and institutional scales described above, the Kaa-Iya Project simultaneously pursues a series of local activities focusing on wildlife and community-based management practices, seeking to integrate tradition with science. Like other indigenous groups, the Izoceños emphasize their role as protectors of nature with traditions of sustainable use that maintain forests and wildlife populations (COICA 1989; Kleymeyer 1994; Robinson and Bennett 2000b). Guaraní culture emphasizes a set of spiritual relations with the environment: the "Tumpa" is a superhuman celestial being that represents a wildlife species and guarantees that the species will always exist on earth. Thus the armadillo-Tumpa protects armadillos, the fox-Tumpa protects foxes, and so on. The principal Tumpa is the rhea (*Rhea americana*), Ñandú, which governs all other animals. Ñandú-Tumpa appears as a constellation in the Milky Way and sends animals to Earth for the benefit of hunters. On Earth the "kaa-iya" or spirit guardian is the immediate keeper of wildlife, the superior of humans, and responsible for ensuring that wildlife is not destroyed or mistreated.

The hunter therefore must request permission of the kaa-iya and the Tumpa before entering the forest to hunt. If he behaves appropriately, he will be rewarded and must then thank the spirit guardians for providing meat and safety. Hunters must not kill for pleasure or cause harm or injury to any wild animal, and they must only take what they need to supply their families. An injured animal will complain to its guardian, who will punish the hunter by not sending him more game, by causing him to become lost in the forest, or even with death (Riester 1984; Combès et al. 1998).

These traditions and their contribution to sustainable use and conservation, again similar to other examples of indigenous peoples, have been undermined by socioeconomic changes over the past decades (Vickers 1994; Brandon 1996; Stearman 1999; Robinson and Bennett 2000b; Stearman 2000):

1. New hunting technologies (firearms and nylon fish nets, along with horses and dogs) and changes in employment patterns with seasonal emigration to sugar cane and cotton harvests. Although migrants do not exploit wildlife in the Izozog during part of the year, upon their return they may depend on wildlife more than do their neighbors, who as permanent residents can maintain livestock. In addition, seasonal migration disrupts community solidarity and the implementation of long-term community projects including wildlife management (Beneria-Surkin 1998; Noss and Cuéllar 2001).

2. Reduction of the area accessible to Izoceño hunters with the installation of private cattle ranches and Mennonite colonies on "fiscal land" now being claimed as part of the Izoceño indigenous territory.

3. A growing population that has tripled since 1930 (Combès 1999), reaching a current level of 9,000 inhabitants in twenty-three communities of the Izozog and a population density of $2.8/km^2$ in the actively exploited hunting range.

4. Rise in the actual standard of living, creating new expectations in terms of basic needs (health and education) as well as material welfare. Wildlife exploitation can provide important economic benefits either directly through the sale of skins and pets or indirectly because wildlife meat consumption allows Izoceño hunters and fishers to save domestic animals (chickens, goats, and cattle) for sale.

Through the Kaa-Iya Project, CABI and WCS sought to address these issues in the context of the regional and institutional frameworks described above. In order to ensure full participation in all aspects of planning and implementation, CABI and WCS formed a technical team composed of non-Izoceño biologists working together with Izoceño "parabiologists." The parabiologists were selected by CABI and their communities for their interests and abilities, though they lacked formal academic training beyond the high school level and in biology. In addition to the technical team, the Project has emphasized scientific research that actively involves local hunters in research and in community discussions in order to collectively design and implement community wildlife management activities in the Izozog (Cuéllar 1999; Noss 1999; Ayala 2000; Leaños and Cuéllar 2000; Painter and Noss 2000; Noss and Cuéllar 2001).

Research focuses on the two sides of the sustainable use equation: the exploitation of the resources on one hand and the availability and productivity of the resources on the other. Research also emphasizes such aspects of the biology of exploited species as population densities, population structure, reproduction, diet, and health . An understanding of the biology could help improve the sustainability of current exploitation patterns through appropriate management plans.

HUNTER SELF-MONITORING

Beginning in 1996, the team established a self-monitoring program with hunters in all Izoceño communities (Townsend 1996c, 2000a). The purpose of the self-monitoring was to identify the principal prey species (for subsistence as well as commercial purposes) on which to focus subsequent biological research in both hunted and nonhunted areas. The comparative information, combined with the mapping of over 300 hunting locations listed by hunters, permits an evaluation of the sustainability of current hunting patterns and pressure and elicits recommendations for management practices that can further promote sustainable use (Noss 2000; Noss and Cuéllar 2001).

Together, the team and the hunters developed, tested, and revised data sheets that hunters could carry with them on hunting trips to record information on hunted animals (appendix 5.1). Hunters also provide specimens from hunted animals: skulls for age classification, stomach contents for diet studies, and fetuses for reproductive studies. Although all hunters participate on a voluntary basis, in 1997 the project hired individuals in eleven communities who are responsible for providing materials to hunters—data sheets, tape measures, and spring scales—and who collect information from hunters once a month. These monitors were selected by their communities and by CABI.

Over 700 Izoceño hunters have provided data on hunting activities to date. Although some provided information on only a single recorded kill, others have recorded more than 100 hunted animals. In total they have reported over 5,000 mammals (31 species), 3,000 birds (15 identified species), and 280 reptiles (5 identified species) (E. Cuéllar 1999; Noss 1999; R. L. Cuéllar 2000c; Leaños and Cuéllar 2000; E. Cuéllar 2002). The most important species for subsistence purposes are mammals: ungulates, i.e., gray brocket deer (*Mazama gouazoupira*), collared peccary (*Tayassu tajacu*), white-lipped peccary (*Tayassu pecari*), and tapir (*Tapirus terrestris*), and armadillos: nine-banded (*Dasypus novemcinctus*), three-banded (*Tolypeutes matacus*), large hairy (*Chaetophractus villosus*), small hairy (*C. vellerosus*), and yellow (*Euphractus sexcinctus*). Birds are also consumed, in particular doves (*Zenaida auriculata, Columbina picui, Columba picazuro,* and *Leptotila verrauxi*) and the only local cracid, the Chaco chachalaca (*Ortalis canicollis*) (Mamani 2000, 2001).

Hunters rarely recorded data on animals hunted for commercial, medicinal, or artisanal purposes, and the team obtained details of these uses through interviews (R. L. Cuéllar 2000a, 2000b) and focused studies. Most important are the commercial harvests of the three abundant psittacids—blue-fronted Amazon (*Amazona aestiva*), blue-fronted parakeet (*Aratinga acuticaudata*), and monk parakeet (*Myiopsitta monachus*) sold as pets (Guerrero et al. 2000; Guerrero and Arambiza 2001; Saavedra 2000)—and the red tegu lizard (*Tupinambis rufescens*), whose skin is marketed (Montaño 2000, 2001).

BIOLOGICAL RESEARCH: HUNTERS AND MONITORS

Parallel to hunting self-monitoring, the team has established several lines of research activities, focused on the species most important for subsistence and trade. A first set of studies depended on specimens of hunted animals voluntarily provided by hunters. Community wildlife monitors collect, prepare, and superficially analyze the specimens. More detailed analyses are performed by undergraduate or graduate biology students at laboratory facilities in Santa Cruz.

AGE STRUCTURE

Wildlife monitors clean and dry skulls provided by hunters, then classify each ungulate skull into three to four age categories according to molar growth and wear (Townsend 1996b). A more accurate age analysis involves counting dental annuli on cross-sections of incisors extracted from the same skulls (Maffei and Becerra 2000; Maffei 2001, 2004).

DIET

Wildlife monitors record the common names of plants and other items encountered in the stomach contents of ungulates, then wash and dry the specimens and store them in formalin for laboratory analysis in Santa Cruz (Caballero and Noss 2000; Lama 2000). Armadillo stomach contents are stored directly in formalin for laboratory analysis (Bruno and Cuéllar 2000).

REPRODUCTION

Wildlife monitors record number, biometrics, and characteristics of fetuses collected by hunters. These permit estimates of fecundity (proportion of females that are reproducing), seasonality of reproduction, and number of young per female (Chávez 1999; Rojas 2001).

HEALTH

Hunters have also collected digestive tracts from armadillos for laboratory analysis of parasites. Additional research on wildlife and domestic animal health, carried out in conjunction with the WCS Field Veterinary Program, has focused on samples (serum, tissue, ectoparasites, and feces) collected while accompanying hunters during their normal subsistence hunting excursions (Villarroel 2000; Parada and Villarroel 2001; Deem et al. 2002, 2003).

BIOLOGICAL RESEARCH:
MONITORS AND PARABIOLOGISTS

A second set of research activities is conducted at a nonhunted site on the boundary of the Izoceño TCO and the KINP using a system of twenty 5-km transects opened throughout the Izozog hunting area. The field camp also opened a system of trails in a rectangle roughly 4 km wide by 12 km long.

TELEMETRY

In order to determine home ranges, activity patterns, and habitat use, the team captured and attached radio transmitters externally (in the case of tapir, brocket deer, collared peccary, and red-legged tortoise [*Geochelone carbonaria*] or internally (in the case of tegu lizard and three-banded armadillo). Tracking ungulates consisted of standard telemetry triangulation procedure. Compass bearings toward the signal from three fixed points (marked every 100 m along study trails) were taken each hour on the hour. Locations were recorded throughout the 24-hour daily cycle, from 8 hours per week to 8 hours per day depending on the number of animals being tracked and the number of parabioligists avialable for as long as the animal and transmitter permitted (Barrientos and Maffei 2000; Ayala 2002; Miserendino 2002; Barrientos and Cuéllar 2003). In the case of the reptiles and armadillos, parabiologists tracked the signal directly to the animal or its burrow and recorded the location directly using a GPS receiver (Soria, Mendoza, and Ayala 2001; Mendoza and Noss 2003; Soria 2003).

Tapirs were tracked for over 2 years (N=5), brocket deer for 6 to 24 months (N=2), collared peccaries for 2 to 24 months (N=9), tortoises from 6 to 20 months (N=8), tegu lizards from 2 to 18 months (N=10), and armadillos from 6 to 12 months (N=2). By combining telemetry data on home range and overlap with group size, it was possible to estimate density at a nonhunted Chaco site.

LINE TRANSECT CENSUSES

The monitors and parabiologists respectively conducted weekly line transect censuses along the transects near the communities and at the field camp respectively.

They used standard methods of walking at a pace of 1 to 1.5 km per hour and recording perpendicular distances from the trail to any animal sighted. In 2001 the team abandoned the method, finding it inappropriate for estimating densities of mammals in the Chaco, given the poor visibility and study of animals that are small, cryptic, solitary, burrowing, and/or nocturnal (Ayala and Noss 2000).

TRACK COUNTS

Along the same study trails, the monitors and parabiologists established 1 x 2 m track plots every 200 m in order to estimate relative abundance from track counts. Track plots were cleared one day then checked the following day for new tracks, with all track plots cleared and checked every week or two weeks (Noss and Cuéllar 2000). An alternative track-based method involved sweeping 5 km or more stretches of trails and roads and checking for new tracks on the following day. This was done daily during 10-day periods (Cuéllar and Noss 1997). Despite biases in the layout of the transects, the track counts at both hunted and nonhunted sites permited comparisons of relative abundance in order to evalute the impacts of hunting and livestock near the Izoceño communities.

DRIVE COUNTS

With the participation of hunters and community members, the team conducted drives to estimate brocket deer density both near Izoceño communities and at Cerro Cortado field camp. For each drive the group of beaters walked from the trail or road approximately 200 m into the forest and formed a line roughly 200 m long and parallel to the trail. Then they began shouting and beating while walking back to the trail and the waiting observers. All deer crossing among the observers or beaters were recorded at the end of each drive (Noss, Cuéllar, and Ayala in press).

PLANT REPRODUCTIVE AND VEGETATIVE PHENOLOGY

Using ten of the same trails in the hunting area near the communities, and two trails at Cerro Cortado, the monitors and parabiologists marked five individuals, separated by 100 m or more, of thirty fruiting plant species important for game species in the area. The plant species were selected during a field course with Izoceño hunters and by examining stomach contents of hunted animals. Each month the plants are checked, with the availability and development of fruits, flowers, and leaves according to a scale of 0–4 (R. L. Cuéllar 2000a; Martínez and Cuéllar in press).

Not all biological research supported by the Kaa-Iya Project has direct management implications, for example fruiting plant phenology, tortoise telemetry, and diet, but it is nevertheless valuable in at least two important ways. It provides baseline data on little-known species and ecosystems, and it supplies training opportunities to local technicians who can later apply their skills in research design, investigation, and data analysis to the full range of natural resource management and environmental monitoring scenarios that CABI faces.

SOCIOECONOMIC RESEARCH

A third and complementary set of research activities depends on interviews and surveys in the Izoceño communities. These interviews are conducted by monitors, parabiologists, and biologists.

MEAT CONSUMPTION

In order to evaluate the relative importance of domestic versus wild sources of meat, a parabiologist conducted a set of surveys. She either administered the survey directly or demonstrated it to a member of the household who then assumed responsibility for recording on a daily basis whether meat was consumed in the household and, if so, what type of meat. Results confirm the importance of wildlife and fish for Izoceño communities: one-third of meat consumed in Izoceño households derives from wild game, one-third from fish, and one-third from domestic animals (Parada and Guerrero 2000).

ACTIVITY PATTERNS

In order to determine the importance of hunting relative to other activities (wage labor, farming, etc.) and the participation rate in the self-monitoring program, the monitors maintain a monthly activity record for all potential hunters (men and boys) in their community. Roughly one-fourth of potential hunters (all teen-age boys and men) actively hunt and fish during any given month of the year, again confirming the importance of these activities for Izoceños.

NATURAL RESOURCE CONSUMPTION

Focusing in particular on firewood consumption, a further survey involved daily visits by monitors in five communities to between eighteen and twenty-five households. This survey recorded the type and quantity of firewood, construction materials, fish, game, honey, etc., brought into the household that day. Information on consumption was compared with data on productivity to determine whether firewood use is sustainable (by species and by habitat type) and to propose management measures (Navarro 1999).

COMMUNITY MEETINGS

In the context of the Kaa-Iya Project, community meetings serve a variety of purposes, through oral presentations in Spanish and Guaraní as well as slide presentations. In part they serve to inform and educate community members with respect to research activities and objectives, legislation, the park management plan, and the TCO management plan, among others. They also serve to train the Izoceño technicians (monitors, parabiologists, and hunters) who have been directly involved in activities by giving them the opportunity to explain what they do and why to their communities.

Finally, the meetings serve to engage the communities in a discussion of the issues described above, particularly wildlife and natural resource management issues. For example, in January and February 1998 the team posed several questions intended to elicit opinions and beliefs regarding wildlife and wildlife management: (a) what activities do you pursue? (b) what purpose does wildlife serve? (c) what problems does wildlife cause? and (d) is it possible to care for wildlife? The team did not provide any options beforehand and recorded all community responses. In response to the final question, the team compiled a list of management proposals:

1. establish hunting zones or a hunting rotation system;
2. hunt only adult animals;
3. hunt only male animals when females are pregnant;
4. hunt only what the family needs without exaggerating;
5. hunt animals that are abundant and protect those that are rare;
6. conserve plants that are important food sources for wildlife;
7. prohibit hunting in the Izozog by outsiders.

Supported by considerable data from the field research, the round of meetings in November and December 1999 reviewed the wildlife management ideas proposed by the communities in the earlier meetings. For each proposal the team provided concrete examples from our experience with Izoceño hunters and from our field research on the game species of what the management idea meant in practice and how it could be applied. In each community the team asked which of these ideas were valuable and which were feasible in that hunters were willing to implement them, recording for each proposal the community's overall response. Responses generally favored the development of management plans but ranged from outright rejection as impossible or irrelevant (proposals 2, 3 , 4, 5), explanations of how current hunting practices in fact represent management (proposals 1, 3, 4, 5, 6), and full acceptance (7). Communities with a stronger tradition of livestock use and with less participation in seasonal emigration to the sugar cane harvest tend to favor active scientific management, whereas other communities emphasized traditional belief systems wherein supernatural forces ensure that wildlife resources will not be depleted (Noss and Cuéllar 2001).

More specific community meetings were also held to unite individuals and community representatives interested in the hunting of a particular species, such as parrots, tegu lizards, and peccaries, in order to discuss and develop conservation or management plans for those species. On the basis of the extensive research described above on wildlife as well as on hunting practices and traditions, the Kaa-Iya Project has developed conservation or management plans for subsistence as well as commercial use, including the following elements depending on the species: annual harvest quotas, age class or sex-based harvest quotas, distribution of quotas among interested hunters and communities, hunting seasons, and hunting methods. Data suggest that the principal prey species persist under current hunting pressure, with the exception of tapir and white-lipped peccary.

The framework of a general management plan for the TCO, together with a zonification of the TCO, was based on the following: delimitation of livestock, hunting, conservation areas; livestock management and veterinary care; prohibition of hunting by outsiders; and the strict conservation of the endangered guanaco (*Lama guanicoe*) and Chacoan peccary (*Catagonus wagneri*). The species and TCO management plans have been presented to the Izoceño General Assembly for a decision on how CABI wishes to proceed (Noss 2000), although formal implementation depends on the titling of the TCO, which is needed to provide the legal framework for any management plans.

TRAINING

In addition to the community meetings described above, practical experience is also an important training method, particularly for parabiologists and monitors, but also for community members participating in hunting self-monitoring as well as in other field activities. Furthermore, the Kaa-Iya Project has provided a series of formal field courses for parabiologists, monitors, community members, and Kaa-Iya National Park guards (table 5.1). Most courses have taken place in Izoceño communities or at the field camp. Aside from the specific technical content of the training courses, the emphasis on research design and the ability of each course participant to develop and carry out his or her own research project, based on the schoolyard ecology methodology (Feinsinger, Margutti, and Oviedo 1997), has been fundamental to the training program. As a result, all of the Izoceño parabiologists and monitors have participated in meetings with other ethnic groups in Bolivia (Chiquitano and Tsimane) and have presented papers, which are based on research projects they themselves have designed and/or implemented, at national and international wildlife management conferences. The participation in international congresses has been particularly motivating for Izoceño technicians, helping build confidence in themselves and in CABI and helping establish professional contacts with others working on similar issues in different contexts across Latin America. In 2002 several individuals assumed positions of responsibility for natural resource management projects undertaken by CABI within the TCO: guanaco conservation, parrot conservation, honey production by native stingless bees, and commercial harvest of tegu lizards and peccaries.

ENVIRONMENTAL EDUCATION MATERIALS

A further means to encourage discussion and implementation of natural resource management issues in the Izozog has been a strong environmental education program directed at both the formal education system and at community members

TABLE 5.1 Training Courses for Parabiologists, Wildlife Monitors, Hunters, and Park Guards

TOPIC	DATE	LOCATION
Field methods in biology I	August 1996	Yapiroa,Izozog
Field methods in biology II	August 1997	Ibasiriri, Izozog
Phenology	November 1997	Iyobi, Izozog
Radiotelemetry	April 1998	Cerro Cortado
Driver training	1998–2000	Santa Cruz
Data analysis	May 1998	Santa Cruz
First aid	May 1998	Santa Cruz
Necropsy and wildlife health	September 1998	Kuarirenda, Izozog
Herpetology	December 1998	Kopere Brecha, Izozog
Wildlife capture and handling	January 1999	Cerro Cortado
Basic administrative management	March–July 1999	Santa Cruz
Ornithology	April 1999	Karaparí, Izozog
Biology and ecology	February–March 2000	La Brecha, Izozog
Research design	April 2000	La Brecha, Izozog
Environmental impact assessment	October 2000	Tucavaca, Kaa-Iya
Herpetology	January 2001	Ravelo, Kaa-Iya
Research design	March 2001	Natividad, Chiquitanía
Protected area management	September 2001	OTS, Costa Rica
Research design	February 2002	Tucavaca, Kaa-Iya

and the public in general. The formal environmental education program includes a curriculum that has been implemented in all schools in the Izoceño communities (Combès et al. 1997) as well as a pair of teachers' guides to plants of the Chaco (Bourdy 2001, 2002) and a guide to wildlife of the Izozog (Cuéllar et al. 1998). The curriculum is currently being extended to the municipal school district. Nonformal education materials derive from Kaa-Iya research activities and include bulletins for general distribution on the following topics: armadillos; uses of wildlife; parrots and parakeets; guanacos; wildlife research at Cerro Cortado; jaguar (*Panthera onca*) and puma (*Puma concolor*) predation on livestock; Izoceño folk tales (Kaa-Iya Project 2000); wildlife health; and fishing in the Izozog. Additional written materials include billboards, posters and leaflets promoting conservation of the guanaco, while a series of radio broadcasts in Spanish and Guaraní have addressed guanaco conservation as well as broader natural resource management issues in the TCO.

The environmental education program emphasizes drawing upon and systematizing Izoceño environmental knowledge and fits in with CABI's efforts to ensure that Izoceño values and lifeways are not implicitly associated by the formal education system with backwardness and other negative characteristics. In addition, the concepts and information help prepare future generations of leaders to assume the technical and administrative challenges of managing this large area that includes the TCO and National Park.

CONCLUSION

The CABI-WCS experience in the Bolivian Chaco suggests that community-based conservation can be an important mechanism to ensure long-term biodiversity conservation. However, the CABI-WCS relationship is based on the explicit understanding that conservation and development are not synonymous and on an explicit definition of the specific interests and objectives guiding each institution's actions. The result is a relationship with clear and limited objectives, minimizing erroneous expectations or perceptions that could destabilize the relationship. At the same time, each institution's interests complement the other's in unexpected ways. CABI's interest in the KINP is not as a territory that Izoceños can occupy but principally as a buffer for the TCO from external pressures. The demand for the TCO itself is a public declaration by an indigenous people of its right to pursue an alternative development path, consistent with its culture and values, and that improvements in quality of life cannot be reduced to economic growth. Thus, CABI supports conservation objectives both in the KINP and the TCO. Meanwhile, WCS sees institutional strengthening and training as fundamental to CABI's long-term ability to effectively manage the KINP and the TCO land titling as the basis for any sustainable resource management. The same institutions, trained personnel, and land titles are essential elements for any development scenarios CABI wishes to pursue.

In order for the relationship to be successful, it must simultaneously address issues on regional and local scales and focus on institutional as well as biodiversity concerns. In fact, all are cross-cutting: for example, community meetings and training at the local level strengthen institutions for regional level action. Likewise, biodiversity concerns cannot only be addressed on a local scale through community management plans for particular species but also depend on landscape scale activities such as the establishment and defense of protected areas. The following lessons derived from the Bolivia experience can provide guidelines for community-based conservation efforts elsewhere in Latin America, including areas much smaller than the enormous expanse of the Bolivian Chaco (Painter et al. 2003).

1. The project created a vision of territory as a space for the long-term survival of a people. Here, protected areas and indigenous territories are complementary—not overlapping—elements. The principle of complementarity that is implicit in the vision becomes real in terms of management practice through the participatory approach to preparing the zoning proposal.
2. The project provided a strategic vision that united communities or indigenous groups. It also allowed consideration of long-term regional (including protected

areas, indigenous territories, and their surroundings) environmental and socio-economic impacts as a basis for negotiation with government and private institutions responsible for development policies and programs.

3. The project required long-term financing mechanisms, such as trust and development funds, that were independent and complementary to government or short-term project funds.

4. The project strengthened community organizations that were responsible for the management of supracommunal territories and resulted in an emphasis on transparency and accountability in the distribution of benefits and internal control.

5. The project promoted strategic relations with important neighbors, such as municipalities and other government agencies, as well as forestry and hydrocarbon concessions.

6. The project required a multidisciplinary focus that addressed not only the biological aspects of conservation but also the socioeconomic needs of local actors.

7. The project demonstrated that conservation actions and ventures must derive from the communities or these enterprises will be doomed to failure.

ACKNOWLEDGMENTS

This publication was made possible by the financial support of the United States Agency for International Development (USAID, Cooperative Agreement No. 511-A-00-01-00005). The opinions expressed are those of the authors and do not necessarily reflect the criteria of USAID. We also thank the Wildlife Conservation Society, the Capitanía del Alto y Bajo Izozog, our colleagues in the Kaa-Iya Project, and most importantly the communities and hunters of the Izozog for their participation and support.

Hunter self-monitoring data sheet with English translation

CACERIA   Comunidad _____
Quienes salieron? _____
_____

Cuántas horas o días duró la salida _____
Dónde cazó? _____ Clima_____
Qué cazó? _____
No. de etiqueta _____
Fecha _____Hora_____Tipo de monte _____
Cómo lo consiguió? Montados_____Cuántos perros _____
   Con qué arma _____
Sexo: Macho_____Hembra_____Tiene leche _____
   Cuántas crías en la barriga _____
Peso: Con tripas_____Sin tripas _____
Medidas: Total_____Cola_____Pata trasera_____Oreja _____
Edad: Juvenil_____Adulto _____
Animales heridos pero no cazados: _____
_____

Otros animales encontrados pero no cazados _____
_____

Por qué? _____
Observaciones: _____
_____

HUNTING Comunity _____
Who hunted? _____
_____

How many hours or days did the trip last? _____
Where did you hunt?_____Weather _____
What did you hunt? _____
Specimen no. _____
Date_____Hour_____Habitat type _____
How did you hunt? Horseback_____no. of dogs _____
   With what weapon? _____
Sex: Male_____Female_____Lactating _____
   Number of fetuses _____
Weight: Whole _____Cleaned _____
Measurements: Total_____Tail_____Hindlimb _____Ear _____
Age: Juvenile_____Adult _____
Animals injured but not captured: _____
_____

Other animals encountered but not hunted _____
_____

Why? _____
Observations: _____
_____

6

Fisheries in the Amazon Várzea

——————

HISTORICAL TRENDS, CURRENT STATUS, AND FACTORS AFFECTING SUSTAINABILITY

WILLIAM G. R. CRAMPTON, LEANDRO CASTELLO,
AND JOÃO PAULO VIANA

The várzea floodplain flanking the sediment-rich whitewater rivers of the Amazon basin is a mosaic of seasonally inundated rain forests, lakes, and winding channels. This ecosystem is exceptionally productive and species rich, with a large proportion of endemic taxa adapted to the prolonged annual floods. Várzea floodplains cover about 180,000 km², or approximately 2.6%, of the 7 million-km² area of the Amazon basin (Bayley and Petrere 1989). This figure does not include the less productive floodplains of nutrient impoverished blackwater and clearwater rivers. Junk (1997) estimated that the total area of seasonal floodplains in the Brazilian Amazon basin is 307,300 km², of which 40% (106,000 km²) is typical whitewater várzea (Bayley and Petrere 1989). An additional one million km² of the Brazilian Amazon's terra firme forest (above the seasonal floodplain) are periodically inundated by the flash flooding of streams (Junk 1997).

Várzeas support an astonishingly diverse fish fauna (Henderson, Hamilton, and Crampton 1998) and highly productive fisheries (Goulding, Smith, and Mahar 1996). Fish are unquestionably the most economically important of all natural resources in the várzea (Batista 1998; Queiroz and Crampton 1999a). They provide the main source of income for the rural settlers (*ribeirinhos*), underpin entire regional economies, and provide the main source of protein for the Amazon's rural and urban population. The fisheries of the Brazilian Amazon currently employ around 25,000 professional and 70,000 subsistence fishermen (Batista 1998). In the early 1990s fishing in the Central Amazon generated profits exceeding US$ 200 million/year (Batista 1998). The main theme of this article is that with appropriate management, várzea fisheries could provide a major contribution to economic growth and rural development in the Amazon basin, as well as a powerful incentive to conserve the habitats that várzea fishes rely upon.

Although there is growing evidence for the overfishing of some species (Bayley and Petrere 1989; Goulding, Smith, and Mahar 1996; Araujo-Lima and Goulding 1997; Barthem and Goulding 1997; Crampton 1999a, 2001), studies suggest that the overall fish productivity of intact Amazonian várzeas can comfortably support contemporary levels of exploitation (Bayley and Petrere 1989; Crampton and Viana 1999). The main threat to the fish stocks of the várzea is not overfishing but habitat loss (Goulding 1999). Floodplain fishes depend upon flooded forests and floating meadows for sustenance, refuge, and breeding sites (Goulding 1980; Crampton 1999b). Because the alluvial soils of the várzea support outstanding agricultural production in comparison to terra firme forests, most of the natural várzea forests and meadows of the middle to lower course of the Amazon have already been cleared or severely degraded by livestock ranching, agriculture, and predatory logging (Smith 1999). More than 70% of várzea forests may have already disappeared (Alexander 1994), whereas the loss of Amazonian terra firme rain forest is estimated at 10 to 14% (Ayres and Fonseca 1997; Schwartz 2000). Relatively intact várzea forests and meadows are still found in the Upper Amazon regions of Brazil and Peru, but their future is uncertain (Laurance 2000).

A major challenge to fisheries management in the várzea is restricting access and economic benefits to independent groups of fishermen. The federal government owns all the várzeas of the Brazilian Amazon. Fishing is permitted in any water body accessible by boat, and there are no formal regulations defining exclusive fisheries rights for any group of stakeholders, including the local populations (McGrath et al. 1999). The only exceptions to this situation occur in some conservation units and in a rising number of local lake-protection schemes that have received official recognition by Brazil's Institute for the Environment and Renewable Resources, IBAMA (Instituto Brasileiro do Meio Ambiente e dos Recursos Naturais Renováveis). For the most part, fisheries legislation in the Brazilian Amazon comprises state-imposed restrictions that are so poorly enforced as to be effectively nonexistent (Isaac, Rocha, and Mota 1993; Crampton and Viana 1999; McGrath et al. 1999). These conditions of almost unrestricted access and impotent regulation define the fish stocks of várzeas as a typical "open access resource," meaning that few stakeholders have exclusive rights of access or the incentive to control their own activities.

Following the collapse of several major fisheries around the world (McGoodwin 1990), planners are losing faith in state-imposed regulation and devoting serious attention to community-based models of fisheries management (Anderson 1986; Berkes 1989). In some ways várzea fisheries are ideally suited to community-based management. First, várzea settlers have evolved a semicommunal organization with cooperative labor (Lima 1999). Second, the outstanding economic value of floodplain fisheries provides the incentive for fishermen to defend their resources and undertake management. Third, várzea settlements are often located at strategic positions such as the entrance to lake systems. Finally, várzea fishermen have an excellent understanding of fish ecology.

In other ways várzea fisheries are less amenable to management. Many species of várzea fishes have complex migratory life histories, and their abundance in any given location is primarily a function of exploitation levels elsewhere. Local management has negligible effects on the stocks of these migratory species.

Self-motivated lake vigilance schemes began to appear spontaneously around the Amazon basin in the mid-1970s in response to invasions of predatory fishing fleets from the growing cities. These schemes had limited success, mainly because of political weakness and lack of infrastructure (Lima 1999). More recently, many self-motivated social movements (e.g., lake-protection schemes and rubber-tapping syndicates) have formed alliances with state or nongovernmental conservation organizations. Such alliances are intended to strengthen community projects by providing funding, training, and technical cooperation (Lima 1999). These kinds of partnerships may represent the most promising direction for fisheries management in the várzea (Ruffino and Isaac 1994; McGrath et al. 1999; Ruffino 1999).

This article brings together information from a variety of disciplines to provide a broad review of the history and status of várzea fisheries in the Brazilian territories of the Amazon. We emphasize the polarization of contemporary management strategies toward (1) government-based regulation and (2) community-based lake-management schemes, including those working in alliance with government authorities or NGOs.

VÁRZEAS: A LONG HISTORY OF HUMAN SETTLEMENT

The reports of early expeditions (Fritz 1922; Acuña 1942; Hemming 1978) and the abundance of archaeological sites along the Amazon suggest that a large indigenous population existed before European colonization (Palmatary 1939; Lathrap 1970; Smith 1980; Porro 1981; Meggers and Evans 1983; Porro 1983; Costa et al. 1986; Roosevelt 1987, 1991; Denevan 1996; Roosevelt 1999). Smith (1999) speculated that there might have been as many as fifteen million Amerindians in the basin before Europeans arrived. Archaeological studies reveal a pattern of large settlements closely spaced along the Amazon and located mostly on terra firme bluffs near várzea (Lathrap 1970; Denevan 1996; Roosevelt 1999). Here, Amerindians benefited from the resources of both várzea and terra firme ecosystems but avoided the flooding and biting insects of the várzea. We do not know when humans first arrived in the Amazon, but there is evidence for settlements near várzeas dating back to 11,000 or perhaps even 16,000 years ago (Roosevelt et al. 1991; Roosevelt 1999; Roosevelt et al. 1996).

How did várzeas provide such a large indigenous population with a sustainable supply of protein? The answer seems to be that several protein sources were exploited, with fish being less important than today. Roosevelt (1999) described the preponderance of turtle carapace fragments at sites dating from 11,200 to 9,800 years near Santarém, Pará. Roosevelt (1999) also found the bones of many fish species, including small characiforms, catfishes, and the pirarucu (*Arapaima gigas*), at paleo-indian sites (from 11,200 to 8,000 years ago) and in more recent pre-

Colombian sites. Roosevelt et al. (1991) reported that freshwater mussels were an important food source near várzeas of the Central and Lower Amazon.

Accounts of travelers and naturalists until the beginning of this century suggest that turtles provided the main protein supply for Indians along the Amazon (Spix and Martius 1822–1831; Wallace 1853; Bates 1863; Coutinho 1868; Fontes 1966; Ferreira 1971, 1972). The most important species at the time was the giant turtle (*Podocnemis expansa*), which emerges onto nesting beaches at low water. Indians harvested turtles for meat and collected eggs to make lard and lamp oil. Manatees (*Trichechus inunguis*) were also abundant before mass slaughters occurred in the seventeenth century (Vieira 1925–1928) and perhaps also represented a significant food supply.

Soon after the arrival of Europeans, the indigenous population of the Amazon suffered a massive decline, mainly because of lack of resistance to Old World diseases (Hemming 1978). Few European immigrants took their place and for the next two centuries much of the basin became almost devoid of human population and commerce. Verissimo (1895) reviewed fishing in the Amazon between the sixteenth and eighteenth centuries. From 1667 until 1827 the imperial administration established Royal Fisheries for commerce in the states of Amazonas, Pará, and Maranhão. It is clear from Verissimo's accounts that fishing pressure in the Amazon was generally low during this period. However, during the late nineteenth and early twentieth centuries, turtles were exploited recklessly for the manufacture and export of an oil made from the eggs. The populations of *P. expansa* and some other species were almost obliterated (Sternberg 1995), and today turtles form a negligible (although prized) component of the Amazonian diet. In the late twentieth century caiman, which were previously abundant in várzeas, were reduced drastically by skin hunters. The black caiman (*Melanosuchus niger*) was particularly affected. However, hunting restrictions have since resulted in a wide-scale recovery of populations (R. da Silveira, INPA, pers. comm.).

The rubber boom from the 1870s to the 1920s (and a minor boom after World War II) brought a wave of immigrants into the Amazon, but most of the latex tapping took place in terra firme forests or in seasonally inundated blackwater igapó forests away from the várzeas. Following the final collapse of the rubber boom in the 1920s, many unemployed workers settled in várzea floodplains. Settlements grew around the outposts of the patrons who controlled trade in the area through the aviamento system of debt bondage. The major economic activities in the early twentieth century were the cutting of firewood for steam ships and the commercial extraction of pirarucu, manatees, and turtles. During the 1940s and into the 1970s, jute growing provided a substantial source of income in the várzea and, along with agriculture and fishing, attracted even more rural Amazonians away from the less productive blackwater and terra firme systems (Goulding, Smith, and Mahar 1996).

By the late 1960s the rural population of the Amazon had grown steadily but was still low in comparison to the estimated pre-Columbian population (Smith 1999). The cities of the Amazon were only just beginning to grow rapidly, and the pressure on várzea fisheries was still light. However, most of the Amazon's rural population had become concentrated in várzea floodplains. As the jute industry collapsed in the early 1970s and as the urban market for fish began to grow at about the same time, many várzea farmers turned to fishing (McGrath, Silva, and Crossa 1998; Smith 1999). With turtle, manatee, and caiman populations almost extirpated, fish remained as the only major natural source of protein. In the next chapter of Amazonian history—the explosive growth of the urban population—várzeas and their fish stocks were poised to play a new and central role in the regional economy.

OPERATION AMAZÔNIA:
URBANIZATION AND COMMERCIAL FISHING

The status of várzea fisheries changed radically with the implementation of Operação Amazônia in 1966. In just three decades this package of resettlement and development projects resulted in more changes to the ecological fabric of the Amazon than in all previous human history (Kohlhepp 1984; Goulding, Smith, and Mahar 1996). The human population of the Brazilian portion of the Amazon basin (considering Acre, Amapá, Amazonas, Pará, Rondônia, Roraima, and Tocantins) increased from 1.8 million in 1960 (Costa 1992) to 9.2 million in 1991 (IBGE 1991) and to 12.9 million in 2000 (IBGE 2001). Most of the growth occurred in urban areas. Manaus grew from just 320,000 in 1970 (Costa 1992) to 1.4 million in 2000 (IBGE 2001) and continues to grow by 6% per year (Costa 1992). In Amazonas, 74.8% (2.10 million) of the total population (2.81 million) lives in towns with over 800 inhabitants (IBGE 1991).

In the 1970s and early 1980s fish represented the primary protein source for people in large Amazonian cities. Estimates of average per capita fish consumption of Manaus at this time varied from 102 g/day (Amoroso 1981) to 155 g (Shrimpton and Giugliano 1979). These figures represent fish consumption of between four and seven times the world average. The dominance of fish continues to prevail in the towns of the Upper Amazon, although in Manaus fish is declining in importance because of imports of cheap poultry and meat from Southern Brazil.

Rapid urban growth and dependence upon fish protein in the 1970s and 1980s created a demand for fish above that supplied by rural fishermen. The response was the development of commercial fishing fleets. Cheap credit and engines were easily available at the time, and the number of (inboard) motorized boats in the Amazonas interior increased from around 70 to 1,700 between 1970 and 1988 (Costa 1992). In 1995 the commercial fishing fleet of the Central Brazilian Amazon (encompassing the Amazon and its tributaries from Tabatinga, Amazonas, to Ilha Tupinambarana, Pará) comprised around 2,500 fishing vessels with inboard engines (Batista 1998).

Until the 1970s salting and the production of a fishmeal called *piracui* were the main preservation methods for fish. The wide scale introduction of ice not only shifted the market toward fresh fish but also allowed fishing to occur at much greater distances from port. The modern fleet of *geleiras* (boats with ice holds) can travel to fisheries hundreds of kilometers away. Várzea lakes were targeted from the onset due to the ease with which tambaqui (*Colossoma macropomum*), pirarucu, and other premium quality fish could be harvested using seine and gill nets. These techniques can be devastatingly indiscriminate, and the geleira crews often discard the lower-value species. The expansion of predatory commercial fishing practices throughout the várzeas of the Brazilian Amazon introduced not only unprecedented pressure on key commercial species but also a new era of conflicts over fishing rights with the local ribeirinho communities.

FISHERIES OF THE MODERN VÁRZEA

The commercial and subsistence fisheries of várzeas and adjacent whitewater river channels are multispecific, seasonal, and dependent on several types of equipment (Meschkat 1961; Batista 1998). They focus on three ecological categories of fishes:

Resident várzea species spend their entire life cycle inside the várzea. They include pirarucu, aruanã (*Osteoglossum bicirrhosum*), tucunaré (*Cichla* spp.), many other cichlids, and some armored catfishes (Loricariidae) (Crampton 1999b).

Migratory characiforms undertake most or at least the initial part of their growth phase in várzea floodplains and then disperse upstream to colonize whitewater and low-nutrient floodplain systems up to several hundred kilometers away. These fishes stay upstream and eventually spawn along the edge of whitewater river channels. Their juveniles are recruited into várzea floodplains adjacent to and downstream of the spawning sites (Goulding 1980). Commercial species with this type of complex migratory life history include tambaqui, pirapitinga (*Piaractus brachypomus*), pacus (Myleinae), curimatá (*Prochilodus nigricans*), jaraquis (*Semaprochilodus* spp.), aracus (Anostomidae), and the matrinchãs (*Brycon* spp.).

Migratory catfishes of the family Pimelodidae (*peixe-liso*) undertake long-distance migrations up the Amazon's main whitewater rivers. At least two species, the dourada (*Brachyplatystoma flavicans*) and the piramutaba (*B. vaillantii*), migrate from the Amazon's estuary to headwater tributaries thousands of kilometers away (Barthem and Goulding 1997). Other species such as the surubim (*Pseudoplatystoma fasciatum*) and the caparari (*P. tigrinum*) are thought to undertake shorter migrations, but the distances involved are unknown (Goulding, Smith, and Mahar 1996).

The capture of characiform fishes and catfishes in the main river channels is undertaken on an almost completely commercial basis. Fishermen distinguish between the long-distance upstream movement of migratory characiform fishes, the *arribação*, the timing of which varies among species, and local spawning runs, the *piracema*, which occur during the rising water period. Piracemas usually involve movements out of whitewater or black/clear water floodplains into spawning grounds along whitewater river margins. Migrating characiform fishes are sold to urban fish markets, while catfishes from the main rivers and their side branches (*paranás*) are mostly sold to *frigoríficos* (freezer storage plants) for export to Colombia, Southern Brazil, and Peru. Fishing within the floodplain is undertaken both by visiting professional fishermen and by ribeirinhos for subsistence and local sale. Falabella (1994) and Batista (1998) describe fishing methods in the Amazon, updating earlier accounts by Petrere (1978 a,b), Goulding (1979), and Smith (1979).

MODELS OF FISHING SUSTAINABILITY IN THE AMAZON VÁRZEA

Várzea floodplains, because of their relatively nutrient-rich waters and annual deposits of alluvium, are much more productive than the floodplains of nutrient-impoverished blackwaters and clearwaters (Schmidt 1973a,b; Fittkau et al. 1975; Schmidt 1976; Goulding, Carvalho, and Ferreira 1988; Henderson and Crampton 1997; Saint-Paul et al. 2000). Few studies have attempted to quantify the fish productivity of Amazon floodplains. Using carbon flow-analysis in várzeas near Manaus, Bayley (1989) estimated that around 1% of the carbon fixed annually by photosynthesis is assimilated by fishes, representing a total fish biomass production of between 174,000 and 523,000 kg/km^2/year. Bayley (1980) previously calculated a maximum fishing yield of 12,000 kg/km^2/year for the same area—around 17% of total annual fish production.

Bayley and Petrere (1989) reviewed potential yield estimates from tropical floodplains and concluded that a conservative sustainable yield of fish from typical whitewater floodplains is probably closer to 5,000 kg/km^2/year. This is within Welcomme's 1979 estimated range of sustainable tropical floodplain fish yields of from 4,000 to 6,000 kg/km^2/year. Quantitative estimates of actual yields for várzeas include 2,000 kg/km^2/year near Manaus in the late 1970s (Bayley and Petrere 1989) and 1,800 kg/km^2/year near Iquitos, Peru, in 1981 (Bayley et al. 1992).

The total area of whitewater floodplain in the Brazilian Amazon basin is an estimated 106,400 km^2 (Bayley and Petrere 1989). Assuming a minimum production of 4,000 kg/km^2/year, a per diem consumption of 100 g of fish biomass by the entire population, and a conservative estimate that half of this fish biomass is edible (Batista 1998), the whitewater várzeas of Brazilian territory should theoretically be able to provide 11.7 million people (almost the entire population of the Brazilian Amazon) with the World Health Organization recommended minimum daily protein requirement of 50 g.

However, fishing pressure is not evenly applied to standing fish biomass. Of the estimated 2,500 or more species of fishes in the Amazon basin (Val and Almeida-Val 1995), of which perhaps around 700 frequent várzeas, only a few dozen are eaten in appreciable quantities. Fewer than ten species provide more than three-quarters of the fish biomass extracted from várzeas by commercial fishing (Goulding 1979; Smith 1979b; Gerrits and Baas 1997; Barthem 1999a). Today's várzea fisheries are characterized by the overexploitation of a very small number of key species.

There is no evidence of overexploitation for the majority of other species (Crampton and Viana 1999).

Goulding, Smith, and Mahar (1996) speculated that urban centers with more than 5,000 inhabitants account for more than three-quarters of the total fish consumption of the Amazon basin. However, Batista (1998) argues that, even though rural Amazonians are a minority, they eat so much more fish than urban Amazonians that they create a greater overall demand. Batista (1988) reported per capita/day fish consumption in several Amazonian towns: Manacapuru (34 to 104 g), Itacoatiara (160 g), Parintins (60 g), and Santarém (28 g). Batista (1998) and Cerdeira, Ruffino, and Isaac (1997) recorded per capita/day fish consumption of from 400 to 800 g in rural settlements or small towns (less than 5,000 people) in Pará and Amazonas. Batista's argument for a larger rural consumption of fish seems compelling when one considers the following calculations. In the states of Amazonas and Pará combined, 44% (4.01 million) of the total population (9.00 million) lives in rural settlements with fewer than 800 inhabitants (IBGE 2001). Assuming a maximum per capita daily consumption of 100 g for the urban population and a minimum per capita daily consumption of 400 g for the rural population, rural Amazonians consume at least 3.2 times more fish than urban Amazonians. Although commercial fishing fleets are driven by the demands of urban Amazonians, rural Amazonians seem to make the greatest demands on fish biomass. This overlooked consideration is important for the planning of fisheries. It implies, for example, that as much as two-thirds of the total fish landings of the Brazilian Amazon cannot be assessed or monitored using the standard method of monitoring market landings.

CONTEMPORARY STATUS OF FISH STOCKS

MAJOR SPECIES

Pirarucu was exploited throughout the colonial period as a substitute for sun-dried salted codfish. Records show a stable supply to Manaus and Belém from 1885 to 1920, with landings exceeding 1,000 tons/year (Fontenele 1948). Following the collapse of the rubber boom in the 1920s, many workers settled in várzeas and took to pirarucu fishing. Landings started to decline during the 1930s (Fontenele 1948), and by the late 1940s only around 300 tons/year were landed in Belém (Menezes 1951). Until the 1970s, pirarucus were mostly harpooned. Now, commercial fishermen employ gill nets, which are far more effective. Few quantitative data are available on the overall status of pirarucu in the Amazon basin, but two trends are clear. The first is that the size structure of pirarucu populations has changed over the last three decades in all but the most remote areas. Pirarucus up to 3 m long were once common but specimens over 2.5 m long are now rare. The second trend is that in some areas pirarucu have reportedly been depleted to the point of commercial extinction (Goulding, Smith, and Mahar 1996).

Tambaqui was once a staple food species. It accounted for around 50% of the total fish catch in the Manaus fish market during the 1970s (Goulding, Smith, and Mahar 1996) and was not considered overexploited until the mid-1980s (Petrere 1983). Merona and Bittencourt (1988) reported declining tambaqui landings in Manaus during the late 1980s. Today, large adults (over 70 cm) are rare throughout most of the Amazon, and the majority of marketed fish are undersized (Goulding, Smith, and Mahar 1996). Evidence for overexploitation of tambaqui has been reported around Tefé (Costa et al. 1999), Manaus (Ribeiro and Petrere 1990; Batista 1998), and Santarém (Isaac and Ruffino 1996).

Pimelodid catfishes are captured in huge numbers along the entire course of the Amazon, mostly for the frigorifico market. Pimelodid catfishes are usually captured during their upstream migrations. The piramutaba is the most important of all fish species exported from the Amazon basin since the 1970s (Goulding, Smith, and Mahar 1996). In the late 1970s 22,000 tons were landed annually in the Amazon estuary, of which three-quarters were exported. By 1990 the harvest had halved. Due to declining yields, the value of the harvest dropped from a peak of US$ 13 million in 1980 to around US $ 3 million in 1986 (Barthem and Petrere 1992). The dourada also appears to be overfished in the Amazon estuary, with the Belém market now dominated by juveniles (Goulding, Smith, and Mahar 1996). Goulding, Smith, and Mahar (1996) predicted that the industrial-scale exploitation of estuarine dourada and piramutaba populations would eventually precipitate a collapse of the inland fisheries of these species. Gerrits and Baas (1997) have already reported declining landings of piramutaba in the Óbidos area of Pará, some 600 km inland. On the basis of market landing data, Isaac, Ruffino, and McGrath (1998) reported overfishing of surubim and caparari in the lower Amazon region of Santarém. The status of other pimelodid catfishes is unknown.

OTHER MAJOR COMMERCIAL SPECIES

In terms of biomass the detritivorous curimatá and jaraquis (Prochilodontidae) are probably now the dominant food fishes in Amazon markets (Batista 1998; Barthem 1999a,b). There is, as yet, no firm evidence for overexploitation, although Batista (pers. comm.) has observed a reduction in jaraqui sizes at Manaus markets over the last decade. Detritus constitutes a major proportion of the biomass of Amazonian aquatic systems (Araujo-Lima et al. 1986; Bayley 1989), perhaps explaining the enormous productivity of these fishes. Likewise, the abundance of newly recruited fishes in várzeas means that the exclusively piscivorous tucunarés represent a direct trophic conversion of a vast but unmarketed protein resource. Goulding, Smith, and Mahar (1996) speculate that this explains the apparent resistance of tucunarés to intense fishing pressure. Several characiform fishes, such as matrinchãs, pacus, and the sardinhas (*Triportheus* spp.), are omnivorous, eating seeds, fruit, insects, and other fish in floodplain forests. The generalist nature of these fishes may account for their continued abundance, despite heavy fishing pressure.

MINOR SPECIES

In addition to the fish groups discussed above, some 150 other várzea fish species are eaten (Goulding 1979; Smith 1979b; Barthem 1999a; Crampton and Viana 1999; Crampton et al. this volume). Some of the previously less popular food species, such as piranhas (*Serrasalmus* and *Pygocentrus*), are now marketed in increasing quantities to compensate for shortages of other fishes.

MONITORING FISH LANDINGS

The assessment of fish stocks in the Amazon basin is based almost entirely on market landing data. The regional planning of várzea fisheries is limited by the paucity of such data. It is impossible to tell, for example, what the current total landings of fish are in the Brazilian Amazon. The markets of Manaus were monitored from 1976 to 1978 (Petrere 1978b), for a few years in the early 1980s (Merona and Bittencourt 1988), and then from 1993 to date (Batista 1998). Landing data have been collected at Tefé by the IDSM (Instituto de Desenvolvimento Sustentável Mamirauá) since 1992 (Barthem 1999a) and at Santarém since 1994 by Instituto Iara (Instituto Amazônico de Manejo Sustentável de Recursos Naturais, i.e., Amazon Institute for Sustainable Resource Management) (Ruffino, Isaac, and Milstein 1998; Ruffino 1999). Short-term landing data were collected in the late 1970s at Porto Velho (Goulding 1979) and Itacoatiara (Smith 1979b). Data from these studies and some governmental statistics from Manaus and Belém constitute just about all that is known about fish landings in the Brazilian Amazon.

Even the best market surveys are only partially informative about total landings. In the first place two-thirds or more of fish consumption may be on a subsistence basis (see above). Also, the trade in controlled species (such as pirarucu) and undersized or closed-season catches is usually diverted from public markets. Finally, the sale of large catfishes to frigorificos and the transport of fish to distant markets via passenger boats are notoriously hard to monitor. Despite these difficulties, landing data probably represent a more realistic means of assessing the (relative) status of stocks of Amazonian fishes than direct stock assessments from wild populations. Researchers from the University of Amazonas, the Federal University of Pará, the Mamirauá Sustainable Development Institute (IDSM), the Iara Institute, and Projeto Várzea in Santarém are joining forces with IBAMA to form a linked web of data collectors throughout the Brazilian Amazon.

ALTERNATIVE PROTEIN SUPPLIES

The management of wild fish stocks should continue to be a regional priority, but the protein demands of the Amazon's expanding population will inevitably need to be met by other forms of production. Cattle and water buffalo ranching are ecologically unacceptable forms of protein production in várzeas (Goulding, Smith, and Mahar 1996). Promising and more acceptable options are fish and poultry farming. Several species of Amazonian fishes, in particular pirarucu and tambaqui, respond well to domestication, and fish farming is now becoming lucrative business in the Amazon and in southern Brazil (Cerri 1995; Smith 1999). Cerri (1995) reports that farmed pirarucu and tambaqui can yield up to 4,560 kg/ha/year and 2,800 kg/ha/year respectively of marketable flesh. This compares very favorably with the production of water buffalo (225 kg/ha/year), cattle (203 kg/ha/year), and sheep (144 kg/ha/year) (Cerri 1995).

Chicken is cheaper than all premium quality fishes and represents a growing supply of animal protein in the Brazilian Amazon. Most is imported frozen from Southern Brazil, but the Amazonas State Livestock Development Institute, IDAM (Instituto de Desenvolvimento Agropecuário do Estado do Amazonas), is promoting poultry production in Amazonas state (E. Nunes de Sá, IDAM, pers. comm.). The captive production of Amazonian turtles is also expanding, and tagged and IBAMA-certified captive-raised turtles are now sold in some Manaus supermarkets. Smith (1999) describes other options for forest-friendly livestock production, including pigs, ducks, and such domesticated game as capybara (see Nogueira-Fiho and Nogueira, this volume).

ACCESS RIGHTS AND FISHERIES LEGISLATION IN THE VÁRZEA

All seasonally flooded land in the Brazilian Amazon is owned by the state. In fact, the semiaquatic nature of várzea places it under the juridical responsibility of the Brazilian Navy. Although the state can concede rights of use to individuals or companies, as it has done in some parts of Pará, várzeas cannot be privately owned (McGrath et al. 1999). Many ranchers and ribeirinhos in várzeas of the lower Amazon hold title deeds that routinely change hands, but these documents have no legal standing. The poorly defined status of land tenure is a major barrier to defining management plans for várzea fisheries.

The formulation and enforcement of inland fisheries legislation in Brazil is the direct responsibility of IBAMA. By law, fishermen are permitted unrestricted access to all waterways under the control of the state (i.e., all várzeas) except those within reserves and national parks. Streams and ponds in the terra firme surrounded by private land are recognized as private property but do not support substantial fisheries.

RESTRICTIVE REGULATIONS

IBAMA policy for inland fisheries regulation is based on a series of legally binding restrictions on fishing activities. The early framework was devised in the late 1960s and covers restrictions on equipment, minimum sizes, and closed seasons. These restrictions were based on fisheries research from southern Brazil and in many cases were inappropriate for Amazonian waters (Isaac, Rocha, and Mota 1993). Closed seasons for pirarucu (fig. 6.1), however, were based on regional studies. Three categories of fishing were defined by this early legislation: commercial, scientific, and sport fishing (although not, to universal surprise, subsistence) (Fischer, Chagas, and Dornelles 1992). In recent years there have been extensive modifications and additions to the laws, including minimum size limits, closed seasons for additional species, and a list of 175 fish species that can be legally exported for the aquarium

RESPEITE O DEFESO

SUPERINTENDÊNCIA DO ESTADO DO AMAZONAS

FIGURE 6.1   1995 IBAMA notice posted in ports and fish markets to remind fishermen of the closed season (defeso) for pirarucu (above) and tambaqui (below).

trade (IBAMA 1996). Some IBAMA regulations are highly restrictive. On the grounds that pirarucu had reached an "advanced stage of over-exploitation," IBA-MA-Amazonas declared a statewide ban on the capture and commercialization of pirarucu in 1996. This ban has been extended without interruption and was still in effect as of July 2002.

An important law in 1994 recognized the authority and competence of regional IBAMA superintendents to enact temporary fishing regulations or closed seasons of up to two months in response to information concerning overexploitation. For instance, a 1997 decree introduced measures to control the total number of commercial fishing boats in Lake Tefé, Amazonas (L. McCulloch, IBAMA, pers. comm.). The devolution of decisions to the heads of regional IBAMA posts is part of a trend toward combating local problems through tactical response rather than a fixed global strategy. Nonetheless, each decision must still be codified as a formal decree (*portaria*) and published in the government's official gazette (*Díario Oficial*) before it can be enacted. These decrees can be subject to long delays, and it is not unknown for them to disappear in a sea of paperwork.

Another advance in IBAMA policy is the recognition of the potential of lake-protection schemes set up by ribeirinhos. Many fishing accords developed by communities around the Amazon have been granted legal backing by IBAMA decrees. IBAMA posts encourage communities to submit management plans prepared by regional fishing councils (Conselhos Regionais de Pesca). Proposals that make provisions for reconciling rights of access with commercial fishermen are supposedly favored. Nonetheless, IBAMA-supported lake-protection schemes are unpopular with commercial fishermen. They argue that IBAMA is awarding privileges to communities, marginalizing professional fishermen, and ignoring the question of how both parties could benefit from joint management (Batista 1998).

THE CHALLENGES OF ENFORCEMENT

Many of IBAMA's restrictive regulations are believed to be both unrealistic and based on insufficient research. Isaac, Rocha, and Mota (1993) provide a critique of the main problems of contemporary fisheries regulations. One recurring criticism refers to the protection of fish during their migrations, when in fact they are often more vulnerable during the lowest water period (Goulding, Smith, and Mahar 1996). Moreover, IBAMA is unable to enforce most of the wide range of measures intended to control fishing in the Brazilian Amazon. IBAMA posts are widely spaced, underfunded, and operated by staffs that are historically undermotivated. Consequently, most of the Amazon basin receives only superficial vigilance (Goulding, Smith, and Mahar 1996; Isaac, Ruffino, and McGrath 1998; Ruffino, Silva, and Castro 1998b; Crampton and Viana 1999; McGrath et al. 1999). Manifestations of IBAMA's failure to enforce regulations include the continued trade in pirarucu following its 1996 suspension in Amazonas and the ubiquitous marketing of undersized tambaqui.

IBAMA REFORM

In response to the difficulties discussed above, IBAMA is currently undergoing considerable reform, including the employment of a new generation of staff (Anon. 2000) and the aforementioned devolution of management decisions to regional offices. One of IBAMA's most innovative recent initiatives is the training of Voluntary Environmental Agents (AAVs), who are expected to complement the activities of IBAMA's field agents and to take responsibility for environmental education in local schools and village meetings (Crampton et al. this volume). IBAMA has also for some time been contemplating the potential for schemes of integrated fisheries management that operate at a regional level (Fischer, Chagas, and Dornelles 1992). The IBAMA-supported Instituto Iara in várzeas near Santarém, Pará, was the first working example of this kind of scheme (Ruffino 1999) and is described later.

These initiatives typify IBAMA's general trend toward decentralized administration and comanagement with the public sector. A seminal internal report (IBAMA 1997) conceded that resource management in the Amazon cannot be resolved through the straightforward enforcement of rules (instructive management). Instead, IBAMA instigated a policy of forging new institutional cooperation and consultation with a wide range of stakeholders (consultative management), such as municipal authorities, fishing and extractive syndicates, environmental NGOs, research institutes, and universities.

COMMUNITY-BASED FISHERY MANAGEMENT

MANAGEMENT PROCEDURES

Várzea communities have practiced basic fishery management techniques for at least the last four decades (Lima 1999). Lakes are usually divided informally into protection, subsistence, and commercialization categories, with the latter often located further away from the community. Subsistence lakes are managed with frequent, low-intensity harvesting and selective fishing techniques. Commercialization lakes are fished infrequently but intensively with less selective techniques (such as. gill netting) and are often left for a period of fallow before being refished. Studies have demonstrated that fish production of várzea lakes increases under community management schemes (IPAM 2000a). McGrath, Castro, and Futemma (1994) demonstrated that a várzea lake near Santarém under community management produced up to double the yield of an unmanaged lake for some commercial species. Crampton et al. (this volume) and Viana et al. (this volume) describe community lake-management schemes in the Mamirauá Sustainable Development Reserve, RDSM (Reserva de Desenvolvimento Sustentável Mamirauá).

Floodplain communities regard commercialization and protection lakes as economic security. During times of difficulty, for example when crops fail due to an unusually protracted flood, the bumper yields of commercialization lakes can provide much needed cash. For the same reasons many families in the várzea like to keep a few head of cattle that can be sold in times of difficulty.

MIGRATORY FISH STOCKS: MANAGEMENT DIFFICULTIES

The migratory life history of many characiform fishes means that their recruitment in any given várzea is primarily a function of exploitation conditions elsewhere. Species like tambaqui are therefore less amenable to in situ management than are resident species like pirarucu. Ribeirinho communities make no attempt to protect or manage the migratory pimelodid catfish stocks of whitewater rivers. The main channels of the Amazon are considered by ribeirinhos to be free for all. Moreover, local management would have negligible effects on their stocks.

CONFLICTS AND VIGILANCE

In the early 1980s campaigns for the defense of community lakes sprung up in várzeas along the Amazon basin. This came chiefly as a response to the growth of predatory fishing by commercial fleets. The early stages involved assistance from the Catholic Church through its Pastoral Land Commission, the Comissão Pastoral de Terra (CPT), and provided some empowerment for communities to expel commercial fishermen (CPT 1992; Lima 1999; McGrath et al. 1999). Around this time villages began to use the term *comunidade* (community) with hierarchical levels of organization and a village committee. These early lake-protection schemes met with some success, but the campaigns were blighted by intercommunity disputes, the weak legal status of várzea residency, and transport or communication difficulties. Moreover, government authorities never endorsed these early lake-protection schemes.

ECOLOGICAL PARTNERSHIPS

Escalating concern about environmental degradation and inequitable development stimulated a recent increase in the number of Brazilian and international conservation-oriented NGOs operating in the Brazilian Amazon. At the same time the growing recognition of the importance of involving local people in biodiversity conservation has encouraged many conservation and development projects to forge partnerships with previously existing social movements (Lima 1999). These alliances can greatly strengthen self-motivated movements by providing the funding, the legal support, and the technical cooperation necessary to formulate management plans, enhance production efficiency, or pursue alternative economic activities. Many ecological partnerships involve programs of environmental education, which seek to increase the general ecological awareness of rural people (Hall 1997). Environmental education programs usually also work with government educational authorities to improve basic educational standards. Illiteracy and innumeracy are perhaps the greatest constraints to the economic independence and self-confidence of rural people.

Many conservation alliances between local people of the Brazilian Amazon and either NGOs or government agencies concentrate on extractive forest products or timber and are based in areas where fishing is not a major economic activity (Lima 1999). There are currently only three projects in the Brazilian várzea that are based on community participation and that have a substantial fisheries component.

PROJETO VÁRZEA

Located at Santarém, Projeto Várzea promotes the sustainable exploitation of fisheries and other natural resources in várzeas of the Middle Amazon (fig. 6.2). This project is run by the (nongovernmental) Institute for Amazonian Environmental Research, IPAM (Instituto de Pesquisa Ambiental da Amazônia), in partnership with local communities and other stakeholders in the Santarém region. The project lists six goals (IPAM 2000b): to develop and strengthen community lake-management programs, to diversify the management strategies of várzea communities, to develop a program of environmental education, to study economic and ecological trends in the fisheries sector, to develop regional fisheries policies, and to provide management and marketing training to local fisheries organizations.

Projeto Várzea is working with várzea communities and the commercial fishing syndicate of Santarém to consolidate a framework for rights of access and regional

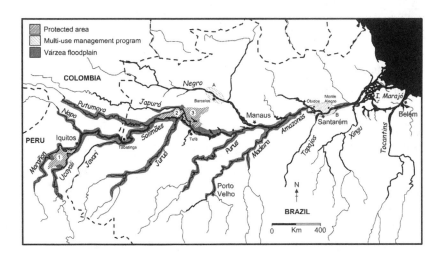

FIGURE 6.2 Location of major fisheries conservation and management programs in the Amazon basin: Pacaya-Samiria National Reserve, Loreto, Peru (1); Mamirauá Sustainable Development Reserve (2); Amanã Sustainable Development Reserve (3); ornamental fish catching initiatives directed by Projeto Piaba (A); and multiple-use fishery management initiatives directed by Projeto Várzea and Instituto Iara (B).

fisheries management. It is also working with communities to promote agricultural activities that reduce the destruction of levee forest and floating meadows and to restore floating meadows damaged by buffalo and cattle ranching (M. Crossa, Projeto Várzea pers. comm.). McGrath, Castro, and Futemma (1994) and McGrath et al. (1994, 1999) review the fisheries management activities at Projeto Várzea.

INSTITUTO IARA

Also based at Santarém, Instituto Iara (previously Projeto Iara) is responsible for the administration of fisheries resources in the Middle Amazon. The Institute's acronym, Iara, is a mythical nymphlike apparition in Amazonian folklore (Smith 1996). Since 1996 Instituto Iara has been "developing, testing and consolidating institutional measures for the sustainable use of fisheries resources in the Middle Amazon that are compatible with the interests and needs of local populations and of the regional and national economy and society" (IBAMA 2000). Instituto Iara is based on technical and financial cooperation between IBAMA and the German technical cooperation agency, GTZ (Deutsche Gessellschaft für Technische Zusammenarbeit). It also involves cooperation with several Brazilian academic institutions. Through a multidisciplinary program of research, training, environmental education, and monitoring, Instituto Iara is working closely with the full spectrum of stakeholders involved in fishing in the Middle Amazon region to define rights of access to fisheries resources and to develop management plans for sustainable use. The training of Voluntary Environmental Agents from local communities and from Santarém's fishing cooperative (Colônia de Pescadores-Z20) forms an important part of these initiatives. Instituto Iara's sphere of influence affects around 250,000 people living in várzeas along a 200-km stretch of the Rio Amazonas between Óbidos and Monte Alegre (fig. 6.2) (M. Ruffino, Instituto Iara, pers. comm.). Many of the activities and results of Projeto Iara are described by Fischer (1995) and Ruffino (1999).

THE MAMIRAUÁ AND AMANÃ SUSTAINABLE DEVELOPMENT RESERVES (RDSM/RDSA)

The RDSM is an 11,240-km^2 area of várzea located at the confluence of the Rios Solimões and Japurá. This Reserve was originally established in 1990 and was given Sustainable Development Reserve (SDR) status in 1996. Here, local people in partnership with the Mamirauá Sustainable Development Institute (IDSM) are mounting an integrated sustainable resource use program. Crampton et al. (this volume) and Viana et al. (this volume) provide a detailed overview of fisheries management in the RDSM. This reserve constituted the first of a new category of Brazilian conservation unit that permits the presence of traditional peoples and allows them exclusive rights of access to the natural resources of the area. In 1999 a second SDR, the 23,500-km^2 Amanã Sustainable Development Reserve (RDSA) was inaugurated (fig. 6.2). This reserve is also administered by IDSM and was established to form a contiguous corridor between the RDSM and the Jaú National Park on the west bank of the Rio Negro. Fisheries management in the RDSA is planned for the future and will follow the RDSM model.

OTHER PROJECTS

The only major conservation and development projects in a várzea floodplain outside Brazilian territory are being developed in the Reserva Nacional Pacaya-Samiria in Peru (see Tello this volume). This 21,508-km^2 state-administrated reserve located at the confluence of the Rios Ucayali and Marañon (fig. 6.2) has an active fisheries management program built around community participation and a multiple-use zoning system (COREPASA 1986; Bayley et al. 1992; Durand and McCaffrey 1999).

One other fisheries-dominated project in the Brazilian Amazon deserves a mention although it is not based in várzea floodplains. Projeto Piaba is based in the town of Barcelos on the blackwater Rio Negro (fig. 6.2) and has investigated the biological and economic sustainability of ornamental fish catching in the area (Chao et al. 2001). The ornamental fish trade involves around 1,600 part-time fishermen in the area and contributes to more than 60% of the economy of the municipality of Barcelos. Commercial food fishing is relatively unimportant in the nutrient-poor blackwaters of the middle Rio Negro. Chao et al. (2001) concluded that current extraction levels of ornamental fishes, including the heavily exploited cardinal tetra (*Paracheirodon axelrodi*), are sustainable. They argue that the trade encourages habitat conservation and contributes positively to local economies and livelihoods. In addition to research Projeto Piaba is developing a program to improve the production and marketing efficiency of rural fishermen.

HABITAT CONSERVATION

Several authorities on Amazonian conservation have stressed that habitat loss through the deforestation of floodplains and river margins or through the construction of hydroelectric dams has a far more devastating and irreversible effect on fish stocks and diversity than does overfishing (Goulding 1983; Leite 1991; Ribeiro, Petrere, and Juras 1995). The two most important fish habitats in the várzea are seasonally flooded forests (Goulding 1980) and floating meadows (Junk 1973, 1983; Crampton 1999b; Henderson and Crampton 1997). Both provide seasonal refuge and sustenance for a huge diversity of fishes. The meadows have an especially important role as nurseries for juveniles of many commercially important species (Crampton 1999b). Of the various types of forest in várzea floodplains (Prance 1979; Ayres 1993) the tall restinga forests found on the higher levees host the highest terrestrial and arboreal biodiversity and provide much of the sustenance for commercially important fruit- and seed-eating fishes, such as tambaqui and pacu (Goulding 1980).

The destruction of levee forest by ranching, agriculture, and predatory logging, as well as the trampling and grazing of floating meadows by livestock, have detrimental effects not just on overall biodiversity of várzeas but also on fisheries production (Goulding, Smith, and Mahar 1996). Although any commercial fishermen in the Amazon will tell you that degraded várzeas offer poor fishing for key commercial species in comparison to intact ones, there is little quantitative data to substantiate this. Ruffino and Isaac (2000) report average catch per unit effort (CPUE) estimates of around 10 to 20 kg/fisherman/day in the Santarém area. Batista (pers.comm.) reports around 20 kg/fisherman/day in the Manaus area. Viana (pers. obs.), on the other hand, estimates that the average CPUE in the Tefé region is from 50 to 80 kg/fisherman/day. The extent to which these discrepancies reflect the diminished and fragmented forest cover in the lower reaches of the Amazon (Santarém and Manaus) versus higher fishing pressure is unknown.

DISCUSSION

Fisheries management in the Brazilian várzea has polarized toward two approaches: state (IBAMA)-imposed restrictive regulations at one end of the spectrum and self-motivated community lake-management programs at the other. The first of these approaches is not working. Like fisheries agencies around the world, IBAMA is unable to adequately enforce its regulations. The second approach, community management, is often extolled as a miracle solution. However, proponents of community management place a great deal of faith in the abilities of rural people to manage and defend resources. We argue that community management is unlikely to work unless five provisos are satisfied:

1. A single group of users is awarded guaranteed rights of access.
2. There are strong economic incentives for defending and managing fish stocks.
3. The users have a general ecological awareness and understand the concepts of resource depletion and management.
4. Management is based on sound scientific grounds and/or traditional ecological knowledge.
5. There are concerted efforts to preserve the intricate mosaic of flooded forests and floating meadows upon which many várzea fish depend.

A worldwide analysis of community-based management programs by Barret et al. (2001) concluded that the capacity of communities to manage resources has been overemphasized and that the success of these programs rarely matches the fanfare. Without financial, legal, and scientific support, local fishing communities are unlikely to be able to satisfy the five provisos above. Alliances between self-motivated social movements and conservation/development-oriented external agencies (governmental or nongovernmental) probably represent the most promising direction for fisheries management in várzeas.

Of the three such alliances with a strong várzea fisheries component currently being developed in the Brazilian Amazon, all approach the challenges of management in different ways. The RDSM is the only one to undertake fisheries management in the context of a protected area in which local people enjoy legally binding rights of access. Projeto Várzea and Instituto Iara operate in the normal context of state-owned várzea. These three projects also differ in their emphasis on habitat conservation, geographical coverage, and the extent to which the interests of the entire spectrum of stakeholders are included.

HABITAT CONSERVATION

The long-term health of várzea fisheries is ultimately dependent upon the conservation of relatively intact forests and floating meadows. The RDSM is unique in prioritizing habitat protection. The reserve effectively compensates for a closed zone of complete biodiversity protection by offering local people a surrounding sustainable use zone. Here, integrated fisheries and forest management, along with a package of economic incentives, promote sustainable exploitation and encourage economically favorable alternatives to such destructive land use as livestock ranching (Crampton et al. this volume; Viana et al. this volume). The RDSM contains a considerable proportion of Brazil's remaining areas of relatively intact várzea floodplain habitats. Assuming two-thirds of the estimated 106,000 km² of Brazilian várzeas (Bayley and Petrere 1989) have already been severely degraded (Alexander 1994), around 32% of the remaining, relatively intact area lies within the RDSM, including its Subsidiary Area (7% within the Focal Area alone).

Instituto Iara is solely concerned with fisheries management. It focuses on the important issues of access rights and market incentives but does not directly address habitat preservation. Projeto Várzea is attempting to promote sustainable forest management and to encourage forms of livestock production and agriculture that are less damaging to várzea habitats. However, it does not emphasize the need for completely protected areas of forest. In fact, there are very few undamaged forests left to protect in the lower Amazon region, making habitat restoration, rather than conservation, the main concern.

GEOGRAPHICAL COVERAGE

Fisheries projects operating on a regional scale are arguably better suited to the management of migratory fish stocks and better prepared for managing the full range of economic concerns and conflicts that decide the fate of management. Instituto Iara is the most expansive fisheries project in Brazil, operating over a 200-km stretch of the axis of the Rio Amazonas and encompassing around 3,000 km² of várzea. This initiative was originally projected to affect around 600,000 people in an area extending from Itacoatiara in Amazonas downstream 1,000 km to Almeirim in Pará. However, this area was subsequently considered to be too large to be effectively managed with available resources. Likewise, because of logistic difficulties and funding limitations, management activities in the 11,240-km² RDSM have been restricted to an area of 2,600 km². The difficulties the Mamirauá and Iara initiatives have experienced in expanding the geographical scale of their operations indicates that single fisheries management programs are unlikely to be effective over areas much larger than 3,000 km² unless they are spectacularly well funded.

STAKEHOLDER INCLUSION

Due to their larger, regional nature, both Instituto Iara and Projeto Várzea seek to reconcile the needs of both várzea residents and commercial fishermen. These projects provide a forum for negotiating mutually acceptable divisions of fishing rights and provide technical cooperation for the production of management plans. For example, the commercial fishing syndicate of Santarém contains many associates who are from the communities of the surrounding várzeas, allowing a balanced forum for discussing fishing agreements and a consolidated front for excluding fishermen from outside the region (e.g., from Belém). Likewise, IBAMA's training program for Voluntary Environmental Agents is providing training for community representatives from both local communities and the urban fishing fleet (M. Ruffino pers. comm.).

In the Upper Amazon region of Tefé, there has always been a much greater separation between várzea fishing communities and the urban fishing fleets of the local towns (Lima 1992). This separation is probably in part a historical consequence of the fact that the rural population of the state of Amazonas is smaller and younger than that of Pará. The rural fishermen of the lower Amazon have for some time been a more forceful political force and have developed a more integrated and mutually beneficial relationship with commercial fishermen.

The fishermen of the RDSM are especially divergent in organization and interests from the nearby urban fishing syndicate of Tefé. Under the area's legal status as an SDR, the resident and user communities are awarded exclusive rights of access to the natural resources of the region. The RDSM is criticized by commercial fishermen from Tefé for providing access rights to 6,000 ribeirinhos while at the same time restricting access rights to almost everybody else in the area. The situation was aggravated by the demarcation of the Amanã Sustainable Development Reserve, which contains large areas of várzea to the east of the RDSM (fig. 6.2). With the formal closure of both the RDSM and RDSA, the Tefé commercial fishing syndicate argues that it has been left with almost no viable fishing grounds. Likewise, ribeirinhos in the remaining unprotected várzeas of the Tefé region are indignant that their lakes are now under much greater pressure.

Anticipating the necessity to concede some access rights to the commercial fishing syndicate of Tefé, the reserve's 1996 management plan made provisions for communities to concede temporary access rights to some lakes for visiting commercial fishermen. However, most communities of the reserve subsequently opted not to provide such concessions (see Crampton et al. this volume). This deadlock has provoked serious debate. How can the managers of a protected area justify developing one part of a regional rural economy at the expense of another?

Protected Areas Act as Fish Supply Areas and Restocking Nuclei The Mamirauá Sustainable Development Institute has argued that managed fisheries and the protection of core no-use zones in the RDSM should, in time, guarantee not only a permanent supply of reasonably priced premium quality fish for the urban markets of the region but also generate surplus stock that will replenish surrounding fisheries (Queiroz and Crampton 1999a). Skeptics argue that there is no evidence that protected areas have the capacity to generate surplus stock and that local communities are in any case likely to maximize harvesting for their own gain. The reluctance of the residents of the RDSM to negotiate fishing concessions with commercial fishermen no doubt fuels this skepticism. So far there is no scientific evidence that the RDSM (or any other protected fishing ground in the Amazon) is restocking adjacent areas. There is evidence, however, for significant increases in the stocks of pirarucu, tambaqui, and caiman in the core protection zone of the RDSM (Crampton et al. this volume; Viana et al. this volume).

The Value of Intact Biodiversity Intact várzea floodplains are among the most diverse and yet threatened ecosystems on earth. The human population is rapidly growing in the Upper Amazon regions where most of the remaining intact várzeas are located (Cincotta, Wisnewksi, and Engelman 2000). Some proponents of biodiversity conservation argue that the economic marginalization of some stakeholders through lost rights of access to protected areas like the RDSM is an unfortunate but necessary price paid by societies that place value on conserving their ecological and genetic heritage (Pimm et al. 2001).

CONCLUSION

Fisheries management is conventionally about providing people with a sustainable supply of fish protein. However, in the Amazon basin the issues are complicated by the multiple interfaces between fish stocks, habitats, and livelihoods. Fisheries managers in the Amazon are concerned with habitat preservation, with sustainable use and development, and with resolving social conflicts.

Of the three contemporary community-based fisheries management models described in this article, all are experimental in nature. They are also young, and results indicating their successes and failures are only just emerging. These three models all have attendant theoretical merits and shortcomings, and strategies for the management of várzea fisheries will in the future probably need to incorporate features of all three. For example, protected areas like the RDSM may in the future need to concede larger areas of várzea for access to outside commercial fishing fleets. Likewise, some areas of the regional fishing grounds in the middle Amazon region may need to include zones of protection. We argue that an increase in the number of fisheries management initiatives would be more effective than expanding the geographical range of existing reserves or multiple-use management programs to much more than around 3,000 km^2, which seems to be about the upper limit for effective administration.

Goulding, Smith, and Mahar (1996) proposed a chain of reserves for "fish forests" and "fish meadows" along the axis of the Amazon River and an integrated regional management program. We envisage a similar model that combines features of contemporary fisheries management schemes by including multi-use regional fisheries management programs and Sustainable Development Reserves, which would need to be arranged in a constellation along the major whitewater rivers of the Amazon (Amazon, Juruá, Purus, Madeira, etc.). Each initiative could be independently managed but coordinated within an overall scheme for national fisheries management within Brazil and Peru (and internationally between the two countries). The administration of such a chain of fishery management initiatives would need to involve the entire spectrum of stakeholders, including communities, businesses, the environmental authorities, commercial fishing syndicates, conservation organizations, and research institutions. Institutional cooperation between the Amazonian countries of Brazil, Peru, Bolivia, Colombia, and Ecuador would also be necessary to formulate management plans for the some migratory catfishes.

Finally, a word of warning: in the drive to establish integrated, socially appropriate models of management, it is important not to forget the biology of the fish. Scientific studies are still needed to define the migratory ranges of commercially important characiform fishes and catfishes, as well as the minimum size of single blocks of várzea necessary to sustain viable populations of commercial fish species.

ACKNOWLEDGMENTS

We thank several authorities cited by personal communication. J. Bampton, S. Gillingham, and G. Watkins commented on the manuscript. The authors were funded by CNPq and DFID.

7

Fisheries Management in the Mamirauá
Sustainable Development Reserve

WILLIAM G. R. CRAMPTON, JOÃO PAULO VIANA, LEANDRO CASTELLO,
AND JOSÉ MARÍA B. DAMASCENO

Fisheries management in the Brazilian Amazon has polarized toward state-imposed regulations at one extreme and community-based management at the other (Crampton, Castello, and Viana this volume). At present there is no overall government fisheries conservation policy for Amazônia, and existing state fisheries restrictions are almost completely ineffective (Hall 1997; Crampton and Viana 1999). Since the 1970s fishing has become an increasingly important source of income for the ribeirinho people of the whitewater várzea floodplain and a growing number of várzea communities have set up lake reserves (reservas de lagos de várzea) to manage fish stocks and to guard them from predatory fishing by the commercial fleets of major towns. The Pastoral Land Commission, Comissão Pastoral da Terra (CPT), of the Catholic Church supported many of these initiatives and reports that up to 15% of all major lakes in Amazonas are inside such reserves (Hall 1997). These self-motivated lake-vigilance schemes met with only limited success due to political weakness, poor infrastructure, and lack of recognition by the state authorities (Hall 1997; Lima 1999). Alliances between local social movements and state or nongovernmental organizations can greatly strengthen the former by providing funding, training, and technical or legal support. These kinds of alliances represent one of the most promising directions for the management of Amazonian fisheries (McGrath et al. 1999; Ruffino 1999; Crampton, Castello, and Viana this volume).

At present, three partnerships between NGOs and local people of Brazilian várzeas involve a substantial fishery component. Projeto Várzea and the Iara Institute (Instituto Amazônico de Manejo Sustentável dos Recursos Naturais) are two multidisciplinary projects designed to promote sustainable fishing at the regional level in the state of Pará (Crampton, Castello, and Viana this volume). These projects seek to reconcile the needs of várzea communities, commercial fishermen, and other stakeholders. The third such alliance, which is the subject of this article, involves a partnership between the people of the Mamirauá Sustainable Development Reserve (or RDSM), Amazonas, and its supporting NGO, the Mamirauá Sustainable Development Institute (Instituto de Desenvolvimento Sustentável Mamirauá, or IDSM).

Crampton, Castello, and Viana (this volume) defined five provisos for successful community-based fishery management:

1. Access rights are restricted to an economically independent group of users.
2. The users understand the concepts of resource depletion and management.
3. Management is accompanied by conservation of the habitats that sustain fish populations.
4. Profits accrued from fish marketing are sufficient to provide economic motives for long-term management and vigilance.
5. Management is based on sound scientific and/or traditional ecological knowledge.

In this article we describe general strategies for fisheries management in the RDSM and how these interlink with biodiversity conservation and the sustainable management of other resources. We emphasize community participation, access rights, restrictive fishing regulations, and the conservation of habitats. These issues correspond to the first three provisos above. Viana et al. (this volume) go on to emphasize the last two of the provisos above by describing an experimental Fish Commercialization Program in the RDSM. We evaluate the progress of ongoing fisheries management activities in the RDSM and conclude with a discussion of the applicability of the RDSM program to general models of fisheries management in the Amazon.

THE MAMIRAUÁ RESERVE

The RDSM encompasses 11,240 km² of várzea floodplain and represents the largest contiguous block of reasonably intact várzea forest left in the Brazilian Amazon (Ayres et al. 1999). Unlike the situation in the Lower Amazon, the várzeas of the Tefé region have not suffered large-scale deforestation or degradation (Goulding, Smith, and Mahar 1996; Goulding, 1999). Research and community-participation in the RDSM have concentrated on a 2,600 km² Focal Area delimited by the Rios Solimões, Japurá, and the Paraná Aranapu (figs. 7.1 and 7.2). Unless otherwise stated, Mamirauá Reserve, RDSM, or just the reserve will henceforth refer specifically to this Focal Area. In 2001 the RDSM contained 1,585 people (0.61 people/km²) who live mostly in twenty-one communities. A further 4,401 people lived in forty-two villages outside the reserve that are classified as user communities because of their dependence on resources from within the RDSM (SCM 1996).

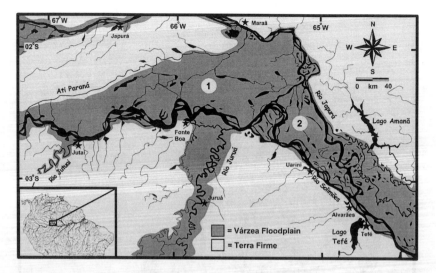

FIGURE 7.1 Map of the Upper Amazon region showing the Subsidiary (1) and Focal (2) areas of the Mamirauá Sustainable Development Reserve. Also shown are regional municipal centers and the extent of the várzea floodplain.

THE FISHERIES STATUS OF THE RDSM AND TEFÉ REGION

The fisheries of the reserve underpin the local economy, contributing at least 79% of the market value of resources extracted from the area (table 7.1). The commercial yield of the Focal Area of the reserve from resident and user communities was estimated at around 320 tons/year in the period 1991–1994, of which 58% was fresh fish, 40% dried and salted fish, and 2% pickled fish (SCM 1996). Commercial fishing boats from the towns of Tefé and Alvarães extracted a further 220 tons/year from the Focal Area of the reserve during the period 1991–1994 (Barthem 1999a). Most commercial fishing is undertaken during the low-water season when fish are concentrated in lakes and river channels (Barthem 1999b; Queiroz 1999). Subsistence fishing, which does not figure in table 7.1, provides around 80% of local animal protein requirements (Howard et al. 1995; Santos 1996). Caiman, game, and turtles provide the rest (Santos 1996). The per capita consumption of fish in the RDSM is as high as 500 g/diem (Queiroz 1999). Ayres et al. (1999) estimated that the total annual subsistence demand for fish in the RDSM was between 240 and 300 tons in the early 1990s.

Added together, the total annual yield of subsistence and commercialization for the RDSM in the early 1990s was in the order of approximately 840 tons/year. This is equivalent to an average extraction of approximately 323 kg/km²/year (840,000/ 2,600). Although probably somewhat of an underestimation, this figure is still an order of magnitude below Bayley and Petrere's 1989 estimated maximum sustain-

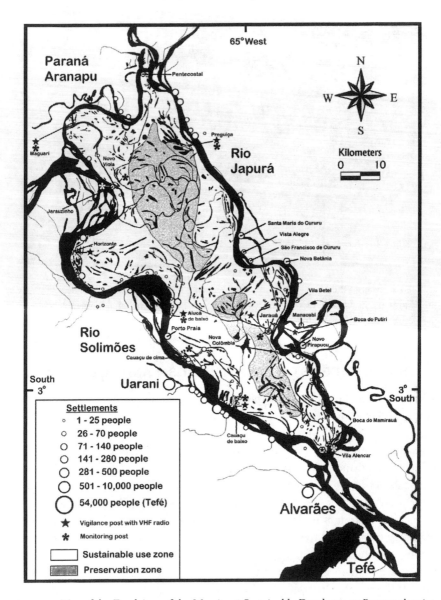

FIGURE 7.2 Map of the Focal Area of the Mamirauá Sustainable Development Reserve showing the core zones of total protection and communities.

able annual yield for Amazonian várzeas of 5,000 kg/km²/year. It is also well below documented levels of production in other Amazonian várzeas (1,800 to 2,000 kg/km²/year) (Bayley and Petrere 1989; Bayley et al. 1992). These calculations imply that in general terms the fish resources of the RDSM are harvested at levels below those that would deplete them. However, as is the pattern throughout the Amazon

TABLE 7.1 Estimated Annual Economic Value of the Principal Natural
Resources of the Focal Area of the Mamirauá Sustainable Development Reserve

RESOURCE	MARKET VALUE
Fisheries	
External fishermen (multispecies)	US $867,000
Internal fishermen	
—Pirarucu (*Arapaima gigas*)	US $329,000
—Tambaqui (*Colossoma macropomum*)	US $240,000
—Other species	US $417,000
Timber and Firewood	US $ 107,000
Agricultural Products (manioc, banana, citrus, etc.)	US $157,000
Hunting (mostly caiman)	US $64,000
Other Resources	US $185,000
Total:	US $2,336,000

Source: Mamirauá Management Plan (SCM, 1996)
Note: Fisheries data is averaged over the period 1993–1995 and exclude the direct sale of large tambaqui to
Manaus via commercial passenger boats and the sale of catfishes to *frigoríficos* (freezer stations).

basin (Crampton, Castello, and Viana this volume), commercial fishing in the
Tefé region is biased toward only a few species.

There are around 600 species of fishes in the Tefé area (Crampton 1999b; W.
Crampton pers. obs.). Of these, around 100 are used for subsistence or commercialization and around fifty regularly appear in the Tefé market (table 7.2). Three
species alone constitute more than half of the annual biomass of fish landed from
RDSM and sold in the Tefé market, and the top ten species represent 84% of the
catch (table 7.3). For brevity, scientific names of all fish species mentioned in this
article are listed in table 7.2.

Research prior to 1995 suggested that three species were being overfished in the
RDSM. Based on size -class and life-table analyses of 1995–1996 data, Queiroz and
Sardinha (1999) concluded that levels of pirarucu exploitation exceeded the maximum sustainable yield. They predicted a halving of stocks within six years in the
absence of management. Costa, Barthem, and Correa (1999) reported overfishing
of tambaqui in the RDSM and documented that 93.5% of tambaqui in the Tefé
market during the low water of 1993 were below the legal minimum length of 55
cm. Crampton (1999b,c) described overfishing of discus in the RDSM by visiting
ornamental fish catchers. There is no evidence for overfishing of aruanã, curimatá,
tucunaré, or any of the other species that are heavily commercialized in the Tefé
area (table 7.3).

TABLE 7.2 Fish Species Consumed by the Rural and Urban
Population of the Tefé Region

Clupeiformes

Pellonidae

—*Pellona castelnaeana*   Apapa-amarela [*3]

—*Pellona flavipinnis*   Sardinhão

Osteoglossiformes

Arapaimidae

—*Arapaima gigas*   Pirarucu [*1]

Osteoglossidae

—*Osteoglossum bicirrhosum*   Aruanã, Sulamba [*3]

Characiformes

Erythrinidae

—*Hoplias malabaricus*   Traira [*3]

—*Hoploerythrinus unitaeniatus*   Jeju [*3]

Anostomidae

—*Leporinus friderici*   Aracu-piau

—*Leporinus fasciatus*   Aracu-flamengo [*3]

—*Rhytiodus microlepis*   Aracu

—*Rhytiodus argenteofuscus*   Aracu

—*Schizodon fasciatum*   Aracu-comum [*3]

Hemiodontidae

—*Anodus elongatus*   Charuto [*3]

—*Anodus melanopogon*   Charuto

—*Hemiodopsis immaculatus*   Orana-branca [*3]

—*Hemiodopsis microlepis*   Orana-flecheira [*3]

—*Hemiodus unimaculatus*   Orana [*3]

Curimatidae

—*Curimata vittatus*   Chorona

—*Curimatella alburnus*   Chorona

—*Potamorhina latior*   Branquinha [*3]

—*Potamorhina altamazonica*   Branquinha [*3]

—*Potamorhina pristigaster*   Branquinha [*3]

—*Psectrogaster rutiloides*   Chorona [*3]

—*Psectrogaster amazonica*   Chorona [*3]

Prochilodontidae

—*Prochilodus nigricans*   Curimatá [*2]

—*Semaprochilodus insignis*   Jaraqui esc. –grossa [*2]

—*Semaprochilodus taeniurus*   Jaraqui escama-fina [*2]

Acestrorhynchidae

—*Acestrorhynchus falcatus*   Peixe-agulhão

Cynodontidae

—*Cynodon gibbus*   Peixe-cachorro

—*Hydrolycus scomberoides*   Peixe-cachorro

—*Rhaphiodon vulpinus*   Peixe-cachorro [*3]

TABLE 7.2 (*Continued*)

Characidae (Characinae)

—*Brycon melanopterus*   Jatuarana [*]

—*Brycon cf. cephalus*   Matrinchã [*2]

—*Triportheus angulatus*   Sardinha-chata [*3]

Triportheus elongatus   Sardinha-comprida [*2]

Characidae (Serrasalminae)

—*Colossoma macropomum*   Tambaqui [*1]

—*Myleus rubripinnis*   Pacu-galo [*3]

—*Myleus torquatus*   Pacu

—*Mylossoma duriventre*   Pacu-comum [*3]

—*Mylossoma aureum*   Pacu-manteiga [*3]

—*Piaractus brachypomus*   Pirapitinga [*3]

—*Pygocentrus nattereri*   Piranha-caju [*3]

—*Serrasalmus elongatus*   Piranha-mucura

—*Serrasalmus spilopleura*   Piranha-jirda

—*Serrasalmus rhombeus*   Piranha-preta [*3]

Gymnotiformes

Rhamphichthyidae

—*Rhamphichthys cf. rostratus*   Sarapó

Siluriformes

Doradidae

—*Centrodoras brachiatus*   Reque-reque

—*Lithodoras dorsalis*   Bacu-pedra [*3]

—*Megalodoras uranoscopus*   Rebeca [*3]

—*Pseudodoras niger*   Cuiu-cuiu [*3]

—*Pterodoras lentiginosus*   Bacu-liso [*3]

Auchenipteridae

—*Trachelyopterichthys taeniatus*   Cangati

—*Trachycorystes trachycorystes*   Cangati

Pimelodidae

—*Brachyplatystoma filamentosum*   Filhote, Piraiba [*2†]

—*Brachyplatystoma flavicans*   Dourada [*2†]

—*Brachyplatystoma vaillantii*   Piramutaba [*2†]

—*Goslinia platynema*   Babão [*]

—*Hemisorubim platyrhynchos*   Braço-de-moça [*3]

—*Hypophthalmus edentatus*   Mapará [*3]

—*Hypophthalmus fimbriatus*   Mapará [*3]

—*Hypophthalmus marginatus*   Mapará [*,*3]

—*Leiarius marmoratus*   Jandiá [*3]

—*Paulicea luetkeni*   Jau, Pacamum [†]

—*Phractocephalus hemioliopterus*   Pirarara [*3†]

—*Pimelodina flavipinnis*   Mandi [*3]

—*Pinirampus pirinampu*   Barba-chata [*3]

—*Platynematichthys notatus*   Mandi

TABLE 7.2 (*Continued*)

—*Platynematichthys sturio*	Mandi
—*Pseudoplatystoma fasciatum*	Surubim *2†
—*Pseudoplatystoma tigrinum*	Caparari *2†
—*Sorubim lima*	Bico-de-pato *3
—*Sorubimichthys planiceps*	Peixe-lenha *3
Ageneiosidae	
—*Ageneiosus brevifilis*	Mandubé *3
Callichthyidae	
—*Hoplosternum littorale*	Tamoatá *3‡
—*Megalechis thoracata*	Tamoatá *3‡
Loricariidae	
—*Glyptoperichthys gibbiceps*	Bodó *3‡
—*Hypostomus carinatus*	Bodó
—*Hypostomus cf. emarginatus*	Bodó
—*Liposarcus pardalis*	Bodó *3‡
Perciformes	
Sciaenidae	
—*Plagioscion squamosissimus*	Pescada *3
—*Plagioscion sp.*	Pescada *3
Cichlidae	
—*Astronotus ocellatus*	Acará-açu *2
—*Chaetobranchus semifasciatus*	Acará-tucunaré *3
—*Chaetobranchus flavescens*	Acará-branco *3
—*Cichla monoculus*	Tucunaré *2
—*Crenicichla gr. lugubris*	Jacundá-vermelho *3
—*Geophagus proximus*	Acará roe-roe *3
—*Heros appendiculatus*	Acará-roxo *3
—*Hypselecara temporalis*	Acará
—*Satanoperca jurupari*	Acará-garrafa
—*Symphysodon aequifasciatus*	Acará-disco *3
—*Uaru amphiacanthoides*	Acará-bararuá *3
Pleuronectiformes	
Soleidae	
—*Achirus sp.*	Soia, Solha
Rajiformes	
Potamotrygonidae	
—*Potamotrygon constellata*	Arraia
—*Potamotrygon hystrix*	Arraia
—*Potamotrygon motoro*	Arraia

Note: Species marked * appear regularly in urban fish markets and are classed by price as 1 (premium quality), 2 (medium quality), and 3 (low quality); † are sold to *frigoríficos* (freezer stations) at Tefé and Alvarães; and ‡ are sold alive. Unmarked species are eaten commonly only in the rural interior.

TABLE 7.3 Average Annual Landings of the Thirty Most Common Species/Groups of Species Originating from the Mamirauá Reserve at the Tefé Market

LOCAL NAME	SCIENTIFIC NAME	WEIGHT (KG)	%	CUMULATIVE %
1. Aruanã	*Osteoglossum bicirrhosum*	38,261	22.3	22.3
2. Curimatá	*Prochilodus nigricans*	28,875	16.8	39.2
3. Tucunaré	*Cichla monoculus*	18,374	10.7	49.9
4. Tambaqui	*Colossoma macropomum*	12,875	7.5	57.4
5. Pacu comum	*Mylossoma duriventre*	10,372	6.0	63.4
6. Pirapitinga	*Piaractus brachypomus*	8,741	5.1	68.5
7. Jaraqui escama grossa	*Semaprochilodus insignis*	7,970	4.6	73.2
8. Jaraqui escama fina	*Semaprochilodus taeniurus*	6,881	4.0	77.2
9. Acará-açu	*Astronotus ocellatus*	5,811	3.4	80.6
10. Matrinchã	*Brycon cf. cephalus*	3,073	3.3	83.8
11. "Salada"	—	5,159	3.0	86.8
12. Branquinha peito-de-aço	*Potamorhina latior*	3,062	1.8	88.6
13. Sardinha comprida	*Triportheus elongatus*	3,046	1.8	90.4
14. Caparari	*Pseudoplatystoma tigrinum*	2,678	1.6	92.0
15. Bodó	*Liposarcus/Glyptoperichthys*	2,313	1.3	93.3
16. Pacu galo	*Myleus rubripinnis*	2,098	1.2	94.5
17. Piranha caju	*Pygocentrus nattereri*	2,021	1.2	95.7
18. Acará-tucunaré	*Chaetobranchus semifasciatus*	1,675	1.0	96.7
19. Branquinha comum	*Potamorhina altamazonica*	1,263	0.7	97.4
20. Pescada	*Plagioscion* spp.	1,016	0.6	98.0
21. Cuiu-cuiu	*Pseudodoras niger*	998	0.6	98.6
22. Pirarucu	*Arapaima gigas*	628	0.4	99.0
23. Sardinha chata	*Triportheus angulatus*	408	0.2	99.2
24. Surubim	*Pseudoplatystoma fasciatum*	403	0.2	99.4
25. Dourada	*Brachyplatystoma flavicans*	206	0.1	99.5
26. Aracu comum	*Schizodon fasciatum*	205	0.1	99.7
27. Jatuarana	*Brycon melanopterus*	158	0.1	99.8
28. Orana	*Hemiodopsis/Hemiodus* spp.	154	0.1	99.9
29. Sardinhão	*Pellona castelnaeana*	141	0.1	99.9
30. Charuto (cubiu)	*Anodus melanopogon*	116	0.1	100.0
Total:		168,975		

Source: Summarized from Barthem (1999a).

Note: Data is averaged over the period 1991–1994. "*Salada*" refers to a mixed catch for which the market data collectors were unable to separate the species by weight. Data are absent or only partial for the following: (1) salted and sundried pirarucu, (2) fresh pirarucu since 1996 when IBAMA introduced an indefinite ban on all pirarucu commercialization, (3) large tambaqui that are transported to Manaus, and (4) large catfishes that are sold to *frigoríficos* (freezer stations) for export to Peru or Colombia.

THE EARLY STAGES OF THE MAMIRAUÁ RESERVE

In the early 1980s a local group, the Movement for Grass-Roots Education, Movimento de Educação e Base (MEB), began to train community leaders in the Tefé region and build upon the CPT's campaign to assist lake-protection schemes (Hall 1997). During the late 1980s the primatologist José Márcio Ayres and the anthropologist Deborah Lima conducted pioneering studies in várzea floodplains at the confluence of the Rios Japurá and Solimões and recognized the outstanding conservation importance of the area. In addition to its relatively intact forest, this region contains a rich fauna and flora, including rare and endemic taxa, such as the white uakari monkey (*Cacajao calvus calvus*) (Ayres 1986). Ayres's and Lima's work led to a proposal for a conservation unit in the area to be established on a philosophy of community participation and to be built on the foundation of existing community-based lake protection schemes in the area.

The Mamirauá Reserve, named after a prominent lake in the area, began in 1990 under the interim status of Ecological Station, a conservation category in which the presence of people and the use of resources for purposes other than scientific research are illegal. Despite these restrictions, early work encouraged the communities of the area to consolidate an organizational structure based on the CPT model in which clusters of nearby communities regularly convene to discuss issues of mutual concern. Nine such clusters, or political sectors, were founded in the reserve. To resolve the irregular status of the Mamirauá Ecological Station and to guarantee defined rights of access for the local people necessitated lobbying for a revision of national conservation policy (Ayres et al. 1999). This was finally achieved in 1996 with the transformation of the Ecological Station into a Sustainable Development Reserve, the first of a new category of Brazilian conservation unit (Amazonas state decree 2.411 of July 16, 1996). The transformation provided local residents with defined access rights and represented a milestone in the inclusion of local people in some protected areas of Brazil. The overall goal of the Mamirauá Reserve is to reconcile biodiversity protection with long-term improvements in the living standard of the local people through three processes:

1. empowering and educating local people to defend the resources of the area from outside interests;
2. encouraging economically motivated sustainable management of these resources;
3. conducting a program of applied research on biodiversity and key natural resources (Ayres et al. 1999; Lima 1999).

RESEARCH AND EXTENSION ACTIVITIES IN THE RDSM

In 1992 Projeto Mamirauá was launched from headquarters in Tefé. Its goals were to formulate a management plan for the sustainable use of natural resources in the area that would become the RDSM and to catalog biodiversity. An international team of biological and socioeconomic researchers and a basic infrastructure of boats and floating research stations were funded by government and overseas aid. This first phase of the Mamirauá Project culminated with the production of the Mamirauá Management Plan (MMP) (SCM 1996) and supporting technical reports (Queiroz and Crampton 1999b). Implementation of the integrated management program outlined in the MMP began in 1997. The program emphasizes the shifting and seasonal nature of resource exploitation in the RDSM and, in addition to fisheries management, covers alternative agricultural practices, timber extraction, and caiman, turtle, and game hunting. (Queiroz and Crampton 1999a).

The Instituto de Desenvolvimento Sustentável Mamirauá (IDSM) was created by presidential decree in 1999 and charged with a mission to expand activities in the RDSM and to develop general models for the sustainable management of tropical forest ecosystems. By 2001 the IDSM hosted around twenty-five professional staff, seventy support staff, twenty interns, and several teams of visiting researchers. This involved an infrastructure comprising six boats, sixteen floating houses, fifty motorized aluminum canoes, and three vehicles. In 2001 the IDSM operated with a core annual budget of US$ 1.3 million, two-thirds of which came from the Brazilian government and the remainder from the U.K. Department for International Development. IDSM is currently expanding a multidisciplinary extension and research program in fisheries, forestry, agriculture, environmental education, and ecotourism. IDSM also runs a microcredit program that provides small loans for residents and users of the RDSM. Limited health and sanitation support beyond the obligations of the municipal authorities are also provided by IDSM. Regularly held meetings, including an annual General Assembly, provide a negotiating forum for the communities of the reserve and for other stakeholders in the region. The IDSM also produces a biweekly radio show and a quarterly newsletter.

BASIC AND ENVIRONMENTAL EDUCATION

Although researchers and extension workers in the IDSM constantly strive to explain the concepts of management to *ribeirinhos* (rural river-dwelling people), a baseline ecological awareness is essential for such concepts to be assimilated. Most fishermen are aware that some fish stocks are under pressure and that the need for preservation and management exists. Nonetheless, it is sometimes difficult for them to appreciate the long-term issues of management. The IDSM runs an environmental education program in which itinerant teachers and guest researchers run practical courses for both adults and children. The IDSM is due to inaugurate a moving (floating) center for environmental and scientific education in 2002 in order to intensify this program and deepen its impact on the young generation of ribeirinhos in the RDSM (E. Moura; IDSM, pers. comm.).

Illiteracy and innumeracy are the archenemies of economic independence and self-confidence in rural people. Just about every aspect of resource management requires a good standard of literacy. Participating in training courses, dealing with the environmental authorities, organizing community associations, and marketing produce, for example, all require reading and writing skills. The proportion of illiterate people over fifteen years old within the RDSM declined from 38% to 31% between 1996 and 2001. However, 55% of people over the age of ten in the RDSM either cannot read, or read except with difficulty (E. Moura pers. comm.). As is the case in most várzeas, schools in the RDSM are small, usually run by just one part-time teacher, and offer education only to around the fourth grade. Thirty-two percent of people who migrate to urban centers from the RDSM do so to continue their schooling. The environmental education team of the IDSM is working closely with the state education authorities to raise standards of reading, writing, and arithmetic in the reserve for both children and adults. In addition to assisting with teacher training, the IDSM is also contributing to the schooling infrastructure, for example by donating solar panels and lights that allow classes to continue into the night (E. Moura pers. comm.).

ACCESS RIGHTS TO THE RDSM

The people of the Brazilian várzea floodplains do not possess exclusive rights of access to fisheries resources. This lack usually represents a major obstacle to the development of community-based fisheries management. Under the legislation supporting the demarcation of Mamirauá as a Sustainable Development Reserve (SDR), the residents are entitled to exclusive access to the natural resources of the reserve, even though they are still not the legal landowners. Therefore, it is illegal for commercial fishing boats to operate inside the RDSM without permission from the residents. If invading fishermen ignore requests to leave the reserve, the residents can request the intervention of agents of the Brazilian Institute for the Environment and Renewable Natural Resources (IBAMA) or the police. The powers and willingness of these authorities to support the population of the RDSM is discussed later. The status of a SDR effectively provides the preconditions for a community property regime in which rights to resources are held by a distinct group of users who exclude outsiders (McGrath et al. 1999).

ZONING IN THE RDSM

In addition to defining general rights of access to the reserve, a system of zoning was implemented. This system divides lakes and forests into areas of no use and sustainable use.

PROTECTION ZONE

A zone of total protection where fishing and all other forms of exploitation are unconditionally prohibited was demarcated in the central areas of the RDSM (fig.

7.2). This demarcation provides, in theory, a refuge for fish stocks, nursery grounds for many resident and migratory species (Crampton, Castello, and Viana this volume), and an area in which the habitats and biodiversity of the ecosystem are preserved intact. The boundaries of protection zone evolved during five years of negotiations with the reserve's residents and were designed to include: (a) substantial areas of all the major lake systems, (b) a variety of relatively intact forest ecosystems, and (c) acceptable divisions of territory between the organizational sectors of the reserve.

SUSTAINABLE USE ZONE

A sustainable use zone designated for multiple resource management by the resident and user communities of the reserve surrounds the protection zone (fig. 7.2). Lakes in this zone are divided into two categories: community lakes (*lagos comunitários*) and town lakes (*lagos de sede*).

Community lakes are reserved for the resident and user communities of the RDSM but closed to fishermen from outside. Each community has a territory of lakes and forest, the boundaries of which are negotiated with neighboring communities. The use of these lakes is left to the discretion of each community, but as described later, systems of active management are being implemented or encouraged. Basic management involves the division of community lakes into three categories: (a) subsistence lakes (*lagos de subsistência*), designated for supplying food; (b) commercialization lakes (*lagos de comercialização*), reserved for commercial operations; and (c) preservation lakes (*lagos de preservação*), set aside for permanent preservation or very occasional use (not to be confused with lakes in the zone of total protection). In the first phase of the Mamirauá Project, many conflicts were precipitated by neighborly incursions into community lakes. Most communities have since agreed on mutually satisfactory territorial borders. However, a considerable proportion of the Mamirauá Institute's efforts and resources were and continue to be expended in appeasing a wide range of these internal conflicts.

Town lakes form a category designated in the MMP for exclusive access by commercial fishing boats from the main towns of the area (Tefé, Alvarães, Uarani, Fonte Boa, and Maraã) but not from distant cities, such as Manacapuru, Manaus, and Itacoatiara. The original idea was that this concession was necessary to secure the cooperation of local urban fishing fleets after the closure of lakes within the RDSM. During the first years of the Mamirauá Project, attempts to provide controlled fishing rights in the RDSM for boats from Tefé's commercial fishing syndicate floundered. Commercial fishermen refused to acknowledge the limitation of fishing rights in what were previously more or less free-for-all areas. They refused to negotiate a settlement and vowed to continue invading the reserve (even with the threat of arrests by IBAMA following the 1990 transformation of the area into a conservation unit). In response, the communities of the RDSM reached a unanimous decision in the 1997 General Assembly to close all lakes to outside users.

Dialog between representatives of the Tefé fishing syndicate and RDSM fishermen was reopened tentatively in 1998 but achieved nothing until a small but symbolic agreement was reached in January 2000. The communities of Maguari and Barroso established three lakes as town lakes. Here, fishing was opened to associates of the Tefé fishing syndicate with stipulations that the communities would receive a share of the fishing profits in the form of fuel (which is used by the communities for lake vigilance). This agreement represented the first step toward establishing good faith between the antagonists of a decade-long deadlock. Reconciling the needs of urban fishermen with the needs of the population of the RDSM continues to be one of the Mamiraué Institute's central concerns.

VIGILANCE

The system of zoning and exclusive access rights described above would be weakened without an accompanying system of vigilance—the desire and motivation for which should ultimately come from the local communities. In the lake-protection schemes that preceded the implementation of the Mamiraué Reserve, the traditional procedure was for communities to intercept invading boats and request that they leave. These requests were usually ignored. To take advantage of the new legal status of the reserve and to strengthen the position of local fishermen, the IDSM is working closely with federal authorities and supports the training of voluntary environmental agents by IBAMA. When invading fishermen refuse to respect requests to leave the reserve, a network of VHF radios installed at strategic locations of the reserve is used to call the IDSM headquarters in Tefé. A formal denunciation is then delivered to the local IBAMA office. IDSM deals not just with incidents involving fishermen from outside the RDSM but also infractions by residents and users of the reserve (e.g., fishing illegally in the protection zone or in another sector's fishing grounds). When necessary, two or three authorized IBAMA agents are sent to resolve the incident, sometimes with the support of a police officer. IBAMA agents have the authority to make arrests and expel fishermen from the reserve. They can also confiscate equipment and illegal catches if they are intercepted in flagrante delicto.

A quantitative system of monitoring invasions of the RDSM by fishing boats was implemented in 1999 by the IDSM. Voluntary agents at the ten monitoring stations marked in figure 7.2 record the origin and motives of each attempted invasion. During the period February to December 1999, 94% of 304 attempted invasions by boats were by fishermen, 5% by professional hunters, and just 1% by timber extractors. Most of the invasions occurred during the peak fishing season at low water (Reis and Souza 2000). During eight IBAMA missions to the RDSM in 1999, 5 canoes, 1.9 tons of salted and sun-dried pirarucu flanks, and 178 tambaqui were confiscated (L. McCulloch, IBAMA-Tefé, pers. comm.). Following a violent encounter between community and commercial fishermen in which one community member was seriously injured, media attention led to Operação Mamiraué. This large-scale mission involved a sweep through the reserve in the peak fishing period preceding Christmas 1999 and was undertaken by a team of IBAMA agents from Manaus and Tefé in collaboration with officers of the federal and local police. Twenty-four illegal fishermen, three boats, sixteen wooden canoes, three motorized canoes, and ninety gillnets were apprehended during the operation. The list of confiscated catches included 6,755 kg of fish, 40 kg of game and 560 live turtles. A total of 17,200 Brazilian Reais in fines were applied (Pantoja 2000).

Similar but smaller scale operations were conducted through the low water season of 2000 as a partnership between the IDSM, voluntary agents from the RDSM, the Amazonas State Institute for Environmental Protection, IPAM (Instituto de Proteção Ambiental do Amazonas), IBAMA, the Tefé police force, and the army. Between August and November around 35,000 kg of fish and over 650 kg of game were apprehended along the margins of the RDSM. Over 330,000 Reais of fines were applied. These operations gave some indication of the scale of clandestine activity in the RDSM and surrounding areas.

Enforcement, however, is often difficult. The Tefé IBAMA post is responsible for an area of 251,000 km² and yet is staffed by just eight field agents. Until 1999 it did not even possess a boat. To address the paucity of field agents throughout the Amazon, IBAMA recently began training Voluntary Environmental Agents (Agente Ambiental Voluntário, or AAV). With additional support from the Catholic Church, a total of 330 voluntary AAVs have already been trained to operate in the Tefé region (Reis and Souza 2000). By August 2001 eighty-five AAVs from the RDSM had been trained and thirty-seven were active (P. Souza, Mamirauá Institute, pers. comm.). The AAVs assume a largely educational role and are trained to give courses in environmental education at schools and village meetings. They are also trained to guard lakes from intercommunity invasions by fishermen of the reserve itself and to confront invaders from outside. AAVs in the RDSM receive small stipends and rations of gasoline and are issued with a field kit including a flashlight and a jacket emblazoned with an AAV logo in IBAMA livery.

LOGISTIC AND INFRASTRUCTURE SUPPORT FOR VIGILANCE

Most of the 500 or so lakes in the RDSM are accessible to large boats only via channels effectively guarded by the presence of communities (fig. 7.2). However, illegal fishermen can easily carry canoes along trails that lead to some lakes. The Mamirauá Institute is strengthening the vigilance of lake systems by expanding a network of floating or fixed houses equipped with VHF radios. These posts also serve as bases for research, monitoring, and extension activities (fig. 7.2). The reserve currently has fifteen floating and two fixed posts equipped with radios. Voluntary agents conduct routine nocturnal forays during the low-water season. By 2001 seven organizational sectors of the RDSM had been provided with speedboats and rations of fuel to conduct these forays. The IDSM contributes 50% of the maintenance costs of the engines (P. Souza pers. comm.).

PARTICIPATORY MANAGEMENT

LAKE MANAGEMENT PROTOCOLS

Subsistence lakes are usually located near communities. Fishing is traditionally controlled in these lakes by using species-specific techniques (table 7.4) and by limiting effort to small and frequent catches. Commercialization lakes are fished infrequently but with a much greater effort and using generalist techniques, usually gill nets, for maximum yield. Similar patterns of gear use and catch frequency/intensity have been observed in várzeas of the Lower Amazon (McGrath, Silva, and Crossa 1998) and Central Amazon (Smith 1981; Merona 1990).

Commercialization lakes in the RDSM are often far from the communities, and camps are set up for fishing expeditions that last two or more days per lake. The management of commercialization lakes involves a *rodízio* (rotation) system of leaving the lakes to fallow for a period of four to six years between exploitation (McGrath et al. 1999). The logic of the fallow period is twofold. First, it encourages the reproduction and recruitment of species that breed within the várzea (see Crampton, Castello, and Viana this volume). Second, the lack of disturbance encourages all fishes, including those that breed outside (e.g., tambaqui and matrinchã), to take low-water refuge in fallow lakes over successive years. Communities suspend fishing and disruptive activities like logging in or near the entrances to favorite fallow lakes during the flood ebb period, a time when fishes move into lakes to seek low-water refuge. Rather than being formally preplanned, the timing of fallow periods and the total catches of these lake rotation schemes usually depend upon community needs and the abundance of fish in a given year.

MANAGEMENT SUPPORT FROM SCIENTIFIC RESEARCH

Scientific research at the Mamirauá Institute is beginning to strengthen traditional management by defining conservative sustainable yields for commercial species. One example of cooperative planning between researchers and fishermen is a community management and stock assessment system for pirarucu. This forms part of the experimental Fish Commercialization Program described by Viana et al. (this volume). A separate management plan for tambaqui included provisions for the protection of spawning sites along the Rio Solimões (Costa, Barthem, and Correa 1999). Nonetheless, in situ management of this species is unlikely to be effective because tambaqui undertake long upriver migrations from their natal sites before colonizing other floodplain areas. A management plan was also prepared for the exploitation of discus for the ornamental fish trade (Crampton 1999c).

RESTRICTIVE FISHERIES REGULATIONS

The fisheries component of the MMP (SCM 1996) included a series of restrictive regulations. These regulations were circumscribed by current IBAMA legislation,

TABLE 7.4 Selective, Semiselective, and General Fishing Techniques Used in Floodplain Lakes and Flooded Forests Within the Mamirauá Reserve

SELECTIVE TECHNIQUES	SPECIES
Arpão (robust single-head harpoon)	Pirarucu
Flecha/Flechão (small single-point harpoon launched from a bow or attached to a spear, usually used by day for fish swimming near the surface)	Aruanã, Tucunaré, Acará-açu, Pirapitinga, Jatuarana, Matrinchã, Traira, Surubim, etc.
Zagaia (spear with a trident head, used with a head lamp or oil lantern for night fishing)	(As above)
Caniço/Caponga (The caniço is a rod and line with, in this case, a hook baited with a seed such as from a latex tree. The caponga is a rod and line with a seed or weight tied onto the end of the line that is splashed onto the water surface to mimic seeds falling from overhead branches.)	Tambaqui, Pirapitinga, Pacus
Espinhel (multihook longline baited with fruit)	Tambaqui, Pirapitinga, Pacus
Currico (hand line with retrievable metallic lure or dead bait)	Tucunaré, Pirapitinga, Aruanã
Pinauaca (rod and line with red cloth attached to hook as a lure)	Tucunaré, Pirapitinga, Jatuarana

SEMISELECTIVE TECHNIQUES	SPECIES
Tarrafa (throw net)	Bodó, Tamoatá, various characiform fishes
Caniço (rod and line with hook baited with berries, insects, cubes of fish, etc.)	Sardinha, Traira, Jeju, Piranha, Jatuarana, Matrinchã, Pirapitinga, Acará-açu, etc.
Espinhel (multihook longline baited with insects, frogs, meat, etc.)	Aruanã, Acará-açu, Piranha, Traira, Jatuarana, Matrinchã, Sardinha, etc.

GENERAL TECHNIQUES	SPECIES
Small *malhadeira* or *miqueira* (gill net with monofilament netting, set in flooded forest or along lake edges)	Most small and medium-sized fishes
Large *malhadeira* (gill net with multifilament netting, set passively in lakes. Large fishes such as pirarucu and tambaqui are often driven into gill nets by *batição* [beating] in the manner of a driven game shoot.)	All large fishes. Small size classes are avoided by using large mesh.

Note: See Barthem et al. (1997) for additional techniques used exclusively in whitewater river channels and paranás.

but extra rules and recommendations were incorporated to accommodate local ecological and economic factors (Queiroz and Crampton 1999a). The following section describes fisheries restrictions in the MMP and how they have changed during the period 1996–2001. The Mamirauá Institute sympathizes with the fact that rural fishermen are often driven by economic circumstances to disobey restrictions and zoning regulations. Some tolerance is exercised in the case of subsistence fishing, but fishing for economic gain is enforced as carefully as possible. IBAMA regulations on size limits are reiterated in the MMP, and extra regulations are included for specific fisheries (Queiroz and Sardinha 1999; Viana et al. this volume).

TACKLE RESTRICTIONS

Gill Nets The MMP banned the use of gill nets throughout the reserve for pirarucu and recommended banning all other kinds of gill netting during low water. These restrictions were lifted through unanimous agreement in the 1997 General Assembly because it was agreed that gill nets are appropriate for the rotation harvesting of commercialization lakes in a well-run management program.

Seine Nets IBAMA laws include complex rules on which types of seines can be deployed in different habitats. The main types (purse and beach seines) were unconditionally banned in the MMP because they cause large-scale mortality of non-target species (Barthem 1999a). This ban has been maintained and well respected since its imposition; very few residents of the reserve own seine nets.

CLOSED SEASONS

Pirarucu The MMP prohibited pirarucu fishing during the period December 1 to May 31. This prohibition is a reiteration of IBAMA policy before pirarucu fishing was indefinitely banned in 1996. This closed season is still applied to pirarucu harvested with special IBAMA authorization by the Fish Commercialization Program (Viana et al. this volume).

Tambaqui IBAMA's closed season for tambaqui usually extends from December to February (dates vary from year to year) and corresponds to the spawning period. The MMP recommended extending the closed season to begin earlier on October 1, which corresponds to the beginning of the low-water season when tambaqui are sensitive to exploitation. However, this recommendation was never approved by the general assemblies of the RDSM because of the economic value of this species.

HABITAT PROTECTION

The intimate dependence of floodplain fishes on seasonally flooded forests and floating meadows (Goulding 1980, 1993; Pires 1996; Henderson and Crampton

1997; Crampton 1999b) means that fisheries management can only work in the long term if integrated with habitat conservation. Floating meadows act as nursery grounds for a variety of commercially important species and provide low-water refuge for almost all fishes (Junk 1984; Crampton 1999b). Of the forest ecosystems, *restinga alta* forest growing on the high levees supports the highest diversity of trees and richest terrestrial and arboreal biota (Goulding, Smith, and Mahar 1996). It is also an important source of sustenance for seed- and fruit-eating commercial species, such as tambaqui, pirapitinga, and pacu. Levee forests are flooded by one to three meters of water for up to four months each year (Ayres 1993) and cover around 12% of the RDSM. A further 50% of the reserve is made up of transitional *restinga baixa* forests, the back-slopes from high levees down to low-lying pioneer chavascal forest (Ayres et al. 1999). Flooded chavascal forests support enormous areas of floating meadows during the high-water season.

At present there are no immediate threats to floating meadow habitats in the RDSM. However, restinga forests are threatened by the clearing of *roças* (gardens) for manioc and banana production. The high levees are always chosen for roças because they remain inundated for less time than lower-lying land. The greatest threat to the várzea habitats of the Amazon is unquestionably large-scale cattle or water buffalo ranching, which involves the complete destruction of várzea forests and the degradation and trampling of floating meadows. Ranching has begun in the Tefé region but does not occur in the RDSM, in part because there are no terra firme areas into which livestock can be driven during the high-water period. A major priority of the Mamirauá Institute is to provide the population of the RDSM with ecologically and economically acceptable alternatives to ranching. Profitable fisheries, integrated with forest management and forest-friendly agricultural activities, provide local communities with strong economic incentives to preserve restinga forests and floating meadows. Effective fisheries management should in this sense promote a self-reinforcing cycle in which habitat conservation and fisheries management are reciprocally beneficial.

FOREST-FRIENDLY AGRICULTURE

Researchers at the Mamirauá Institute are introducing new seed stock and teaching techniques for the cultivation of beans, corns, rice, peanuts, and melons on exposed beaches (J. Inuma, Mamirauá Institute, pers. comm.). These crops have no impact at all on the forest ecosystems of the várzea. Methods for extending the duration of roças or using secondary-growth roças instead of new forest are also being explored. Finally, the cultivation of understory trees for agro-forestry production is being evaluated. Cacao (*Theobroma cacao*), *açai* palm (*Euterpe oleracea*), and some other species already grow naturally in the várzea but are not very economically attractive. One promising species for commercialization is the camu-camu tree (*Myrciaria dubia*), the fruits of which are used to make a vitamin C-rich juice (SCM 1996).

FORESTRY

Várzea forests have outstanding economic potential due to the fast growth rates of a variety of commercially important trees, low harvesting costs (logs can be floated to the market), and constant market demand (Albernaz and Ayres 1999). With careful management, restinga forests (where most of the valuable timber grows) have the potential to provide a long-term supply of timber, the economic value of which exceeds that of agricultural production. With low harvest rates (approximately 5 trees/ha/year) and selective felling of trees, managed restinga forests are expected to retain most of their biodiversity and continue to sustain fish stocks during the high-water period (J. Bampton, DFID, pers. comm.). A forest management program at the IDSM is building the capacity of local communities to undertake sustainable management and to market wood through the formation of formal community associations. As with fisheries the definition of rights of exclusive access to timber resources is fundamental.

DISCUSSION

The primary aim of the Mamirauá Reserve is to reconcile wildlife conservation with long-term improvements in the living standards of the local people. So far, after almost ten years of activities, the partnership between MSDI and local people continues to be expanded with great enthusiasm by the resident and user population of the RDSM. The spending power of many communities has increased, and some indices of general standard of living such as infant mortality, literacy, and parasite infestation levels have improved over the last decade (IDSM 2001; E. Moura pers. comm.). Moreover, the results from the Fish Commercialization Program described by Viana et al. (this volume) give a clear indication of the magnitude of financial benefits that can accrue from sustainable fisheries management.

Is there evidence that participatory management, community vigilance, and zoning are also having the desired effects of restricting access rights to local users and promoting the conservation of resources? Below, we summarize some lines of evidence to suggest that for fish resources the answer in both cases seems to be yes:

MONITORING LANDINGS

Fish landings from the Focal Area of the RDSM have been monitored at the Tefé market since October 1991 (fig. 7.3) (Barthem 1999a,b). Although much of the variation in landings illustrated in figure 7.3 is related to seasonal effects, mean monthly landings from the Focal Area of the RDSM are 58% lower in the second half of the time series (5.40 tons, SD 4.14) than in the first half (12.74 tons, SD 8.87). This disparity is strongly significant (two-tailed t-test, n = 51, T = 5.35, P = 0.001). Comparing the same periods, there was a smaller but significant decline (14%) in mean monthly landings from outside the Focal Area of the reserve from 159.25 tons (SD

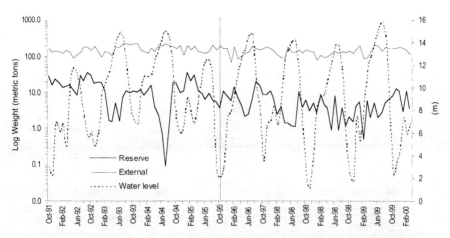

FIGURE 7.3 Commercial landings of fish at the Tefé fish market originating from inside and out-side the Mamirauá Sustainable Development Reserve. Left axis shows landings on logarithmic scale. Right axis shows water level measured near Tefé at the beginning of each month. The da-tum at 10 m corresponds to the point at which high levee forest of the várzea become flooded. The vertical solid line divides the time series in half. The data excludes or is incomplete for the cate-gories of fish listed in table 7.3. "Reserve" refers to the Focal Area of the RDSM (Mamirauá Sus-tainable Development Reserve) where fisheries management and vigilance are undertaken. "Ex-ternal" refers to the sum of three categories of data: (1) Subsidiary Area of RDSM (Mamirauá Sustainable Development Reserve), (2) outside the reserve, and (3) without information. The without information category represents 3.4% of the total data set and refers primarily to landings from outside the reserve.

36.94) to 136.99 tons (SD 34.01) (two-tailed t-test, n = 51, T = 3.17, P = 0.005). These analyses indicate a substantial decline in the proportion of fish landings deriving from the Focal Area of the RDSM.

Landings from different classes of fishing boat were also discriminated (fig. 7.4). Fish brought to Tefé from the RDSM in canoes with long-shaft (*rabeta*) engines belong almost exclusively to residents and users of the reserve. On the other hand, commercial boats with inboard engines and with or without fixed iceboxes are ex-clusively owned by fishermen outside the RDSM and mostly belong to the urban fleet of Tefé. Between 1991 and 2000 there was a clear decline in (invasive) fishing in the RDSM from the most important category of commercial fishing boat—those with fixed iceboxes. At the same time there was a distinct rise in the proportions of landings at Tefé by the rabeta canoes.

Our feeling is that, despite some shortcomings, these data and observations indi-cate a decline in invasive fishing in the Focal Area of the RDSM and a concomi-tant increase in landings from residents and users. On a smaller scale, a similar trend was also observed in várzeas of the Amanã Sustainable Development Re-

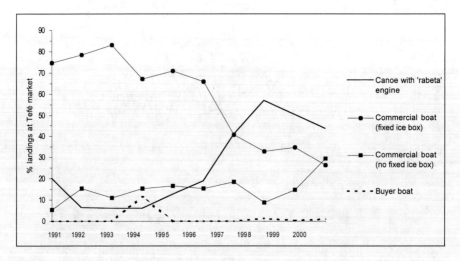

FIGURE 7.4 Proportional contribution of different categories of fishing boats to the commercial landings of fish at the Tefé fish market that originate from the Focal Area of the RDSM (Mamirauá Sustainable Development Reserve).

serve, a new protected area created in 1999, and in várzeas along the Rio Solimões within 50 km of Tefé. Both of these areas developed self-motivated community vigilance schemes over the last decade with support from the Tefé-based Preservation and Development Group (Grupo de Preservação e Desenvolvimento, or GPD) and from MEB of the Catholic Church (Hall 1997). Ongoing work at IDSM is attempting to describe more precisely who is fishing in the Tefé region and when.

INVASIONS BY THE *GELEIRA* FLEETS

One of the Mamirauá Reserve's most resounding successes has been the almost complete elimination of commercial *geleira* boats (with large ice holds) originating from the cities of Manaus, Manacapuru, and Itacoatiara (Ayres et al. 1999). The evidence for this elimination comes mostly from the reports of fishermen who recall a series of conflicts with geleira crews in the years preceding the designation of the area as a reserve. The geleira boats travel over distances of up to 700 km, and the closure of the Mamirauá Reserve affected only a small proportion of their potential fishing grounds. In contrast, a substantial portion of the fishing grounds of the Tefé commercial fishing fleet has been affected. This difference probably explains why there has been a more marked reduction in invasions by the geleira boats than by commercial boats from the Tefé fleet.

PERCEPTIONS OF LOCAL FISHERMEN

Interviews with thirty-eight fishing families from ten communities in the RDSM documented a general observation of increasing stocks for several commercial species (table 7.5).

POPULATION GROWTH OF KEY COMMERCIAL SPECIES

There is mounting evidence that populations of several key commercial species are increasing in the protection zone of the MSDR. Pirarucu populations are higher in the protection zone of the Jarauá Sector than in the surrounding sustainable use zone (Viana et al. this volume). Costa, Barthem, and Correa (1999) recorded consistently higher densities of tambaqui in protection-zone lakes of the RDSM than in lakes used for community subsistence or commercialization. The population of black caiman (*Melanosuchus niger*) has also increased dramatically within the protected zone of the Mamirauá Sector since the early 1990s (Ronis da Silveira, INPA—National Institute for Amazonian Research, pers. comm.).

OVERVIEW

The results above provide early evidence for the growth of key commercial species within the core protection zones of the MSDR and for a decline in invasions by outside users. The extent to which the two are linked is impossible to quantify, but

TABLE 7.5 Results of Interviews with Thirty-eight Fishermen in the RDSM to Characterize Perceived Changes in the Abundance of Key Fish Stocks

	NUMBER OF REPLIES				
SPECIES	Major Decrease	Minor Decrease	No Change	Minor Increase	Major Increase
Pirarucu	3	4	7	16	8
Tambaqui	2	2	8	13	13
Tucunaré	0	3	6	14	15
Aruanã	0	6	1	10	21
Pimelodid catfishes	4	8	13	8	5
Discus	2	2	17	8	9

Note: The interviews were conducted in September 2000 using a multiple-choice questionnaire. All of the communities are located on or near the banks of the Rio Japurá within twenty kilometers of the Jarauá community (fig. 7.2). The ten communities are Jarauá (4 fishermen interviewed), Santa Maria do Cururu (4), Vista Alegre (4), São Francisco do Cururu (4), Nova Betânia (3), Nossa Senhora da Fátima (4), Vila Betel (3), Manacabi (4), Novo Pirapucu (4), and Nova Colômbia (4).

presumably there has been an overall decline in pressure on key resources within the reserve (and especially within the core protection zones) that is promoting the recovery and growth of previously overexploited species. Community management and vigilance are presumably the main explanations for these patterns. Nonetheless, anecdotal evidence suggests that the mere presence of scientists and extension workers of the IDSM has an appreciable effect on the extent to which local users are likely to break zoning rules and on the extent to which outsiders are likely to invade.

Despite the incomplete and early nature of results emerging from the Mamirauá Reserve, it is evident that the kind of partnership between local people and a supporting NGO that is being developed in the RDSM can create the conditions for successful management, vigilance, and economic gain that are required to set up a self-reinforcing cycle of sustainable management.

These emerging results are welcome, especially because just five years ago it was unclear as to whether the substantial costs of establishing a Sustainable Development Reserve would be rewarded with any evidence for simultaneous improvements in ecosystem health, livelihoods, and access rights. Crampton, Castello, and Viana (this volume) suggest that future models for fisheries management in the Amazon basin will need to look carefully at the early results of contemporary experiences in fisheries management—both in the zoned reserve context of the RDSM and in the open multiuse context currently being developed by the Instituto Iara and Projeto Várzea in the lower Amazon. The early successes of the RDSM indicate that zoned reserve nuclei offer a tangible and potentially effective means of reconciling the conservation of fish stocks and habitats with sustained economic growth. The model offered by Crampton, Castello, and Viana (this volume) proposes a chain of such nuclei within regional zones of multiuse management. Each one of these nuclei will require the intervention and support of outside agencies or NGOs like the IDSM. Whether funding and expertise will be available for this proposal stands as one of the major challenges for environmentalists and politicians of the coming century.

ACKNOWLEDGMENTS

We thank J. M. Ayres and the staff of the Instituto de Desenvolvimento Sustentável Mamirauá. J. Bampton, G. Castro, M. Harrison, and S. Gillingham commented on the manuscript. The authors were funded by CNPq and DFID.

8

Hunting Effort as a Tool for Community-Based Wildlife Management in Amazonia

PABLO E. PUERTAS AND RICHARD E. BODMER

Wildlife hunting is a major activity of Amazonian inhabitants, both in seasonally flooded várzea forest and nonflooded tierra firme forest (Beckerman 1994; Bodmer 1994). Local people in Amazonia preferentially exploit large and mid-sized mammals as sources of protein and cash income through meat sales (Redford and Robinson 1991; Bodmer et al. 1997b). In the western Amazon, hunting patterns are strongly influenced by meat values, and wildlife conservation strategies must incorporate local people's needs for wildlife meat. Community-based wildlife management allows people to obtain subsistence and cash benefits from hunting, while at the same time promoting conservation. Community-based strategies are apparently successful at conserving wildlife species in western Amazonia, in large part because human populations are relatively low (less than 1 person/km^2) and because an intact habitat is still abundant.

In a community-based system local people must make management decisions about access rights and levels of hunting. Thus, local communities must have a mechanism to evaluate the impact of hunting on wildlife species. Most models that evaluate the impact of hunting combine information on hunting pressure and some estimation of species populations (Bodmer and Robinson this volume). Information on hunting pressure is relatively easy for local communities to collect since hunters usually bring back animals to villages. However, information on animal populations requires great effort and in the Neotropics usually involves line transect censuses. These censuses are very time consuming and require methods that use unhunted trails with no hunting activity being conducted during censuses (Rabinowitz 1993). Local people must take time away from such other important activities as small scale farming, fishing, or subsistence hunting to do censuses. These other demans make it difficult for local people to carry out line transect censuses, especially if they do not receive financial incentives from outsiders.

Catch per unit effort (CPUE) analysis is an alternative method that can be used to evaluate the abundance of wildlife species and to measure trends in wildlife populations. CPUE is assumed to indicate whether a species is overhunted or not overhunted. A decrease in the CPUE would suggest overuse (a decreasing population), a constant CPUE would suggest a stable population, and an increase in CPUE would suggest an increasing population (Vickers 1991).

CPUE methods do not interfere with the activities of local people since they do not compromise other work, and CPUE data are relatively easy for community members to collect. Further, unlike line transects, CPUE is also relatively easy to analyze and can potentially be analyzed by local people. In turn, local people can make management decisions using CPUE.

This study evaluated the effectiveness of CPUE as a tool for community-based wildlife conservation. First, we determined whether CPUE could be used in Amazonia to evaluate wildlife species abundances. We did this by comparing CPUE with an independent measure of abundance using line transect censuses. Next, we determined whether CPUE data could be collected easily by local people, and could be used as a tool for community-based wildlife management.

The effectiveness of CPUE as a community-based wildlife management strategy was studied in the Tamshiyacu-Tahuayo Community Reserve (Reserva Comunal Tamshiyacu-Tahuayo, or RCTT) in northeastern Peru. Community-appointed wildlife inspectors collected data on CPUE. These inspectors were community members who were responsible for the vigilance of hunting, and they formed part of the community-based wildlife management program of the reserve. This inspection system was already in place prior to the start of the project. CPUE data were collected during 1994, 1995, and 1996 in the Tamshiyacu-Tahuayo community reserve. The data examine the relationships between effort and yield, and in this case the relationship is presented as animals per hunter-days.

We evaluated changes in CPUE both annually and seasonally and compared differences in CPUE inside and outside the reserve. The comparison between seasons allowed us to test CPUE against an independent measure of abundance and to test whether CPUE reflects abundance. The comparison of CPUE inside and outside of the reserve showed us what species are appropriate for CPUE analysis and why CPUE analysis does not work for certain species. The comparison between years was used to test whether CPUE could be used as a measure of the sustainability of hunting.

STUDY AREA AND METHODS

THE RESERVA COMUNAL TAMSHIYACU-TAHUAYO

The RCTT, located in the northeastern Peruvian Amazon, comprises 322,500 ha of continuous, predominantly upland forest (75%) with a lesser amount of flooded forest (Bodmer et al. 1997b). (fig. 8.1). The city closest to the reserve is Iquitos, located

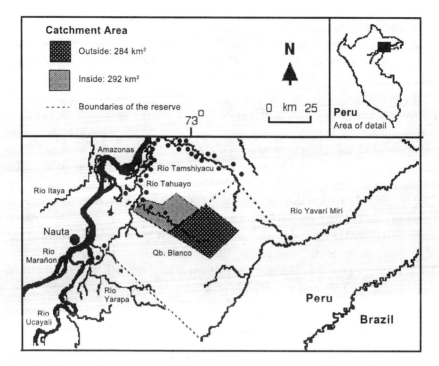

FIGURE 8.1 Location of the study area in the Reserva Comunal Tamshiyacu-Tahuayo, northeastern Peru, showing the catchment areas inside and outside the reserve.

about 100 kilometers northwest of the reserve, with a population of approximately 400,000 inhabitants. The reserve is bordered to the west by the upper Tahuayo and Blanco rivers, to the south by the upper Yarapa River, to the east by the upper Yavarí Mirí River, and to the north by the upper Tamshiyacu River.

The RCTT is divided into three distinct land use zones: (a) a protected source area of approximately 160,000 ha, (b) a zone of subsistence use of approximately 160,000 ha, and (c) an area of permanent human settlement that has no defined boundaries (Bodmer et al. 1997b). The subsistence use and source areas are within the official limits of the reserve and have no people living inside them. The area of subsistence use is for extraction of natural resources by local residents of the permanent settlement zone. Residents cannot set up houses or clear land for agricultural uses within the boundaries of the subsistence use or source areas. The area used for permanent settlements along the Tamshiyacu, Tahuayo, Yarapa, and Yavarí Mirí rivers is adjacent to the reserve. This area includes villages and is for intensive land use, such as agriculture. The human settlement zone was not officially included in the reserve in order to avoid conflict over land-use practices, but it is an important part of the RCTT management plans (Bodmer et al. 1997b).

The RCTT has an extraordinarily high diversity of faunal and floral groups (Castro 1991; Puertas and Bodmer 1993), due in part to the juxtaposition of tierra firme (upland) forest with rich soils and várzea (flooded) habitats and to its location in a biogeographic region of high species diversity in western Amazonia (Gentry 1988). At least fourteen species of primates are found in the RCTT, the greatest diversity of primates reported for any protected area in Peru (Puertas and Bodmer 1993).

The majority of rural inhabitants who use the reserve are nonindigenous people known as *ribereños* (Bodmer et al. 1997b). The major economic activities of these people include fishing, agricultural production, game hunting, small-scale lumber extraction, and collection of minor forest products, such as fruits, nuts, and fibers (Coomes 1992). Ribereños, like Amazonian Indians, have a great knowledge of forest plants, agriculture techniques, and hunting and fishing methods. Many have only recently abandoned their indigenous heritage in order to claim themselves as Peruvians. Ribereños have an intricate involvement in the market on both regional and international levels and harvest products, such as spices, rubber, and furs, that have traditionally been marketed in European countries and North America. These rural Amazonians are renowned for their ability to switch harvest patterns as markets change, a reason for their wide geographic mobility (Padoch 1988).

DATA COLLECTION, SAMPLING, AND HUNTING

Information on CPUE was collected from hunters who live in the middle and upper sections of the Blanco river and who use both the human settlement and subsistence areas of the Reserva Comunal Tamshiyacu-Tahuayo (fig. 8.1). The forests that hunters used in the human settlement area cover 284 km^2 and will be referred to as the "human settlement catchment area" or the "area outside the reserve." The forests that hunters used in the subsistence area are 292 km^2 and will be referred to as the "subsistence catchment area" or the "area inside the reserve."

Information on CPUE was collected from both direct observations and through hunting registers. Hunters began registering harvests in 1991 as part of the participatory involvement of community-based comanagement. Hunter participation relies on building interest in community-based wildlife management by having researchers work with hunters when evaluating the impact of harvests (Bodmer and Puertas 2000). Hunting registers involve hunters and their family in data collection. This participatory method helps researchers, extension workers, and hunters find common ground to discuss wildlife issues. These registers also provide information on CPUE and can be analyzed as the number of kills per person-day of hunting per year. Hunting registers were analyzed for this article using data from 1994, 1995, and 1996.

Three hunters and their families were trained as recorders for the registers in 1993. However, only two families collaborated effectively with the project, and we only used their data. The recorders' homes were strategically located along the banks of the Blanco river (Quebrada Blanco) so that they could easily note which hunters went up or came down the river. Hunting registers were checked continually during the first six months of the study to evaluate their accuracy and to make adjustments. Later, registers were checked monthly and compared for uniformity. Recorders' wives continued to register hunting activity when their husbands were away hunting or went to Iquitos to sell products. If hunters were not registered immediately after their return from the forest, they were registered indirectly through information provided by other local inhabitants.

Hunting registers included information about the number of animals hunted, the species, sex, and location of the kill, and dates of departure and return of the hunter. This last item was used to determine effort. The majority of the animals were identified by direct observation. In some cases, identification was made by comparison with specimens at the Zoology Museum of the National University of the Peruvian Amazon or the Peruvian Primate Project Manuel Moro Sommo in the city of Iquitos.

Line transects censuses were used as an independent measure of animal abundance. While over 3,000 km of line transects have been conducted in the area, we only used a portion of the data set to compare CPUE with abundance. We used data from 1997 collected during the low- and high-water seasons from the upper Qb. Blanco site inside the reserve. A total of 92 km of census was used for the low-water season and 170 km for the high-water season. The encounter rates of animal groups were used as a measure of abundance since individuals are not independent in social species.

RESULTS

COMPARISON BETWEEN SEASONS

There was no observed difference in CPUE between the high- and low-water seasons from 1994 to 1996 in the subsistence area inside the reserve (one-way Anova, $F = 0.009$, $p = 0.931$, $df = 1$) (fig. 8.2a). Likewise, there was no difference in the abundance of animal groups sighted on line transect censuses between the high- and low-water seasons when all species are considered together (Kruskal-Wallis, $p = 0.1$; $df = 7$) (fig. 8.3). The total harvest was greater in the high-water seasons than in the low-water (one-way Anova $F = 10.443$, $p = 0.032$, $df = 1$) (fig. 8.2b). This finding reflects a difference in access to the hunting zones between seasons. Hunters can reach the subsistence area inside the reserve more easily in the high-water season since high-water levels facilitate access with canoes and small boats. This access was reflected in a greater number of hunter-days in the high-water season than in the low-water.

There was no observed seasonal difference in CPUE between the high- and low-water seasons from 1994 to 1996 in the human settlement area (one-way Anova, $F = 0.168$, $p = 0.703$, $df = 1$) (fig. 8.4a). Thus, the collective abundance of species was similar during both seasons in the human settlement area. In addition, there

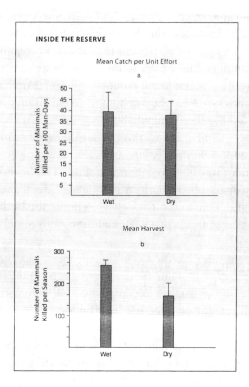

FIGURE 8.2 Mean (SD) CPUE (catch per unit effort) and hunting pressure inside the reserve between the high/(wet)-water and low/(dry)-water seasons.

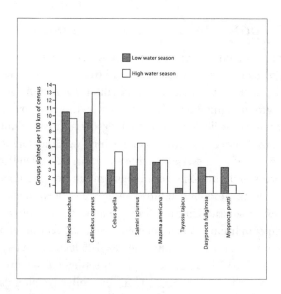

FIGURE 8.3 Relative abundance of large mammals between the high- and low-water seasons in the Tamshiyacu-Tahuayo Community Reserve in 1997.

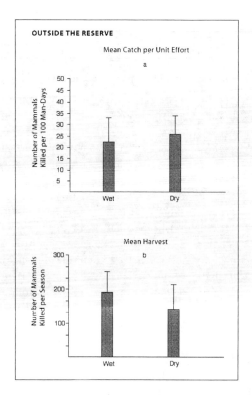

FIGURE 8.4   Mean (SD) CPUE (catch per unit effort) and hunting pressure outside the reserve between the wet and dry seasons.

was no observed difference in the harvests of mammals between the seasons (one-
way Anova F = 0.748, p = 0.436, df = 1) (fig. 8.4b).

COMPARISON BETWEEN HUNTING AREAS

A total of 3,358 man-days of hunting were recorded during 1994, 1995, and 1996 in
the subsistence area inside the reserve. On a few occasions, hunting dogs were observed accompanying novice hunters inside the reserve (table 8.1A). Artiodactyls,
rodents, perissodactyls, primates, and edentates were the most important orders for
local people hunting in the reserve. The ten most commonly hunted species, in order of importance, were: *Tayassu pecari, Tayassu tajacu, Agouti paca, Mazama
americana, Dasyprocta fuliginosa, Tapirus terrestris, Dasypus novemcinctus, Pithecia monachus, Mazama gouazoubira*, and *Tamandua tetradactyla*.

A total of 4,200 person-days of hunting were recorded during 1994, 1995, and
1996 in the human settlement area outside the reserve. People were not observed
using dogs when hunting in the human settlement area. Artiodactyls, rodents, primates, carnivores, and perissodactyls were the most important orders for local peo-

TABLE 8.1 Number of Individuals Hunted

	A	B	C	D	E	F
Species	In (SD)	Out (SD)	In (SD)	Out (SD)	In (SD)	Out (SD)
Tayassu pecari	42 (2.1)	21 (5.9)	11.3 (1.5)	4.3 (1.5)	3.46 (0.69)	1.393 (0.472)
Tayassu tajacu	30 (3.8)	14 (1.5)	7.7 (1.5)	2.7 (2.1)	2.037 (0.343)	0.727 (0.487)
Mazama americana	9 (1.5)	8 (0.6)	2.3 (0.6)	1.8 (1.9)	0.73 (0.115)	0.6 (0.575)
Mazama gouazoubira	2 (0.6)	2 (0.6)	0.5 (0.1)	0.1 (0.1)	0.097 (0.029)	0.107 (0.006)
Tapirus terrestris	3 (2.0)	3 (1.5)	0.7 (0.4)	0.7 (0.3)	1.123 (0.544)	0.6 (0.575)
Pithecia monachus	2 (2.0)	6 (5.0)	0.5 (0.5)	1.1 (0.9)	0.01 (0.01)	0.026 (0.022)
Callicebus cupreus	1 (0.2)	6 (1.2)	0.2 (0.1)	1.0 (0.1)	0.002 (0.001)	0.01 (0.0001)
Lagothrix lagothricha	1 (1.0)	2 (1.0)	0.5 (0.2)	0.4 (0.1)	0.053 (0.025)	0.043 (0.012)
Cebus apella	1 (1.4)	1 (1.4)	0.2 (0.1)	0.3 (0.1)	0.007 (0.004)	0.01 (0.009)
Cebus albifrons	1 (0.5)	2 (0.6)	0.2 (0.2)	0.5 (0.2)	0.005 (0.005)	0.016 (0.006)
Ateles paniscus	0.3 (0.6)	0.3 (0.2)	0.1 (0.2)	0.1 (0.1)	0.01 (0.017)	0.004 (0.004)
Cacajao calvus	0.4 (0.2)	0.5 (0.2)	0.1 (0.1)	0.1 (0.1)	0.005 (0.001)	0.004 (0.001)
Alouatta seniculus	0.7 (0.4)	0.8 (0.3)	0.2 (0.1)	0.2 (0.1)	0.016 (0.012)	0.016 (0.008)
Saimiri spp.	0	2 (1.3)	0	0.4 (0.3)	0	0.003 (0.002)
Aotus nancymae	0	0.1 (0.2)	0	0.02 (0.05)	0	0.0002 (0.0003)
Saguinus spp.	0	0.5 (0.5)	0	0.09 (0.1)	0	0.0004 (0.0005)
Agouti paca	44 (12.2)	22 (12.9)	17.0 (4.0)	4.7 (2.9)	1.137 (0.189)	0.393 (0.257)
Dasyprocta fuliginosa	4 (2.1)	8 (5.3)	1.1 (0.8)	1.8 (1.3)	0.057 (0.02)	0.083 (0.056)
Myoprocta pratti	0	2 (2.3)	0	0.5 (0.5)	0	0.004 (0.004)
Hydrochaeris hydrochaeris	0	0.1 (0.2)	0	0.03 (0.5)	0	0.01 (0.017)
Coendou bicolor	0	0.1 (0.2)	0	0.1 (0.1)	0	0.003 (0.005)
Sciurus spp.	0.2 (0.2)	4 (0.1)	0.1 (0.05)	0.1 (0.1)	0	0.0002 (0.0003)
Dasypus novemcinctus	2 (2.1)	3 (3.3)	3 (4.4)	0.5 (0.5)	0.047 (0.015)	0.039 (0.037)
Tamandua tetradactyla	1 (0.6)	2 (2.4)	0.4 (0.1)	0.5 (0.5)	0.017 (0.007)	0.022 (0.024)
Myrmecophaga tridactyla	0.1 (0.2)	0.9 (0.9)	0.03 (0.1)	0.2 (0.2)	0.01 (0.017)	0.057 (0.064)
Priodontes maximus	0	0.3 (0.2)	0	0.05 (0.04)	0	0.01 (0.012)
Didelphis marsupialis	0	0.1 (0.2)	0	0.03 (0.05)	0	0.0002 (0.006)
Chironectes minimus	0.1 (0.2)	0	0.03 (0.05)	0	0.002 (0.003)	0
Puma concolor	0.2 (0.4)	0	0.05 (0.04)	0	0.063 (0.065)	0
Nasua nasua	0.5 (0.4)	5 (3.0)	0.1 (0.1)	1.1 (0.9)	0.159 (0.27)	0.033 (0.021)
Leopardus spp.	0.3 (0.1)	0.6 (0.3)	0.1 (0.01)	0.1 (0.1)	0.007 (0.003)	0.01 (0.009)
Eira barbara	0	0.5 (0.5)	0	0.1 (0.1)	0	0.004 (0.005)
Potos flavus	0.1 (0.2)	0.4 (0.1)	0.03 (0.05)	0.1 (0.01)	0	0.002 (0.0006)
Total	150 (34.7)	115 (54.3)	46.4 (15.3)	23.7 (15.85)		

Note: (A) Annual mean number of individuals hunted in the subsistence area inside the reserve (In) and (B) in the human settlement area outside the reserve (Out); (C) CPUE (per 100 person-days) of mammals hunted in the subsistence area (In) and (D) in the human settlement area (Out); (E) CPUE (per 100 person-days) of mammalian biomass extracted in the subsistence area (In) and (F) in the human settlement area (Out). Values are for 1994, 1995, and 1996, and provided with (SD).

ple hunting outside the reserve. The ten most frequently harvested species, in order of importance, were: *T. pecari*, *T. tajacu*, *A. paca*, *M. americana*, *D. fuliginosa*, *P. monachus*, *Callicebus cupreus*, *Nasua nasua*, *T. terrestris*, and *D. novemcinctus* (table 8.1B).

Registries of hunters were greater in the subsistence area than in the human set-tlement area of the reserve (one-way Anova F = 663.1, p < 0.001, df = 1). Hunters living in communities close to the reserve did most of the hunting, while people living further from the reserve hunted less. Hunters living in communities further from the reserve tended to use the subsistence area inside the reserve. In contrast, hunters living in the communities closest to the reserve hunted more frequently in the human settlement area.

CPUE was greater in the subsistence area of the reserve than the human settle-ment area outside the reserve (one-way Anova, F = 7.708, p = 0.037, df = 1) (fig. 8.5). This finding suggests that animals are more abundant inside the reserve than outside.

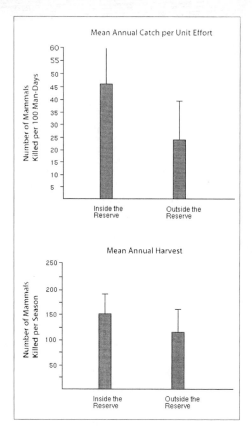

FIGURE 8.5 Mean (SD) annual CPUE (catch per unit effort) and hunting pressure inside and out-side the reserve.

CPUE was greater in the subsistence area than the human settlement area for economically valuable species, such as *Tayassu pecari* (one-way Anova, $F = 31.5$, $p = 0.005$, $df = 1$) and *T. tajacu* (one-way Anova, $F = 11.25$, $p = 0.028$, $df = 1$), reflecting a greater abundance of peccaries inside the reserve. However, for such smaller, noneconomically valuable species as the titi monkey (*Callicebus cupreus*). CPUE was greater in the human settlement area than the subsistence area (one-way Anova, $F = 700$, $p < .0001$) (table 8.1C and D). Interestingly, noneconomically valuable species are used for household food and are usually hunted close to the villages. In contrast, hunting in the subsistence area focuses more on large animals whose meat can be used for subsistence or sold to purchase other basic subsistence goods (Bodmer et al. 1997b).

CPUE can also be analyzed using measures of biomass extracted; such an analysis looks at the amount of meat hunted from an area. The species that had the greatest biomass extracted included *T. pecari*, *T. tajacu*, *Tapirus terrestris*, and *A. paca*. Total CPUE of biomass extracted differed between the subsistence and human settlement areas (one-way Anova, $F = 9.296$, $p = 0.038$, $df = 1$) (table 8.1E and F). *T. pecari* (one-way Anova, $F = 17.59$, $p = 0.014$), *T. tajacu* (one-way Anova, $F = 14.52$, $p = 0.019$, $df = 1$), and *Agouti paca* (one-way Anova, $F = 16.24$, $p = 0.016$, $df = 1$) had greater CPUE inside the reserve than outside, suggesting greater abundance of these species in the subsistence use area than in the human settlement area. These species are also the ones with the greatest economic value for people who hunt in the subsistence area.

In contrast, *C. cupreus* (one-way Anova, $F = 53$, $p < .0001$) and *Potos flavus* (one-way Anova, $F = 49$, $p = 0.002$, $df = 1$) had greater CPUE of biomass extracted in the human settlement area than the subsistence area. Smaller species like *Callicebus* and *Potos* are exclusively hunted for household use and have little economic value.

These results show that CPUE should be divided among species. The effort for nonpreferred species is lower than for the preferred species in some sites, and CPUE analyses should be partitioned per species. For example, a low CPUE for agouti may be false if very little time was actually spent on hunting agouti at a site. If more time was then spent on hunting agouti at another site, then CPUE is not comparable for agouti. Thus, one can only really compare CPUE between sites for preferred species or for overall hunting yields, without a division into species.

COMPARISON OF CPUE BETWEEN YEARS

Comparing the annual values of total CPUE suggests that the animal populations are increasing in both hunting areas, although more data are needed to demonstrate this statistically (fig. 8.6). Overall, CPUE is showing similar trends between the subsistence area inside the reserve and the human settlement area outside the reserve. This positive trend in CPUE suggests that the community-based comanagement of wildlife is working for the conservation of most species. However, some

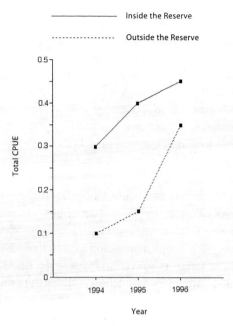

Inside the Reserve

Outside the Reserve

FIGURE 8.6 Trend in the total CPUE (catch per unit effort) per year inside and outside the reserve.

species of carnivores, primates, and edentates had decreasing CPUE over the years, indicating a decrease in their abundance.

CPUE AND COMMUNITY-BASED WILDLIFE MANAGEMENT

CPUE appears to be a reliable method for assessing the populations of wildlife in the Tamshiyacu-Tahuayo community reserve. Results show that, when independently comparing each hunting area, there are no statistically significant differences between high- and low-water seasons. This constant CPUE suggests that the abundance of species was similar between seasons both inside and outside the reserve. This similarity provides evidence of CPUE being an accurate measurement of mammal abundance since one would not expect changes in densities between seasons in Amazonian mammals that are found in upland (tierra firme) forests and that generally do not migrate.

One weaknesses of using CPUE as a comparative index of abundance between sites is that it only works well with economically important species, such as peccaries, deer, tapir, and large-bodied rodents. These mammals are hunted whenever they are encountered, either in the human settlement area or the subsistence area. However, with nonpreferred species, such as *Callicebus* and *Potos*, hunters really only kill these animals in the settlement area close to their homes since they are used almost exclusively for household consumption. These smaller, nonpreferred species are rarely hunted in the subsistence area since hunters visit the subsistence area for economically important species. Thus, the higher CPUE of *Callicebus* and *Potos* in the human settlement area compared to the subsistence area does not necessarily reflect a difference in abundance. Rather, it reflects a difference in hunter preference.

Community-based conservation is potentially a very important means of achieving conservation in tropical regions because people living in tropical forests often need to use resources for subsistence and market sale. If they are involved with managing those resources through community-based mechanisms, then there is potential for sustainable use and, in turn, conservation.

For community-based resource management to be truly participatory, local people must have a means of evaluating the impact of their resource use. For wildlife hunting this need means local people must have some way of evaluating the status of animal populations. Most simple population models that have been used to evaluate hunting in tropical forests rely on abundance or density estimates, usually using line transect counts. Abundance or density estimates are then entered into the models.

Line transect counts involve visual, auditory, or indirect counts (tracks) along trails that have been cut in the forest. Line transects only work well if trails are not used by hunters and if no hunting is occurring during censuses. For many tropical forest species, approximately 1,000 km need to be censused per site to obtain appropriate sample sizes. People conducting the censuses should walk at an average speed of 1 km/hour. It would then take an estimated 1,000 hours to conduct line transects or 125 person/days (using 8 hours/day), not taking into consideration the time it takes to open unhunted trails.

The effort needed for local people to conduct appropriate line transects as part of a participatory wildlife management program is likely to be impractical. People would need to take considerable time away from other activities to conduct line transects. If these transects were not financed it would be difficult for most local people to participate in wildlife censuses.

Alternative ways are therefore required for people to evaluate the impact of hunting that do not involve line transects. One of these might be using age structure analysis. Skulls of animals can be collected easily by local people, especially if they cook skull-brain soup, which is often the case in Amazonia. The animal skulls can then be used to assess the age of individuals, often using simple techniques such as tooth wear. These age data can be easily plotted into age structures. The problem of using age structure as a measure of hunting impact is in the interpretation of the results. Age structures of a hunted sample from randomly hunted populations might not differ significantly between slightly hunted and overhunted populations. Thus, using age structure of a hunted sample might give a false impression of sustainability when in fact the population is heading toward extirpation.

Another alternative is using CPUE as a measure of animal abundance and then using comparative abundances as a way of evaluating the impact of hunting. For example, in this study the annual values of total CPUE were increasing for each hunting area. Preliminary results indicate that the overall wildlife population in both study areas was healthy and that the populations were not declining. The increase in the CPUE between 1994 and 1996 might be due to the comanagement programs in the reserve. However, this finding does not apply for the specific species of carnivores, primates, and edentates. The results of this study suggest that wildlife is more abundant inside the reserve than outside and that the reserve zone conserves wildlife populations.

One great advantage of CPUE is that it agrees with the activities of local people and does not take much time away from other activities. Local people in the Tamshiyacu-Tahuayo community reserve were able to collect CPUE data easily by setting up a system of wildlife inspectors. Hunters would record the number of animals hunted and the time they spent hunting upon their return to the villages. This system meant that they could continue their resource use activities and collect the CPUE information at the same time. Similar to skull collections, CPUE information is compatible with community-based management systems. It is also more reliable than age structure analysis in the interpretation of the results, and it gives a clearer picture of the impact of hunting. Indeed, local hunters can evaluate CPUE results with minimal assistance from wildlife extensionists. This ability makes CPUE particularly valuable as a community-based technique. Furthermore, hunters could potentially work out abundances and see if animal populations are increasing, stable, or decreasing in their hunting areas. Local hunters could then make management decisions on hunting according to the CPUE results.

For CPUE to be reliable, hunting technology must be constant during the monitoring period. If technology changes then the CPUE does not give accurate results. In the Neotropics there are examples of indigenous groups converting from bow and arrows to shotguns (Hames 1980; Yost and Kelly 1983). This type of drastic change will influence the effort needed to hunt and, in turn, will make comparisons of CPUE between technologies difficult to interpret.

ACKNOWLEDGMENTS

We are grateful to Dr. José M. V. Fragoso and Dr. F. Wayne King for their critical comments on and suggestions for this study. This field work was partially supported by the Wildlife Conservation Society, the Biodiversity Support Program, Rainforest Conservation Fund, Consejo Nacional de Ciencia y Tecnología, and the Fondo de Desarrollo Universitario of the Universidad Nacional Mayor de San Marcos. We are also grateful for the institutional support of the University of Florida. Finally, we are indebted to the invaluable support provided by the local communities of the Reserva Comunal Tamshiyacu-Tahuayo.

PART II

Economic Considerations

9

Economic Incentives for Sustainable Community Management of Fishery Resources in the Mamirauá Sustainable Development Reserve, Amazonas, Brazil

JOÃO PAULO VIANA, JOSÉ MARIA B. DAMASCENO, LEANDRO CASTELLO, AND WILLIAM G. R. CRAMPTON

The increasing demand for and degradation of limited resources by the rising human population of the Amazon basin has precipitated a great deal of discussion about the sustainable use of its natural resources (Hall 1997; Ayres et al. 1999). For the most part the fisheries of the Amazon basin are underexploited, and fishing pressure is concentrated on only a few species (Bayley and Petrere 1989; Crampton and Viana 1999). Crampton et al. (this volume) provide a detailed account of the history and current status of floodplain fisheries in the Brazilian Amazon basin. Until the 1970s tambaqui (*Colossoma macropomum*) and pirarucu (*Arapaima gigas*) represented staple protein supplies in the Amazon basin. With the growth of commercial fishing fleets, these species began to show clear signs of overfishing and today are luxury food species (Petrere 1986; Costa 1992; Goulding, Smith, and Mahar 1996). In the last two decades the detritivorous curimatá (*Prochilodus nigricans*) and jaraquis (*Semaprochilodus* spp.) have become the staple food species. As yet, there is only anecdotal evidence for overfishing of these species (jaraqui sizes are decreasing in Manaus markets, V. Batista pers. comm.). The major challenge in Amazon fisheries management is to avert a situation in which one species after another is depleted.

This article describes the principal activities and results of an experimental community-based fisheries management initiative in the Reserva de Desenvolvimento Sustentável Mamirauá (Mamirauá Sustainable Development Reserve, or RDSM), a protected area of várzea floodplain in the Brazilian state of Amazonas. This reserve covers 1,124,000 hectares and is delimited by the rivers Solimões, Japurá, and Uatí-Paraná. The work reported here is centered on a smaller, 240,000-ha Focal Area, bordered by the Japurá and Solimões rivers and by a connection between them, the Paraná Aranapu (fig. 9.1). Resident and user communities in or near this reserve have exclusive access rights to fishing resources granted by its status as a conservation unit. In addition to providing strong economic incentives for the sus-

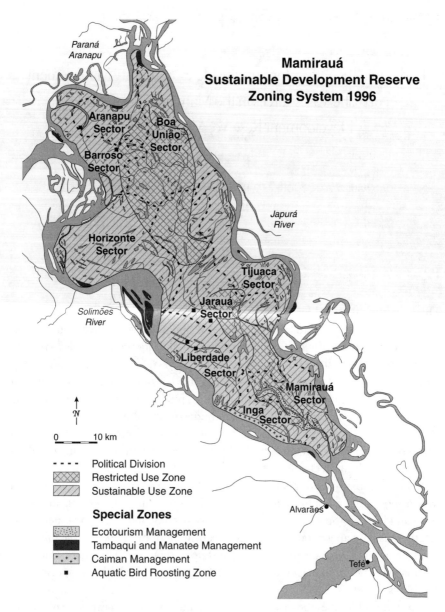

FIGURE 9.1 Organizational (political) sectors and Focal Area of the Mamirauá Sustainable Development Reserve.

tainable management of fisheries, the program described here supplied scientific and technical support in the definition of sustainable quotas. The program strove to improve the production efficiency of an existing fishery and to diversify its species base—both important strategies for reducing the pressure on key commercial stocks (Crampton and Viana 1999).

The RDSM was created in 1990 by the state of Amazonas and is one of the largest areas of relatively intact várzea floodplains in the Brazilian Amazon. The lakes, channels, and seasonally flooded forests of this unique ecosystem offer fishes a rich supply of vegetation, detritus, seeds, fruits, and invertebrates (Junk, Bayley, and Sparks 1989). The immense productivity and size of this reserve probably means that the area is a regionally important nursery ground for commercial food species. Such resident species as pirarucu and tucunaré (*Cichla* spp.) pass their entire life cycle in the floodplain. Many species of migratory characiform fishes, such as the detritivorous jaraquis and curimatá, spend the first years of their lives in the várzea before migrating upstream to colonize other areas.

Crampton, Castello, and Viana (this volume) describe the human population of the RDSM. The principal economic activities of the area are fishing, agriculture, and timber extraction, which are undertaken seasonally and on a communal or familial basis. The average annual income of families in the reserve is around US$ 900, of which 53% is spent on the purchase of food and basic supplies. Fishing represents by far the most lucrative source of income in the reserve, forming 72% of the average domestic income (SCM 1996).

The management of natural resources in the RDSM is based upon an alliance between the local resident and user population and a multidisciplinary research and extension project carried out by the Instituto de Desenvolvimento Sustentável Mamirauá (Mamirauá Institute for Sustainable Development, or IDSM). The local people of the reserve are involved in making management decisions that reconcile sustainable economic development with the conservation of biodiversity (Howard et al. 1995; SCM 1996; Crampton, Castello, and Viana this volume). Because of their economic importance, fisheries resources represent a major focus of management initiatives in the reserve.

Research undertaken in the Mamirauá reserve and negotiations with local communities resulted in the development of a series of restrictive and advisory measures designed to regulate fishery resource use in the area. These measures are outlined in the reserve's management plan (SCM 1996) and summarized by Crampton, Castello, and Viana (this volume). In addition to restrictive measures, the reserve's management plan established a zoning system. This comprises a core no-use zone surrounded by a zone designated for sustainable resource use by the residents and user communities of the Reserve (fig. 9.1).

To promote the conservation of biodiversity, to stimulate the sustainable use of natural resources, and to improve the economic well being of the residents and users of the Reserve, the IDSM is undertaking several extension programs in partnership with the local communities. One of the most important programs, named *Novas Alternativas Econômicas* (novel economic alternatives), aims to promote alternative economic activities, diversify income sources, and enhance existing activities by either increasing production efficiency and/or by forging more lucrative marketing arrangements. This initiative involves the organization and empowerment of production groups, technical and administrative training, the construction and implementation of new infrastructure, and the development of systems to keep track of the dynamic market conditions. The experimental fish commercialization initiative described in this article is a central component of this Novas Alternativas Econômicas program.

THE FISH COMMERCIALIZATION PROGRAM

In 1998 an experimental Programa de Comercialização de Pescado (Fish Commercialization Program, or PCP) was implemented in one of the nine organizational sectors of Focal Area of the reserve. The Jarauá Sector, which comprises around two hundred inhabitants in four communities, was selected for this experimental program for four reasons:

1. Its communities are located strategically at the entrance to the largest lake system in the Focal Area.
2. Fishing represents the main economic activity.
3. The communities have a relatively well organized social structure.
4. The communities have a history of cooperation with the IDSM throughout the installation of the reserve.

The program began with an economic viability study prepared by outside consultants. This study projected the scale of fisheries production in the area and proposed an operational structure and chronogram (Bostock 1998). It assisted with the early planning of the program but underestimated the time and infrastructure that would be required.

Several meetings and consultations with the participating communities were held at the onset of the program in order to define the logistic and administrative organization necessary to stage a new fisheries production program. Training courses in fish processing were also given in order to improve the quality of fish for marketing. Also, visiting instructors ran workshops on the structure and management of commercial associations and cooperatives. As far as possible, the traditional systems of fish production were maintained, or altered only so as to improve productivity without drastically altering the community organization.

There was, however, room for consensual experimentation. In the beginning one attempt was made to conduct communal pirarucu exploitation in which labor was divided among the participants into fishing, transporting, and processing, and the income then equally divided. This strategy was proposed because of the relatively small fishing quota allowed (see below) and the unequal distribution of fishing equipment (canoes, engines, ice chests, and tackle) among fishermen. Most fishermen and the supporting staff felt that a communal organization with division of labor was more appropriate for such conditions. However, this approach was unsuccessful for several reasons. Friction developed around the division of duties and unequal division of labor. More importantly, pirarucu fishing is traditionally organized on an individual or family level or by small teams composed mostly of relatives. In view of the failure of communal fishing arrangements, the organization of pirarucu fishing expeditions reverted to individual or small group affairs organized by the fishermen themselves.

Following capture, fish are transported from the managed area to a processing base located at the Jarauá community. The base comprises a covered floating raft with a large icebox and water purifying system. This facility was built in partnership with the fishermen and serves as a base for eviscerating, cleaning, and chilling fish, as well as an administrative base for the fishing operations. The base allows up to 4 tons of fish and ice to be stored. When this limit is reached, the fish are transported to the nearby city of Tefé in a boat dedicated to this program and equipped with an ice hold. The fish are either sold in Tefé or sent to Manaus or other cities. The IDSM invested (at no cost to the fishermen) approximately US$ 15,000 capital funds into the construction of the floating fish processing base plus the purchase of the boat and some equipment and supplies (a scale, office supplies, ice chests, etc.). The participating fishermen contributed manual labor and wood for the base.

The most important feature of the PCP is the elimination of intermediaries (*atravessadores*) between the fishermen and the market. Intermediary purchasers travel around the interior buying fishes at prices way below the market values. Most of the transactions are based on an exchange of fish for household supplies, such as cooking oil, salt, sugar, and coffee, all of which are bartered at inflated prices. By removing intermediate purchasers and improving production quality of the fish products, the fishermen of the Jarauá Sector are able to achieve a far greater economic return with smaller landings of fishes. Fishing expenses are the responsibility of the fishermen, while commercialization expenses are discounted proportionally from the sales. The profits are then divided according to the amounts and species of fishes captured by different fishermen. For team efforts the leader is responsible for paying the others according to the arrangements made with his partners.

The original proposal of the PCP was for fishermen of the Jarauá Sector to avoid capturing species which are prohibited by IBAMA bans and to capture fish only above legal minimum sizes. The idea was to divert fishing pressure toward species without capture restrictions and toward those for which there is no evidence of current overfishing (Bostock 1998). However, from the onset it became clear that only a few species, including pirarucu and tambaqui, were capable of generating sufficient profits to make the PCP an economically viable operation.

THE PROBLEMS OF *PIRARUCU* AND *TAMBAQUI*

Pirarucu is the most economically important species of fish for the residents and users of the RDSM (Queiroz and Sardinha 1999). It made up around 40% of the total weight of fish landed for sale and local consumption in the early 1990s. In six communities between 1993 and 1995, an average annual catch of between 1.4 and 1.6 tons was registered. On the basis of this information, the annual capture of pirarucu in the Focal Area of the reserve was estimated to be in the order of 110 to 150 tons. The production of pirarucu occurs mostly during the low-water months of September to December. Queiroz and Sardinha (1999) showed that only 30% of the landed pirarucu were larger than the legal minimum total length (TL) of 1.5 m. This shortfall indicated that the species was being exploited beyond a maximum sustainable yield in some parts of the Reserve.

These studies led to the establishment of pirarucu fishing guidelines in the reserve's management plan (SCM 1996). The closed season and minimum capture size follow those established by federal legislation (IBAMA decrees 480/1991 and 8/1996). However, other guidelines imposed by IDSM were more restrictive. IBAMA decree 14-N/1993 established legal minimum sizes for the salted and sundried flanks (*mantas*) of pirarucu at 1 meter, effectively allowing the landing and commercialization of pirarucu well below 1.5 m TL. Instead, the Mamirauá management plan set limits of 1.15 m for salted and dried mantas and 1.25 m for fresh mantas. Given the economic importance of pirarucu, these restrictions on the commercialization of mantas were predictably unpopular with the reserve's residents and created some animosity towards the IDSM. Nonetheless, they were considered necessary to regulate pirarucu fishing, along with other measures proposed by reserve residents and users. These measures included a minimum gill net mesh size for pirarucu fishing (29 cm measured across opposed angles) and the prohibition of fishing soon after the water level starts to drop, extending until the floodplain lakes become isolated.

From 1996 onwards the IBAMA representation in Amazonas declared pirarucu stocks to be in a critical stage of overexploitation and established a two-year statewide ban on the capture and commercialization of this species. This ban was enacted by prohibiting fishing from June through November because a 1996 decree by the IBAMA head office in Brasília had already banned fishing from December through May every year throughout the Amazon basin (a period that roughly corresponds to the species reproductive season). So far this statewide suspension of pirarucu fishing has been renewed twice without interruption, and there are no indications that it will be lifted in the near future.

The ban ruled out the possibility of legal pirarucu fishing within the RDSM, in principle resulting in a potentially significant decline in income for the resident communities. Tambaqui fishing is still permitted by IBAMA, with a minimum landing size of 55 cm TL (IBAMA decree 8/1996) and a closed season, which lasts from three to four months and varies in exact dates from year to year (IBAMA decrees 6/1996 and 142/2001). The minimum size of 55 cm TL drastically complicates the capture of tambaqui in the reserve. Tambaqui spend the first five years or so of their life in floodplain lakes and forests and then, as adults (above around 55 cm TL) undertake upstream migrations along main whitewater river channels (Goulding 1979; Costa, Barthem, and Correa 1999). Only around 5% of the tambaquis in the floodplain lakes of the RDSM are larger than the legal minimum size of 55 cm TL (Costa, Barthem, and Correa 1999).

RESOLVING THE PROBLEM OF *PIRARUCU*

From the onset of the PCP the biggest difficulty was related to the restrictions placed on the species with highest commercial value. The participating fishermen made it clear that the initiative would not generate financial returns unless pirarucus were involved. A monitoring system was implemented in early 1998 with the objectives of evaluating local fish production and identifying alternative species for extraction. The first results made it clear that the fishermen were right. Without pirarucu and tambaqui the yields of other species when weighed against production expenses would not generate profits.

The IBAMA decrees banning pirarucu since 1996 do allow for the controlled capture and commercialization of the species provided that they derive from *bona fide* managed fisheries. The PCP therefore proposed a system of managed pirarucu extraction involving a rotation system in thirty-one of the eighty lakes in the sustainable use zone of the Jarauá Sector. This proposal, requesting an initial quota of 3 tons of pirarucu for the late 1999 season, was submitted to IBAMA-Amazonas and accepted in June 1999. The quota was based on a previously published estimate of pirarucu production in várzeas of the Peruvian Amazon at around 0.3 kg/ha/year (Bayley et al. 1992). The Jarauá Sector comprises 56,300 ha of floodplain lakes and forests, allowing a crude production estimate of around 15 tons/year of pirarucu in the sector (assuming that the area is under low-fishing pressure). Considering an average weight of 40 to 50 kg for a 1.55 m pirarucu, this production corresponds to an annual harvest of 375 fish. The quota was set at a conservative one-third of this number in order to err on the side of caution during the beginning of this program and to review the results thereafter.

The first year's quota of 3 tons was significantly below the previous levels of pirarucu fishing in the Jarauá Sector. Monitoring data in the peak production months of September to December 1998 showed a total landing of around 600 pirarucu (12 tons of mantas), of which at least 95% were below the legal minimum length of 1.5 m. We later learned from the fishermen themselves that the number was actually higher, maybe around eighteen to twenty tons of mantas, because the fishermen smuggled part of the production. Of course, due to the continuing IBAMA ban on pirarucu fishing, all these fish were illegally captured, including those above the minimum size. Even though the 3-ton quota was well below previous harvesting levels, the fishermen of the Jarauá Sector were satisfied to reduce production and to follow all of the RDSM fisheries guidelines in return for being able to sell IBAMA certified legal pirarucu and to avoid intermediary purchasers.

In the PCP's second year (late 2000 season), a new tool became available to monitor pirarucu stocks. This tool was a direct counting method conducted as a collaboration between local pirarucu fishermen and a researcher from IDSM (Castello in press). Pirarucu are obligatory air breathers and betray their presence to fishermen when they rise to gulp air from the surface. Using a combination of auditory and visual cues, experienced fishermen are able to distinguish between several fish surfacing at the same time and are able even to estimate the approximate size of the fish. These abilities, when used in the context of quantitative methods, yield replicable information about pirarucu numbers in floodplain lakes. This direct count methodology was compared to independent estimates of pirarucu population sizes from mark-release-recapture studies in floodplain lakes (at low water) and proved to be very accurate for fish larger than 1 m TL (Castello in press). Once the accuracy of this method had been established, it was then employed to monitor pirarucu stocks and to calculate sustainable landing quotas in the Jarauá Sector.

POLITICAL INTEGRATION OF THE PCP IN THE JARAUÁ SECTOR

Each organizational sector of the RDSM operates as a separate entity with its own local coordinator responsible for organizing bimonthly meetings. These meetings provide a forum to discuss and resolve community-related issues and disputes. This organizational structure is modeled on the rural community projects of the Catholic Church. Working at the sector instead of community level helped to distribute the benefits of the PCP to more people and also to increase the number of people trained to undertake duties for the PCP. In the beginning management was set up in an informal manner and consisted of a technical coordinator (held by an IDSM technician) and a community coordinator (elected by the participating fishermen). A formal terms-of-agreement document was drafted in which the fishermen agreed to support the PCP work by maintaining the infrastructure (boat and floating fish processing base), guarding the lakes selected for rotation, and enforcing the rotation system. The final, formal, organizational structure and the necessary delegations of responsibility were left for the fishermen to develop.

As the discussions evolved, the communities opted to register the PCP as part of a formal production association (Associação de Produtores do Setor Jarauá), marketing not just fish but also agricultural produce and artisanal products, such as pottery and basketwork. A decision was made to extend the PCP infrastructure to the storage and transport of agricultural and artisanal products. The formation of a production association also allowed the formal inclusion of women. The presence of women was significant from the administrative point of view because it increased the possibility of finding people with skills to assume such tasks as record keeping and accounting. Most fishermen are illiterate, whereas literacy rates are higher among women. By the time the production association was formally registered in July 2001, the associates had already assumed most of the administrative and technical duties originally assumed by extension workers and community assistants from IDSM. Several women were elected to administrative posts, among them secretary and accountant. Currently, technicians from IDSM support the association mainly by establishing contacts with fish buyers outside Tefé (because of communication difficulties from the RDSM).

The legal registration of the production association permits marketing outside the state of Amazonas. In 2001 a contract was signed to supply a chain of restaurants in Brasilia with a major part of the annual pirarucu quota at an excellent price of 8.00 Brazilian Reais (R$) per kilo (exchange rate in 2001 was approximately $US 1 = R$ 2.3–2.7). The restaurant chain in turn was able to provide its customers with environmentally friendly pirarucu purchased legally and derived from managed fisheries in an area protected for biodiversity conservation.

ECONOMIC, SOCIAL, AND ECOLOGICAL RESULTS

The results of the PCP's first three years of operation have proven very satisfactory from the economic, social, and ecological points of view (table 9.1). In 1999 most of the harvested pirarucu were sold in Tefé. For the first batch of fish brought from the RDSM to Tefé in 1999, the possibility of sending the fish by passenger boat for sale in Manaus was investigated. However, the sale price would not have justified the expenses (freight costs and expenses for somebody to accompany the produce to its final destination). At that time it was impossible to find a buyer in Manaus who was willing to pay a premium for legally commercialized pirarucu. The trade in illegal pirarucu has continued more or less unabated throughout Amazonas state despite the IBAMA 1996 ban on commercialization. IBAMA enforcement of this ban has completely failed and most traders are not in the least bit worried about breaking the law.

In 1999 400 kg of pirarucu from the PCP were sold to a food processing company in Manaus at R$ 4.00/kg. The rest was sold in Tefé via an intermediary who agreed to pay the going rate for Manaus (R$ 3.00–3.40/kg). The food processing company in Manaus placed another order, but by the time it was received, the an-

TABLE 9.1 Production Statistics for the First Three Years of Operation of the Fisheries Commercialization Program of the Jarauá Sector, Mamirauá Sustainable Development Reserve

	1999	2000	2001
Number of participating fishermen	42	46	67
Number of species commercialized	7	13	12
Total production (tons)	6.2	9.9	15.0
Pirarucu production (tons)	3.0	3.5	5.3
Average sale price for pirarucu (R$/kg)	3.85	6.00	7.96
Total sales (R$)	16,903	29,209	56,687
Mean income per fisherman (R$)	402	635	846

Note: Data refer only to the three-month production season of late September through early December (including trimester incomes). Exchange rates for US$1 are approximately: 1999 (R$1.6–2.0), 2000 (R$1.7–2.3), 2001 (R$2.3–2.7).

TABLE 9.2 Total Capture (TC, in kg) and Average Sale Price (ASP, in R$) for the Fish
Species Commercialized by PCP from 1999 through 2001

		1999		2000		2001	
SPECIES	SCIENTIFIC NAME	TC (kg)	ASP (R$)	TC (kg)	ASP (R$)	TC (kg)	ASP (R$)
Pirarucu	*Arapaima gigas*	3000.0	3.60	3377.0	6.00	5285.0	7.96
Aruanã	*Osteoglossum bicirrhosum*	–	–	1042.0	0.22	4380.0	0.46
Tambaqui	*Colossoma macropomum*	2166.6	2.34	2524.0	2.07	2921.00	2.87
Pirapitinga	*Piaractus brachypomus*	–	–	–	–	40.0	1.00
Caparari	*Pseudoplatystoma tigrinum*	73.4	1.89	754.8	1.92	401.0	2.33
Surubim	*Pseudoplatystoma fasciatum*	–	–	30.5	1.46	5.0	0.60
Dourada	*Brachyplatystoma flavicans*	–	–	217.5	1.39	96.0	2.15
Filhote	*Brachyplatystoma filamentosum*	–	–	219.0	1.44	551.5	2.43
Pirarara	*Phractocephalus hemiliopterus*	–	–	116.0	0.71	211.0	0.56
Pacamum	*Paulicea lutkeni*	–	–	21.0	0.85	–	–
Pescada	*Plagiocion squamosissimus*	11.5	0.85	–	–	2.0	0.80
Tucunaré	*Cichla monoculus*	879.0	0.92	1188.5	0.93	565.0	2.19
Acará-açu	*Astronotus ocellatus*	84.0	1.00	230.5	0.52	586.0	0.64
Acará-branco	*Chaetobranchus flavescens*	–	–	84.5	0.69	–	–

nual quota of 3 tons had already been reached. The next year, with a renewal of
IBAMA authorization for PCP pirarucu landings, the same company purchased
the entire PCP quota of pirarucu at R$ 6.00/kg. The other species captured in 1999
and 2000 were sold mostly in Tefé. The complete list of fish species exploited by
PCP is presented in table 9.2.

Socioeconomic monitoring in one of the communities of the Jarauá Sector, São
Raimundo do Jarauá, showed that, despite the restrictions in fishing (all fishes were
captured in accordance with minimum size regulations and closed seasons established by law and by the Mamirauá Management Plan) and despite the overall reduction in pirarucu landings, there was no attendant reduction in annual family incomes. On the contrary, average family incomes increased from around R$ 1,900
in 1995 to R$ 2,700 in 1998/1999 and to R$ 4,100 in 2000. Using as a reference point
a *cesta básica* (a standardized shopping basket of household supplies used by
Brazilian social scientists for economic surveys), the buying power of São Raimundo do Jarauá doubled from 1995 to 2000 (table 9.3). Unfortunately, due to the way
the data were collected, it was not possible to calculate the proportion of the annual wage derived from fishing. However, we believe that the contribution of fishing
was significant because no other major revenue-generating activity was introduced
in Jarauá during this period.

All fish commercialized by the PCP were above the legal minimum size limits

TABLE 9.3 Mean Annual Family Incomes from a Socioeconomic Monitoring Survey at the Community of São Raimundo do Jarauá

	1994–95	1998–99	2000
Number of families	16	20	19
Mean annual family income (R$)	1,939	2,721	4,142
Cost of a *cesta básica* (R$)	43.68	44.14	46.98
Buying power (number of *cestas básicas*)	44	61	88

Source: Edila Moura

TABLE 9.4 Mean Total Lengths (TL) for Species Exploited by the PCP for Which Size Limits Have Been Established by IBAMA

SPECIES (MINIMUM TL)	1999 Mean ± SD (n)	2000 Mean ± SD (n)	2001 Mean ± SD (n)
Pirarucu (150 cm)	162.9 ± 21.5 (126)	157.8 ± 10.3 (143)	165.8 ± 13.4 (188)
Tambaqui (55 cm)	61.3 ± 5.1 (455)	59.2 ± 4.0 (610)	63.4 ± 4.6 (582)
Caparari (80 cm)	89.5 ± 8.4 (12)	86.5 ± 6.8 (147)	88.0 ± 8.4 (69)
Surubim (80 cm)	—	86.0 ± 9.4 (6)	85.8 ± 7.0 (5)
Aruanã (44 cm)	—	—	68.1 ± 4.9 (495)
Tucunaré (25 cm)	—	—	36.4 ± 3.0 (134)

Note: TLs for pirarucu were estimated from lengths of filleted flanks.

established by IBAMA (table 9.4). These sizes contrast with the situation immediately before the PCP project began when 95% of landed pirarucu from the Jarauá sector were below the legal size limit (fig. 9.2). There was also an encouraging increase in the diversity of commercialized species (table 9.1), indicating that fishermen were beginning to divert fishing efforts toward previously underexploited species (table 9.2).

Stock assessments in the sustainable use zone of the Jarauá Sector showed a 300% increase in the number of pirarucu between 1999 and 2001 (table 9.5). These assessments utilized the direct count methods described earlier (Castello in press) and discriminated between juvenile fish (1 to 1.5 m TL) and adult fish (1.5 m TL). The number of adult pirarucu counted in the 1999 stock census was used to calculate the quota submitted to IBAMA-Amazonas for the subsequent low-water season of 2000. The requested quota represented the removal of approximately 30% of the number of adults counted. This percentage corresponded to a total of 3 tons of mantas, or around 120 adult fish (assuming a mean capture size of 1.55 m, 40 to 50 kg of total weight per fish, and 20 to 25 kg of saleable meat per fish). A stock census

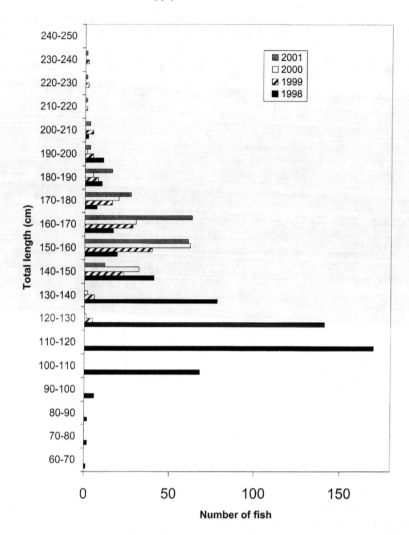

FIGURE 9.2 Histogram of size classes for pirarucu captured by communities in the Jarauá Sector of the Mamirauá Sustainable Development Reserve in the low water periods (September through December) of 1998–2001.

(based on the same methodology) in the low-water season of 2000 was used to plan the quota for the subsequent harvest in 2001. For 2001 a quota of 8 tons, corresponding to around 300 fishes, was submitted to IBAMA.

Before the 2001 season began, the fishermen proposed to relax the lake rotation system rules in favor of choosing lakes on a year-to-year basis. Fishing and access conditions to any given lake vary considerably from year to year, making it difficult to stick to a rigid schedule. In fact, some lakes are almost impossible to enter during very dry low-water periods. Such was the rate of increase of pirarucu stocks in

TABLE 9.5 Estimates of the Number of Pirarucu in Floodplain Lakes of the Jarauá Sector of the Mamirauá Sustainable Development Reserve from 1999 to 2001

YEAR	JUVENILES (1–1.5 M)	ADULTS (> 1.5 M)	TOTAL
1999	2,149	358	2,507
2000	2,984	994	3,978
2001	5,901	1,441	7,342

Note: Data from censuses involving a new direct counting technique (see text).

the sector that the fishermen thought it might be possible to capture the established quotas by fishing in the floodplain channels alone and without ever needing to enter the lakes. IBAMA agreed with the changes in management but unjustifiably refused to allow an increase in the pirarucu quota to the 8 tons requested for the 2001 low-water season; the quota was based upon the pirarucu stock assessment method. Instead, only 5 tons was granted. This tonnage corresponded to around 200 adult pirarucu.

In contrast to previous years, the IBAMA harvesting permit for 2001 established not only a quota for the total weight of mantas but also a limit to the total number of fish that could be landed. This alteration in the proceedings resulted in considerable changes in the way in which fishing was organized in 2001. On the basis of the numbers of fish, the pirarucu quota became more concrete and manageable. It began to be treated as a share and split among the associates according to specific rules. Also, each shareholder tried to make the best use of his share.

In 1999, the first pirarucu fishing season, no criteria were established by the PCP members for the distribution of pirarucu quotas. Those fishermen who benefited most were the best equipped (in terms of canoes, ice boxes, and tackle) and those with cash to finance fishing trips. In 2000 the PCP members decided to divide equitably the total pirarucu quota among the four communities, taking into account the number of fishermen in each community. In 2001, with the production association already legally registered, criteria were established by the directors to distribute the fish to associates according to their relative contribution to the different aspects of management required to sustain the fishery. These contributions included participation in lake-vigilance excursions, respecting the fishing rules (such as minimum sizes and closed seasons), and participating in meetings. The production association leaders viewed these criteria as a way to increase the number of people involved in fisheries management. During an open meeting the distribution of the quota (in numbers of pirarucus instead of weight) was then established at between zero and three fish per associate.

This new system of quota distribution was well received by the associates and is expected to be maintained for the future. The results from 2001 demonstrate that each fisherman made the best of the quota he received by selecting and culling larger fishes that would give a better financial return. The average pirarucu yielded 28.2 kg of mantas in 2001, as compared with 23.8 kg in 1999 and 24.5 kg in 2000. The average size of captured pirarucu reached 1.6 m in 2001 (fig. 9.2), its highest level since the onset of the PCP.

FINAL CONSIDERATIONS

In just three years of operation, the Fish Commercialization Program of the Jarauá Sector has proven to be a viable model for the management of várzea fisheries. During this time many operational problems were identified and resolved. One of the greatest sources of difficulties was the complex infrastructure and the maintenance of a boat. For example, the Port Authority requires specific training for boat crews, including at least one skipper and one mechanic. Four fishermen volunteered to take the required courses offered by the navy and gained the necessary diplomas and documents. Another problem was that the administrative affairs of the PCP require the dedicated full-time attention of at least two people. The association consequently provided two salaried positions for two organizers and paid a salary from R$ 150 to 250 per month (depending upon the number of active fishermen since each contributed with a share of R$ 5 to make up the salary). In the end a small number of people assumed multiple tasks, including administration, boat piloting, and participating in sales trips. In the future more people are expected to become involved and the workload to be spread more comfortably.

Administrative delays at IBAMA proved to be one of the greatest sources of difficulties in the development of the Jarauá PCP. Large amounts of complex paperwork attended almost every stage of the proceedings and the resulting permits and documents were sometimes issued late. In 2001, for example, IBAMA permits for capture and commercialization of pirarucu arrived at the end of October, already two months into the low-water fishing season. This delay left only one month for the fishing quotas to be met, in contrast to the two to three months of the previous two years. The short fishing season made it impossible for the association to meet the quota, and only 188 of the 200 authorized pirarucu were caught.

The most important contribution of this experimental Fisheries Commercialization Program is that it proves the feasibility of exploiting high-value species at good prices while at the same time allowing stocks to increase through a program of combined management, vigilance, and monitoring. The community-based production association was able to manage its own affairs with initial—and thereafter occasional—technical assistance from the outside, demonstrating that communities can incorporate new systems and build upon them.

The general model developed in Jarauá may be a powerful tool for the management and conservation of fishery resources in other parts of the Amazon basin. The direct counting method for assessing pirarucu stocks was also considered to be vital to the success of this program. This tool represents the integration of traditional knowledge with scientific methodology, and helped tremendously with the introduction of fisheries management principles to the communities. This method is already being successfully taught to fishermen from other areas of the Amazon basin, such as the Santarém region in the Brazilian state of Pará, the Pacaya-Samiria National Reserve in Peru and Guyana. In fact, it is currently under consideration by the Guyanese government as a tool in the country's strategy to promote the recovery and sustainable use of the pirarucu stocks.

The positive results of the PCP in the Jarauá Sector have attracted much interest from fishermen in other sectors of the Mamirauá Reserve and of the adjacent 2.3 million-ha Amanã Sustainable Development Reserve. Several communities of these other areas have requested technical assistance to develop similar programs. Nonetheless, the PCP system in the Jarauá Sector was expensive to set up and took a long time to be implemented by and integrated with the communities. We conclude that the ideal solution would be to implement similar but less complex, low-cost systems adapted to local ecological and socioeconomic conditions. These initiatives could, for example, rely on hiring boats for the transport of fish rather than investing in the purchase and maintenance of vessels. Replicating simplified and less expensive systems based upon the Jarauá model of fisheries production and management would allow more communities to benefit in a shorter time frame.

In 2001 a simpler fisheries production program was initiated in the Tijuaca Sector of the Mamirauá Reserve. The initial results were promising, but delays in the liberation of IBAMA capture and commercialization permits for pirarucu meant that in its first year of operation the Tijuaca PCP only had time to land 40 of a total quota of 120 pirarucus before the season closed. In 2002 two other simplified programs will be launched, one targeting the Fishermens Association of the nearby town of Maraã (around 160 members) and the other assisting the seven communities in the Coraci Sector of the Amanã Reserve (benefiting around 400 people). Both projects have been submitted to IBAMA and are currently being appraised.

Since 2001 IBAMA-Amazonas streamlined its internal procedures. The permits for the Jarauá sectors (500 pirarucus or approximately 15 tons of mantas) and Tijuaca (120 pirarucus or approximately 3 tons of mantas), with the full quotas requested by the fishermen through the stock assessment method, were issued in late June 2002. We expect that the results for Maraã and Coraci projects (120 pirarucus each) will be out well before the beginning of the fishing season, allowing time for the fishermen to plan ahead for their first pirarucu management experience.

The Jarauá fisheries management system, which began in 1999 with merely 3 tons of pirarucu, is now ready to harvest 15 tons of legally caught and sized fish. On the basis of ongoing monitoring of pirarucu stocks, we expect that the total production of the four ongoing systems will double next year. This significant increase is expected because only a fraction of the full potential of the Maraã and Coraci systems will be harvested in 2002. On the basis of the direct counting method, the fishermen from Maraã, for example, could start harvesting 400 fish in the first year.

However, they preferred to begin on a smaller scale and review the situation afterwards. In the case of Coraci the first quota was based on a stock assessment survey using the direct count method in a representative subset of the available lakes made in 2000. After this survey, and in the expectation of having a pirarucu management system for this area, the fishermen decided to completely ban fishing for commercialization purposes. They wanted pirarucu stocks to recover and to start harvesting them in a different, sustainable manner. In a four to five year time frame and with the addition of new community fisheries management systems, we expect that the total production of wild pirarucu in the Mamirauá and Amanã reserves will reach 100 to 150 tons per year.

Today, the Mamirauá Institute is working on the development of a large-scale model for the implementation of this new generation of simplified community-based fisheries management programs around the várzeas of the Brazilian Amazon. Simplicity, technical practicability, and the ability to mould new initiatives around preexisting local conditions are fundamental principles for the successful planning and implementation of such initiatives. The role of technicians will be mostly limited to training fishermen in procedures for stock assessment (for pirarucu and other species), methods for setting quotas, basic management principles, fish processing techniques, and association management. The commercialization of the harvests, and the day-to-day running of production associations will be the responsibility of the associates themselves. Due to communication difficulties, communities at great distances from markets are placed at a considerable disadvantage in establishing commercial contacts. These cases would probably need the intervention of technicians from such organizations as IDSM. Likewise, the technical reports required from community fisheries programs by IBAMA would probably need to be compiled by experienced fisheries technicians for most communities.

ACKNOWLEDGMENTS

We thank Dr. José Márcio Ayres for the invitation and support to work in the Instituto Mamira04. Funding was provided by Conselho Nacional de Desenvolvimento Científico e Tecnológico and the U.K. Department for International Development. Support also came from the Instituto de Proteção Ambiental do Estado do Amazonas and the Instituto Brasileiro do Meio Ambiente e dos Recursos Naturais Renováveis do Amazonas. Finally, special thanks to Saide Pereira, Antonio Martins, and the fishermen of the Jarauá Sector who over the last four years have taught us the realities of fishing and living in the Amazon floodplain.

10

Community Ownership and Live Shearing
of Vicuñas in Peru

EVALUATING MANAGEMENT STRATEGIES
AND THEIR SUSTAINABILITY

CATHERINE T. SAHLEY, JORGE TORRES VARGAS,
AND JESÚS SÁNCHEZ VALDIVIA

In 1994, after years of being listed under Appendix I of the CITES convention, vicuña (*Vicugna vicugna*: Camelidae) populations in Peru were reclassified as an Appendix II species. Under this classification Peru obtained permission to export fiber from animals captured in their wild state, partially shorn, and then released. In 1995 the Peruvian government passed Law # 26496, "System of regulation of property, commercialization and sanctions for the hunting of vicuñas, guanacos, and their hybrids." The purpose of the law was to give Andean *campesinos* a direct interest in the conservation of the vicuña and to motivate them to participate in conservation efforts. This law includes the following components:

1. It gives usufruct rights of vicuñas to communities if such animals are found within community boundaries and also the rights to a portion of profits from the sale of vicuña fiber
2. It gives the communities the responsibility of management and "rational" use of the vicuña
3. It enacts legal penalties for poaching that range from two to twenty-five years imprisonment, depending on the gravity of the crime committed (El Peruano, July 11, 1995).

This law represents a new focus for conservation efforts of the vicuña in Peru. *Campesino* communities are legally recognized entities, which by and large, are made up of Andean indigenous peoples. As of 1998, 5,666 campesino communities were legally registered, and 3,956 have legal title to their land (Velasquez 2001). With the granting of exclusive usufruct and management rights to communities, the vicuña has in effect ceased to be a totally public resource. The capture, live-shearing, and release program (called the *chaku* in Quechua) offers many possibilities for the sustainable utilization of the vicuña in Peru and other South American countries. However, many aspects of the current management program need to be carefully evaluated, and an urgent need exists for studies that generate information on the biological and economic costs and benefits of the program.

In the present article we review the most common management systems that are being utilized in Peru and examine the results of the program in terms of fiber harvests and economic gain. We also present data obtained for the communities of Tambo Cañahuas, and Toccra in Arequipa, Peru, with the objective of highlighting some of the current limitations of the management program. Finally, we present some conclusions and recommendations regarding the implementation of the live-shearing program in Peru.

BACKGROUND

PRESENT MANAGEMENT PLANS

Several management plans proposed by the government are currently being implemented. The first is the capture, live-shearing, and release program of free-ranging vicuñas; a program that has been approved by CITES. Once a year, each community implements the capture/live-shearing/release activity that should occur between May and November. The actual scheduling of the chaku varies greatly among communities and is influenced by a variety of factors, such as community activities that may or may not conflict with optimal capture dates, logistic constraints regarding the availability of appropriate capture equipment, and the organizational capacity of each community.

The primary method utilized for the live capture and shearing of vicuñas is the use of a temporary capture corral made of fishnetting 2 m in height. Community members hide behind bushes or in trenches and run behind the vicuñas carrying a colorfully flagged rope once the vicuñas have entered the capture zone. The vicuñas try to escape but are hampered by the funnel-shaped capture area. Finally, if the roundup is successful, the vicuñas are herded into a smaller mesh-shaped corral from which they cannot escape. They are then captured by hand, shorn (either with scissors or electric shears), and released.

The second plan, which is being pursued aggressively by the state and its extension workers, is the capture of vicuñas in the wild, followed by their subsequent transfer to corrals that are generally 500-1000 ha in size. The size of corrals varies, and currently, there are no published standards about management requirements of vicuñas that are kept in corrals. A third plan consists of a repopulation program, wherein animals caught in more populated areas are transferred to communities that have few vicuñas. These animals are not released but are maintained in corrals.

All participating communities are required to formally organize vicuña management committees and to register with the government in order to participate in the fiber-shearing and marketing program. They are also required to organize antipoaching patrols, reinvesting a portion of proceeds from the sale of fiber to this end.

POTENTIAL COSTS AND BENEFITS OF THE
CURRENT MANAGEMENT PLANS

Some concerns have been expressed about the techniques employed (Bonacic 1996; Wheeler and Hoces 1997), especially regarding the levels of stress experienced by the animals and its effect on their subsequent well-being. Few data exist on the effect of capture and shearing on rates of mortality, fertility, and social structure of vicuñas in the wild. Bonacic (1996) studied levels of stress in animals captured and shorn in Chile. Working only with males and following them for eight days in captivity, he showed that capture increased cortisol secretion for two days but that, in and of itself, stress from capture did not contribute directly to vicuña mortality. Instead, Bonacic's 1996 results suggest that perhaps a more important factor affecting mortality is lowered body temperature because of shearing, which leads to decreased immunological response. His data show that total and partial body shearing caused respiratory disease in some cases, with completely shorn animals suffering higher mortality than partially shorn animals. Captured, but not shorn, animals did not suffer any mortality during the study.

Bonacic's work highlights the importance of including physiological stress as one factor in vicuña management plans. Stress could be especially important for pregnant females, which in many cases are gestating and lactating at the same time and have high energetic requirements. Thus, time of year and climate is likely an important factor to consider when scheduling shearing. Presently in Peru the capture and shearing of vicuñas starts in winter (June) and continues until spring (November). This long shearing season is due to the lack of sufficient capture netting, which necessitates preparing a timetable for transporting and sharing equipment.

Since the widespread commercial exportation of fiber products began in 1995, management of the vicuñas has moved in the direction of animal husbandry proposals and projects to the detriment of investment in wildlife management initiatives. In 1996 a new management proposal was devised via an agreement between the National Breeders Society, the Ministry of Agriculture, and CONACS (Consejo Nacional de Camelidos Sudamericanos, or the National Council for South American Camelids). This proposal, "Convenio de Coóperacion Interinstitucional Para el Uso Sustentable de la Vicuña," significantly changed the scope of the free-ranging/capture/live-shear program previously in place. It outlined a new Ministry of Agriculture project, Programa de fortalecimiento de la competitividad comunal en la crianza de vicuñas (Program to strengthen community competitiveness in the breeding and care of vicuñas).

One of the major premises of this program was that "sustainable use modules", or large corrals, were more efficient, productive, and profitable than the free-ranging program currently in place. Furthermore, the proposal stated that the use of corrals constitutes the "best alternative for the rational management of the species". The specific objectives of the program call for: (a) protection and conservation of the vicuña (no reference is made to wild vicuñas), (a) reintroduction of vicuñas into appropriate habitats, (c) production of fiber from live-shorn animals, (d) legal commercialization of fiber, and (e) generation of a productive activity for Andean campesinos. The state has been actively promoting these management alternatives and is implementing them on a wide scale throughout Peru.

The program called for the implementation of corrals in 600 campesino communities by the year 2000 (Ministerio de Agricultura 1997), a task that has been completed for approximately 260 communities (Hoces 2000). It has been estimated that 250 vicuñas are needed for the corral scheme to be economically viable; an economically viable system would therefore require 150,000 vicuñas to be in what is euphemistically called semicaptive management, a number that has not been reached because it is greater than the current population of vicuñas in Peru. This program, unfortunately, completely neglects the management of free-ranging vicuñas. In fact, no free-ranging management plan currently exists in Peru.

The plan's assumption that the so-called sustainable use modules are more efficient, productive, and profitable was untested at the time that the program was implemented on a national level. Unfortunately, although under certain well-defined circumstances it could be argued that corrals might be necessary for vicuña management, the implementation of this program has caused confusion regarding appropriate management practices. It can be argued that implementation of this program has been detrimental in some respects to the overall conservation program of the vicuña in Peru. For example, the implementation of corrals to date has cost Andean communities approximately US$ 2,500,000.00, while little has been spent on strengthening antipoaching efforts, which has been one of the primary concerns of campesino communities in Peru (Sociedad Nacional de la Vicuña 2000). It is evident that a coherent management plan must be formulated to maintain viable wild populations of vicuñas in Peru and that at present the term sustainable use modules has led to confusion about what sustainable use of the vicuña means, both at the professional and the community level.

The maintenance of vicuñas in large corrals has been the least studied option from both technical and socioeconomic perspectives. Initially, this modality was proposed by the state to facilitate monitoring of vicuñas and consequently to lower rates of poaching. Another presumed advantage is greater capture efficiency and consequently greater economic benefits for campesinos (Zuñiga 1997).

This option has an initial economic disadvantage because it requires an upfront high capital investment; each corral costs US$23,000.00, not including labor, which campesinos provide for free. For many communities this quantity is more than the income received from the sale of vicuña fiber. Communities with few vicuñas that become involved in this program become immediately indebted to the state. A cost-benefit analysis conducted by Lichtenstein et al. (2001) compared two communities that utilized free-range management with two communities that utilized semicaptive management. By the time the study ended, the two communities that utilized semicaptive management had not received any short-term benefits, and long-term benefits were considered low income and high risk. In contrast, the two communities that utilized free-range animals had received cash that was used for community improvement projects. Lichtenstein et al. (2001) concluded that management of free-ranging vicuñas, while being a moderately high-risk venture, also had high chances of being profitable.

Biological impacts of maintaining vicuñas in corrals or in semicaptive management can be significant, both in the short and long term. By increasing densities within fenced-in areas, reproductive rates can decrease if there are density-dependent effects. Also, an increase in density can facilitate the transfer of disease and parasite load. An immediate effect is change in dispersal behavior and movement patterns, changes that can ultimately influence genetic structure at the metapopulation level as well as for animals within corrals.

As for the repopulation program, which could arguably be socially, economically, and biologically necessary, Wheeler et al. (2000) have identified four genetically distinct populations of vicuñas within Peru. They recommend caution with regards to the repopulation program, and suggest that repopulation efforts occur within the four distinct sub-populations. They also recommend caution regarding corrals and the potential for inbreeding effects. From a population dynamics perspective, if no interaction is permitted between vicuñas in the wild and vicuñas placed in corrals , these animals may as well have been harvested from the population. If too many animals from a wild population are placed within corrals, these de facto harvests may be unsustainable (Sahley 2000; see table 10.1).

Another factor is that animals within corrals may receive more vigilance and attention than animals left in the wild, thus putting animals in the wild at greater risk for poaching. Because emphasis is placed on shearing animals within corrals because of purportedly increased efficiency, free-ranging vicuñas may not be shorn. Shearing vicuñas has been proposed as a disincentive to poachers. In fact, the original motto surrounding the vicuña management program has been "a vicuña sheared is a vicuña saved."

An additional concern is that animals in semicaptive management may be more susceptible to predators (Sociedad Nacional de la Vicuña 2000). This may not only jeopardize vicuñas within the corrals but also create negative attitudes toward predators, such as the puma (predator control for vicuñas in corrals was a concern voiced by several campesino delegates at the SNV 2000 conference). Further, at the interface between the vicuña and its habitat, the removal of animals from the wild and into the corrals has immediate effects on the interaction of vicuñas and the landscape of the Andean puna.

Finally, no standards exist as to management of vicuñas within corrals. For example, animals in the "repopulation program" currently do not undergo quarantine before they are transferred to less populated areas. While in the short-term corrals for repopulation programs could be justified, in the long-term, if animals are

TABLE 10.1 Vicuña Census Data for Animals Within and Outside of Corrals, as Well as Fiber Production for the Year 1999

COMMUNITY	DISTRICT	PROVINCE	VICUÑAS WITHIN CORRAL	VICUÑAS OUTSIDE CORRAL	FIBER PRODUCTION 1999	INCOME GENERATED
San Juan de Tarucani	S. J. Tarucani	Arequipa	106	158	12.99	3,553.09
Salinas Huito	S. J. Tarucani	Arequipa	11	65	1.59	441.36
Tambo Cañahuas	Yanahuara	Arequipa	0	560	16.49	4,449.09
Toccra	Yanque	Caylloma	144	62	18.71	5,239.39
Total			261	845	49.79	13,682.93

Source: Data from CONACS (Consejo Nacional de Camelidos Sudamericanos)

Note: All communities are located within the National Salinas Aguada Blanca Reserve. The Toccra corral is located just outside the park border, which is a highway.

kept behind fences, these translocations cannot be considered legitimate reintroductions of the vicuña. Currently, no timetables are set for the eventual release of repopulated animals, and no conditions have been established for eventual releases into the wild should animals in semicaptive management reach carrying capacity within corrals. Extremely limited data exist on the biological effects corrals are having on vicuña populations.

STUDY AREA AND METHODS

We utilized information from CONACS, the governmental agency in charge of camelid management in Peru. We also utilized data and reports from the National Society of Vicuña Breeders, which represents the Andean communities and which is in charge of commercializing vicuña fiber and distributing profits.

Field data were obtained from within a 75-km^2 zone pertaining to the campesino community of Tambo Cañahuas, located within the boundaries of the Salinas-Aguada Blanca National Reserve in Arequipa, Peru. The study site is primarily a large plain surrounded by mountainous terrain. The area belongs to the subalpine, subtropical desert scrubland as defined by the Holdridge system (INRENA 1995). Data on habitat use, density, and fertility of vicuñas were obtained from the end of 1996 through 2000, using fixed-width repeated transects as well as opportunistic observation. When a vicuña was observed during a transect count, we noted activity, number of vicuñas, social group, and position of group or individual using GPS. In 1999 and 2000 wild-caught vicuñas were marked with plastic colored and numbered ear tags, which allowed us to identify animals by individual, age, and sex categories. Tagging caught and shorn animals allowed us to experimentally compare the effects of capture and shearing on female fertility with uncaptured and unshorn vicuñas. Additional data regarding community management of the vicuña were obtained during informal interviews with state officials and members of the community of Tambo Cañahuas, attendance at community meetings, regional meetings, workshops, and personal observations.

RESULTS

VICUÑA POPULATIONS, FIBER HARVESTS, AND PROFITS

Information obtained from CONACS (1996) indicates that at a national level, community interest in participating in the live-capture and shearing program is high. Since 1994 more than 300 communities registered with CONACS in order to participate. By 1996 more than 600 communities had formed conservation committees and registered with the government to participate in the program (fig. 10.1). This number has increased to approximately 780 communities and over 250,000 families (Sociedad Nacional de la Vicuña 2000). However, many communities are not yet participating because of a lack of infrastructure and organizational capacity.

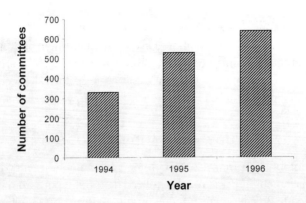

FIGURE 10.1 Formation of vicuña management committees, 1994–1996 (Data from CONACS).

The level of interest indicated by the formation of vicuña management committees and the animated and passionate debates at local and regional meetings (pers. obs.) is a positive development, especially given the participation of community members in the process of decision making and commercialization. However, the politics of vicuña management are still mainly organized in a top-down fashion, with the state maintaining a large influence in the decision making and implementation process. In addition, community interest does not necessarily lead to commercial success, as we shall see in the next section.

There is one question of extreme importance regarding the utilization of vicuña fiber by campesino communities: does giving communities a direct interest in conservation of the vicuña through economic incentives have a positive impact on vicuña populations? No published studies exist that show a direct link between fiber commercialization and vicuña conservation, although convincing circumstantial evidence exists. Government censuses conducted at a national level indicate a continuous increase in the vicuña population (fig. 10.2), even though a black market still exists for vicuña pelts. Nonetheless, a total of 30,391 vicuñas were shorn and contributed a total of 6.5 tons of fiber between 1994 and 1997, and by 1997 279 campesino communities had participated in the chaku (Hoces 2000; table 10.2).

Perhaps the most vexing problem is the economic and social sustainability of the current system, at least in the short term. Unlike hunting, vicuña capture requires capital investment and a high degree of community organization and technical skills for the monitoring, capture, and shearing of animals. For example, the approximate cost of a capture net is US$ 2,700 (Wheeler and Hoces 1997). Communities must first program the dates of the chaku in a regional assembly, and thereafter community meetings need to be organized and held in order to plan the assembly of the capture nets, monitoring of vicuñas, and finally capture and shearing. In many communities, the ultimate limitation in this system is the low number

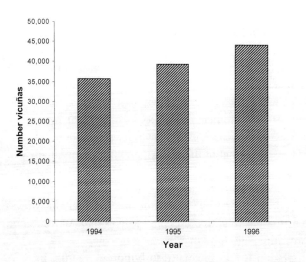

FIGURE 10.2 Government census data for vicuñas, 1994–1996. The same areas were not censused every year; therefore only those areas for which data were available in all years are included in the graph. Thus, this data represents a trend, not an actual total for Peru.

TABLE 10.2 Summary of Capture/Live-shear Fiber Production, 1994–1997

| DEPARTMENT | TOTAL: 1994–1997 | | | |
	Captured Vicuñas	Shorn Vicuñas	Kg of Fiber	No. of Campesino Communities
Ayacucho	39,474	17,597	4,028.66	76
Arequipa	682	350	69.35	8
Apurimac	2,100	1,573	245.28	22
Cajamarca	95	52	13.00	1
Cusco	298	202	41.185	8
Huancavelica	744	517	112.20	12
Ica	1,018	657	125.80	7
Junin	3,396	2,442	466.54	15
Lima	2,208	1,552	362.278	24
Puno	10,118	5,449	1,076.94	106
Total	60,133	30,391	6,541.15	279

Source: CONACS (taken from Hoces 2000)

of vicuñas, which are still recuperating from several poaching cycles. The last of these cyles occurred in 1992–93, just prior to the vicuñas' downgrading to Appendix II in CITES.

Compounding the problems of low vicuña densities in many areas of the Peruvian Andes is the fact that fiber production per vicuña is low, with an average of 250 g/individual. Vicuña fiber grows so slowly that animals can be shorn only once every two years. At current prices (US$ 300.00/kg fiber) an average vicuña will produce, at best, US$ 42.00/year. Despite low fiber production, however, the price of vicuña fiber far exceeds that of the alpaca, another luxury fiber (which has an extremely low price of US$2.00/kg) and that of sheep, US$0.15/kg (prices correspond to those of 2001–2002). Thus, the addition of vicuña fiber as an additional economic option for Andean campesinos is significant. This benefit is real even if prices were to drop further, as might be expected when other Andean countries, Bolivia, Chile, and Argentina, enter the market.

Another limitation is the unequal distribution of vicuñas in Peru. The departments of Ayacucho, Lima, Puno, and Apurimac (INRENA 1994) have the highest populations of vicuñas, and these numbers have led to higher fiber production in these zones (table 10.2). Large differences exist in the production of fiber in different geographic localities, as well as the number of campesino communities involved. Compare, for example, the 4,028 kg of fiber harvested in Ayacucho in 1994–1997 with the 69 kg of fiber harvested in Arequipa. These differences in the production of fiber and income can affect power relations of the communities with respect to management and organizational methods at regional and national scales and will surely influence the attitudes of people toward vicuñas in different zones. Indeed, signs of such changes are apparent. In the year 2000 some representatives from the community of Lucanas in Ayacucho (an area of high vicuña density and fiber production) expressed their desire to commercialize their fiber independently from the SNV (Sociedad Nacional de la Vicuña 2000). The political ecology of community interaction and relationships with the fiber industry is one that merits additional attention.

TAMBO CAÑAHUAS, AND TOCCRA, AREQUIPA: A CASE STUDY

Human census data indicate that approximately 115 families inhabit the 50,000-ha, legally recognized communal territory of Tambo Cañahuas. Approximately 50% of the families dedicate themselves in large part to raising livestock, principally alpacas and llamas and, on a smaller scale, sheep (Torres 1998). Censuses indicate that within community lands there are approximately 400 vicuñas.

Once a year the community of Tambo Cañahuas plans and implements the chaku. The base price for fiber is approximately US$ 300.00 /kg, lower than the US$ 500.00 /kg offered by the International Vicuña Consortium in previous years. In addition to paying for capture nets, the community is also discounted the price for additional equipment, as well as 10% for the Vicuña Breeders Society (SNV).

The 10% discount for the Sociedad Nacional de la Vicuña goes to the regional delegates and contributes to their administrative and travel costs.

It is evident that, although the shearing program is generating income, the quantities currently received will probably not drastically change the economic conditions of the community. However, even a few thousand dollars per year distributed at the community level can make a difference in terms of financing small community-based projects or businesses (table 10.3). By the year 2000, for example, the town of Tambo Cañahuas had received enough income to pay for equipment costs and to purchase a large truck as an investment not only for vicuña-related necessities but also for use in other community-based enterprises. Because the town residents were able to capture and shear vicuñas in 1995, they perceived immediate economic benefits (although because of bureaucratic delays, they did not receive money for two years).

Thus, the live shearing of wild vicuñas produced short-term gains (small debt and immediate returns). These short-term gains will last as long as the community continues to capture and shear animals because data indicate that the capture and live-shearing process is biologically sustainable, and the population is growing (table 10.4). Although the actual income of the capture/shearing/release program is not high, it can cover the cost of capture equipment and provide a surplus within a short time frame. From the point of view of wildlife management, the capture/release program has the added benefit of maintaining populations of wild vicuñas. This maintenance is especially important in Tambo Cañahuas, which falls within a National Reserve, and has populations that are still recuperating from a severe poaching event in 1992–1993.

In the case of Tambo Cañahuas, which began shearing vicunas in 1995, a ten-

TABLE 10.3 Fiber Production and Projected Income for the Community of Tambo Cañahuas, Arequipa, Peru

YEAR	NO. ANIMALS CAPTURED	FIBER	PROJECTED INCOME (US$)[a]
1995	48	8.185 kg ($434.00/kg)	$2,863.82
1996[b]	21	5 kg ($434.00/kg)	$1,953.00
1997[b]	16	1.5 kg ($434.00/kg)	$650.00
1998[c]	0	0 ($300.00/kg)	$ 0.00
1999[c]	55	16.5 kg ($300.00/kg)	$4,449.09

Source: Data from CONACS (Consejo Nacional de Camelidos Sudamericanos)
[a]These figures do not add up if the price is US$ 300 or US$ 434, as per contract for clean fiber; thus the communities must have been discounted for not selling the fibers previously cleaned, or for another rule. There is a convoluted system of discounts during the processing of the gross fiber and the clean fiber to the final product, and the actual discounts are not made available to the community in the reports.
[b]Second lot
[c]Third lot

TABLE 10.4 Vicuña Census Data for Catchement Area, Tambo Cañahuas, Arequipa

YEAR	TOTAL NUMBER	λ	AVERAGE DENSITY/KM^2	NUMBER YOUNG	PROPORTION YOUNG-FEMALE
1997	187		2.8 (SE = 0.6)	38	0.48
1998	242	1.22	1.27 (SE = 0.24)	53	0.70
1999	264	1.09	2.8 (SE = 0.33)	35	0.33
2000	284	1.07	2.0 (SE = 0.20)	55	0.48

dency (table 10.3) toward diminished capture success and fiber harvests existed from 1996 to 1998. The reasons for diminished success are various. The decrease was not due to the reduction of vicuña populations from poaching or the shearing program. Census data indicate the population is increasing at an average rate of r = (0.14), which is within normal range of the species (Sanchez 1984; Cattan and Glade 1989).

Observations conducted since 1995 indicate that two significant factors were influencing the success of vicuña capture. First is the lack of sufficient equipment to overcome the logistic difficulties of capture and the second is lack of capacity in several aspects of community organization. For the community it is difficult to organize meetings to plan communal activities. Tambo Cañahuas makes up an area of approximately 54,000 ha and many inhabitants have to walk for several hours to reach the community center. Planning a meeting can take weeks, as meeting announcements have to be individually given by a courier traveling by bicycle and on foot. Delays in the scheduled chaku can occur because of bad weather, but more significantly they occurred because of lack of sufficient logistic assistance or necessary equipment. Often, members of the community do not have the necessary means to transport the wood posts and netting to areas of highest vicuña densities. For example, due to the lack of trucks to transport equipment in 1996 and 1997, temporary capture corrals had to be set up in an area near the village and not in the area of highest vicuña density.

Since 1995, when the community relied heavily on state assistance for transportation and planning, people have become more organized and the capture process has become more streamlined and almost routine. Although the state agency CONACS has tried to persuade the community to install permanent corrals, the people of Tambo Cañahuas have not yet agreed to construct them.

It is important to note that in 1995, the first year the program was implemented in Tambo Cañahuas, the local state agency was involved in planning and logistics. In that year vehicles were loaned to transport people and equipment to a high-density area of vicuñas, resulting in a successful chaku. Subsequently, conflicts between the local state agency and the community led to a decrease in communication and collaboration between the two groups. The disillusionment with the poor capture success in 1996 was another factor that contributed to low yields in 1997 and 1998.

It is evident that, with the absence of telephones, radios, and vehicles that would permit communication and transport between community members during the planning and implementation of the chaku, at the outset of the program it was necessary to have some degree of assistance form the state or from other sources. The capture and shearing program was not easy to implement in the first few years. Nonetheless, persistence on the part of the community, sporadic state assistance, and logistic support from NGOs has made the wild-caught vicuña shearing enterprise an independently economically sustainable one within a few years.

Data collected during our study indicate that partial shearing of animals in the spring (September-October) has no negative effect on the vicuña population (Sahley, Torres, and Santos unpublished data). Our data thus far indicate that, if properly managed, capture, shearing, and release of free-ranging vicuñas is a biologically sustainable option. Moreover, the experience in Tambo Cañahuas has shown that the wild-caught system originally proposed is feasible and results in better short-term economic gains than the corral system.

Beyond economics, maintaining wild vicuñas on communal lands may not only serve to promote vicuña conservation but also of the Andean landscape and its associated flora and fauna. An additional benefit is the strengthening of the cultural identification of Andean people with an iconic wildlife species in Peru. The strengthening of pre-Columbian tradition in communities and their passionate attachment to the new vicuña enterprise is a fascinating combination of the meeting of tradition, human interaction with the wild landscape, and the global marketplace. As such, the case study in Tambo Cañahuas illuminates myriad complex interactions that have yet to be revealed. Thus far it seems that it is possible to combine Western wildlife management with a traditional system of resource use that can serve to strengthen cultural and historical ties with the landscape and also work with an international market. We believe that the information and data collected in Tambo Cañahuas can serve as a model for other communities in the Andes.

TOCCRA: AN ABBREVIATED CASE STUDY OF THE CORRAL SYSTEM

Adjacent to the campesino community of Tambo Cañahuas is the village of Toccra. Located at a slightly higher elevation (approximately 4,300 meters above sea level) and more sparsely populated by both humans and vicuñas, this village was chosen as the site of a vicuña "repopulation" effort. With strong support from the state and vigorous promotion by the Inca Group (the Peruvian company that has sole rights to the production of vicuña textiles and is part of the International Vicuña Consortium), a corral was installed a few meters outside the boundaries of the Salinas-Aguada Blanca Reserve but within its buffer zone.

In 1997 ninety-five vicuñas were brought from Pampa Galeras to Toccra and placed in the corral. Shortly thereafter, these animals became part of a highly publicized international vicuña festival, which attracted several hundred American tourists. A staged capture was enacted, in which tourists were allowed to participate, but animals were subsequently not shorn. Members from neighboring communities were asked to help with the capture. Although some community members did help, others did not, because of a lack of any compensation for labor (even though festival organizers charged tourists for the spectacle). Due to pressure from the industry, local state officials focused on organizing this event and virtually ignored other communities in the area such as Tambo Cañahuas, which subsequently had low capture success that year.

Efforts to study the biological effects of the corrals on vicuñas by comparing biological parameters of the captive animals with those of the wild populations in Tambo Cañahuas were unfortunately hampered by vicuña politics. Nevertheless, we did obtain census data in 1998, 1999, and 2000. Through interviews we learned that wild vicuñas were caught when possible and placed in corrals. Thus data on population dynamics and sources of vicuña births and mortality are impossible to quantify precisely. What our census data showed was that in 1998, an El Niño year, production of young was significantly lower in Toccra than in Pampa Cañahuas (Sahley 2000). After three years population growth of animals in the corrals was similar to that in the wild, even though animals had been introduced into corrals to apparently compensate for undocumented mortality or low birth rates. Fiber production for Toccra in 2000 was only 2 kg more than that for Tambo Cañahuas, and the meager economic gain (US$ 780.00 table 10.1) in comparison to Tambo Cañahuas did not compensate for the expense of the corral and veterinary treatment for vicuñas that had contracted mange or had brought mange with them from Pampa Galeras. While Tambo Cañahuas saw an increase in vicuña populations, as well as direct profit from the shearing of wild vicuñas, Toccra continues to pay off the debt incurred by the corral expenditure.

CONCLUSION AND RECOMMENDATIONS

Data obtained from this study indicate that live shearing of vicuñas in Peru has much potential for the sustainable utilization of this species and can also benefit Andean communities that traditionally have been marginalized in Peruvian society. While the "privatization" of the vicuña may cause social and economic conflict by promoting the construction of corrals, it remains unclear how the economic and social sustainability of this process will be influenced. However, it is apparent that vicuña management needs to be conducted in collaboration with technicians familiar with wildlife management theory and techniques. Although implementation of the program is still fairly recent, we believe it is more prudent to develop methods to improve capture efficiency of wild vicuñas and to protect them from poachers instead of changing this methodology for the more risky and costly option of maintaining vicuñas under conditions of captivity. We suggest emphasis be placed on the following efforts:

1. Measure biological and socioeconomic costs for each community or region that will have different advantages and disadvantages with respect to vicuña management.

2. Proceed with great caution with the implementation on a large scale of permanent corrals. Research is urgently needed to compare health, fertility, social behavior, movement patterns, and effects on the landscape between wild and captive vicuñas.

3. Strengthen community organization and basic wildlife management techniques at local levels. If communities are going to be legally responsible for vicuña management and conservation, they must receive training in basic concepts of monitoring and management of vicuñas.

4. It is critical for communities to have complete information on the commercialization process and sale of fiber (for example, what percent of profits are they receiving?) and also to be able to understand the full implications of any contract or agreement presented to them by state or private institutions.

5. While community management may have several advantages for the vicuña, one potential disadvantage is that management on regional or national scales could be overlooked. This oversight could have a negative effect on the structure of the metapopulation of the vicuña or have important effects on the Andean landscape. It is important to maintain regional and national perspectives regarding management of wild vicuñas. This need requires collaboration and cooperation with state and nongovernmental organizations, as well as the creation of a national management plan for the vicuña.

Fortunately, the interest of Andean campesinos, who are important protagonists in this saga, is high. We hope that management practices based on scientific principles will lead to a conservation success story in Andean Peru.

POSTCRIPT

Since the completion of this article, a new presidential decree was signed by former President Fujimori. This decree grants individuals, as well as communities, the right to own vicuñas that occur on their land. Thus, privatization of the vicuña has, for the time being, been increased in scope. This expansion has resulted in a protest by the Society of Vicuña Breeders, which represents communities, and raises important issues regarding the privatization of a wild mammal and the future social, economic, and biological implications this decree might have.

In 2002, under the new government of President Toledo, a vicuña working group was established. This working group will examine the biological, legal, commercial, and social aspects of vicuña management and current legislation. Recommendations from the working group are as still pending.

ACKNOWLEDGMENTS

We are profoundly grateful to the community of Tambo Cañahuas, which for several years has granted us permission to conduct our research on their land and to observe the chaku. In addition, we thank Caroline Sahley who revised an early draft of the manuscript and gave comments that greatly improved its organization. INRENA has granted us permission to work in the Salinas Aguada Blanca Reserve for several years, and CONACS has also granted us permission to conduct vicuña research and to have access to their fiber harvest data. The Society for Vicuña Breeders has also granted us interviews and provided us with information. The study has been supported by USAID through the Biodiversity Support Program (a consortium of the World Wildlife Fund, The Nature Conservancy, and The World Resources Institute), the Wildlife Conservation Society of the New York Zoological Society, the International Vicuña Foundation, and Atlantic Hospitality.

11

Captive Breeding Programs as an Alternative for Wildlife Conservation in Brazil

SÉRGIO LUIZ GAMA NOGUEIRA-FILHO
AND SELENE SIQUEIRA DA CUNHA NOGUEIRA

The incorporation of wild species into the meat production industry has attracted the interest of Brazilian farmers in the last few years because of the increased demand for products and by-products of wild mammal origin. The use of new protein sources for the human population is also of great social interest. Rational use of native resources can be a beneficial process, resulting in economic and social advantages, while at the same time reducing damage to wild animal populations caused by irrational hunting and habitat destruction.

Additionally, there is a great demand in the international market for wild animal leathers. Germany, Italy, and the United States are the main importers of collared peccary (*Tayassu tajacu*) leather from Peru (Bodmer and Pezo 1999). In Argentina, where leather is the main product of capybara (*Hydrochaeris hydrochaeris*) exploitation, skins are used for such luxury products as shoes, jackets, and gloves (Nogueira-Filho 1996). This demand for wildlife leather has always been met by commercial hunting in several South American countries, especially in Brazil. From the 1960s and 1970s on, commercial hunting became illegal in many of these countries. Nevertheless, traffic in wildlife products continued. In the 1980s, for example, one German importer purchased 36,000 collared peccary skins from Paraguay (Redford and Robinson 1991).

Commercial hunting is forbidden in Brazil. However, wildlife farming in full or semicaptivity is permitted, and Brazilian law includes a provision that enables this use of wildlife. The Brazilian forest code—one of the most important laws for the protection of Brazilian forests—requires farmers to protect a part of their property as an Área de Reserva Legal, or legal reserve area. This is an area inside the agricultural proprerty required for the protection of wildlife and biodiversity conservation. Depending on the region, the legal area ranges from 20 to 80% of the property. In the Amazon region the Área de Reserva Legal can comprise up 80% of rural properties, and in the Cerrado ecosytem, up to 35%. The sustainable use of natural resources is allowed in these protection areas, and some Brazilian farmers are using part of their legal protection areas for the semi-intensive production of wildlife. Unfortunately, landowners do not always support or respect this law, and a draft proposal is currently in discussion in the Brazilian congress that will reduce the size of the protection areas across the country.

Currently IBAMA, the Brazilian Environmental Agency responsible for wildlife use and protection, allows capybaras, rheas (*Rhea americana*), and other wild animals to be raised, on an experimental basis, in a low management-intensity system such as ranching. IBAMA classifies this kind of wildlife exploitation as an extensive production system (Francisco Neo, pers. comms.). Such extensive production gives farmers the option of conserving forests by making wildlife an economically valuable resource. The hope is that wild animal production, whether in full captivity or through ranching, will meet the meat and leather demands on wild animals and, in consequence, decrease illegal hunting.

In this article we describe the biological characteristics, products, and by-products of capybara and collared peccary, as well as the current production systems adopted in Brazil to produce these two species commercially. Analyses of economic parameters such as inputs, outputs, and financial return of the three production systems are also compared.

PART I. CAPYBARA: SPECIES CHARACTERISTICS, PRODUCTS, AND BY-PRODUCTS

BIOLOGY

Among Brazilian wild mammals, the capybara is the species that has the highest productive potential in captive, intensive small-scale farm settings, as well as in natural, extensive ranch settings (Parra, Escobar, and González-Jiménez 1978; Lavorenti 1989; González-Jiménez 1995; Nogueira-Filho 1996). It is herbivorous, grows rapidly, and adapts easily to captivity. There is also a great potential market for its meat and leather (Ojasti 1991; González-Jiménez 1995; Andrade et al. 1996; Nogueira-Filho 1996).

Capybara habitat consists of a combination of water bodies (rivers and lakes), pasture, and available natural cover. Its semiaquatic habit and its reliance on wetlands led some authors to propose capybara ranching as an alternative in order to enhance wetland benefits for local communities (Ojasti 1973; González-Jiménez 1995). The ranching system allows rural inhabitants of wetland regions, such as the Brazilian Pantanal, to use this species as a source of protein and income as long as they safeguard its natural habitat. Alternative uses, such as cattle ranching and agriculture, require extensive land draining.

MEAT CHARACTERISTICS

In Venezuela and Colombia capybara meat is traditionally eaten during Easter. In the sixteenth century, in response to a petition from the people, the Pope decreed the capybara a fish (Mukerjee 1994). Consequently, its meat can be eaten by Catholics during periods of abstinence from red meat.

In the biggest Brazilian cities, such as São Paulo, Rio de Janeiro, and Belo Horizonte, capybara meat is sold for up to US$ 7.5 per kilogram. To introduce capybara meat to the market, the following strategy was used in São Paulo. Initially, the meat was offered only at a few luxury restaurants, at which it was presented as an exotic product and listed at the highest price on the menu. This process increased the demand for wild meat in urban centers, resulting in incentives for commercial captive breeding (Nogueira-Filho 1996).

Some rural peoples, however, reject capybara meat as having a disgusting taste, probably because they have consumed meat from hunted animals. In general, hunted animals are slaughtered without care, and under inadequate hygienic conditions, resulting in meat contaminated by intestinal contents. On the other hand, people that tested the meat of captive-reared animals in a double-blind experiment found no differences in scent and flavor among capybara, pig, and cow meat (Frasson and Salgado 1990). The captive animals used in the test had been fed a diet of elephant grass (*Pennisetum purpureum*), supplemented with corn and soybean meal ration and slaughtered under adequate conditions. Nevertheless, more studies of the fatty acid composition of capybara meat are necessary to establish how diet affects its organoleptic characteristics.

Capybara meat contains on average only 4.5% of fat and 24.5% of crude protein (Lavorenti et al. 1990). It has a digestibility of 89.4% and a biological value of 61.3% (Frasson and Salgado 1990).

ECONOMIC CONSIDERATIONS

In addition to high quality meat, capybaras produce light and resistant leather. This leather can be used for footwear, glove, jacket, and bag production (González-Jiménez 1995). In Buenos Aires, jackets made from capybara leather sell from US$ 400 to 720, while a pair of gloves sells for US$ 80 (Nogueira-Filho 1996).

Brazil exported 3,500,000 tons of capybara skins between 1956 and 1969, an amount that brought around US$ 10,700,000 to the country (Mourão 1999). However, because of the high market price of capybara meat, currently in Brazil the skin is sold incorporated to the carcass (Nogueira-Filho and Nogueira 2000). Capybara leather will only be marketed separately when there is sufficient production to guarantee a constant supply to the industry.

Capybara oil, derived from the body fat, is popularly used for such illnesses as asthma, bronchitis, rheumatism, and allergic diseases and as a growth tonic

(Nogueira-Filho 1996). A recent study showed that it functions to control choles-terol in mice fed a diet rich in saturated fat (Fukushima et al. 1997). Capybara body fat contains high levels of polyunsaturated fatty acids, including omega fatty acids (Fukushima et al. 1997) that may be responsible for its therapeutic action. The con-jugated linoleic acid (CLA), which has anticarcinogenic properties (Bauman et al. 1998), may also be present in capybara body fat. The possibility that capybara meat and fat contain substances that can bring benefits to human health can be used for marketing purposes.

CAPYBARA SOCIAL BEHAVIOR AND CAPTIVE MANAGEMENT

Capybaras live in social groups ranging in size from one male and two females up to 100 individuals; the larger groups include several adults of both sexes with their offspring (Ojasti 1973; Macdonald 1981; Schaller and Crawshaw 1981). Groups are cohesive with a complex social structure characterized by a dominance hierarchy and individual specialization of functions (Azcarate-Bang 1980). A dominant male protects the group, is possessive of the females, and through threats and attacks maintains subordinate males at the periphery (Ojasti 1973). Satellite males some-times copulate with young females in estrus (Ojasti 1973; Macdonald 1981; Schaller and Crawshaw 1981). Offspring remain in the original group until they reach sexu-al maturity, at which time females may be incorporated and males expelled follow-ing a hierarchical dispute with the dominant male (Ojasti 1973). Expelled mem-bers often form new groups (Ojasti 1973; Schaller and Crawshaw 1981).

Some authors have suggested the existence of cooperative care or alloparental behavior among capybaras (Ojasti 1973; Macdonald 1981). In the wild, capybara fe-males have been seen with twelve to fourteen infants (Macdonald 1981; Schaller and Crawshaw 1981; Alho, Campos, and Gonçalves 1987), although the maximum litter size is seven (González-Jiménez 1995; Nogueira 1997). In captivity, juveniles spend more time with one female than with other group members (Nogueira 1997). Nogueira et al. (2000) studied the alloparental behavior of two capybara fam-ily groups in captivity and concluded that females did not discriminate between young when suckling because each suckled her own infant as often as those of oth-er group females.

Despite the occurrence of alloparental behavior, breeding capybara in full or semi-intensive captivity can be very problematic because of the high rates of infan-ticide (Nogueira 1997). The hypothesis that infanticide is caused by lack of experi-ence in primiparous females has been discarded (Nogueira et al. 1999). Rather, in-fanticide was associated with reproductive groups containing females who had not been together since weaning. Unfamiliar females lived together without apparent conflict until the occurrence of births when they killed their pen-mates' offspring (Nogueira et al. 1999). When a capybara group is set up with females that have been living together since weaning, infanticide does not occur, and females do not have to be isolated to farrow either in intensive or semi-intensive breeding (Nogueira et al. 1999).

CAPYBARA FEEDING HABITS

Capybaras feed mostly on grasses (Lavorenti 1989; González-Jiménez 1995) and therefore do not compete directly with humans for food (Nogueira-Filho 1996). They enhance their nutrient intake from food by increasing the efficiency of their digestion by extensive mastication and hindgut fermentation (Mones and Ojasti 1986). The cecum represents 74% of the whole gastrointestinal tract of capybaras (González-Jiménez and Parra 1972). Cecum pH is around 6.0 (0.3 (González-Jiménez 1977), providing an optimal environment for microbial fermentation of structural carbohydrates (cellulose and hemicellulose) and liberation of volatile fatty acids that are an energy source for the host (González-Jiménez 1995).

Capybaras may be able to use microbial protein from the cecum through either (a) reingestion of the cecal content (Herrera 1985; Mendes 1999; Mendes et al. 2000; Hirakawa 2001) or (b) digestion and absorption of microbial protein in the large intestine (González-Jiménez 1995). This ability suggests that expenses for supplements of essential nutrients, such as amino acids and vitamins, can be decreased for adult animals in captivity. Nevertheless, the best growth results were obtained when the diet of young animals was supplied with protein supplement (Silva-Neto 1989; Andrade, Lavorenti, and Nogueira-Filho 1998). Under these conditions young animals gain up to 110 g per day and reach slaughter weight more quickly than when just fed grass, which results in 60 grams of daily weight gain (Ojasti 1973).

REPRODUCTION

Although capybaras can breed throughout the year, in the wild they do so most frequently at the onset of the rainy season (Mones and Ojasti 1986). Capybara gestation lasts approximately 150 days (López-Barbella 1984), and females usually produce one litter per year in the wild (Mones and Ojasti 1986).

Nogueira (1997) analyzed data obtained from the experimental captive-breeding project at the University of São Paulo at Piracicaba (ESALQ/USP). From November 1984 through December 1995, 243 young were born in 80 litters. Births peaked in the rainy months of November, December, and January. Mean litter size was 3.3 (range 1–7) with a male-female sex ratio of 53:47. Mean birth weight (±SD) was 1,988.0 g (± 346.7) for males and 1,991.0 g (± 356.8) for females.

Females could reach sexual maturity at seven months of life, but there was some individual variation in the age at which females produced their first litter (1,449.9 ± 744.2 days of age). The mean interbirth interval was 380.0 days (± 163.7). However, females could reinitiate reproductive activity just twenty days after farrowing. The potential for postpartum estrus with ovulation, unlike what was observed by

López-Barbella (1984), allows capybaras to produce young at a faster rate than animals that are unable to ovulate soon after parturition. There was a significant difference in interbirth interval among females of different ages. The number of young per litter decreased, and the interbirth interval increased after six years of age (Nogueira 1997).

Data from this study also showed that intensive breeding and handling procedures led to stress and poor condition in captive animals, with negative impact on productivity. It is therefore necessary to improve husbandry practices through behavioral studies and to implement a selection program to improve reproductive parameters and to reduce the enormous variability observed. Nogueira (1997) concluded that the variability in reproductive parameters could be partially explained by individual temperament as a consequence of adaptation to captivity. Nogueira et al. (2000) proposed that subordinate females take better care of juveniles than do dominant females. This observation may be one possible explanation for the variability in reproductive parameters showed by capybara females in captivity.

Lack of adapted husbandry practices for this species in captivity may also have consequences for reproductive performance. For example, the delay in entrance into reproductive life was due partly to the fact that young females and males of nearly the same age made up the reproductive groups. Capybara males reach sexual maturity between fifteen and twenty-four months of age (Ojasti 1973), females between seven and twelve months (López-Barbella 1982; Nogueira 1997). Groups composed of young females with adult males should be tested to verify the possibility of reducing the age at which capybara females produced their first litter.

CAPYBARA CAPTIVE-BREEDING SYSTEMS

INTENSIVE PRODUCTION SYSTEM

In the intensive breeding system family groups comprising a male and up to eight females are maintained in small enclosed reproduction paddocks (120 to 400 m², depending on group size), with at least 20 m² per adult animal (Nogueira-Filho 1996). Paddocks should contain a water tank, feeders, and a sheltered area. Additional paddocks, with the same facilities and space per animal, are used for rearing juveniles (Andrade, Lavorenti, and Nogueira-Filho 1998).

Females introduced to each of the reproduction paddocks must belong to the same family group in order to avoid fights and infanticide (Nogueira et al. 1999). The male can be from a different origin than the females, allowing the exchange of genetic material from different populations. If wild-reared females are used, adaptation to captivity is facilitated by placing them in a paddock already containing a captive-born adult male (Nogueira 1997). An alternative reproductive group composition is to place in one paddock juveniles no more than sixty days old from several groups. Because they will grow to maturity together, fighting and infanticide will not occur (Nogueira et al. 1999).

Adult animals are fed supplementary ration (12% CP) at 1.0% of live weight on a dry matter basis and grass ad libitum. Young animals receive grass ad libitum and a protein supplement at 1.5% of live weight on a dry matter basis. Since all live grass is removed from the confinement area, grass must be cut and brought to the paddock. The forage crop of choice is the elephant grass (*Pennisetum purpureum*), which has high productivity (Nogueira-Filho 1996).

Infants born in the reproduction paddock will stay with the group until they are sixty days old. Then they are weaned and transferred to a juvenile growth paddock until they reach the weight/age of 20 kg/6 months or 35–40 kg/12–18 months, depending on local market preferences (Nogueira-Filho 1996).

Space allowances for captive capybara are much higher than those used for domestic animals and easily meet the animals' behavioral needs. Even farmers who adopt the intensive production system use at least up to 20 m² per individual, whereas pig farming typically allocates 2 m² or less per animal (Nogueira-Filho 1996; Andrade, Lavorenti, and Nogueira-Filho 1998). In cases where appropriate handling practices, feeding requirements, enclosure facilities, and social environment are not supplied, low productivity, low growth rates, infanticides, or even lethal fights will occur, leading to breeding failure (Nogueira 1997; Andrade, Lavorenti, and Nogueira-Filho 1998; Nogueira et al. 1999).

SEMI-INTENSIVE PRODUCTION SYSTEM

Over 90% of the 137 capybara farming facilities in Brazil use the semi-intensive production system. Females are farrowed and juveniles raised to maturity in fenced areas, ranging from 5,000 to 40,000 m², including a lake or impoundment of at least 200 m² surrounded by arborescent vegetation (Nogueira-Filho 1996). One or more traps are built in the breeding area. The trap must allow the capture of multiple animals in order to reduce the costs of trapping and handling. Inside the traps mineral salts, grain corn, or supplementary ration are supplied in sheltered feeders. The number of traps and their size depend on the number of animals in the breeding area, which in turn depends on the size of the area and the farm's capacity for food (grass) production.

A family group comprising a male and up to nine females is captured from the wild or purchased from another breeder and introduced into the paddock. If the animals were captured in the wild, only animals caught in the trap at the same time should be used in order to avoid the inclusion of a nongroup member (Nogueira 1997). If the capybaras were purchased from another breeder, the group individuals should have been living together at least since weaning (Nogueira et al. 1999). Inside the breeding area the group will increase only by reproduction — no more animals are added from the wild or captivity in order to avoid fights and infanticide (Nogueira 1997).

Appropriate density studies are still required, but at densities of 100 m² per adult animal, weaning should take place at sixty days of age. Juveniles are ear-tagged, and young males placed in a growth paddock with an area ranging from 200 to 800 m² — at least 20 m² per young (Andrade, Lavorenti, and Nogueira-Filho 1998) — where they will grow more quickly without competition from the adult animals. Young females can be maintained with the adults. When a breeder needs to replace an adult female, he can select among those presenting faster growth; because they already belong to the group, there will no difficulty with group incorporation (Nogueira-Filho 1996).

At lower densities, on the other hand, weaning occurs naturally and young males grow up with the adults until they reach the weight/age of 20 kg/6 months old. After this age young males are subject to aggressive behavior from adult males. At this time they should be slaughtered or transferred to a juvenile growth paddock until they reach 35–40 kg/12–18 months old, depending on local market preferences (Nogueira-Filho 1996).

Normally, the adult diet is grass and mineral salt, while the young receive a protein supplement in a creeper feeder. As in the intensive system, cut elephant grass is placed in the trapping devices. The grass growing inside the area is not considered in the feeding plan because capybaras are selective feeders who choose forage plants with high protein content, such as plant seedlings, using their incisors to clip vegetation at ground level. Given this behavior, large areas and a pasture rotation system would be required to ensure adequate natural pasture supply. This practice is not cost effective because of the high prices of the stout galvanized chain link wire fences used in capybara farming (Nogueira-Filho 1996).

We consider this breeding system as semi-intensive, despite food supplementation, because the groups are maintained in large areas and because there is no individual control on reproductive parameters as in the intensive system. When analyzing reproductive parameters, the breeder considers only the average group reproduction data, such as interbirth intervals or number of young produced per female each year.

EXTENSIVE PRODUCTION SYSTEM

In some portions of Brazil, especially in the states of São Paulo and Minas Gerais, capybaras are shot as agricultural pests because they harm corn, rice, sugarcane, and other crops and may compete with cattle and other domestic livestock for food during the dry season (NRC 1991; Moreira and Macdonald 1997). An alternative way to control and use these animals is through extensive production or ranching. This production system in particular has the ecological benefit of preserving wetlands rather than draining them as would be done for cattle ranching. Currently IBAMA allows this kind of exploitation system only on an experimental basis.

In this system capybara population density is assessed, and a selective quota determined, ranging from 10 to 30% of the population; the range depends on annual population variations. This system is used in the Venezuelan Llanos (Ojasti 1973); however, there are some differences between the experimental system in Brazil and the better-known Venezuelan harvest. In Brazil the animals are live-captured, allowing ranchers to select the animals that will be slaughtered. Furthermore, animals can be slaughtered before they reach sexual maturity. IBAMA requires that all ranched animals be identified with electronic microchips, and only young born after ranching is implemented may be slaughtered. Therefore monitoring is easier than in the Venezuelan harvest system, and it is possible to avoid the overharvesting that has occurred in the Venezuelan Llanos (Ojasti 1991).

An example of the extensive production system in São Paulo state is the Perobal farm, whose main activity is beef cattle production. There, a population of 800 capybaras shares the pasture with cattle. On alternate days grain corn is placed inside the traps that allow animals to be handled. These trapping devices, with an ingenious system of automatic doors, are set one day before the planned capture and allow the rancher to hold and select the animals that will be slaughtered (Nogueira-Filho 1996).

The main problem with this production system is that capybaras can quickly degrade the pasture because of their selective feeding on seedlings. Animals thus have to be moved out of some areas while the pasture recovers, potentially resulting in confrontations when transferred groups encounter resident groups in new areas. Therefore specific husbandry practices must be developed for this production system, especially those that allow an increase in the extraction quota.

ECONOMIC ANALYSIS OF THE THREE CAPYBARA PRODUCTION SYSTEMS

An essential practical consideration in evaluating a productive activity is its cost in terms of the return obtained for the product. While it is obvious that the financial aspects of breeding operations cannot be neglected, the expression of the results in terms of dollars, unless properly interpreted, may obscure rather than clarify the facts because monetary statements are not experimental results. They are based upon factors that are not under experimental control, the same combination of which may never happen again. The relative prices of feeds and the selling prices of product vary from time to time and from place to place, according to market conditions.

Let us consider a hypothetical comparison, based on data obtained from commercial breeders, regarding the farm expenses and income of a dairy farmer who wants to breed capybara through either intensive, semi-intensive, or extensive production systems. The farm has an available area with arborescent vegetation and a 600-m^2 lake, and in another area a large group of capybaras is harming a corn crop and grass pasture used by the cows.

We simulated the expenses and income for intensive, semi-intensive, and extensive breeding systems on this farm (table 11.1). In the intensive system the breeder will need to build six paddocks, each with an area of 400 m^2, a 20-m^2 water tank, and a 40-m^2 sheltered area. Three reproduction groups, each comprising a male

TABLE 11.1 Initial Expenses for the Maintenance of Twenty-four Female and Three Male Adult Capybaras in Intensive, Semi-intensive, and Extensive Breeding Systems

INITIAL EXPENSES (US$)	INTENSIVE	SEMI-INTENSIVE	EXTENSIVE
Consultant[a]	500.00	500.00	500.00
IBAMA[b]	100.00	100.00	100.00
Purchase of animals	1,350.00	—	—
Materials and labor for facilities building	6,820.00	4,500.00	1,500.00
Equipment	200.00	200.00	200.00
Total	8,970.00	5,300.00	2,300.00

[a]The breeder needs to contract a consultant to develop the project.
[b]The breeder must pay an initial charge to IBAMA to legalize the project.

and eight females, will occupy three of those paddocks. The other three paddocks will be designated for juvenile growth. In the semi-intensive system the breeder will need to surround two hectares (20,000 m^2) with a 1.5-m high wire fence around the lake to contain a breeding group comprising three males and twenty-four females, which will be captured on the farm. It will also be necessary to build a trap for animal handling. For the ranching system the breeder will need to build two panel traps, establish electrified fences around crops to avoid damage, and increase the carrying capacity of the habitat by enriching the habitat with supplementary food—e.g., elephant grass—that capybara groups can use throughout the year in order to increase their production.

To estimate farm expenses (table 11.2) and incomes (table 11.3) from the three breeding systems, we used a spreadsheet for computing costs and expenses designed by researchers at the Economics Department at ESALQ/USP. Intensive and semi-intensive productivity data are from Nogueira (1997) and Nogueira-Filho (1996), respectively. Currently, there are no data available on the productivity of capybara ranching in Brazil; therefore we used the natural productivity data from González-Jiménez (1995). We considered an extraction quota of 30%, the experimental extraction quota currently allowed by IBAMA.

In spite of the apparent high annual potential net income (table 11.3), the small scale of commercialization limits actual profits. Only a few big producers can sell their production directly to the consumers, mainly restaurants, and so obtain the highest incomes. Most capybara farmers sell their production to intermediaries who pay only US$ 1.5 per kilogram of animal live weight. Therefore, farmers must organize themselves to sell their joint production directly to the consumers and to obtain additional income through the industrialization of capybara leather and fat (Nogueira-Filho and Nogueira 2000).

Ranching incurs the lowest expense per kilogram of animal produced (table 11.2) but also the lowest financial return (table 11.3). Contrary to the situation in the

TABLE 11.2 Analysis of the Total Expense (per kg) of Live Animal Weight in Intensive, Semi-intensive, and Extensive Breeding Systems for Capybara

EXPENSE ITEMS	INTENSIVE (US$)	%	SEMI-INTENSIVE (US$)	%	EXTENSIVE (US$)	%
Feeding[a]	1.19	64.3	1.17	75.0	0.36	52.9
Labor	0.22	11.9	0.10	6.4	0.10	14.7
Capital interests[b]	0.25	13.5	0.14	9.0	0.10	14.7
Fuel, electricity, and telephone	0.03	1.6	0.02	1.3	0.05	7.4
Veterinarian and medicines	0.04	2.2	0.04	2.6	0.02	2.9
Other expenses[c]	0.12	6.5	0.09	5.8	0.05	7.4
Total expense per kg of live animal weight	1.85	100	1.56	100	0.68	100

[a]US$ 7.00/t of *Pennisetum purpureum* grass, crushed and placed in the feeder (labor included) and US$ 0.15 per kg of supplementary growth ration—both used in intensive and semi-intensive system. US$ 5.00/t of grass grazed by the animals itself (expenses for pasture recovery) in the extensive system.
[b]Includes interests (6% per year) on total initial costs
[c]Includes expenses with interests (6% per year) on operational capital, taxes, and commercialization expenses

TABLE 11.3 Projection of the Production, Costs, and Annual Incomes for Twenty-four Adult Female and Three Male Capybaras in Intensive, Semi-intensive, and Extensive Breeding Systems

ANNUAL INDICES	INTENSIVE	SEMI-INTENSIVE	EXTENSIVE
Mean number of young	118.6	144	204.3
Number of commercial animals sold[a]	110.2	134.4	69.4
Total weight of commercial animals (kg)	3,306.0	4,032.0	2,082.0
Total weight of discarded reproducers (kg)	173.3	173.3	—
Labor per kg of commercial animal produced (hours)	0.4	0.1	0.1
Kg of food consumed/kg of commercial animal[b]	58.8	55.6	126.2
Facilities and equipment expenses per female (US$)	292.50	195.83	70.83
Hypothetical annual total production expenses (US$)	6,438.00	6,568.00	1,992.00
Hypothetical annual net income (US$)[c]	5,743.00	8,151.00	5,295.00

[a]Currently, IBAMA adopts an annual extraction quota of 30%.
[b]In the extensive system the farmer will need to feed all animals but only a percent of the young will be slaughtered.
[c]Selling the production at US $ 3.5/kg of animal live weight.

Venezuelan Llanos where ranchers incur only slaughter expenses, Brazilian breeders incur expenses for supplementary food and landscape management to maximize the capybara carrying capacity. Furthermore, IBAMA does not allow breeders to slaughter the oldest animals. The rancher thus incurs expenses for all animals but obtains economic benefit only from animals born after ranching is implemented. Because natural predators are controlled in ranching, capybara density will greatly increase after a few years. Therefore IBAMA extraction quotas should be increased in order to make this production system economically feasible. Breeders will incur further costs carrying out the studies required to win approval of a higher extraction quota.

PART II. COLLARED PECCARY: SPECIES CHARACTERISTICS, PRODUCTS, AND BY-PRODUCTS

The collared peccary occurs throughout Brazil in habitats ranging from arid environments to tropical rainforest (Sowls 1997). Adult collared peccaries measure from 80 to 90 cm in length, 25 to 45 cm in height, and attain weights of up to 27 kilograms in captivity (Emmons 1999; Nogueira-Filho 1999). Currently, peccary meat is sold for up to US$ 10 per kilogram in the city of São Paulo. Unlike capybara meat, peccary meat is a preferred wildlife product throughout Brazil (Nogueira-Filho 1999) and is also the game meat most frequently sold at the urban market of Iquitos, Peru (Bodmer and Pezo 1999). The meat has 22.7% of crude protein and less than 1.5% of fat in its composition (Carrilo 1999). The species also produces excellent quality thin, soft, durable, and stain-resistant leather, characteristics rarely found combined in any one leather (Bodmer and Pezo 1999).

The Province of Salta, Argentina, exported 12,600 peccary skins between the years of 1988 and 1992, even though the trade was illegal. In Salta hunters receive the equivalent to US$ 3 for each skin (Barbarán 1999), while tanned leather is offered to footwear manufacturers at US$ 45 per m^2, and a pair of boots is offered to the public for around US$ 150 (Barbarán 1999). The market is relatively stable because Europeans traditionally use products made from peccary leather (Bodmer and Pezo 1999; Bodmer, Pezo, and Fang this volume). The United States provides another important market; a pair of peccary gloves at Barneys, New York, retails for up to US$ 195.

Today the demand for peccary leather is met primarily by the harvest of natural populations in Amazonian Peru (Bodmer and Pezo 1999) and by illegal hunting in several other South American countries, especially Brazil (Nogueira-Filho 1999). The collared peccary adapts well to captivity (Nogueira-Filho 1999), can support high indices of extraction in natural populations (Bodmer 1999), and would be an ideal animal for incorporation into agroforestry projects (Sowls 1997). For these reasons commercial projects are being developed in several South American countries in order to exploit the species in a sustainable way (Bodmer 1999; Nogueira-Filho 1999).

SOCIAL BEHAVIOR AND CAPTIVE MANAGEMENT

Collared peccaries live in stable and cohesive herds containing from six to thirty-four individuals with an approximately 1:1 sex ratio (Emmons 1999; Fragoso 1999). The smaller herd size may be characteristic of hunted areas or areas of low productivity (J. Fragoso pers. comm.). Herd members eat, sleep, and forage together (Sowls 1997). In captive groups comprising individuals from different sources, however, feeding and sleeping subgroups frequently form according to their origin, and females are more cohesive than males (Nogueira-Filho, Nogueira, and Sato 1999). Therefore, in order to avoid the occurrence of conflicts during feeding, several feeders must be available in the enclosures (Lima-Neto, Nogueira, and Nogueira-Filho 2001). Conflicts occur mainly because certain individuals retain definite dominance over others, but dominance ranks are ill defined (Diaz 1978; Sowls 1997; Nogueira-Filho, Nogueira, and Sato 1999).

Captive groups tolerate the presence of several adult males without noticeable conflicts, even when receptive females are present. Unlike the situation with capybara groups, male and female subadult peccaries are also tolerated (Nogueira-Filho, Nogueira, and Sato 1999). Even though collared peccary groups usually reject the introduction of solitary animals into an established colony (Lochmiller and Grant 1982), they tolerate the introduction of groups of three or more related individuals, which leads to the observed subgroup composition (Nogueira-Filho, Nogueira, and Sato 1999).

If the group contains unrelated or unfamiliar females at high densities, pregnant females should be isolated to farrow in order to avoid infanticide (Packard et al. 1990; Nogueira-Filho 1999). However, this practice increases breeding expenses due the need for special facilities and the lengthening of the interbirth interval (females can have postpartum estrus; López-Barbella 1993). One solution is to breed peccaries at low densities of about 250 m² per adult animal (Engel 1990), but this requirement implies the need to enclose large areas, a very expensive procedure for small producers because of the high prices of the wire fences and concrete or rock base used in peccary farming (Nogueira-Filho 1999). The best solution, if the rural producer decides to adopt the semi-intensive production system, may be to build small paddocks and set up groups only with related females. These paddocks will range from 200 to 400 m² each, according to group sizes and the breeder's financial resources. Young animals produced from these groups can then be raised together from weaning to make up familiar groups in a larger area; this practice helps prevent later infanticide (Nogueira-Filho 1999).

FEEDING AND NUTRITION

Collared peccaries are largely frugivorous and also eat a wide variety of roots, tubers, greens, bulbs, and rhizomes (Kiltie 1981; Bodmer 1989; Sowls 1997; Fragoso 1999). In captivity they adapt easily to different kinds of food, including cassava,

cassava hulls, pumpkin, grain corn, sorghum silage, corn silage, crushed sugar cane, and commercial ration for pigs (Liva et al. 1989; Nogueira-Filho 1999).

Collared peccaries have a forestomach with active fermentation (Langer 1979; Cavalcante-Filho et al. 1998), which has given rise to considerable speculation regarding their ability to utilize coarse roughage by transforming the dietary fiber to usable volatile fatty acids (VFA) (Sowls 1997). Lochmiller et al. (1989) proposed that the VFAs found in collared peccary forestomachs arise from noncellulose components of the diet. Shively et al. (1984) concluded that fiber digestion in the collared peccary is considerably lower than in true ruminants. Strey and Brown (1989) suggested that collared peccaries might be concentrate selectors, selecting highly digestible forages and plant parts over forages with higher fiber contents. In other studies Gallagher, Varner, and Grant (1984) and Comizzoli et al. (1997) showed that collared peccaries could digest forage like true ruminants, while Nogueira-Filho (1990) determined that this species could digest even low quality roughage with high lignin contents and that peccaries could handle up to 30% of roughage in their diet. This characteristic should reduce farm expenses and the dependence on external products through the use of coarse products like grasses and leaves from farm and agroindustry by-products.

The protein demand for adult collared peccary maintenance is relatively low, around 0.82 g N/kg$^{0.75}$ a day, or 6.9% of crude protein (CP; Gallagher, Varner, and Grant 1984; Gary and Brown 1984). Part of this protein requirement could be furnished by nonnitrogen sources, such as urea (Leite et al. 2001). Daily digestible energy needs are 148.5 kcal/ kg$^{0.75}$ (Gallagher, Varner, and Grant 1984).

Growing young, on the other hand, have higher protein requirements. Different levels of crude protein (12, 15, or 18% CP) were tested in the diet of twelve collared peccaries in the initial phase of growth. Animals that received a higher protein ration (18% CP) showed superior weight gain and better feed:gain ratio (F:C; Nogueira-Filho et al. 1991). These results are preliminary, and more experiments are needed to determine the species' nutritional demands in the growth phase.

In tests with an experimental growth ration with 16% of crude protein and 4,100 kcal/kg of gross energy, an average daily weight gain of 80 g from 60 days to 120 days of age was achieved with 380 g/day of dry matter intake, resulting in a F:C of 4.7:1. Between 120 and 300 days of age, an average daily weight gain of 55.1 g was obtained with 634 g/day of dry matter intake, for a F:C of 11.5:1. Above ten months of age, when juveniles weighed 18.5 kg on average, the average daily gain weight was 19 g with a F:C of 21.7:1. These data were used to estimate a mean slaughter weight for collared peccaries of around 18.5 kg of animal live weight at ten months of age (Nogueira-Filho and Lavorenti 1997).

REPRODUCTION

From April 1986 through December 1991, forty-three young were born in twenty-six collared peccary litters at ESALQ/USP. Reproduction took place during the whole year, and gestation ranged from 140 to 148 days. Mean litter size was 1.6 (range 1–3), with a male:female ratio of 43:57. Mean birth weight was 710 g and 617 g for males and females, respectively. Females could reach sexual maturity at seven months, but on average females produced their first litter when they were 416 days old (± 88.9). The mean interbirth interval was 215.1 days (± 57.1) (Nogueira-Filho and Lavorenti 1997). López-Barbella (1993) established that collared peccaries can come into estrus at 46.0 ± 24.8 days postpartum (range from 14 to 92 days) and estimated 224.2 ± 63.8 days for the farrowing interval. However, he also found high fetal mortality and the possibility of estrus without ovulation or low fertility in the first postpartum estrus. Studies that provide data on the reproductive physiology of the collared peccary are necessary in order to improve management plans and productivity.

COLLARED PECCARY CAPTIVE -BREEDING SYSTEMS

INTENSIVE PRODUCTION SYSTEM

Currently only scientific breeders at universities or other research centers adopt the intensive production system in Brazil (e.g., Universidade Estadual de Santa Cruz, the Universidade Federal do Pará, and Centro de Multiplicação de Animais Silvestres, that is, CEMAS). In this breeding system family groups, comprising a male and up to three females, are maintained in small pens that range from 12 to 40 m². Alternatively, peccaries are maintained in small herds of up to fifteen individuals (maintaining a male to female ratio of 1: 4–5) in paddocks ranging from 100 to 600 m². Paddocks should contain a watering trough, feeders, and a sheltered area. Additional pens with 20 to 40 m² are required for growth of young.

Females introduced in each one of the reproduction pens must belong to the same family group in order to prevent fights and infanticide (Nogueira-Filho, Nogueira, and Sato 1999). The male could be from a different origin than the females, ensuring genetic diversity. Adult animals are fed grain corn or cassava at around 3.5% of live weight on a dry matter basis and ad libitum mineral salts. Pumpkins, bananas, or other fruits or foodstuffs available from farms are offered ad libitum. Meanwhile the young receive grain corn and a protein supplement (16% of crude protein) at 5% of live weight on a dry matter basis until they reach 12 kg of live weight. After that the young receive the same diet as adult animals.

Usually, if the peccary herds are maintained in small pens, females are isolated to farrow in an available free pen. In larger paddocks of at least 200 m², the females can farrow inside the confinement area. In both situations, infants will stay with the mother until they are sixty days old. Then they are weaned and transferred to a juvenile growth pen until they reach the weight/age of 18.5kg/10 month (Nogueira-Filho 1999).

SEMI-INTENSIVE PRODUCTION SYSTEM

Currently there are twenty-one collared peccary commercial breeders in Brazil, most of them using the semi-intensive production system. Some breeders use areas of their farms that cannot be used for traditional agriculture because of edaphic and/or topographic factors. Other farmers use parts of their legal protection areas. The semi-intensive production system thus provides an alternative for the integral use of the farms.

In semi-intensive production areas ranging from 1,000 to 50,000 m^2 and including arborescent vegetation are enclosed with wire fence on a concrete or rock base. These areas are used for farrowing females and growing young (Nogueira-Filho 1999). Peccaries are purchased from another breeder or captured in the wild in places in which they are considered agricultural pests and introduced into the paddock in a proportion of one male to four to five females. The captive group will increase through reproduction or by the introduction of new members from the wild or from captivity, up to a density of 250 m^2 per adult animal. At this density infanticide does not occur (Engel 1990). One or more trapping devices are built in the breeding area, and mineral salts, grain corn, cassava, or supplementary ration are supplied in sheltered feeders within the traps. The number and size of trapping devices and feeders depends on the number of groups put together inside the area (Nogueira-Filho 1999; Lima-Neto, Nogueira, and Nogueira-Filho 2001).

Normally, the animals are fed cassava or grain corn, supplemented with pumpkin, banana, other fruits, and protein supplements, as in the intensive system. Fruits produced by trees inside the area are not considered in the feeding plan. As in the case of capybara farming, this breeding system is considered semi-intensive because the group is maintained in a large area and because there is no individual control on reproductive parameters. The infants are ear-tagged at approximately sixty days of age. Weaning is natural and young grow up with the adults until they reach the weight/age of 18 kg/10 months, at which time they are designated for slaughter or for the formation of new reproduction groups.

The main problem with this production system is that, at densities of less than 250 m^2 per adult animal, peccary rooting damages natural vegetation and can provoke erosion, primarily near the capture areas (Nogueira-Filho et al. 1991). Such damage can be minimized with the use of mobile panel traps, like those used to trap and handle feral pigs in the U.S.A. (Sweitzer et al. 1997).

EXTENSIVE PRODUCTION SYSTEM

Collared peccary herds harm cassava and other crops in some areas of Brazil (e.g., southeastern Bahia) and are shot as agricultural pests (Nogueira-Filho and Nogueira 2000). As with capybara, ranching allows the use and control of the species and provides an alternative to destructive use of the land. However, to date there is no experience in Brazil with collared peccary ranching. To develop a rational ranching system, we require information on the ecology of natural populations. Currently, several research institutions, UESC, EMBRAPA-Pantanal, CIRAD-EMVT and DICE (Durrell Institute of Conservation Ecology, University of Kent), are studying peccary ecology in northeastern Brazil and in the Pantanal as part of the project Development of Different Production Systems for the Sustainable Exploitation of the Collared Peccary in Latin America, which is financed by the European Commission. Research on feeding ecology, habitat use, population size, reproductive productivity, and limiting and keystone resources during periods of low food production is aimed at determining sustainable levels of collared peccary harvest.

As with capybara IBAMA will allow ranching only on an experimental basis. All ranched animals must be identified with electronic microchips, and only young born after ranching is implemented will be slaughtered. Once a selective quota is determined based on local population density, individuals will be live-captured in order to allow selection of animals for slaughter. As in capybara ranching this production system requires specific husbandry practices, such as traps that capture multiple animals, thus reducing the costs of trapping and handling.

ECONOMIC ANALYSIS OF THE THREE COLLARED PECCARY PRODUCTION SYSTEMS

We carried out a hypothetical comparison, based on data obtained from commercial breeders, of the farm expenses and income of a beef cattle farmer who wants to breed collared peccaries using any of the three production systems. The farm has an available area with arborescent native vegetation and abandoned pig facilities comprising fourteen pens with 12 m^2 each and four pens with 24 m^2. In another part of the property, a group of twelve female and twelve male peccaries is harming a cassava crop meant for subsistence use by the farm workers.

In the intensive system the farmer will need to increase the height of the walls in the old pig facilities with wire mesh to prevent peccaries from escaping. Eight reproduction groups comprising one male and three females each will occupy eight of the 12-m^2 pens. The other six pens will be used for farrowing, while the four 24-m^2 pens will be used for juvenile growth. In the semi-intensive system the breeder will need to surround 7,500 m^2 with a 1.5-m high wire fence around the native vegetation to create a breeding group of six males and twenty-four females purchased from another breeder. At least two panel traps will be built to handle the animals. For ranching the breeder will need to build two panel traps, establish electrified fences around crops to avoid damage, furnish supplementary food to animals at the bait station, and increase the carrying capacity by enriching the habitat with supplementary foods, such as abandoned crops of corn, sweet potato, and cassava that peccaries can use throughout the year (table 11.4).

We used the intensive productivity data from Nogueira-Filho and Lavorenti (1997) and semi-intensive productivity data from Nogueira-Filho (1999) to estimate farm expenses and incomes from breeding systems (table 11.5). There are no available data on productivity of Brazilian collared peccaries through ranching; therefore we used the reproductive parameters in the wild from Bodmer (1999). Presently, there is no annual extraction quota established by IBAMA. Therefore we used a harvest level of 30% that was considered sustainable by Bodmer (1999).

TABLE 11.4 Initial Expenses for Maintenance of Twenty-four Female and Eight Male Adult Collared Peccaries in Intensive Breeding System, Twenty-four Females and Six Males in Semi-intensive System, and Twelve Females and Twelve Males in Extensive Breeding System

INITIAL EXPENSES (US$)	INTENSIVE	SEMI-INTENSIVE	EXTENSIVE
Consultant[a]	500.00	500.00	500.00
IBAMA[b]	100.00	100.00	100.00
Purchase of animals	1,600.00	1,500.00	—
Materials and labor for facilities building or adaptation	2,000.00	5,756.00	1,500.00
Equipment	400.00	200.00	200.00
Total	4,600.00	8,056.00	2,300.00

[a]The breeder needs to contract a professional consultant to develop the project.
[b]The breeder must pay an initial charge to IBAMA to legalize the project.

TABLE 11.5 Analysis of the Total Expenses (US$) (per kg) of Live Animal Weight of Collared Peccaries in Intensive, Semi-intensive, and Extensive Breeding Systems

EXPENSE ITEMS	INTENSIVE	%	SEMI-INTENSIVE (US$)	%	EXTENSIVE (US$)	%
Feeding[a]	1.47	42.5	0.94	39.7	0.37	15.9
Labor	0.47	13.5	0.13	5.5	0.50	21.5
Capital interests[b]	0.56	16.3	0.77	32.5	0.84	36.0
Fuel, electricity, and telephone	0.18	5.2	0.10	4.2	0.19	8.1
Veterinarian and medicines	0.36	10.4	0.21	8.8	0.19	8.1
Other expenses[c]	0.42	12.1	0.22	9.3	0.24	10.3
Total expense per kg of live animal weight	3.46	100	2.37	100	2.33	100

[a]US$ 0.08/kg of peccary ration
[b]Includes interests (6% per year) on total initial costs
[c]Includes expenses with interests (6% per year) on operational capital, taxes, and commercialization expenses

As is the case for capybaras, ranching yields the lowest financial return (table 11.6) because there will be expenses with all animals but economic return only from young born after ranching is implemented. Productivity is further limited by the peccaries' 1:1 sex ratio.

In spite of the apparently highest annual potential net income (table 11.6), if our

TABLE 11.6 Projection of the Production, Costs, and Annual Incomes for Twenty-four Female and Eight Male Adult Collared Peccaries in the Intensive Breeding System, Twenty-four Females and Six Males in the Semi-intensive Breeding System, and Twelve Females and Twelve Males in Extensive Breeding System

ANNUAL INDICES	INTENSIVE	SEMI-INTENSIVE	EXTENSIVE
Mean number of weaned young	27.6	51.5	23.0
Number of commercial animals sold[a]	24.4	47.5	14.0
Total weight of commercial animals (kg)	451.7	879.4	259.0
Total weight of discarded reproducers (kg)	100.5	88.5	—
Labor per kg of commercial animal produced (hours)	0.9	0.2	0.8
Kg of food consumed/kg of produced commercial animal[b]	18.4	11.7	4.6
Facilities and equipment expenses per female (US$)[c]	100.0	250.00	250.00
Hypothetical annual total production expenses (US$)	1,909.00	2,291.00	985.00
Hypothetical annual net income (US$)[d]	300.00	1,581.00	52.60

[a]In the extensive system we used an annual extraction quota of 30%.
[b]In the extensive system only supplementary food furnished to animals at the bait station was considered.
[c]Remember that for the intensive system only expenses with facilities adaptation were considered. If we considered building expenses (US$ 75 per m²), the facilities and equipment expenses per female in intensive system will increase to US$ 850.00.
[d]Selling the production at US$ 4.00/kg of animal live weight

hypothetical beef cattle farmer chooses the semi-intensive production system to produce collared peccaries, he will have some problems because of small-scale commercialization. Most peccary farmers sell their production to intermediaries who pay at most US$ 2.0 per kilogram of animal live weight, resulting in economic loss to the farmers. Thus farmers will need to market their production jointly to obtain better prices. They could also seek additional income through the industrialization of peccary leather.

CONCLUSION

Wildlife farming has been indicated as a possible source of animal protein for poor populations in developing countries because native mammals have specific advantages over domestic species (Lavorenti 1989; NRC 1991; González-Jiménez 1995). We suggest that wildlife farming should also be considered an alternative source of income for Brazilian farmers since wild animal meat is sold at high prices as an exotic product in Brazil's largest cities (Nogueira-Filho and Nogueira 2000).

Relative to species of domesticated livestock, wild animals may thrive as meat producers in their native habitat because they possess heat tolerance, resistance to local diseases and parasites, and overall hardiness and tolerance for poor nutritional conditions (Nogueira-Filho and Nogueira 2000). Another advantage is the lower environmental damage they cause relative to domestic animal production—consider the environmental pollution and deforestation caused by pig or cattle production, respectively (Nogueira-Filho 1996; Nogueira-Filho 1999).

Wildlife captive breeding, in either full captivity or semi-confinement, is an alternative use for unproductive areas of agricultural holdings, which for one reason or another are unfit for any other agricultural production. Also, in particular for Brazilian farmers, it is an economic alternative use for the legal protection area (Nogueira-Filho 1996, 1999). Semi-intensive captive breeding is not a high-tech or capital-intensive agrobusiness. However, it is not always practical because of the reality of small-scale commercialization. Therefore, farmers must become organized to sell their production jointly and directly to the consumers. Furthermore, the industrialization of subproducts (e.g., leather) will encourage more farmers to consider wildlife breeding as an alternative for the integral use of their property and as a way to diversify production.

On the other hand, the extensive system or ranching is frequently considered the preferred conservation/management tool because it gives farmers an alternative to conserve forests and wetlands by making wildlife a valuable resource. This activity could be especially beneficial in places where wild animals are considered very abundant or agricultural pests, as is the case for capybara in the Pantanal and collared peccaries in southeastern Bahia. Nevertheless, the projection of production, expenses, and annual incomes show that the ranching system for both capybara (table 11.3) and collared peccary (table 11.6) will only be economic feasible if harvest levels are increased. Such a change in policy will only be possible through studies of the ecology of natural populations, aimed at improving landscape-scale management to maximize carrying capacity. However, these studies are very expensive, and they must be developed in each and every place where the breeders intend to establish an extensive system.

Because of the high cost of the intensive system and the lack of knowledge in the extensive system, Brazilian farmers are choosing the semi-intensive production system to breed wild animals. Nevertheless, it is still necessary to improve husbandry practices to increase production and to lower costs in this system. We believe that the establishment of an efficient semi-intensive exploitation system of wild animals will help create an economic alternative for land use that in turn will help to conserve the natural ecosystem and its wildlife.

12

Economic Analysis of Wildlife Use
in the Peruvian Amazon

RICHARD E. BODMER, ETERSIT PEZO LOZANO,
AND TULA G. FANG

Successful conservation programs often depend on practitioners and researchers integrating the biological limitations of species with the social and economic realities of people (Barbier 1992). Indeed, there has been considerable dialog about the need to extend the reach of conservation biologists into the realms of social and economic analysis (McNeely 1988). The need to integrate biology with social and economic considerations is particularly relevant in tropical countries (Plotkin and Famolare 1992).

Conservation efforts must deal with all levels of wildlife use from local hunters to international trade. However, wildlife conservation efforts will only be successful if they focus on the level of wildlife use that ultimately influences hunting pressure. Economic analyses help determine what level is most critical. For example, there is debate whether conservation efforts should be directed at the level of national and international trade and their corresponding policies, or at the local level and community-based actions (Swanson 1992). This debate is particularly relevant in tropical countries where funds for wildlife conservation are limited. The scarce funding must be used efficiently to have an impact on conservation.

This article uses economic analyses of wildlife use in the Peruvian Amazon to help determine where conservation efforts should be directed. Should the focus be directed toward rural hunters who harvest wildlife, toward the national meat trade in urban markets, or toward the international pelt trade? Economic analyses also identify factors that drive overhunting and strategies that can be used to reduce overhunting of economically valuable species.

Wildlife and wildlife products in tropical forests are used for subsistence food, local meat markets, and national and international trade. Subsistence and commercial use of wildlife has traditionally been part of the economy in the Peruvian Amazon (Pinedo-Vasquez 1988; Dourojeanni 1990). Currently, wildlife is an important resource for the regional economy in terms of subsistence food, local meat sales, and international pelt exports. People from both the rural and urban sectors are involved with the commercial use of wildlife, with some uses being legal and others illegal. In addition, the peccary (*Tayassu pecari* and *T. tajacu*) pelt trade in the Peruvian Amazon is economically important both nationally and internationally (Bodmer et al. 1990).

Conserving wildlife in the Peruvian Amazon will have socioeconomic consequences because of the importance of wildlife to the rural sector, regional economy, and international trade. Restricting wildlife use will have negative economic consequences to the people who subsist on and commercialize wildlife products. Likewise, overhunting will result in negative economic returns and a loss of biodiversity. This paper is based on the premise that the well-managed use of wildlife will result in continued subsistence benefits and economic returns, and will contribute to biodiversity conservation (Freese 1997a).

BACKGROUND: LEGAL STATUS OF WILDLIFE HUNTING IN THE PERUVIAN AMAZON

Due to excessive hunting during the professional pelt period between 1940 and 1973, the Peruvian Ministry of Agriculture enacted a national management law in 1973 that prohibited professional pelt hunting in the Peruvian Amazon. This legislation permitted the use of certain wildlife species for subsistence by rural Amazonians (Bodmer 1994). Skins obtained from these species could be commercialized if the pelt originated from an animal killed by subsistence hunters. The law prohibiting professional pelt hunters was apparently successful in curbing the pelt trade (COREPASA 1986; Bodmer, Fang, and Moya 1988b).

Subsequently, in 1976 the Ministry of Agriculture noted increasing sales of wildlife meat in the city markets of Iquitos. To curb professional meat hunting, the Ministry of Agriculture enacted a management law in 1979 that restricted sale of wildlife meat to cities under 3,000 inhabitants. Again, only animals listed as sources of subsistence wildlife meat could be commercialized. While the professional meat law appears to have curbed hunting, its implementation was fraught with difficulty. Management authorities could not effectively control small unlicensed meat vendors in city markets. The demand of wild game meat from urban populations added to the problem of controlling meat sales. The 1979 law restricting wildlife commercialization in city markets has had little effect on actual wildlife meat sales, and such meat is currently openly sold in city markets.

More recently there have been a number of legislations in Peru dealing with wildlife use. Some promote captive breeding, while others are apply to hunting in timber and conservation concessions.

METHODS

Four representative sites in the Department of Loreto, northeastern Peru, were used to evaluate hunting by rural people (fig. 12.1). The major landscape features of

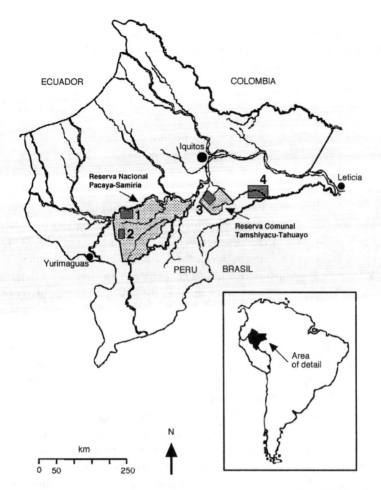

FIGURE 12.1 Map of Loreto, Peru, showing the four representative sites: flooded forest with heavy hunting (1); flooded forest with light hunting (2); upland forest with heavy hunting (3); and upland forest with light hunting (4). The map also shows the city of Iquitos.

Loreto include the seasonally flooded (*várzea*) forests and the nonflooded upland (*terra firme*) forest. Two representative sites were in seasonally flooded forests and two in upland forests. The two upland sites were located in and around the Reserva Comunal Tamshiyacu-Tahuayo in the forests that divide the Amazon and Yavari valleys (fig. 12.1). The principal forms of habitat in upland forest include vegetation on hill tops, in stream valleys, on inclines, and in backswamps.

The two flooded forest sites were in and around the Pacaya-Samiria National Reserve. Water level in the flooded forests of the Pacaya-Samiria National Reserve vary by approximately 11 meters between June and October. The principal forms of vegetation consists of two forested areas (high and low *restingas*), one area intermediate between forest and open habitat (*chavascal*), and palm and nonpalm swamps.

Data on hunting pressure were collected in the two flooded forest sites and the two upland forest sites. To cover the range of hunting pressures in Loreto, each forest type had one site with greater hunting (heavily hunted) and another site with lesser hunting (lightly hunted).

HUNTING PRESSURE

Information on hunting pressure for all mammalian species was obtained by involving hunters in data collection in all four representative areas, totaling approximately 100 hunters and their families. Studies were run for a minimum of three years at each site. Information was collected continuously by the hunters and recorded by wildlife extensionists every few months. Hunting pressure was determined by recording the number of animals hunted in each area from skulls collected by hunters and from hunting registers. An error margin was added to the calculation of hunting pressure to account for animals that were hunted but not recorded either by skulls or registers. The error was calculated by determining which local hunters were not participating in the study. The error margin varied between study sites.

The total number of mammals hunted annually in Loreto was estimated using wildlife harvests from the four representative sites and the annual peccary pelt harvests. We assumed that the annual peccary pelt harvests were a realistic estimate of the total number of peccaries harvested in Loreto, an assumption that appears valid (Pacheco 1983). The numbers of peccary pelts discarded by Peruvian hunters appears to be balanced by the number of pelts entering Peru illegally from Colombia, Ecuador, and Brazil. We used the proportional harvests in the four representative sites to estimate the annual harvests of other mammalian species solving PH/PP = SPH/SPL, where PH is the peccary harvest in each of the four representative sites, PP the peccary pelt exports from Loreto, SPH the harvest in each of the four representative sites of the species being estimated, and SPL the estimated annual harvest in Loreto of that species. The peccary pelt harvest in the Department of Loreto was estimated by government records of peccary pelt exports to Lima. This trade is closely monitored since peccary pelts exported out of Peru must have a Departmental certificate before being issued a CITES certificate.

ECONOMIC ANALYSIS

The economic analysis of wildlife use was done for both the rural and urban sectors. The rural sector consisted of the transactions of wildlife meat prior to being sold by market vendors. This sector includes hunters, carriers, and intermediaries. The urban sector consists of market vendors and urban consumers. Market vendors usually buy meat from carriers or intermediaries and then sell meat to the public in city markets.

To evaluate the economic value of wildlife meat in the rural areas, prices of meat sales in twenty towns and villages were determined from informal interviews at representative sites along the Rio Yavari, Rio Tigre, Rio Marañon, and Rio Amazonas. There are numerous ways that wildlife meat is transacted by the rural sector; these methods are outlined in Bendayán (1991). Hunters usually have to pay transportation costs and/or intermediaries in order to sell wildlife meat in Iquitos, the largest city in Loreto. The transportation and intermediary costs for hunters varies primarily depending on (a) distance to markets and (b) knowledge hunters have about markets. From informal interviews conducted at the four representative sites, we estimated that transportation and intermediary costs to hunters were between 40 to 80% of the income earned from meat sales, with an estimated average of 60%.

To analyze the economic importance of wildlife meat in the urban sector, we conducted a year-long survey of wildlife meat sales in the city markets of Iquitos, the capital of the Department of Loreto with a population of around 274,759 inhabitants (INEI 1993). For comparison, in rural Loreto, which covers an area of 368,851 km^2, there are around 412,523 inhabitants, of which 123,663 live in towns (INEI 1995). Wildlife meat sales were surveyed in the Mercado Belén and Mercado Modelo in Iquitos. Bendayán (1991) showed that 91% of meat sales in Iquitos took place at these two markets, with the other markets in Iquitos only selling small amounts. These sales were recorded from January through December 1996 for all months expect May. Sales for May were extrapolated from the averages of April and June.

Surveys were conducted through interviews with market vendors. All market vendors were surveyed between 6 and 11 A.M. daily Monday through Saturday. Markets were closed on Sundays. During some months surveys were not conducted every day. During these months the averages from days surveyed were used for days not surveyed.

Market vendors participated willingly in the study. They were informed of the purpose of the study and that their names would be kept confidential. During the year of the survey there were no decommissions of wildlife meat from the markets, and market vendors did not fear loosing their products.

Information obtained during the surveys was recorded separately for each species and included the kilograms sold daily and the prices of sales. Prices were determined for meat bought and for meat sold in the markets of Iquitos. Meat bought represents the price paid to the rural sector, including the hunters, carriers, and middlemen. The difference between the price bought and sold represents the profit to market vendors. The number of vendors selling wildlife meat during the day was also recorded.

The number of individual animals sold at market was estimated by converting the kilogram of meat sold to individuals. Wildlife meat is sold in four distinct forms in Iquitos: *fresco* (fresh), *ahumado* (smoked), *seco salado* (dry salted), and *fresco salado* (fresh salted). Dried salted and smoked meat had a conversion of 40% of live weight, whereas fresh and fresh salted had a conversion of 60% of live weight. Average body weights were used for each species and the total number of individuals of each species sold annually in Iquitos was then estimated.

RESULTS

HUNTING PRESSURE

The site with greatest annual hunting pressure was the heavily hunted upland forest (255 individuals/100 km^2), followed by the heavily hunted flooded forest (133 individuals/100 km^2). The flooded forest with lighter hunting had slightly more mammals harvested per year than the upland forest with lighter hunting (73 and 54, respectively). In both upland forests and the lightly hunted flooded forest, ungulates were the most frequently hunted group (Bodmer and Pezo 2001). In the heavily hunted flooded forest, rodents were the most frequently hunted group.

Ungulates had much greater amounts of biomass extracted than the other species in all four sites and were the most important species for meat. In terms of biomass extracted, rodents were the next most important, followed by primates, marsupials, edentates, and carnivores. White-lipped peccary, collared peccary, and lowland tapir (*Tapirus terrestris*) had the greatest amount of meat extracted from the representative forest sites (Bodmer and Pezo 2001).

ECONOMICS OF WILDLIFE FOR THE RURAL SECTOR OF LORETO

There are an estimated 113,000 mammals hunted in Loreto annually. The average annual white-lipped peccary harvests are 14,000; collared peccary, 20,000; brocket deer (*Mazama* spp.), 5,000; lowland tapir, 4,000; paca (*Agouti paca*), 17,000; agouti (*Dasypracta fuliginosa*), 10,000; and primates, 28,000. Using these estimates, the value of wildlife meat for the rural sector is estimated at US$ 1,131,910 annually (Bodmer and Pezo 2001). Lowland tapir has the greatest annual meat value for the rural sector of US$ 291,235, followed by collared peccary at US$ 268,853, white-lipped peccary at US$ 237,512, primates at US$ 104,617, and paca at US$ 75,447.

THE URBAN MEAT MARKET

The number of individuals of each species annually sold in the Iquitos meat markets was estimated from the market surveys. A total of sixteen mammal species were sold during the study. The greatest number of individuals sold in the Iquitos market were from collared peccary with an estimated 2,542 individuals sold annually, followed by white-lipped peccary with 2,316 individuals, and paca with 1,860 individuals (table 12.1). There were fewer individuals of the other large wildlife species sold in the Iquitos meat markets with an estimate of only 232 red brocket deer (*M. americana*), 110 wooly monkeys (*Lagothrix lagothricha*), 76 grey brocket deer (*M. govazoubira*), 76 capybara (*Hydrochaeris hydrochaeris*), and 43 lowland tapir.

An estimated 72,972 kg of wildlife meat was sold in the Iquitos meat markets during 1996. By weight, the most frequently sold meat was white-lipped peccary, contributing 42.6% of all wildlife meat sales. The next most frequently sold was col-

TABLE 12.1 Estimated Number of Individual Mammals Sold in the Iquitos Markets During 1996

SPECIES	FRESH MEAT KG (INDIVIDUALS)	FRESH SALTED KG (INDIVIDUALS)	SMOKED KG (INDIVIDUALS)	DRY SALTED KG (INDIVIDUALS)	BODY WEIGHT KG	NUMBER OF INDIVIDUALS
Tayassu pecari	338 (17)	207 (10)	1,925 (146)	28,295 (2,143)	33	2,316
Tayassu tajacu	461 (31)	149 (10)	2,277 (228)	22,735 (2,273)	25	2,542
Mazama americana	341 (17)	57 (3)	428 (32)	2,373 (180)	33	232
Mazama gouazoubira	75 (8)	42 (5)	79 (13)	300 (50)	15	76
Tapirus terrestris	110 (1)	208 (2)	83 (1)	2,540 (40)	160	43
Agouti paca	1,115 (206)	162 (30)	2,354 (654)	3,491 (970)	9	1,860
H. hydrochaeris	560 (31)	354 (20)	165 (14)	134 (11)	30	76
Dasyprocta fuliginosa	18 (6)	8 (3)	5 (2)	101 (51)	5	62
Lagothrix lagothricha	0	15 (2)	25 (6)	451 (102)	11	110
Ateles spp.	3 (1)	0	0	13 (4)	7.8	5
Alouatta seniculus	0	0	0	8 (2)	8	2
Dasypus novemcinctus	45 (15)	13 (4)	36 (18)	17 (9)	5	46
Didelphis marsupialis	0	5 (8)	0	0	1	8
Tamandua tetradactyla	5 (2)	7 (2)	0	0	5	4
Nasua nasua	0	0	0	10 (8)	3	8
Potos flavus	0	0	0	3 (2)	3	2
Total						7,392

Note: Number of individuals was estimated from (meat factor) (body weight). A factor of 2.5 or +0% of live weight was used for dried salted and smoked meat, and 1.66 or 60% for fresh and fresh salted meat (Bendayán 1991).

lared peccary, contributing 35.8% of sales. Combined, peccaries dominated the wildlife meat market contributing 78.4% of all sales. Other species that were commonly sold included paca (9.5%), red brocket deer (4.4%), lowland tapir (4%), and capybara (1.8%). Other species contributed less than 1% of wildlife meat sales.

Fresh meat made up 4.2% of wildlife meat sales; smoked, 10.2%; dry salted, 83.8%; and fresh salted, 1.7%. Paca and capybara were the species most frequently sold as fresh meat and paca, collared peccary, and white-lipped peccary as smoked meat. The two peccary species dominated the dry salted sales, and capybara was the species most frequently sold as fresh salted.

Total profits for the rural sector from meat sales in Iquitos were estimated at US$ 156,040 annually. Of this amount, approximately US$ 93,624 were profits for hunters, with the remainder going to carriers and intermediaries. The white-lipped peccary brought the greatest profits to the rural sector with an estimated earnings of US$ 69,206, followed by collared peccary at US$ 57,694, paca at US$ 14,464, brocket deer at US$ 8,080, and lowland tapir at US$ 4,493 (table 12.2). In terms of meat types the greatest profit to the rural sector were from the sale of dried salted meat, which had an annual earning of US$ 132,439, followed by smoked meat at US$ 17,700, fresh meat at US$ 4,649, and fresh salted at US$ 1,252.

Total profits to the urban meat vendors were estimated at US$ 94,228 annually. There were twenty meat vendors in the two markets surveyed, resulting in an aver-

TABLE 12.2 Annual Profits and Consumer Values of Wildlife Meat Sales in the Markets of Iquitos

SPECIES	PROFIT FOR RURAL SECTOR	PROFIT FOR URBAN SECTOR	CONSUMER VALUE
Tayassu pecari	69,206	37,837	107,043
Tayassu tajacu	57,694	32,382	90,076
Tapirus terrestris	4,493	3,022	7,515
Mazama americana	7,021	4,375	11,396
Mazama gouazoubira	1,059	721	1,780
Agouti paca	14,464	13,795	28,259
Dasyprocta fuliginosa	121	220	341
Hydrochaeris hydrochaeris	993	1,001	1,994
Dasypus novemcinctus	166	219	385
Lagothrix lagothricha	789	630	1,419
Ateles spp.	21	16	37
Alouatta seniculus	13	10	23
Total	156,040	94,228	250,268

Note: Profits for the rural sector include hunters, carriers, and middlemen. Profits for the urban sector include the market vendors. Consumer value or total value is the monetary value consumers paid for wildlife meat. All values are in $US.

age annual income per vendor at US$ 4,711 or US$ 393 per month. White-lipped peccary brought the greatest annual profits for the market vendors at US$ 37,837, followed by collared peccary at US$ 32,382, paca at US$ 13,795, brocket deer at US$ 5,096, and lowland tapir at US$ 3,022. The greatest profit to the market vendors was from the sale of dried salted meat, which had an annual earning of US$ 71,024, followed by smoked meat at US$ 14,736, fresh meat at US$ 6,885, and fresh salted meat at US$ 1,583.

Consumers spent US$ 250,268 annually on wildlife meat in the Iquitos markets. This amount reflects the total value of wildlife meat sales in Iquitos. White-lipped peccary had the greatest value, followed by collared peccary, paca, brocket deer, and tapir. Dried salted meat had the greatest total value followed by smoked meat, fresh meat, and fresh salted meat.

The amount of wildlife meat sold in Iquitos has increased over threefold during the ten-year period from 1986 to 1996, amounting to an increase of 48,911 kg of meat (1986 survey of the markets is from Bendayán 1991) (table 12.3). The largest increase in meat sales was for white-lipped peccary, which has increased by 27,447 kg/year, followed by collared peccary by 14,888 kg/year, brocket deer by 2,468 kg/year, and paca by 2,047 kg/year.

COMPARING THE RURAL AND URBAN SECTORS

Wildlife harvests in the rural sector were compared to wildlife sold in Iquitos. Overall, only 6.5% of the mammals hunted in Loreto were sold in the markets of Iquitos (table 12.4). The remainder was used in the rural sector as subsistence food or was

TABLE 12.3 Differences in the Sale of Wildlife Meat in the Markets of Iquitos Over a Ten-year Period

SPECIES OR SPECIES GROUP	KG SOLD IN 1986	KG SOLD IN 1996	DIFFERENCE
Tayassu pecari	3,654	31,101	+27,447
Tayassu tajacu	11,211	26,099	+14,888
Mazama spp.	1,305	3,773	+2,468
Tapirus terrestris	1,584	2,905	+1,321
Agouti paca	4,855	6,902	+2,047
Dasyprocta fuliginosa	298	130	−168
Hydrochaeris hydrochaeris	572	1,332	+760
Primates	315	536	+221
Marsupials and Edentates	180	22	−158
Carnivores	36	15	−21
Total	24,010	72,921	+48,911

TABLE 12.4 Percent of Mammals Hunted in Loreto That Are Sold in Iquitos Markets Versus the Percent Used in the Rural Sector for Subsistence Food or for Sale in Villages and Towns

SPECIES OR SPECIES GROUPS	PERCENT OF HARVEST SOLD IN IQUITOS	PERCENT OF HARVEST USED FOR SUBSISTENCE OR SOLD IN RURAL AREAS
Tayassu pecari	16	84
Tayassu tajacu	13	87
Mazama spp.	6	94
Tapirus terrestris	1	99
Agouti paca	11	89
Dasyprocta fuliginosa	0.6	99.4
H. hydrochaeris	4	96
Primates	0.4	99.6
Marsupials and Edentates	1	99
Carnivores	0.1	99.9
Total	6.5	93.5

sold in villages and towns. White-lipped peccary had the greatest proportion of its harvest sold in Iquitos with 16% going to market, followed by collared peccary with 13%, paca with 11%, and brocket deer with 6%. The lowland tapir had only 1% of its harvest sold in Iquitos, while primates only had 0.4%.

SUPPLY AND DEMAND OF MEAT TYPES

The type of meat sold was used to examine supply and demand relationships. For hunters there was no supply and demand relationship between the price paid to the rural sector and the supply of meat types (fig. 12.2a). In contrast, market vendors were driven by supply and demand of meat types (fig. 12.2b). Fresh and smoked meat has a small supply and prices sold to consumers were high. On the other hand, supply of dried salted meat was large, and prices sold to consumers were lower.

Hunters and market vendors see the economics of wild game meat from different perspectives. Hunters are concerned with the number of animals killed in terms of individuals because animals are hunted as individuals. In contrast, market vendors are interested in the kilogram of meat and the type of meat since their earnings depend on the amount of meat bought and sold.

For example, the price paid to hunters for peccary and deer meat is about the same in terms of the kilogram per live animal whether it is fresh, smoked, or dried salted. This relationship is important for the hunters since the price paid for an animal hunted is about the same. Therefore hunters consider effort and logistics involved in preparing the type of meat they sell. Fresh meat has little effort in terms

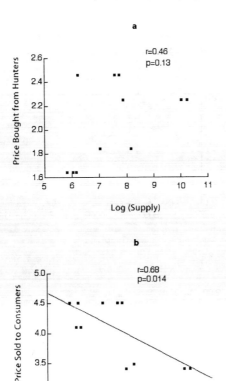

FIGURE 12.2 Supply and demand relationship of meat bought and sold in the Iquitos meat markets.

of preparation but can only be sold if the animals are hunted close to the city. Smoked meat has little financial input but takes much effort in time required to smoke meat and maintain it in a sellable condition. Dried salted meat takes financial input for salt but much less effort in time spent preparing meat. Also, salted meat usually lasts longer than smoked.

Most peccary and deer meat is prepared as dried salted since these species are rare in forests close to the city and are usually hunted in remote areas at distances too far from the city to sell as fresh meat. Hunters receive considerably more money for fresh paca than for smoked or dried salted paca. Thus, hunters prefer to sell fresh paca meat. Paca can still be hunted in areas close enough to the city to be sold fresh.

SEASONAL CHANGES IN SUPPLY AND DEMAND

Supply and demand relationships can also be examined between low- and high-water seasons for the sale of peccary meat in Iquitos. Substantially more peccary meat

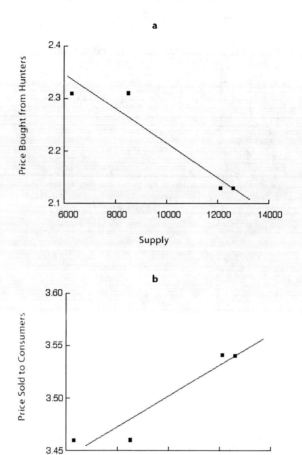

FIGURE 12.3 Supply and demand relationships of peccary meat bought and sold in the Iquitos meat markets between different seasons.

was sold in Iquitos during the high-water season, with 24,719 kg sold during high water and 14,796 kg sold during low water.

The seasonal changes in the prices paid to the rural sector for peccary meat appear to be driven by supply and demand relationships. When the supply of peccary meat decreases, the price paid to the rural sector increases, and when the supply increases, the price paid to the rural sector decreases (fig. 12.3a).

In contrast, the urban sector does not show this supply and demand relationship. As the supply of peccary meat increases, the price sold to the consumer also increases, and when the supply of peccary meat decreases, the price sold to the consumer also decreases (fig. 12.3b). What appears to be driving the seasonal price paid by the consumer for peccary meat is the total amount of meat products available at market. During the dry season when peccary meat supplies are low, the total supply of meat products is at its greatest because the supply of fish is at its highest. During the flooded season the supply of peccary meat is at its greatest, but the total supply of meat products is at its lowest because of a decrease in the supply of fish.

VALUE OF THE PECCARY PELT TRADE

Pelts from species listed as subsistence wildlife can be legally exported for commercial profits from Peru if the species is not listed on Appendix I of CITES. However, in Peru peccary pelts are the only legally exported mammalian pelt that has any significant economic importance. Peccary pelts are sold principally to West Germany, Italy, Japan, and the United States (table 12.5). Peccary leather is used mainly for gloves, shoes, belts, and watchbands. The peccary pelt market is relatively stable because peccary products have traditionally been used by continental Europeans. Peccary leather is prized for its softness and durability, two qualities that are rarely found in a single leather.

The peccary pelt trade has economic value to rural hunters. Rural hunters in Loreto usually sell peccary pelts between US$ 2 to US$ 5 and obtain approximately US$ 74,500 annually. However, the total value of peccary meat for hunters is around US$ 633,265 annually, including the value of both subsistence food and market sales. Thus, pelts only contribute around 11% of the total value of peccaries for rural hunters.

Peccary pelts are bought by intermediaries who air cargo the skins to tanneries in Lima and Arequipa. The tanneries carry out the initial tanning process, which involves rehumidifying, degreasing, and chrome tanning. Peccary hides are then exported at approximately US$ 16 each. The national intermediaries and tanneries have substantial costs because of air transportation, tannery equipment, supplies, labor, government taxes, and CITES permits. Profits for the pelts are approximate-

TABLE 12.5 Exports of Collared and White-lipped Peccaries from Peru

COUNTRY	PRODUCT	COLLARED PECCARY	WHITE-LIPPED PECCARY
Germany	Skins	163,175	83,572
	Gloves	449	1,540
Italy	Skins	69,239	38,693
Japan	Skins	9,822	6,080
USA	Skins	615	200
	Specimens	20	15
Uruguay	Skins	256	64
France	Gloves	150	—
Total		243,726	130,164

ly US$ 4 to US$ 7 per hide, resulting in total profits to the national peccary pelt industry of around US$ 187,000 annually and a total value of US$ 544,000.

A substantial number of peccary pelts are exported to Germany, Italy, and Japan each year. Peccary is an expensive leather, with a pair of peccary gloves currently retailing in Europe at around US$ 125 a pair. A single pelt is usually used for a pair of gloves. There are substantial costs involved in preparing a pair of gloves. These costs include the costs of finishing the hides, cutting and sewing the gloves, transporting the hides and finished products, and the costs of CITES permits and customs. The net value of a peccary pelt is estimated at around US$ 40 after subtracting all the costs. Profits made by the international leather industry on peccary pelts from Loreto are around US$ 1,360,000 annually, with a total value of around US$ 4,250,000.

Combined, profits of US$ 1,621,500 are earned annually from peccary pelt sales in Loreto, with 5% earned by rural hunters, 12% by the national pelt industry, and 83% by the international leather industry. The total value of the peccary pelt trade is estimated at US$ 4,868,500, with 1.5% contributed by the rural sector, 11.1% by the national pelt industry, and 87.3% by the international leather industry.

TOTAL ECONOMIC VALUE OF THE WILDLIFE HARVESTS IN LORETO

The total annual economic value of wildlife harvests from Loreto is estimated at around US$ 6,250,678 for all three socioeconomic sectors combined; rural, urban and international (table 12.6). The rural sector obtains an estimated 21.8% of the to-

TABLE 12.6 Total Annual Economic Value of Wildlife Harvests in Loreto, Peru, by Socioeconomic Sector

SOCIOECONOMIC SECTOR	ECONOMIC VALUE	PERCENT
Rural sector		
Subsistence meat	1,131,910	18.1
City market meat sales	156,040	2.5
Peccary pelt sales	74,500	1.2
Subtotal	1,362,450	21.8
Urban sector		
City market meat sales	94,228	1.5
Peccary pelt sales	544,000	8.7
Subtotal	638,228	
European sector		
Peccary pelt sales	4,250,000	68.0
Total	6,250,678	100

Note: All values are in $US.

tal value, the urban sector in Peru 10.2% and the international sector obtains the greatest proportion at 68%.

The meat markets in Iquitos only account for 3% of the total estimated value of wildlife harvests. Subsistence food and sales in small villages and towns in the rural sector account for 18.1% of the total estimated value. The greatest value is for the peccary pelt trade, which accounts for 79%. However, most of the value of the peccary pelt trade is in the international sector, not in Peru.

DISCUSSION

Economic value of different wildlife species varies considerably between the rural and urban sectors. In the Iquitos markets the lowland tapir and primates made up very little of the annual value. In contrast, in the rural sector, lowland tapir and primates are very important sources of wildlife meat.

Results from this study clearly show the importance of the rural sector in the use of wildlife meat. The results also show that the Iquitos wildlife meat market only consumes a small part of the wildlife harvests in Loreto. Totally prohibiting the wildlife meat markets in Iquitos would only result in a 6.5% reduction of the total harvests of Loreto. Legalizing the wildlife meat markets might promote a further uncontrolled and unmanaged hunt, which would lead to greater overharvesting and more local extinctions. In Iquitos there has been a threefold increase in wildlife meat sales over the past ten years.

These results show the importance of managing wildlife hunting with a focus on the rural sector, not the Iquitos meat markets. Management programs directed at hunters, villages, and towns of rural Loreto are imperative for the success of wildlife management in the Peruvian Amazon.

Management of wildlife in Loreto must reduce overhunting. Previous studies in the four representative areas of Loreto have shown that primates and lowland tapir are usually overhunted, while peccaries, deer, and large rodents are usually not (Bodmer, Eisenberg, and Redford 1997). Therefore wildlife management programs need to reduce the hunting of primates and lowland tapir and maintain a sustainable harvest of peccaries, deer, and large rodents.

There will be short-term economic costs if overhunting is reduced in Loreto. These economic costs can be estimated by examining what happens if the hunting of primates, lowland tapirs, carnivores, edentates, and marsupials is stopped, and the current harvest levels of peccaries, deer, and large rodents are maintained (Bodmer and Pezo 2001).

This management approach would have only minor economic costs to the meat markets in Iquitos, with these costs estimated at 3.6% of the economic value of the meat markets. In contrast, the short-term economic costs to the rural sector would be significant, with the rural sector having a 36.4% decrease in the economic benefits from wildlife hunting, or an annual loss of US$ 412,978 (Bodmer and Lozano 2001).

If management programs are not set up, there will be further overhunting and an increase in local extinctions. It is likely that these local extinctions would result in species extinctions and an overall loss of biodiversity to Peru. Thus, the economic costs to the rural sector must be weighed against the biodiversity loss to Peru.

ECONOMICS OF THE CURRENT PELT TRADE

Professional pelt hunting is prohibited in Peru, and peccary pelts exported from Loreto should only be collected by subsistence hunters. Pelt hunters obtained a relatively good income from peccary skins in the 1950s, 1960s, and early 1970s, and peccary pelt exports from the Peruvian Amazon exceeded 200,000 skins/year (Grimwood 1969). Since the pelt trade has become both less lucrative for hunters and more strictly controlled, exports have fallen to the current level of around 34,000 skins/year.

Peccary pelts from the Peruvian Amazon are usually of poor quality and can not be used for such large leather products as jackets but only for such smaller products as gloves, shoes, belts, and watchbands. The poor quality of peccary pelts results from a combination of causes that include (a) epidermal parasites, especially ticks, infecting free-ranging animals; (b) scars from intraspecific aggression; (c) shot holes produced by the 16-gauge shotguns used by most rural hunters; (d) cuts and holes caused by the rough skinning by rural hunters; (e) blisters caused by drying skins in direct sunlight; and (f) mold and pest damage caused by storing pelts in the humid Amazonian climate.

Rural hunters, however, have little interest in improving their processing techniques because the price paid to hunters for pelts does not make it profitable for them to improve their methods. The international tanneries and leather manufactures, on the other hand, are very interested in getting better quality pelts.

Increasing the price paid to hunters for pelts as a strategy to increase the quality of pelts might add value for hunters and lead to improved pelt quality. However, this increase can only be done sustainably if peccary hunting is well managed. If pelt prices are increased without improved management programs, then hunting pressure on peccaries might exceed sustainable levels and cause overhunting of peccary populations. On the other hand, if added value for peccary pelts is provided to hunters who manage their hunting practices, especially through community-based approaches, this added value could be an incentive for better wildlife management practices. Indeed, added value for peccary pelts could act as a broad incentive for community-based wildlife management.

HUNTING PRESSURE AND ECONOMIC VALUE

In order to focus wildlife management programs, it is imperative to understand what factors are influencing hunting pressures. In the case of the Peruvian Amazon, there is a clear disjunct between the realized hunting pressure and the economic value of wildlife. Hunters are harvesting species primarily for the value of subsistence meat, secondarily for the value of meat sales in the urban markets, and lastly for the value of peccary hides. In contrast, the greatest value of the wildlife is in the international market of peccary hides. Still, the international pelt trade is not a major influence in determining hunting pressure of wildlife in the Peruvian Amazon. Likewise, the urban meat market is influencing hunting pressure to a much lesser degree than the use of wildlife meat in the rural sector.

The success of conservation efforts in the Peruvian Amazon will depend on the success of working with the rural hunters. This effort includes the sale of wildlife meat in villages and towns. One promising management strategy that focuses on wildlife management with rural hunters is community-based wildlife management (Bodmer and Puertas 2000). Community-based management is focused at the level that can actually influence hunting pressure.

ACKNOWLEDGMENTS

We are indebted to the tremendous support provided by the communities of the Reserva Comunal Tamshiyacu-Tahuayo and the Reserva Nacional Pacaya-Samiria; to Pablo Puertas, Rolando Aquino, and César Reyes who helped with the fieldwork; and to Julio Curinuqui and Gilberto Asipali for their dedicated field assistance. We thank the Wildlife Conservation Society, the Chicago Zoological Society, Instituto Nacional de Recursos Naturales-Peru, the Universidad Nacional de la Amazonía Peruana, The Nature Conservancy, the Mellon Foundation, ProNaturaleza, AIF-WWF/DK Programa Integral de Desarrollo y Conservación Pacaya Samiria, and US-AID-Peru for logistical and financial support of the project.

PART III

Fragmentation and Other Nonharvest
Human Impacts

13

Mammalian Densities and Species Extinctions in Atlantic Forest Fragments

THE NEED FOR POPULATION MANAGEMENT

LAURY CULLEN, JR., RICHARD E. BODMER, CLAUDIO VALLADARES-PADUA, AND JONATHAN D. BALLOU

The Brazilian Atlantic Forest (Mata Atlântica) is one of the most threatened ecosystems on earth, currently at risk of large-scale destruction. The forests in this ecosystem have been fragmented and reduced to about 7% of their original extent (SOS Mata Atlântica and INPE 1993). The Mata Atlântica also harbors one of the greatest levels of biological diversity in the world, containing nearly 7% of the world's species, many of which are endemic to this region and threatened with extinction (Quintela 1990).

Currently, most of the remaining forest cover in the Mata Atlântica is found on the hillsides along the coast. Very little forest remains in the interior region because agricultural and industrial expansion has resulted in the loss of more than 98% of these forests. As a consequence of deforestation, most of the remaining interior forests are scattered in a mosaic of forest fragments. Today a combined area of only about 280,000 ha of these forests remains, and nearly all of the interior forests that still exist are found in the Pontal do Paranapanema region located in the western part of the state of São Paulo. This region alone comprises 84% of the remaining interior forest cover and is considered one of the poorest and most underdeveloped areas of the state (SOS Mata Atlântica and INPE 1993).

In landscapes dominated by humans, one of the major challenges for conservation biologists is remedying the long-term deleterious consequences of population fragmentation and extinctions. This is the scenario in the interior region, where patches of the original habitat are increasingly being encroached upon by new human settlements, and plant and animal populations are being extirpated by illegal timber extraction and poaching. In addition to selective logging and poaching, genetic, demographic, and environmental forces can harm small and isolated populations (Soulé and Wilcox 1980; Ralls and Ballou 1983; Soulé 1987; Malcolm and Ray 2000; Beissinger and McCullough 2002). For their long-term persistence, isolated populations will most probably require monitoring, effective law enforcement, protection, and management (Gibbs, Snell, and Causton 1999). For protection and management, information on population size and long-term viability is essential.

Few studies have analyzed density and population size of mammals in remnants of the Brazilian Atlantic Forest (Chiarello 2000). In this study we estimate population sizes for some medium- and large-bodied mammalian species living in different forest remnants of the Atlantic Forest in the state of São Paulo. With this information we discuss the viability and the need for management of these mammalian populations. We also analyze the sudden extinctions observed among white-lipped peccaries (*Tayassu pecari*) and tapirs (*Tapirus terrestris*) in some heavily hunted sites and discuss the possible long-term consequences of their absence on the interior forest remnants. We conclude by recommending some research priorities that should contribute to the future of wildlife conservation in the Brazilian Atlantic Forest.

METHODS

STUDY SITES

During the period of June 1995 through December 1996, five forest fragments were studied within the interior forests range. Three of the sites are in the western part of the state of São Paulo (Fazenda Tucano, Fazenda Mosquito, and Morro do Diabo State Park) (fig. 13.1), and two sites are in the central part of the state (Fazenda Rio Claro and Caetetus Ecological Station). The greatest distance between sites is 300 km. Sites were categorized as slightly, moderately, and heavily hunted based on several indicators of human activity (Cullen, Bodmer, and Valladares-Padua 2000; table 13.1). Climatic descriptions of the region can be found in Valladares-Padua (1987) and geomorphologic descriptions in Setzer (1949).

Most of the interior forests are considered a transitional ecosystem, bordered in the east by the Tropical Evergreen Broadleaf Forest, which originally covered most of the Atlantic coastline (Eiten 1974; Ab'Saber 1977; Alonso 1977). At the other extreme, most of the western and northern range of the interior forests are bordered by the dry *cerrado* vegetation of Mato Grosso do Sul state and northern São Paulo state. Cerrado is a "tall dense semideciduous xeromorphic savanna vegetation" (Redford 1983:126).

Morro do Diabo State Park is located right on the edge of the cerrado, and accordingly, the best classification of the park's forest would be an "upland semideciduous Atlantic Forest interspersed with some areas of Cerradão" (Baitello et al. 1988; cited by Valladares-Padua 1993). The region is characterized by a pronounced dry season: the park annually receives an average of 1,131 mm of rain, of which 30% falls between April and September (Valladares-Padua 1993). Most of the emergent trees lose their leaves during the dry months (Hueck 1972). The region is also known for its generally nutrient-poor sandy soils (Setzer 1949).

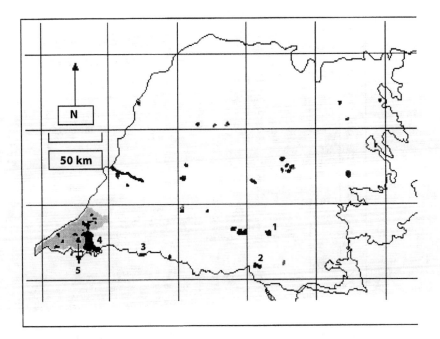

FIGURE 13.1 Map of forest fragments in the Plateau region of São Paulo state, Brazil. The sites used in this study are Caetetus Ecological Station (1), Fazenda Rio Claro (2), Fazenda Mosquito (3), Morro do Diabo State Park (4), and Fazenda Tucano (5). Shaded area is the Pontal do Parana-panema Region.

DENSITY ESTIMATES

Abundances of species were determined by censuses along line transects (Emmons 1984; Caro 1999; Cullen, Bodmer, and Valladares-Padua 2000). Four to eight transects were established in each forest fragment, away from hunting trails and in such a way as to incorporate the known diversity of landscape features (Cullen, Bodmer, and Valladares-Padua 2000). Transects ranged from 0.5 to 8 km in length. The cumulative distance censused at each fragment ranged from 161 to 618 km (mean = 381), with a total of 2,287 km censused. Transects were censused between 6 and 10 A.M. and again between 3 and 7 P.M. Sampling effort was greatest for diurnal species but also included crepuscular and some nocturnal animals.

Transects were walked slowly (approximately 1 km/hour following Emmons, 1984) by an experienced biologist or park ranger, with brief stops every 50 to 100 m (Cullen, Bodmer, and Valladares-Padua 2000). The time, species, location along the trail, group size, and perpendicular distance from the trail were recorded for each sighting. Observations were ended after a maximum of 15 minutes, and only accurate counts were used to estimate average group size. For social species the distance was recorded to the closest individual. Densities were estimated using the

TABLE 13.1 Estimates of Group Size

SITES		M. D. S.PARK		C. E. STATION		FAZ. MOSQUITO		FAZ. TUC 95		FAZ. TUC. 96		FAZ. R. CLARO	
Hunting Intensity		Unhunted		Slight		Moderate		Heavy		Heavy		Heavy	
		N/SD	Min. Avg. Max.	N/SD	Min. Avg. Max.	N/SD	Min. Avg. Max.	N/SD	Min. Avg. Max.	N/SD	Min. Avg. Max.	N/SD	Min. Avg. Max.
Tayassu pecari	Min	(4)	38.00	(3)	30.00	(3)	5.00	(3)	1.00	EXT	EXT	EXT	EXT
	Avg	4.89	42.00	7.50	37.50	8.00	12.00	3.50	4.30				
	Max		48.00		45.00		20.00		8.00				
Tayassu tajacu	Min	(6)	6.00	(3)	4.00	(4)	4.00	(5)	1.00	(4)	2.00	(8)	2.00
	Avg	14.48	16.50	5.90	13.50			5.40	6.40	2.10	4.00	1.70	6.37
	Max		43.00		15.00				14.00		7.00		7.00
Sciurus aestuans	Min	(1)	1.00	(26)	1.00	(34)	1.00	(2)	1.00	(1)	1.00	(5)	1.00
	Avg			0.40	1.80	1.83	1.64	1.41	2.00			0.54	1.40
	Max				2.00		11.00		3.00				2.00
Nasua nasua	Min	(1)	6.00	(12)	1.00	(5)	1.00	(3)	1.00	(4)	1.00	(5)	1.00
	Avg			4.70	6.30*	3.50	5.80*	4.10	5.70*	2.10	4.00*	5.50	5.40*
	Max				11.00		14.00		9.00		6.00		14.00
Cebus apella	Min	(24)	1.00	(25)	1.00	(44)	1.00	(7)	1.00	(9)	1.00	(15)	1.00
	Avg	4.23	10.20	3.30	7.40*	3.30	5.95*	4.90	8.50*	3.70	8.00*	3.50	6.46*
	Max		25.00		14.00		12.00		17.00		13.00		15.00

TABLE 13.1 (Continued)

SITES	M. D. S.PARK		C. E. STATION		FAZ. MOSQUITO		FAZ. TUC 95		FAZ. TUC. 96		FAZ. R. CLARO	
Hunting Intensity	Unhunted		Slight		Moderate		Heavy		Heavy		Heavy	
	N/SD	Min. Avg. Max.	N/SD	Min. Avg. Max.	N/SD	Min. Avg. Max.	N/SD	Min. Avg. Max.	N/SD	Min. Avg. Max.	N/SD	Min. Avg. Max.
Alouatta fusca	(6) 1.47	1.00 3.25 4.00		4.0	(78) 1.90	1.00 4.06* 10.00	(12) 1.30	1.00 4.58* 6.00	(23) 2.40	2.00 4.78* 10.00	(33) 1.80	1.00 3.33* 7.00

Note: N equals number of group counts considered accurate; SD, standard deviation; Min., smallest group observed based on (N); Max., largest group observed based on (N); Avg., average group size; and EXT, locally extinct.

*No significant difference in pairwise comparisons (P > 0.05)

DISTANCE sampling method and computer software developed by Buckland et al. (1993) and Laake et al. (1993).

DISTANCE selects a model (or density estimator) that best fits the detection function produced by the data. AIC (Akaike's Information Criterion; Akaike 1973) was used to aid in selecting the best model, especially in cases of small sample sizes and cases in which most of the sightings were concentrated at or near zero distance from the trail (Buckland et al. 1993). AIC was usually used for ungulates that were rarely encountered at the hunted sites. In some cases small sample sizes led to distortions of the detection curves and resulted in unrealistic density estimates, usually from the hazard rate model (Buckland et al. 1993). In these cases the analysis was done using the uniform model or the half-normal model. The model with the greatest AIC was selected, and its calculated density considered. Density values from data sets with small samples (less than thirty sightings) were still used for comparative analysis.

Of all mammals observed during the censuses, only four species of ungulates (*Tapirus terrestris, Tayassu tajacu, Tayassu pecari,* and *Mazama sp*), three primates (*Alouatta fusca, Cebus apella,* and *Leontopithecus chrysopygus*), one edentate (*Dasypus novencinctus*), one carnivore (*Nasua nasua*), and two rodents (*Dasyprocta azarae* and *Sciurus aestuans*) were considered in the analysis. *Mazama gouazoubira* and *M. americana* were pooled in the analysis due to difficulty in visual identification; track evidence, however, indicates that the gray brocket deer (*M. gouazoubira*) is the more abundant of the two species at our study sites.

POPULATION ESTIMATES

In the case of social species, group density was estimated first, and then transformed to individual density by multiplying by the average group size. Data for each species were analyzed separately, giving a density value for each species at each study site. To estimate population sizes for each species, the individual density data were multiplied by the total forest habitat available at each study site.

POPULATION VIABILITY

Estimating viability of these populations is not possible without population viability analyses based on extensive life-history, ecological, and threat data specific to each population (Beissinger and McCullough 2002). This estimate is not yet possible with the populations included in our study. However, an effective population size of fifty has been suggested as a size sufficient to mitigate the deleterious effects of inbreeding (Franklin 1980; Allendorf and Ryman, 2002). Effective population sizes are on average about 11% of census population sizes (Frankham 1995). Therefore, the number of 500 individuals was considered as a rough estimate of the minimum number of individuals required for avoidance of significant inbreeding effects over the short term in these population fragments.

When considering the interactions of multiple population fragments at the metapopulation level, longer-term (i.e., evolutionary) considerations are also needed. Franklin (1980), Lande (1995), and Frankham and Franklin (1998) suggest that populations with effective sizes of 500 to 5,000 are needed to balance the loss of genetic diversity caused by drift with the gain in genetic diversity due to new mutations. This requirement translates to actual population sizes on the order of at least 5,000 individuals (Frankham 1995). We therefore used 5,000 as a rough guide to the number of individuals required for metapopulation persistence.

While these guidelines are based on genetic considerations, it is likely that estimates of the number of animals needed to maintain population viability when taking into consideration all types of threats (genetic, demographic, environmental, and catastrophic) will be at least several thousand (Beissinger and McCullough 2002). This number does correspond to our use of 5,000 as a guideline for determining viability at the metapopulation level.

RESULTS

GROUP SIZES

White-lips (*Tayassu pecari*) exhibited larger average group sizes at Caetetus Ecological Station than at all other sites (37.50 versus 7.50) (table 13.1). Smaller groups were found at sites with some degree of poaching. Collared peccary (*T. tajacu*) herd sizes were more constant across sites, with Caetetus Ecological Station having on average larger herds (13.50 versus 5.96) than all other sites. All four herds of collared peccaries observed at Fazenda Mosquito had four individuals each. At the other hunted sites, collared peccaries were observed with greater frequency with group sizes having approximately half the numbers of individuals in herds observed at Caetetus Ecological Station. Coati (*Nasua nasua*) groups seemed to be constant across all sites ($\bar{x} = 5.44$, P = 0.926), as were the capuchin monkey (*Cebus apella*) ($\bar{x} = 7.20$, P = 0.165), howler monkeys (*Alouatta fusca*) ($\bar{x} = 4.15$, P = 0.086), and black lion tamarins (*Leontopithecus chrysopygus*).

DENSITY ESTIMATES ACROSS SITES

Mammalian densities and population estimates are presented in tables 13.2 and 13.3, respectively. On the basis of average density estimates, Morro do Diabo State Park met the requirements for maintaining local populations of at least 500 individuals for most of the species evaluated (shaded areas in table 13.3). The only other area holding a short-term viable population was Fazenda Mosquito for *Alouatta fusca*. Species without any short-term viable population in any of the surveyed areas include tapirs, *Mazama spp.*, *Dasypus*, and *Sciurus*. The black lion tamarin is a very endangered endemic species that occurs in low densities, and local populations at these 2,000-ha fragments are far below the minimum recommended. The

TABLE 13.2 Density of Groups (individuals/km^2)

SPECIES/SITES	MORRO DO DIABO ST. PARK (35,000 HA), PROTECTED	CAETETUS ECOLOGICAL STATION (2,178 HA), SLIGHTLY HUNTED	FAZ. MOSQUITO PRIVATE (2,100 HA), MODERATELY HUNTED	FAZ. TUCANO PRIVATE (2,000 HA) (1996), HEAVILY HUNTED	FAZ. RIO CLARO PRIVATE (1,700 HA), HEAVILY HUNTED
Tapirus terrestris[a]	0.20–0.41–0.84 (8)	0.24–0.47–0.91 (9)	0.12–0.30–0.97 (6)	EXT	EXT
Tayassu pecari[a] / G	3.49–6.94–13.79 (10)	3.87–6.30–10.25 (12)	1.08–3.60–15.22 (3)	EXT	EXT
Tayassu tajacu[a] / G	3.19–5.67–10.06 (17)	3.71–6.41–11.22 (12)	0.15–0.81–3.22 (4)	0.97–2.21–5.05 (5)	3.62–5.63–8.79 (14)
Mazama spp.[a]	0.31–1.13–4.12 (6)	0.88–1.82–3.75 (13)	0.50–1.75–1.13 (14)	P	P
Dasyprocta azarae[a]	18.41–26.80–35.47 (82)	P	0.11–0.39–1.23 (3)	2.53–4.10–6.32 (32)	0.22–0.91–4.81 (2)
Dasypus novencinctus[a]	0.12–0.80–5.10 (1)	8.29–23.63–67.33 (19)	4.11–16.97–37.92 (14)	2.07–9.60–26.32 (6)	P
Nasua nasua[a] / G	0.89–4.20–11.30 (3)	2.84–4.47–8.01 (14)	1.72–3.40–5.97 (15)	0.33–3.11–7.46 (2)	1.56–5.20–17.98 (3)
Sciurus aestuans / G	0.18–0.37–0.78 (5)	6.13–11.31–28.40 (28)	17.31–20.20–23.23 (37)	0.10–3.91–21.02 (1)	0.97–2.71–5.92 (5)
Cebus apella / G	7.44–9.96–13.17 (48)	14.67–17.64–21.16 (122)	6.27–8.31–11.02 (56)	5.53–8.57–13.29 (31)	6.59–10.18–15.71 (28)

TABLE 13.2 (*Continued*)

SPECIES/SITES	MORRO DO DIABO ST. PARK (35,000 HA), PROTECTED	CAETETUS ECOLOGICAL STATION (2,178 HA), SLIGHTLY HUNTED	FAZ. MOSQUITO PRIVATE (2,100 HA), MODERATELY HUNTED	FAZ. TUCANO PRIVATE (2,000 HA) (1996), HEAVILY HUNTED	FAZ. RIO CLARO PRIVATE (1,700 HA), HEAVILY HUNTED
Alouatta fusca / G	2.94–15.66–82.32 (6)	0.60[b]	27.42–36.30–48.07 (90)	7.26–10.91–16.20 (30)	10.60–16.27–24.95 (36)
L. chrysopygus / G	2.66[c]	0.52–1.71–2.75 (5)	0.25[d]	1.00[c]	3.23–3.66[e]–4.10

Note: Left and right values represent 95% confidence intervals. G equals species where the value of the number of group counts represent sightings of groups; EXT, locally extinct; and P, present in the area but not recorded during transects.

[a] Considered game animals in the study region.

[b] Density value is derived from the minimum number of three groups known to occur in the area.

[c] Density estimates is taken from Valladares-Padua (1993).

[d] Represent the density from the only five individuals translocated to the area in 1996.

[e] Density estimate from Valladares-Padua and Cullen Jr. (1992).

TABLE 13.3 Population Estimates of Groups (individuals/site)

SPECIES/SITES	MORRO DO DIABO ST. PARK (35,000 HA), PROTECTED	CAETETUS ECOL. STATION (2,178 HA), SLIGHTLY HUNTED	FAZ. MOSQUITO PRIVATE (2,000 HA), MODERATELY HUNTED	FAZ. TUCANO PRIVATE (2,100 HA) (1996), HEAVILY HUNTED	FAZ. RIO CLARO PRIVATE (1,700 HA), HEAVILY HUNTED	METAPOPULATION
Tapirus terrestris[a]	[70–143–294]	[5–10–18]	[2–6–19]	EXT	EXT	[77–159–286]
Tayassu pecari[a] / G	[1,221–2,429–4,826]	[78–127–206]	[22–72–304]	EXT	EXT	[1,321–2,628–5,336]
Tayassu tajacu[a] / G	[1,116–1,984–3,521]	[74–130–226]	[3–16–64]	[19–44–101]	[62–96–148]	[1,274–2,270–4,060]
Mazama spp.[a]	[109–396–1,442]	[17–36–75]	[10–35–23]	P	P	[136–467–1,240]
Dasyprocta azarae[a]	[6,443–9,380–12,414]	P	[2–8–25]	[51–82–126]	[4–15–82]	[6,500–9,485–12,647]
Dasypus novencinctus[a]	[42–280–1785]	[167–476–1,358]	[82–339–758]	[41–192–526]	P	[332–1,287–4,427]
Nasua nasua[a] / G	[312–1,470–3,955]	[57–90–161]	[34–68–119]	[7–62–149]	[27–88–306]	[437–1,778–4,690]
Sciurus aestuans / G	[63–130–273]	[124–228–573]	[346–404–464]	[2–78–420]	[16–46–101]	[551–886–1,831]
Cebus apella / G	[2,604–3,486–4,610]	[296–356–426]	[125–166–220]	[111–171–266]	[112–173–267]	[3,248–4,352–5,789]
Alouatta fusca / G	[1,029–5,481–28,812]	[12]c	[548–726–961]	[145–218–324]	[180–277–424]	[1,914–6,714–24,533]
Leontopithecus chrysopygus / G	[931]b	[10–34–55]	[5]c	[20]	[55–62–70]d	[1,021–1,052–1,081]

Note: Left and right values represent 95% confidence interval. G equals species where the value of number of group counts represent sightings of groups; EXT, locally extinct; and P, present in the area but not recorded during transects.

[a]Game animals.

[b]Density estimates taken from Valladares-Padua (1993).

[c]Five individuals translocated to the area in 1996.

[d]Density estimate from Valladares-Padua, and Cullen (1992).

[e]Number of three groups known to occur in the area.

confidence intervals of the population estimates are quite large. However, when compared, the estimates using the mean with estimates based on the more conservative lower 95% confidence intervals, only one additional species, *Nasua nasual*, can be considered nonviable in Morro do Diabo. Still, when we sum mean population sizes in different forest fragments and consider a metapopulation scenario (i.e., fragmented populations are considered subpopulation of a regional population), interior fragments still hold short-term viable populations of most mammalian species.

DISCUSSION

MAMMALIAN POPULATION VIABILITY IN THE ATLANTIC FOREST OF THE INTERIOR

Conservation biologists who manage small and isolated forest fragments must ask the question, "How many individuals should be maintained within a particular reserve to ensure that a local population will still be thriving 100 years from now?" Although there has been much discussion over the optimum numbers (Lande 1995; Lynch and Lande 1998), the 50-500 rule is still the most commonly accepted (Hunter 1996). This rule states that a local population of 50 effective individuals (i.e., about 500 actual individuals) is a reasonable minimum viable population size (MVP) required to avoid significant problems of inbreeding in the short term. However, long-term survival requires that the MPV should be at least 500 effective individuals (i.e., 5,000+ actual individuals) so that a population will not lose genetic variability and will be able to cope with and evolve in changing enviorments (Lande 1995; Frankham and Franklin 1998). Accordingly, we use 500 individuals (adults and juveniles) as the minimum number required for the short-term conservation of the species. We examine the potential effectiveness of the last remaining forest patches of interior for conserving faunal species.

The Morro do Diabo State Park (35,000 ha) seems to be large enough to support viable populations in the short term for at least 85% of the species examined. *Tapirus terrestris* and *Mazama* spp are probably below the viable number of 500 at the Morro do Diabo State Park. *Sciurus aestuans* and *Dasypus novencinctus* also do not meet the MVP levels at Morro do Diabo State Park. However, the results for these two species should be considered with caution since the figures were derived from a small number of observations. Morro do Diabo State Park can also be considered as one of the two recommended reserves to maintain core wild populations of around 2,000 individuals of *Tayassu tajacu* and *T. pecari* proposed by the pigs and peccaries specialist group (Taber and Oliver 1993).

Each site can be considered a subpopulation of a metapopulation, occupying a patch of the metapopulation (Hanski and Gilpin 1991; Gilpin 1997). A metapopulation is a population of populations in which dispersion of animals promotes gene flow and recolonization of extinct patches (Hanski and Simberloff 1997). However, dispersion of animals within this metapopulation would not be frequent since these fragments are separated either by long distances (e.g., 300 km) or by unsuitable habitat.

Therefore, long-term conservation would require artificially moving animals from one forest patch to another. Management may include the shifting (reintroduction, translocation, and managed dispersal) of individuals among fragments. The populations of different patches should not be treated as separate populations in terms of genetic isolation in this metapopulation scenario. Some populations are sources (e.g., Morro do Diabo State Park) because they produce a substantial number of emigrants that could disperse to the other sites. Smaller fragments are sinks because they cannot maintain populations without a net immigration of individuals from the other patches. Thus, some of the small, hunted, or locally extinct populations could be recovered by the rescue effect (Brown and Kodric-Brown 1977; Harrinson 1991; Novaro, Redford, and Bodmer 2000).

Neither *Dasypus* nor *Sciurus* currently exist in any fragments in sufficient numbers to maintain short-term viability (N > 500). However, metapopulation management would effectively increase the numbers of individuals to sizes sufficient to meet these objectives. The total number of *Dasypus* and *Sciurus*, if managed under a metapopulation management strategy, would increase to an average estimate of 1,287 and 886 individuals, respectively.

Through regional metapopulation management, wild populations of most of the species examined can still be maintained as genetically healthy populations over the short term, despite the extreme fragmentation of the interior forests, the current forest encroachment, and the illegal poaching. Large cats like jaguars and pumas, and some of the smaller ones (e.g., *Leopardus* spp), are likely to be an exception, and their long-term survival will require the incorporation of other forested and protected areas that occur outside the interior range (Crawhaw et al. this volume).

FOREST FRAGMENTS AND SUDDEN LOCAL EXTINCTIONS

In forest patches poaching could quickly exhaust species populations, especially for large-bodied species that occur at lower densities and are preferred by hunters (Robinson 1996). This argument is based on the following premises: a) game species are usually more exposed and locked inside fragments; b) fragmented populations are less resilient to poaching since fragmentation and isolation hinders recolonization; and c) poachers usually have greater access to fragments surrounded by roads and other means of transportation.

Fazenda Tucano can be used to examine some of the premises stated above by examining the changes in mammalian biomass from 1995 to 1996 (table 13.4). This period coincides with an increase in the numbers of potential poachers living around Fazenda Tucano and with the arrival of new landless families in the region. Total biomass of mammals dropped from 296 kg/km^2 in 1995 to 201 kg/km^2 in 1996, a 35% decrease that represented approximately 1,875 kg of animal biomass. Howev-

TABLE 13.4 Change in Crude Biomass (kg/km^2) from 1995 to 1996 at Fazenda Tucano

SPECIES	BODY WEIGHT (KG)[a]	FAZ. TUCANO PRIVATE (2,000 HA) (1995)		FAZ. TUCANO PRIVATE (2,000 HA) (1996)		CHANGE IN C.B, 1995–96
		D	C.B	D	C.B	+/–
Tapirus terrestris[b]	148.95	0.34	50.64	0.00	0.00	—
Tayassu pecari[b]	31.67	1.22	38.67	0.00	0.00	—
Tayassu tajacu[b]	19.10	2.38	45.45	2.21	42.21	—
Mazama spp.[b]	21.72	1.07	23.24	P	P	—
Dasyprocta azarae[b]	2.84	0.43	1.22	4.10	11.64	+
D. novencinctus[b]	3.54	10.25	36.28	9.60	33.98	—
Nasua nasua[b]	3.88	1.82	7.06	3.11	12.06	+
Sciurus aestuans	0.38	4.54	1.72	3.91	1.50	—
Cebus apella	3.44	6.80	23.39	8.57	29.48	+
Alouatta fusca	6.46	10.44	67.44	10.91	70.47	+
L. chrysopygus	0.60	1.00	0.60	1.00	0.60	ø
TOTAL BIOMASS		%	**295.71**	%	**201.94**	
Total biomass of primates		30.91	91.43	49.79	100.55	+
Total biomass of ungulates		53.43	158.00	20.90	42.21	—
Total biomass of nongame		31.49	93.14	50.53	102.05	+
Total biomass of game		68.49	202.56	49.46	99.89	—

Note: Crude biomass was calculated for each species using the average body weight of adult individuals (BW) and multiplying by the estimated individual densities (D) (BW*D - kg/km^2). Mean body masses were taken from the literature (Robinson and Redford 1986a), with the exception of peccaries and the black lion tamarin for which data were available from the study sites. Percentage contribution (%) refers to a group's relative contribution in relation to the total biomass value of the mammalian species. D equals density (individuals/km^2); C.B., crude biomass (kg/km^2); P, present but not seen during census.
[a]Average of mean body weights taken from Robinson and Redford (1986a).
[b]Game animals.

er, primate biomass increased by 10%, while ungulate biomass had a fourfold decrease. The decrease in ungulate biomass was mainly due the local extinction of tapirs and white-lips. Deer also had lower biomass in 1996. Collared peccary biomass remained about the same.

The sudden extinction of tapir and white-lips suggests that in forest fragments large game species can be overhunted rapidly when poaching pressure becomes too excessive. The future of these isolated game populations will thus probably depend on the poaching pressure imposed on them. Poaching may override other effects of fragmentation (i.e., genetics and demographics) and be the ultimate factor responsible for causing the extinctions observed among the large mammalian fauna (Cullen, Bodmer, and Valladares-Padua 2000; Cullen, Bodmer, and Valladares-Pauda 2001).

RECOMMENDATIONS AND RESEARCH PRIORITIES FOR THE INTERIOR ATLANTIC FORESTS

The results of this study provide some empirical suggestions for developing management and conservation strategies for the threatened interior forests. We recommend areas in which we think immediate actions are needed. Results suggest that the implementation of these recommendations should contribute to wildlife conservation in the Atlantic Forest.

Environmental Education Environmental education and community participation are essential for successful conservation. A successful program already established for the Morro do Diabo region showed that the local community became aware of the importance of the park as a conservation site and has been contributing to its protection (Padua 1991, 1997). We recommend that environmental education initiatives and efforts for the interior region for the next few years focus more intensively on large landowners and the rural people, with the goal of increasing the awareness of the importance of other isolated forests as well. These communities should be provided with the knowledge, attitudes, and skills to conserve nature. Past experience and new results suggest that with a continued and systematic team effort, including local community education and participation, ecological research and management, legislation, and law enforcement, we will be able to conserve wildlife. Conservation at the combined community, ecosystem, and landscape levels probably offers the most promising alternative to biodiversity conservation of the interior forests.

Metapopulation Approaches As our results suggest, despite being highly fragmented and isolated, interior fragments together still sustain a viable population of most of the original biota. Subpopulations should be managed as a metapopulation that may include the reintroductions or translocations of animals among these last remaining forest patches. This approach should especially be emphasized for endangered and endemic species, such as the black lion tamarin. Source and sink habitats need to be included in conservation plans; otherwise, the metapopulation could be threatened. New reserves need to be established, the existing ones protected, and community-based programs implemented. Particular emphasis has to be directed at reducing current poaching and degradation of forest remnants.

Studies on Landscapes and Animal Dispersion In highly fragmented landscape areas like the interior region, the existing reserves and forest remnants are definitely islands in a matrix of agricultural lands. This situation should always be kept in mind during the development of new studies and conservation measures. Interior forest patches are parts of a landscape mosaic, and the presence and survival of species may be a function not only of poaching, patch size, and isolation but also of the kind of neighboring habitat around these forest fragments. Some species that are habitat generalists may survive in very small patches because they can exploit surrounding resources. Future studies should focus on the adaptability of species to the new landscape. New studies should investigate ecological and behavioral attributes of species that might help to enhance their survival in the surrounding matrix. One of the most immediate pieces of information needed is the dispersal behavior of species in a habitat mosaic and thus their likelihood of recolonizing the surrounding fragments. Future conservation measures will have to involve effective monitoring and metapopulation approaches that will make the forest patches functional and linked.

Agroforestry Buffer Zones Agroforestry is a type of land management in which woody perennials are planted on the same land management units as agricultural crops or as animals in either a special arrangement or a temporal sequence and with ecological and economic interactions between the different components (Fernandes and Nair 1986). Local people that live and farm on the borders of the fragments are currently depleting forest fragments. Poaching, trees blown down by wind, vine colonization, desiccation by wind, fires, cattle grazing, the spread of aggressive grasses, and pesticides are some of the processes leading to a gradual and continuous erosion of these forest edges (Laurence 1991). These processes in the long term are likely to affect forest structure and cause the loss of many plant and animal species, mainly by the known consequences of the edge effects.

A case project has been implemented in which an agroforest surrounding a forest fragment is functioning as a benefit zone to supply services, vegetables, fruits, grains, and protein to the local farmers, thus relieving some of the pressure on the forest (Cullen, Bodmer, and Valladares-Padua 2001). This benefit zone is providing an insulative/protective zone around fragments and reducing edge effects. Implicit in this conservation approach is the assumption that stimulating the planting and use of multiple-use trees in these edge areas places a value on the resources, and this ascribed value will help to pave the way to conservation of wildlife. Effective programs often begin by encouraging villagers to establish very simple demonstration experiments or vivid examples and to evaluate and share their results with others.

In places like the Pontal do Paranapanema Region, the economic value of forest fragments is low, and economic, political, and demographic pressures are bringing about overexploitation and unsustainable uses of these fragments. This study shows that, except for the Morro do Diabo State Park, the largest protected area of the interior Atlantic Forest, none of the other remaining forest fragments in the Pontal do Paranapanema sustain viable populations for most of the species evaluated. Hence, we must develop and implement innovative and active management schemes that will represent the marriage between these forests and the local people around them.

ACKNOWLEDGMENTS

This study was funded by a collection of small grants, in the following order of receipt: Scott Neotropical Fund of the Lincoln Park Zoological Society; Tropical Conservation and Development Program, University of Florida; and the Tinker Foundation. Another major contributor was the Biodiversity Support Program, a consortium between USAID, World Wildlife Fund/U.S., the Nature Conservancy, and the World Resources Institute. The Conservation, Food and Health Foundation; the Programa Natureza e Sociedade at World Wildlife Fund-Brazil; 100% Fund from Fauna and Flora International/U.K; the Wildlife Trust, U.S.A.; the Beneficia Foundation, U.S.A.; the Duratex Company, and the Smithsonian Institution also provided support. Institutional support was also provided by IPÊ (Instituto de Pesquisas Ecológicas) and the Instituto Florestal de São Paulo, IF-SMA.

14

Abundance, Spatial Distribution, and Human Pressure
on Orinoco Crocodiles (*Crocodylus intermedius*)
in the Cojedes River System, Venezuela

ANDRÉS E. SEIJAS

Crocodilians in general, and the Orinoco crocodile (*Crocodylus intermedius*) in particular, have been traditionally hunted by both aboriginal and rural people in Venezuela because of their value as a food resource or because of the putative medicinal or magical properties of their teeth and fat (Petrullo 1939; Codazzi 1940; Tablante-Garrido 1961; Gumilla 1963). The first attempt to commercialize crocodile skins in Venezuela was initiated in 1894–1895 by a U.S. company that established its headquarters in El Yagual, in Apure state (Calzadilla 1948; Medem 1983). At that time crocodiles were hunted with firearms during the day, a highly inefficient method in which many dead and injured animals could not be recovered from the river. That early commercial enterprise failed. The expenses of preparing and transporting the hides proved to be so great that the work had to be abandoned (Mozans 1910; Calzadilla 1948). Despite this early commercial exploitation, during the first quarter of the twentieth century the Orinoco crocodile was probably as abundant as it was when Humboldt (1975) and other nineteenth-century naturalists were amazed by its numbers.

A new phase of commercial exploitation started at the end of the 1920s (Medem 1983). New hunting methods (flashlights and harpooning) and an international demand for crocodilians hides combined to bring to the brink of extinction in less than three decades a species that originally could be counted in the millions. The peak of the exploitation occurred in 1930–1931, when between 3,000 and 4,000 skins were traded daily in San Fernando de Apure. From 1933 to 1935 Venezuela exported 900,000 crocodile hides. The large-scale exploitation ended in 1947–1948, due mostly to the scarcity of the resource by that time. Independent hunters persisted in this activity for several years, but the export of *C. intermedius* hides from 1950 to 1963 was minimal (Medem 1983).

Commercial exploitation extirpated the Orinoco crocodile from most of its historic distribution range (Godshalk 1978; Medem 1981, 1983; Thorbjarnarson 1992; Ross 1998). Today, the Orinoco crocodile is one of the most threatened crocodilian species in the world (Ross 1998). The species is listed as critically endangered in the Venezuelan Red Data Book (Rodriguez and Rojas 1995). Although *C. intermedius* has been legally protected both in Colombia and Venezuela for more than thirty years and although international trade has been prohibited by the Convention on International Trade in Endangered Species of Wild Fauna and Flora (CITES) since the middle 1970s (King 1989), little recovery of wild populations has occurred.

Even though commercial exploitation of Orinoco crocodiles in Venezuela is today probably negligible, occasional killings of individuals still occur because they are considered vermin. Also, they are hunted for their meat or fat, and their eggs and hatchlings are collected as food or pets, respectively.

Currently, the most important, and probably the only viable, populations of the Orinoco crocodile (Arteaga et al. 1997) are found in two areas of divergent characteristics in Venezuela. First is the Capanaparo River in the state of Apure (Godshalk 1978; Thorbjarnarson and Hernández 1992), a prime-quality habitat, more than 100 m wide, that is impacted relatively little by human activities and that is in the center of the species' range, where it reached its historically highest densities. Second, in the states of Cojedes and Portuguesa is the Cojedes river system (CRS), a set of highly modified and contaminated narrow river sections (in general less than 20 m wide), near the periphery of the distribution of the Orinoco crocodile and very close to some of the most important agricultural, urban, and industrial centers in the country (Ayarzagüena 1987, 1990; Seijas and Chávez 2000).

Although the survival of the Orinoco crocodile in a river like the Capanaparo is easy to understand, the presence of a dense population of this species in the CRS is somehow paradoxical. One of the factors that may explain the survival of *C. intermedius* in the CRS (Ayarzagüena 1987) is the isolation in which that region remained during the years of intense commercial exploitation of the species (1929–1945). The CRS today is, however, not as isolated as it was in the recent past. Some areas of the river are closer to human population centers, and presumably under greater human pressure, than others. Human population in the CRS is mostly concentrated in the north, close to the piedmont of the Coastal Range. The southern part is sparsely populated by humans, with El Baúl (5,236 inhabitants) the most important town. Is this distinct pattern of human occupation of space a factor that could explain the current distribution of the Orinoco crocodile in the CRS? In this paper I explore that possibility. My hypothesis is that human proximity negatively impacts crocodile survival, and consequently, crocodiles should be found more frequently in river sections far from human settlements.

STUDY AREA

For the purposes of this study, the Cojedes River System is defined as the middle and lower portions of the Turbio-Cojedes River basin. It covers a wide fringe of land along the Cojedes and Sarare Rivers. The study area in the CRS encompasses

FIGURE 14.1   Cojedes River System, Venezuela. Rivers flow toward the south. Major cities are located in the north, whereas the south is sparsely populated. The acronyms indicate the locations
of the river sections surveyed: CON, Cojedes Norte; CAN, Caño de Agua Norte; CAS, Caño de
Agua Sur; SAR, Sarare; CAM, Caño Amarillo-Merecure; SUC, Sucre section; and CUL, Caño
La Culebra.

the cities of Acarigua and San Carlos to the north and extends southeast to the confluence of the main course of the Cojedes River with Caño Amarillo-La Culebra
near the town of El Baúl (fig. 14.1).

In the northern part of the CRS, agricultural lands dominate the landscape and
are interspersed with large- and medium-sized urban centers and cattle ranches.
The southern part of the region (south of the Lagunitas-Santa Cruz road) is a matrix of forested savannas and cattle pastures intermixed with forest relicts, scattered
agricultural lands, wetlands, and other less extensive land-cover categories. The
CRS has zones of relatively high human population densities in the north, where
the cities of San Carlos (> 80,000 people) and Acarigua (~ 200,000 people) are located and where the rivers have been modified by damming, canalizing, dredging,
contamination, and deforestation. In the south the rivers retain more of their original conditions, and El Baúl (6,000 people) is the largest town.

METHODS

Based on two landsat TM satellite images of the study area, taken on January 10 and February 27, 1990 (early dry season), and on data from more than 1,500 GPS locations, I updated the previously existing cartographic information of the region. The basic land cover features considered for mapping were agriculture, pasture lands and open savannas (taken together as a unit), urban areas, water bodies, forests, permanent rivers, and roads. Maps were converted into raster images for Geographic Information System (GIS) analyses (IDRISIS 1997). The initial raster image generated from the classification of the satellite images had a spatial resolution of 32 x 32 m. Because of the extension of the land surface being modeled (9,660 km²) and in order to speed up the GIS analyses, the raster images used for the final analyses had a spatial resolution of 64 x 64 m.

With the GIS I generated a cost-distance (CD) layer for every major city as well as for small towns, villages, and other human settlements located close to the river. Each CD layer modeled the cost of movement from a particular human settlement to any location on the landscape, i.e., it represented the ease with which people could reach every spot in the study area, considering the friction offered by different land-cover types. The value assigned to a pixel in the CD layer was a function of its distance to the human settlement under consideration and of the friction exhibited by the land surface between them.

The friction surface used to calculate the cost-distance layer was generated according to the relative cost shown in table 14.1. Primary roads were assigned a friction of 1. This value, in fact, means that there was no cost for travel by car on that surface and that cost-distances measured along them were equivalent to Euclidean distances. Because it is possible to travel on average at 80 km/h on primary roads, the friction values assigned to other land-cover types were calculated relative to how much longer it takes to travel an equivalent distance on or through them (using the fastest transportation method that can be used on that surface). Friction values assigned to rivers were somewhat arbitrary but larger than the values assigned to most land-cover surfaces in order to reflect the fact that they are important obstacles to human movements (although river sections that are navigable facilitate human movements). The highest friction was assigned to lakes, which were considered barriers to human movement for the purposes of this study.

The CD layer obtained for each town or city was used to model the presumed human pressure exerted by that city on every reach of the Cojedes River system (indeed, on every location within the study area). The human pressure index (HPI) is a value that indicates the strength of the expected impact of a particular urban area on every point (pixel) in its surrounding landscape. The HPI of a particular spot (i.e., pixel in the raster layer) was calculated as a function of its proximity to human settlements and of the human population size of these particular human settlements. That river reaches close to cities and towns were assumed to be under greater human pressure than river reaches located farther from those urban areas.

TABLE 14.1 Relative Cost of Movement (friction) Through
Different Land Cover Types in the Study Area

LAND COVER TYPE	MEAN SPEED (KM/HOUR)	RELATIVE FRICTION
Primary roads	80	1.00
Secondary roads	60	1.33
Improved roads	40	2.00
Dirty roads	20	4.00
Urban areas	35	2.29
Main rivers	—	80
Secondary rivers	—	60
Intermittent rivers	—	40
Agriculture fields	—	20
Savannas	4	20
Forests	2	40
Lakes	—	100

Note: See text for calculation of friction values. The high friction value assigned to river may not apply to navigable river sections.

On the other hand, large cities were expected to exert a higher pressure than small ones. Mathematically,

$$(1) \qquad HPI_i = P_i \cdot CD_i^{-2}$$

where P_i is the population size of a particular human settlement and CD_i is the cost-distance layer obtained for that particular human settlement. Equation 1 is in essence a particular case of a gravity model (Forman 1995), which states that the movement or interactions between two nodes increase with node size but decrease with the square of the distance between nodes. In my case one of the nodes (the pixel for which the *HPI* was been calculated) received, arbitrarily, a unit free value of 1. The *HPI* as expressed in Equation 1 has the units of density (inhabitants/km²).

The urban centers considered in the model are listed in table 14.2. Many other small villages (among them, Retajao, El Estero, and La Palmita) and cattle ranching operation centers (including La Batea, Merecure, and Las Guardias) were used to generate cost-distance surfaces. Because of a lack of precise information on human population size in these human settlements, I assigned a figure of 500 inhabitants to small villages and hamlets and 100 to the cattle ranching operation centers.

Because any particular point on the study area may be simultaneously under the influence of several human settlements, the layer of Total Human Pressure (THP) of the entire study area was obtained adding the HPI_i layers of all these settlements (fig. 14.2). In this way every pixel, including those representing the rivers, had an associated *THP* value. Mathematically,

TABLE 14.2 Towns and Other Human Settlements
in the Cojedes River System, Venezuela, Used to
Model the Human Pressure in the Study Area

TOWNS	NUMBER OF INHABITANTS
Portuguesa state	
Acarigua-Araure	171,850
Agua Blanca	9,393
San Rafael de Onoto	7,206
Pimpinela	4,563
Santa Cruz	4,090
Cojedes state	
San Carlos-Tinaco	68,325
Las Vegas	6,897
El Baúl	5,236
Apartaderos	4,260
Cojeditos	4,911
Lagunitas	3,353
Sucre	1,886
El Amparo	1,105

Note: The number of inhabitants is based on OCEI (1993).

$$(2) \qquad\qquad THP = \Sigma HPI_i$$

In the CRS several river sections, ranging from 5.2 to 16 km, have been repeatedly surveyed since 1991 (Seijas 1998; Seijas and Chávez 2000). In 1996 and 1997 the position of most crocodiles seen in those river sections during nocturnal spotlight surveys was recorded with a GPS. The GPS locations for those surveys with the highest number of crocodile sightings were used to generate a new map layer. The THP of the spot (pixel) in which each crocodile was seen was obtained by overlaying the crocodile locations layer on the THP layer.

For all surveyed river sections the frequency of pixels with a particular THP value was calculated and tabulated in ranges. That gave an indication of the availability of river habitat under different THP values and allowed the calculation of the number of crocodiles expected to be found in each of these THP ranges. Chi-square analyses (G^2, likelihood ratios; Sall and Lehman 1996) were used to compare frequency distribution of THP of crocodile sightings in relation to the frequency distribution of THP for the surveyed river sections.

To assess the importance of other human-related factors in determining the abundance of crocodiles in the CRS, I performed a nonparametric correlation analysis between crocodile density in each river section (table 14.3) and the rela-

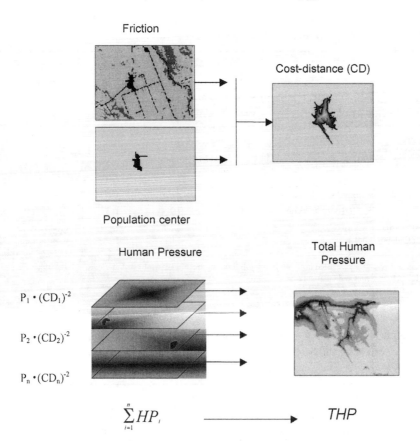

FIGURE 14.2 Flow chart indicating the procedure followed to obtain the Total Human Pressure (THP) over every place in the study area. From a friction surface and each population center a cost-distance (CD) surface was generated using a geographic information system. The population size of each town, divided by the CD^2, allowed the calculation of the human pressure (HP) it exerted over every place in the study area. The Total Human Pressure was calculated adding the HP of all population centers.

tive importance of the following variables: isolation from human populations, navigability, and contamination. I ranked each river section according to the relative importance of these variables (table 14.4). Information on isolation from human population was obtained from data in table 14.3. Ranking according to contamination was based of information presented by Campo and Rodríguez (1997) and Seijas (1998).

Regarding navigability, the only river sections that are navigated on a regular basis are those close to Sucre (SUC) and, to a lesser extent, La Culebra (CUL). For people living in Sucre and El Baúl, the Cojedes River is an essential means of year-round communication with cattle ranches. There is also commercial and subsistence fishing and, presumably, illegal spectacled caiman (*Caiman crocodilus*)

TABLE 14.3 Total Human Pressure (THP) and Mean Crocodile Population Index in Surveyed River Sections of the Cojedes River System, Venezuela

RIVER SECTION	LENGTH (KM)	MEAN THP (INH./KM^2)	RANGE THP	MEAN CROCODILE DENSITY (IND./KM)[*]
Cojedes Norte (CON)	7	15.6	7–56	2.0
Caño de Agua Norte (CAN)	16	7.2	1–100	4.4
Sarare (SAR)	8.4	11.7	1–78	3.1
Caño de Agua Sur (CAS)	5.2	2.0	1–6	7.3
Caño Amarillo-Merecure (CAM)	8.4	1.2	1–3	4.9
Sucre (SUC)	11.6	1.7	1–19	0.6
La Culebra (CUL)	12.8	1.0	1	1.4

Note: River sections are listed from north to south (from upstream to downstream).
[*]Taken from Seijas and Chávez (2000).

TABLE 14.4 Ranks of Crocodile Densities, Isolation from Urban Areas, Contamination, and Navigability of the Different River Sections That Were Surveyed in the Cojedes River System, Venezuela

RIVER SECTION	CROCODILE DENSITY	ISOLATION FROM HUMANS	CONTAMINATION	NAVIGABILITY
Cojedes Norte (CON)	5	7	1	6.5
Caño de Agua Norte (CAN)	3	5	2	6.5
Sarare (SAR)	4	6	5	4
Caño de agua Sur (CAS)	1	4	3	5
Merecure-Caño Amarillo (CAM)	2	2	4	3
Sucre (SUC)	7	3	6	1
La Culebra (CUL)	6	1	7	2

Note: Contamination ranks were based on information presented in Campo and Rodriguez (1997) and Seijas (1998) Navigability is based on personal observations.

hunting around Sucre and in La Culebra. Upriver from Sucre the river section Merecure-Caño Amarillo (CAM), although navigable year round, seems to be navigated only sporadically since only one family with a small canoe was observed there. Caño de Agua Sur (CAS) is difficult to navigate because of obstructions created by fallen trees and urban debris and garbage that drift from upstream towns.

That section seems to be occasionally visited and sporadically navigated by hunters and campers. The Sarare (SAR) river section surveyed are in the same situation as CAM. Caño de Agua Norte (CAN) and Cojedes Norte (CON) are rarely, if ever, navigated by people other than myself and other crocodilian researchers.

RESULTS

A tri-dimensional representation of the THP in the study area is shown in Figure 14.3. As would be expected, the highest THP, represented in the figure as high elevation plateaus, was located in and around the main cities (Acarigua and San Carlos). Consequently, the river reaches flowing through densely human populated areas were under relatively high human pressure (table 14.3). Cojedes Norte (CON) for example, which is very close to the towns of Apartaderos and San Rafael de Onoto, had THP ranging form 7 to 56 (mean 15.6, the highest of all surveyed river sections). At the other extreme, THP was relatively low near Sucre (SUC) and especially so in La Culebra (CUL), where all the pixels representing the river had THP of 1 (table 14.3).

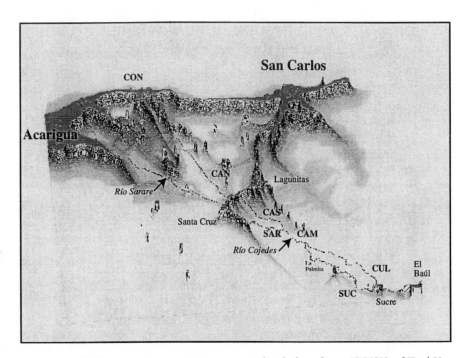

FIGURE 14.3 Tridimensional representation, generated with the software IDRISIS, of Total Human Pressure (THP) on the Cojedes River System, Venezuela. Areas in clear gray or white are under low human pressure (THP < 2). Different tonalities of gray (elevation) represent the intensity of human pressure, with the highest THP (> 100) in dark gray found in towns and cities represented as plateaus. Acronyms indicate the locations of the river sections surveyed (see fig. 14.1).

According to my hypothesis, crocodile sightings should be more frequent in river spots (pixels) under relatively low THP. An analysis of the distribution of 226 nonhatchling crocodiles spotted in 1996 and 1997, according to the THP of the specific spot where they were observed (fig. 14.4, upper), indicated that, contrary to expectations, crocodiles were underrepresented in river spots (pixels) of very low human pressure (THP ranging from 1 to 2) ($G^2 = 23.02$, $P = 0.002$). That was a consequence of low densities of crocodiles in SUC and in CUL, the surveyed river

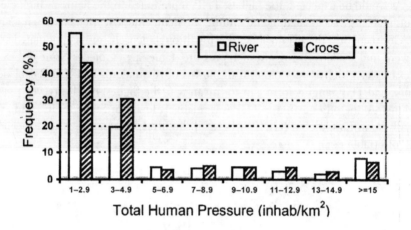

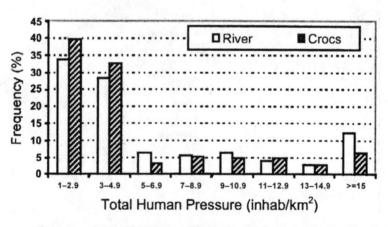

FIGURE 14.4 Frequency of crocodile sightings compared to the frequency distribution of human pressure in different river sections of the Cojedes River System, Venezuela. Bars labeled river represent percentage of pixels (64 x 64 m) in each range of human pressure (HP) along river sections surveyed. Bars labeled crocs indicate the percentage of crocodiles seen in pixels within a particular range of HP. Upper graph includes all river sections; bottom graph excludes navigable sections.

sections under the lowest THP, whereas crocodile densities were highest in such river sections under moderate THP as CAS (mean THP = 2, range 1 to 6). In contrast to other river sections surveyed, CUL and SUC are navigated year round. When data from these river sections were dropped from the analysis, the frequency distribution of THP of locations of crocodile sightings and THP of river reaches differed in the direction predicted by the hypothesis (G^2 = 15.42, P = 0.03) (fig. 14.4b). Within the nonnavigable sections crocodile abundances were negatively related to human pressure (Spearman Rho = –0.9, P = 0.04, n = 4) (fig. 14.5).

Correlation analyses indicate that the variable with strongest relationship to crocodile densities was navigability, but that correlation (negative) was not statistically significant (Spearman Rho = –0.505, P = 0.248). The correlation between isolation and crocodile densities was also negative but not significant (Spearman Rho = –0.11, P = 0.8), in agreement, as would be expected, with the THP analysis presented above.

DISCUSSION

Although the isolation of the Cojedes River may have played an important historical role in preserving a small population of Orinoco crocodiles (Ayarzagüena 1987),

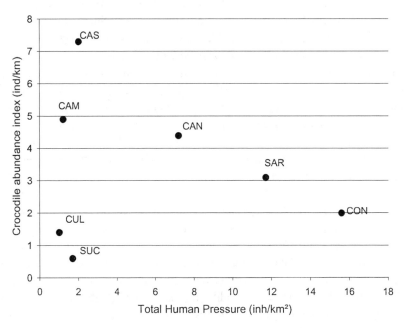

FIGURE 14.5 Relationship between mean Total Human Pressure (THP) and mean crocodile population index in river sections of the Cojedes River System, Venezuela. La Culebra (CUL) and Sucre (SUC) are the only river sections that are navigable.

today the river sections most isolated from human settlements showed the lowest crocodile densities, whereas the areas with the highest crocodile densities were moderately isolated from urban centers. Low densities of crocodiles in parts of the study area, such as those river reaches close to Sucre or in caño La Culebra, suggest that navigability and difficulty of access are probably important factors explaining the current pattern of distribution of the species in the CRS.

A combination of isolation from humans and impossibility of navigation have been used to explain the persistence of other small population of Orinoco crocodile in the Tucupido River (Ramo and Busto 1986; Thorbjarnarson and Hernández 1992). According to Thorbjarnarson and Hernández (1992) crocodiles in the Capanaparo River are protected during the dry season when low-water levels make it unnavigable.

Proximity to towns and cities seems to explain the abundance and distribution of crocodiles in river sections of the CRS, but future studies should include more river sections under relative high human pressure, such as the upper section of CAN near Cojeditos and some of the Sarare River close to Pimpinela and south to Agua Blanca. In some areas with high THP that were not properly surveyed at night, some crocodiles were observed. Seven crocodiles, for example, were seen on January 14, 1993 in Toma Cojedes, an area with a high THP of 46 to 48. In Retajao, a hamlet along the left margin of CAN, a nesting female and a subadult crocodile were observed in 1996 and 1997. The latter individual was sighted just across the street from an elementary school, a spot with a very high THP of 50.

Anecdotal information indicated that downstream from Sucre, and particularly downstream from El Baúl, where the Cojedes river is routinely navigated by two or three dozen small boats and canoes, Orinoco crocodile populations remain as low as they were almost twenty years ago when first evaluated by Godshalk (1978). Young crocodiles are seen occasionally in the Cojedes River near El Baúl (M. González pers. comm.). They probably represent transient individuals or individuals that have been carried downstream by the river during the peak of the rainy season. Some of these crocodiles are taken by people in El Baúl as pets. Others are presumably killed by fishermen, accidentally or deliberately, or move farther downstream toward the Portuguesa River.

Most reproduction of the Orinoco crocodile in the CRS takes place in the middle sections of Caño de Agua and lower Sarare (Seijas 1998), where the species is protected by the relative isolation from towns and difficulty in navigation. These river sections are population sources, in which more than forty females nest every year (Seijas 1998). My data also indicate that reproduction is poor or absent near Sucre and in La Culebra sections, some 50 or 60 km downstream from the previously mentioned sections. No crocodile nest has ever been observed in that area. Under the current circumstances, because of low reproduction and presumably high risk of being killed by people, the later mentioned river reaches are population sinks for the Orinoco crocodile.

ACKNOWLEDGMENTS

This study was made possible by a grant from the Wildlife Conservation Society. I thank John Thorbjarnarson from that institution for his continuous support. Satellite images were generously donated by the U.S. Geological Survey (EROS Data Center). Carlos Chavéz assisted me in the field work.

15

Impacts of Damming on Primate Community Structure in the Amazon

———

A CASE STUDY OF THE SAMUEL DAM, RONDÔNIA, BRAZIL

ROSA M. LEMOS DE SÁ

The importance of tropical rain forests to global biodiversity is clearly appreciated when one realizes that they cover only 7% of the earth's land surface but that they contain more than half the species of the world's biota (Wilson 1988). Despite the importance of tropical forests and the fact that very little is known about their fauna and flora, development of tropical areas is occurring at a rapid pace and will bring about the extinction of species. To avoid mass extinction and to be able to guide developing agencies, a better understanding of tropical forest communities and their responses to environmental changes is needed. One increasingly important source of environmental change in the Amazon is the construction of large hydroelectric dams.

Until 1980 only two small hydroelectric dams were operating in the Brazilian Amazon: Curuá-Una, near Santarém, and Paredão in Amapá state. Each dam impacted an area of less than 100 km^2 (Junk and Nunes de Mello 1987). Since then, three large dams have been added to the region and are operating in the Amazon: Tucuruí, near Belém; Balbina, near Manaus; and Samuel, near Porto Velho. Collectively, these three dams have flooded an area of 5,350 km^2. If Eletronorte (Brazilian Agency for Hydroelectric Power Development in the Amazon region, the government agency responsible for hydroelectric dam constructions in northern Brazil) succeeds in completing all the dams projected for the Amazon in the 2010 plan, an area of roughly 100,000 km^2 will be flooded (Fearnside 1989).

The flooding of such large areas has a tremendous impact on humans and wildlife. The most significant effect is the loss of land with its consequent human and animal displacement and/or death that bring about extinction of species (Liao 1988). Eletronorte greatly improved its rescue operation policy between its first rescue at Tucuruí and the effort that took place at Samuel. The new policies of sending the bulk of animals to research institutions rather than releasing all of them, and of creating protected areas, are commendable changes on the part of Eletronorte. Greater changes, however, must be implemented in order to minimize impact and maximize conservation of the Amazonian region; these changes will require funding for both preliminary and follow-up ecological studies. This article documents the response of mammalian communities to environmental changes resulting from the construction of the Samuel Hydroelectric Dam in the Amazon.

PROJECT DESIGN

The Samuel Dam is located on the Jamarí river in the state of Rondônia approximately 50 km east of the state's capital of Porto Velho (fig. 15.1). Construction on the dam began in 1982, actual filling of the reservoir in 1988, and completion occurred in 1989. The total lake area is 502 km², of which 22 km² remained green in the form of islands (measured from landsat images 1:250,000 by Adolfo de La Pria Pereira, SEDAM-RO).

Two sites were monitored for this study, which I will refer to as the Reserve and Jusante. Prior to the filling of the reservoir, Eletronorte created a 21,000-ha ecological reserve (Estação Ecológica de Samuel) to compensate for the loss of 56,000 ha of forest flooded by the reservoir. This Reserve is located east of the reservoir's embankment, approximately 26 km from the dam, and its forests are continuous with those of the reservoir area. During the rescue operation, from November 1988 to March 1989, 2,374 mammals were released inside the Reserve (Eletronorte 1989).

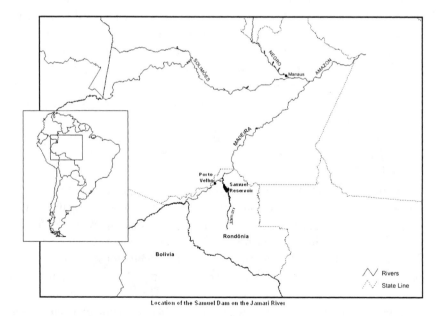

Location of the Samuel Dam on the Jamari River

FIGURE 15.1 Location of Samuel Dam.

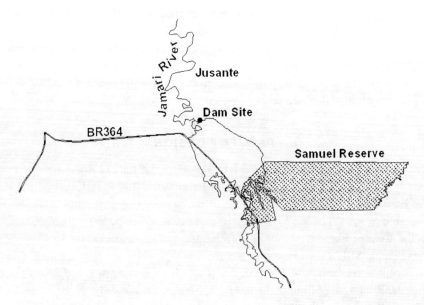

FIGURE 15.2 Location of Reserve and Jusante study sites with reference to Samuel Dam. The size of the reserve was increased in 1994 to 74,300 thousand hectares, up from the original 21,000 ha established at the time the dam was built.

The site was monitored during 1988 (before the flooding of the reservoir), 1989, 1990, and 1991. Monitoring in 1988 was carried out by Eletronorte personnel, not by the author. The second site, Jusante, was an undisturbed area located downstream from the dam on the right bank of the Jamarí river (fig. 15.2). It was monitored during the 1990 and 1991 field seasons. No animals were released into the site.

Before flooding, the area was undisturbed primary forest with little or no hunting pressure. The forests at both sites were continuous with the flooded area but presented some structural and floristic differences. Compared to Jusante, the Reserve had a higher and denser canopy, and greater basal area and tree density for trees greater than or equal to 30 cm dbh. Adult *Bertholletia excelsia* (Brazil nut), and *Orbignya barbosiana* (*babaçu*) were very common at the Reserve but rare at Jusante. On the other hand, *Hevea brasiliensis* (*seringueira*) was very common at Jusante but never seen at the Reserve; the latter observation was expected because this particular species is known to occur along water courses and not in terra firme. Differences in elevation support the idea that the forest at Jusante (at lower elevation) is younger, probably because of disturbance effects related to its proximity to the river. The more open forest at Jusante allows for greater penetration of light, providing an opportunity for shrub and liana species to develop and creating a forest floor more densely covered by vegetation.

My hypothesis at the beginning of this study was that the mammalian community in the Reserve study site could have been affected by the release of rescued animals, by the migration of animals fleeing from the flooded reservoir, or by a combination of the two. The Jusante site, on the other hand, could only have been affected by the migration of animals from the reservoir area or not affected at all. More specifically, I hypothesized that the Reserve would experience animal overcrowding for an undetermined length of time, possibly surpassing the carrying capacity for the area. My hypothesis could be tested by estimating mammalian densities in the Reserve at different points in time and by examining differences in biomass values (Lemos de Sá 1996) for the community.

If my assumptions were correct, the noted responses in density changes, regardless of site, would be immediate in the case of terrestrial mammals (because they would have to flee from the rising water) but possibly delayed for arboreal species (because they could stay on top of trees while the vegetation was still alive). The time frame in which density changes would occur was unknown. To increase the probability of detecting such changes (completely or partially), the sites were sampled repeatedly.

METHODS

Five plots of 1 km^2 were established in the Reserve in 1989, and three plots at Jusante in 1990 (fig. 15.2), creating 4 km of transect lines along each plot's perimeter. Transect surveys were conducted by walking slowly (1 km/h) and by stopping periodically to watch and listen for animals. Transects were conducted from 6:30 or 7:00 A.M. (depending on the time of sunrise) to 10:30 or 11:00 A.M., and from 1:00 to 5:00 P.M. The number of transect samples in each area was divided equally between morning and afternoon surveys. Whenever possible, different transects were walked in the morning and afternoon of the same day. If that were not possible because of logistics, the two daily surveys of a plot always began in the same direction to give an interval of six hours between the morning and the afternoon survey (i.e., the same point in the trail would be traversed in the afternoon six hours after the morning survey). Transect surveys on different days began at opposite ends of the route to reduce potential biases resulting from direction of travel by the observer. Each transect had equal numbers of surveys originating in both directions. The time, transect identification, location on the trail, species, number of individuals sighted, angle of sighting, and distance from the observer to the animal when first seen was recorded for every nonvolant mammal encountered. Surveys were conducted by myself, an undergraduate student as an assistant, and two local field helpers. All medium- to large-sized mammals were recorded (Lemos de Sá 1995), but only primate data are analyzed in this paper.

Transects were surveyed for ten days each month from May to August 1989 and from May to October of 1990 and 1991. These months correspond to the dry season in the region. During 1989, because of logistical problems and the lesser amount of time spent at the study site, data were gathered only at the Reserve and not at Jusante. Two of the five plots sampled in 1989 in the Reserve were abandoned in 1990 and 1991, again because of logistical difficulties. However, observations collected in these areas were included in the 1989 analysis to arrive at a density estimate for the entire site. Data from the various plots at a site were pooled within years to give an overall density estimate.

DATA ANALYSIS

Data were analyzed using the computer program TransAn, version 1.00, which is a flexible computer program that uses a nonparametric, shape-restricted, density estimator (Payne 1992). The shape-restricted estimator, first introduced by Johnson and Routledge (1985) and later modified by Fyfe and Routledge (1991), involves modeling the probability of detecting an individual as a function of its perpendicular distance from the transect line. TransAn requires sightings from at least four independent transect lines to calculate confidence limits. Because I had only three transects per study site, the data were divided between morning and afternoon transects to increase the number of transects to six (for the 1989 data the total number of transects was ten because there were five different plots at the Reserve). Despite inherent biases, the transect censuses are currently the most cost-effective method to evaluate large mammal densities in rainforests (Emmons 1984).

RESULTS

PRIMATE SPECIES OBSERVED

Eight primate species were recorded in all years at both the Reserve and Jusante. They were *Aotus azare, Callicebus bruneus, Pithecia irrorata, Cebus apella, Saimiri ustus,* and *Ateles paniscus* in the Cebidae and *Callithrix emiliae* and *Saguinus fuscicollis* in the Callithrichidae.

PRIMATE DENSITY ESTIMATES PRIOR TO DAMMING

Primate censuses were performed at the Reserve by Eletronorte researchers from September 1987 to February 1988 (table 15.1). Note that *Saimiri ustus* was not recorded at that time. Density estimates were based on 145 km of transect surveys. Techniques used were comparable to the ones used in this study (National Research Council 1981).

Primates represented 48.4% (n = 1,806) of all mammals captured and 47% (n = 1,352) of all animals released during the rescue/release operation (Eletronorte 1989; table 15.2). This was a relatively small number and probably had an insignificant effect on most density shifts during the years following flooding.

PRIMATE DENSITY ESTIMATES AFTER DAMMING

Primate densities (ind/km^2) were estimated by calculating group density and multiplying by mean group size (table 15.3). Figures 15.3 and 15.4 summarize group

TABLE 15.1 Primate Density Estimates at the Samuel Reserve Prior to the Flooding of the Reservoir

SPECIES	D	MGS	IND/KM^2
Ateles paniscus	2.60	5.2	13.5
Cebus apella	4.02	5.4	21.7
Pithecia irrorata	1.38	3.8	5.2
Callicebus bruneus	0.26	2.6	0.7
Saimiri ustus	—	—	—
Saguinus fuscicollis	2.00	8.5	17.0
Callithrix emiliae	0.36	15.0	5.4

Note: D equals group density (km²) and MGS mean group size (Eletronorte 1988). *Saimiri* were not recorded during the 1988 censuses.

TABLE 15.2 Number of Primates Captured at the Samuel Dam Reservoir and Number of Primates Released at the Samuel Ecological Station

SPECIES	NUMBERS CAPTURED	NUMBERS RELEASED
Aotus	104	60
Ateles	35	35
Callithrix	71	42
Saguinus	171	76
Cebus	207	180
Callicebus	348	309
Pithecia	369	324
Saimiri	501	326
Total	1,806	1,352

Source: Eletronorte 1989.

density changes at the Reserve and Jusante, respectively. Densities of *Ateles, Callithrix*, and *Saimiri* at the Reserve were high in 1990 and lower and approximately equal in 1989 and 1991. *Cebus* and *Saguinus* densities were also high in 1990 in the Reserve; however, their densities in 1991 remained high and similar to the 1990 densities instead of returning to values comparable to 1989. The densities of *Callicebus* and *Pithecia* in the Reserve were at their highest in 1989 and decreased steadily through 1990 and 1991 (table 15.3, fig. 15.3). With the exception of *Pithecia*, whose density was similar in 1990 and 1991, all other primate densities decreased substantially from 1990 to 1991 at the Jusante site (table 15.3; fig. 15.4).

TABLE 15.3 Primate Density Estimates at the Reserve and at Jusante Sites for 1989, 1990, and 1991

SPECIES	SITE	YEAR	N	D	MGS	IND/KM^2	95% DCI
Ateles paniscus	Reserve	89	27	3.15	4.2	13.2	1.53–07.43
	Reserve	90	72	6.06	3.9	23.6	3.95–09.91
	Reserve	91	66	3.69	3.0	11.1	2.14–07.06
	Jusante	90	04	0.60	9.8	5.9	0.17–02.63
	Jusante	91	01	—	1.0	—	—
Cebus apella	Reserve	89	54	5.42	3.7	20.1	3.21–09.22
	Reserve	90	71	6.63	4.3	28.5	4.15–10.31
	Reserve	91	72	6.45	4.2	27.1	4.01–10.32
	Jusante	90	32	2.92	6.0	17.5	1.59–06.15
	Jusante	91	27	2.12	3.8	8.1	
Pithecia irrorata	Reserve	89	33	3.43	3.0	10.3	1.76–06.59
	Reserve	90	17	2.07	2.5	5.2	0.75–05.06
	Reserve	91	12	1.17	3.0	3.5	0.35–02.49
	Jusante	90	21	2.60	2.9	7.5	1.16–05.57
	Jusante	91	16	2.38	3.2	7.6	1.08–06.39
Callicebus bruneus	Reserve	89	15	3.61	2.0	7.2	1.57–08.34
	Reserve	90	05	0.83	1.6	1.3	0.20–03.03
	Reserve	91	09	0.62	2.0	1.2	0.24–01.74
	Jusante	90	59	8.33	2.4	20.0	4.77–14.72
	Jusante	91	39	4.85	2.4	11.6	2.49–09.89
Saimiri ustus	Reserve	89	08	0.91	7.6	6.9	0.25–02.43
	Reserve	90	05	3.68	7.8	28.7	0.89–05.68
	Reserve	91	12	1.41	6.3	8.9	0.49–03.64
	Jusante	90	15	1.69	14.1	23.8	0.62–04.55
	Jusante	91	09	0.75	8.0	6.0	0.29–02.58
Saguinus fuscicollis	Reserve	89	14	1.85	3.3	6.1	0.71–03.88
	Reserve	90	24	4.08	3.5	14.3	1.92–09.18
	Reserve	91	20	3.70	3.8	14.1	1.51–06.53
	Jusante	90	34	5.83	3.5	20.4	2.88–11.23
	Jusante	91	28	3.67	3.7	13.6	1.74–07.79
Callithrix emiliae	Reserve	89	10	1.41	2.9	4.1	0.38–03.82
	Reserve	90	15	3.06	3.3	10.1	1.31–06.69
	Reserve	91	09	1.82	2.0	3.6	0.74–06.00
	Jusante	90	20	3.13	2.9	9.1	1.34–05.97
	Jusante	91	07	0.91	3.1	2.8	0.28–03.75
Total	Reserve	1989	161	17.59	3.6	63.3	13.86–22.81
	Reserve	1990	209	23.25	3.9	90.7	18.92–29.02
	Reserve	1991	200	18.03	3.6	64.9	13.71–23.68
	Jusante	1990	185	24.51	4.4	107.8	17.62–33.46
	Jusante	1991	127	13.83	3.5	48.4	8.86–23.14

Note: N equals number of sightings of groups; D, group density; and MGS, mean group size (based on total number of individuals sighted and total number of sightings per year).

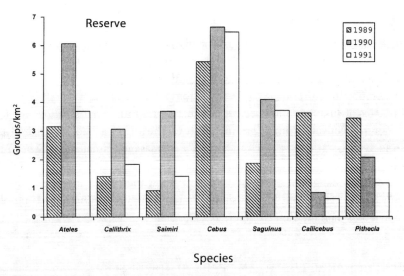

FIGURE 15.3 Primate densities at the Samuel Ecological Station (groups/km²).

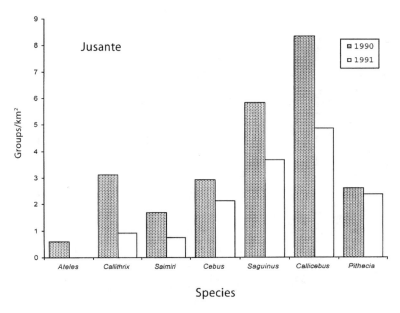

FIGURE 15.4 Primate densities at the Jusante site (groups/km²).

TOTAL PRIMATE DENSITY ESTIMATES

Primates at the Reserve comprised 61% of all mammalian sightings in 1989, 69% in 1990, and 68% in 1991. When data for all primates are pooled and density estimates are calculated for the area as a whole, the result shows a 32% increase in group density from 1989 to 1990 and a 23% decrease from 1990 to 1991 (table 15.3).

Primates at Jusante comprised 71 and 58.5% of all mammal sightings for 1990 and 1991, respectively. Primate group density for the area as a whole decreased 44% from 1990 to 1991 (table 15.3). Even though a decrease in primate density occurred in both areas from 1990 to 1991, the decrease at Jusante was almost twice that of the Reserve. The changes in densities of ind/km² show the same pattern as the group density changes; however, Jusante shows a more abrupt reduction in total number of individuals than the Reserve (table 15.3).

DENSITY CHANGES BETWEEN YEARS AT THE RESERVE

A cluster analysis comparing density results for all four years of data for the Reserve shows that the 1988 densities had only a 25.27 degree of similarity with the 1989 densities (all comparisons excluded *Saimiri* because this species was not recorded during 1988). The 1990 community still only shows a 27.94 degree of similarity with 1988. By 1991 the degree of similarity with the 1988 community increased to 61.92 (fig. 15.5).

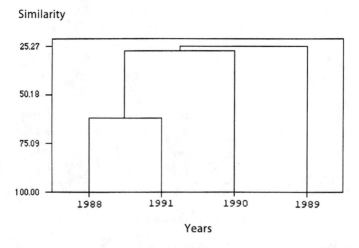

FIGURE 15.5 Cluster analysis for all four years of density data at the Reserve. Similarity levels = 61.92, 27.94, and 25.27, respectively (Minitab 10, Hierarchical cluster analysis of observations).

DISCUSSION AND CONCLUSION

The results suggest that the environmental changes created by the construction of the dam temporarily altered the mammal community in the areas adjacent to the reservoir. Degree of fluctuations in population levels for individual species varied, as did the persistence of the changes in population levels. Some of these differences can be explained by the interactions among the different species as well as by their autoecology.

PRIMATE DENSITY ESTIMATES: RESERVE

Three different patterns can be seen with primate density changes at the Reserve (fig. 15.3). The first one, seen with *Ateles*, *Callithrix*, and *Saimiri*, is a large increase in density from 1989 to 1990 and then a sharp decrease in 1991, returning to density levels similar to those found in 1989. The second pattern, which involves *Cebus* and *Saguinus*, also involves an increase in density from 1989 to 1990; however, the 1991 densities remain high. The third pattern is very different, with the densities of *Callicebus* and *Pithecia* being at their highest in 1989 and declining in 1990, and 1991.

The increase in densities from 1989 to 1990 can be explained by the migration of animals from the reservoir into the Reserve. Sixty percent of the reservoir is, on average, only 3.5 m deep in water, which allowed several arboreal species to survive for at least eight months after the flooding began. I observed flowering and fruiting trees inside the reservoir in August 1989 (five months after the completion of the filling of the reservoir). Child (1968) observed the same phenomenon during the formation of Lake Kariba as a result of the impoundment of the Zambezi river in Zimbabwe. At Lake Kariba species had different survival times that varied from four to twelve months. Child also observed that "most species standing in water came into leaf and/or remained in leaf until they died" (Child 1968:37).

Because primates spend most of their time in the middle to upper forest strata, it is reasonable to assume that the majority of the primate population was still living inside the reservoir when the 1989 survey was carried out at the Reserve. This assumption is supported by the relatively low number (1,806) of primates captured during the rescue operation in a 56,000-ha area (0.03 ind/ha) (Eletronorte 1989). In contrast, at the Tucuruí Dam site (which has a much deeper reservoir), a total of 27,039 primates were captured in a 243,000-ha area (0.11 ind/ha) (Eletronorte 1985).

When I arrived at the Samuel Dam in May 1990 all vegetation inside the reservoir was dead. The only exception was in the higher elevation lands, which formed green islands inside the reservoir. Because the reservoir is bordered by dikes on both sides and by a paved highway on the left river bank, the Reserve constituted a natural escape route for displaced animals (figs. 15.1 and 15.2). Thus the increase in primate densities in the Reserve was likely a result of the natural migration of animals between August 1989 and May 1990, caused by the loss of habitat inside the reservoir.

This scenario does not explain the third pattern of density changes detected in *Callicebus* and *Pithecia*, whose population declined steadily after 1989. However, *Callicebus*, *Pithecia*, and *Saimiri* were the most frequently captured species, with over 300 individuals released (table 15.2). This suggests that the 1989 census in the Reserve documented the density increase in these species as a result of the release operation. This argument is even more convincing when density estimates from the preflooding study done in 1988 (Eletronorte 1988), are examined. Densities for all three species prior to flooding were much lower than the ones found in 1989.

The densities for *Ateles* and *Saguinus* in the Reserve were almost identical in 1988 and 1989. Because only a few individuals were released in the Reserve, there was no reason to expect otherwise. The higher 1989 estimates for *Saimiri*, *Callicebus*, *Pithecia*, and to some extent, *Cebus* reflect the increase in density caused by the released animals. Because my study began only two months after the release was completed, it is reasonable to assume that the animals were still inside the Reserve, and that is why the 1989 densities were higher than those in 1988. As for *Callithrix*, the difference between the 1988 and 1989 densities may be a reflection of the species' characteristics. *Callithrix* are among the most cryptic primates, in both pelage and habitat (Ferrari and Rylands 1994), making them difficult to detect during transect samples. Because the 1988 censuses were not performed by my team, their low density might be a consequence of differences in researchers' detection abilities. Despite having similar body weight, the same detectability differences do not apply to *Saguinus*, which tend to use lower strata of the forest (Ferrari and Rylands 1994) and are therefore more visible during a census.

The decline in density estimates in 1991 for all species is most likely a consequence of the dispersal of animals to adjacent areas or to death. The Reserve is located adjacent to an area of continuous forest without human inhabitation or access roads (fig. 15.2), and dispersal into those areas would be the expected behavior for overcrowded animals. Because the rescue operation only removed a small fraction of the animals, the waters did not cover the tree tops and there was no incidence of high primate mortality inside the reservoir, I can reasonably state that the animals moved into adjacent areas. Together with data on biomass (Lemos de Sá 1995), this dispersal out of the Reserve implies that the Reserve was already at carrying capacity for these primate species. Not only does this information allow us to predict impacts of future reservoirs, it also gives us baseline data on carrying capacities for primates in an unhunted lowland Amazonian site, data which can be used in studies that model sustainable harvests of primates at other Amazonian sites (Bodmer and Robinson this volume).

PRIMATE DENSITY ESTIMATES: JUSANTE

There was a sharp decline in density (between 37 and 71%) from 1990 to 1991 at Jusante for all primates, except *Pithecia* (fig. 15.4). The decline in density is not very apparent for *Cebus* and *Saimiri* if we only consider the number of groups per km^2, however, these two species showed drastic reduction in mean group sizes over the years (table 15.3). Richard-Hansen, Vié, and de Thoisy (2000) documented extensive fragmentation of troops for translocated *Alouatta seniculus* in French Guiana, attributing this reduction in group size to stress, interactions with resident animals, and lack of familiarity with the new environment, among other factors. A similar situation may have occurred with the naturally migrating primate groups in this study. Hence, if we consider the decline in the number of individuals per km^2, both species also show drastic decreases in densities from 1990 to 1991.

Even though there was no sampling at Jusante during 1988 or 1989, it is logical to assume a similar effect on primate communities in both areas as a consequence of the creation of the reservoir, that is, the 1990 densities were artificially high because of an influx of animals after creation of the reservoir. A buffer area for the protection of the dam turbines was created inside the reservoir by clear cutting the forest closest to the dam, which together with several construction projects and the concentration of human activities near the dam could have inhibited animal migration to Jusante. However, migration did occur, probably because of the sheer proximity of the area to the reservoir (animals stranded inside the reservoir near the dam could probably see green forest on the other side) (fig. 15.2).

Despite the lack of data for Jusante in 1989, density estimates for six of the seven primates sampled at Jusante most likely fit the first pattern of density change described for primates at the Reserve (density increase in 1990, followed by a sharp decrease in 1991). The seventh species, *Pithecia*, fits the second pattern of density change (density stays at similar levels from 1990 to 1991). Because no data exist for 1989, it is not possible to determine if the third pattern described at the Reserve (a continued decrease in density) was present at Jusante. However, because the explanation given for this pattern in density change was the active release of animals in the Reserve, such pattern would not be expected to appear at Jusante.

SPECIES-SPECIFIC RESPONSES

Differences in crude densities and in the degree of density decline among species and sites are most likely the result of differences in species behavior and/or habitat requirements.

Ateles *Ateles* are frugivore-herbivores (Eisenberg 1981; Robinson and Redford 1986a, 1989), with 83 to 90% of their diet consisting of fruits and the remainder of other plant parts (van Roosmalen and Klein 1988). Because the distribution of fruits in a forest is widely scattered, *Ateles* density is probably restricted by the availability of this food type (Robinson and Ramirez 1982). Home-range size increases as group weight increases (Eisenberg 1979), and *Ateles* are the largest of all primates in the area requiring, in Suriname, 12.2 ha per individual (van Roosmalen 1980; Robinson and Janson 1987). The new arrivals at the Reserve were most likely displaced to areas outside the Reserve by the resident groups because of the unavailability of fruit crops large enough to maintain the higher population density.

Their almost complete absence from Jusante can be explained by the fact that they are restricted to or occur in higher densities only in primary forest, using upper levels of canopy and emergent trees (Mittermeier and van Roosmalen 1981; Robinson and Ramirez 1982; van Roosmalen and Klein 1988). Because the forest at Jusante has a lower, more open canopy with fewer emergent trees than the Reserve, it is not surprising to find that the Reserve represented a more suitable habitat for the species.

Callithrix *Callithrix* are insectivore-omnivores, with more than 50% of their diet consisting of invertebrates (Eisenberg 1981; Robinson and Redford 1986a, 1989). They are also adapted to feed on plant exudates at certain times of the year in order to compensate for seasonal scarcities in the availability of fruits (Ferrari and Lopes Ferrari 1989; Ferrari 1993; Rylands and Faria 1993). They attain highest densities in second growth forest and edge habitat.

Callithrix species tend to have larger group sizes and smaller home ranges than *Saguinus* species and generally occur at higher densities (Ferrari and Lopes Ferrari 1989; Rylands and Faria 1993). Average group size and densities of *Callithrix* in both of my study sites were lower than those of *Saguinus* (table 15.3), in contrast to previous studies. This difference may be partly a consequence of their cryptic nature, as described earlier.

Saimiri *Saimiri* are classified as frugivore-omnivores, with more than 50% of their diet composed of fruits and the remainder mostly of invertebrates and vertebrates (Eisenberg 1981; Robinson and Redford 1986a, 1989). They are habitat specialists, typical of flooded and riverine forests (Eisenberg 1979; Freese et al. 1982; Rylands and Keuroghlian 1988). The species is known for its preference for more open, secondary habitats, and it is most often encountered in liana forests (Mittermeier and van Roosmalen 1981; Johns and Skoruppa 1987). Neither of the study sites, the Reserve or Jusante, included flooded forests. Although the Jusante site is closer to the Jamarí river and has a more open forest structure, transect censuses started at a distance of 500 m away from the river's edge. Therefore high densities of *Saimiri* were not expected at either site. The high densities in the Reserve in 1990, as well as the very high number of individuals per km^2 at Jusante (due to a larger mean group size; table 15.3), probably occurred when animals living along the Jamarí river inside the reservoir moved to these areas in search of new suitable habitat. Because neither area is suitable habitat, the animals most likely dispersed along the Jamarí river, causing the density decrease seen in 1991.

Cebus *Cebus* are also classified as frugivore-omnivores, with more than 50% of their diet composed of fruits and the remainder mostly of invertebrates and verte-brates (Eisenberg 1981; Robinson and Redford 1986a, 1989). The species has a broad habitat tolerance (Eisenberg 1979). *Cebus* are opportunistic and usually well able to persist in disturbed forest (Johns and Skoruppa 1987), an ability that makes them the most adaptable primate species in the Neotropics (Mittermeier and van Roosmalen 1981). It is not surprising then that they were able to maintain high population density in the Reserve.

The reduction of 54% in the number of individuals estimated per km² at Jusante from 1990 to 1991 (table 15.3) seems inconsistent with their ecology. However, be-cause Palmae species were more abundant in the Reserve than at Jusante and be-cause *Cebus apella* rely heavily on palms in a number of different ways (insect for-aging, fruits, seeds, flowers, and many other plant parts) (Mittermeier and van Roosmalen 1981; Terborgh 1983), it is possible that the carrying capacity for the species is higher at the Reserve than at Jusante.

Saguinus *Saguinus* is classified as an insectivore-omnivore, with more than 50% of its diet consisting of invertebrates (Eisenberg 1981; Robinson and Redford 1986a, 1989). According to Rylands and Keuroghlian (1988), optimal habitat for this species includes secondary forest and forest edge mixed with tall primary forest. Several studies have shown that the species occurs in greater densities in secondary forest near natural clearings than in mature forest (Eisenberg and Thorington 1973; Mittermeier and van Roosmalen 1981; Robinson and Ramirez 1982; Johns and Skoruppa 1987). Emmons (1984) concluded that *Saguinus* density appeared to have increased in some areas where large monkeys had been exterminated. *Saguinus* also overlap with *Cebus* in most habitat and diet categories (Mittermeier and van Roosmalen 1981). The lower densities of *Ateles* and *Cebus* at Jusante and the increased edged habitat at the Reserve created by the reservoir were most likely favorable factors influencing the maintenance of high *Saguinus* densities at both sites in 1990 and 1991.

Callicebus *Callicebus* is also classified as frugivore-omnivore, with more than 50% of its diet composed of fruits and the remainder mostly of invertebrates and vertebrates (Eisenberg 1981; Robinson and Redford 1986a, 1989). The species oc-curs in greatest densities in areas characterized by forest openings with early suc-cessional vegetation, spending more time in the lower canopy levels and understo-ry vegetation (Kinsey 1981; Terborgh 1983; Robinson and Redford 1986a; Robinson, Wright, and Kinzey 1987). The more open vegetation at the Jusante site most likely created a more suitable habitat for *Callicebus* than in the Reserve, a possible explanation for their much higher densities at Jusante (figs. 15.3 and 15.4).

Interference competition with *Cebus* might also affect *Callicebus* densities. Both species are catholic in their diet; however, *Cebus* generally have larger group size and are more aggressive during interspecific encounters, possibly displacing *Callicebus* groups from feeding trees. According to Emmons (1984), troops of larger monkeys, such as *Cebus*, physically prevent access to fruit sources by small ones, such as *Saguinus* and *Callicebus*. The lower densities of *Cebus* at Jusante might benefit *Callicebus*, affecting their population positively.

Pithecia *Pithecia* are also frugivore-omnivores, with more than 50% of their diet composed of fruits and the remainder mostly of invertebrates and vertebrates (Eisenberg 1981; Robinson and Redford 1986a, 1989). They are usually found in the understory and lower to middle parts of the canopy (Mittermeier and van Roosmalen 1981), and they occur in gallery and both primary and secondary forest (Robinson and Ramirez 1982). *Pithecia* are always rare (Mittermier and van Roosmalen 1981; Robinson, Wright, and Kinzey 1987; Rylands and Keuroghlian 1988), despite the fact that they have no distinct habitat preference. However, their rarity may indicate that they are specialists within the forest they occupy or at least dependent on certain floristic communities (Rylands and Keuroghlian 1988). According to Johns and Skoruppa (1987), *Pithecia* are able to feed on fruits from some of the early colonizing trees, an ability that might explain their higher densities at Jusante.

DENSITY CHANGES BETWEEN YEARS AT THE RESERVE

The low degree of similarity between the 1988 and 1989 densities (fig. 15.5) was most likely due to the increased densities of *Pithecia* and *Callicebus* as a result of both the release of captured animals and the movement of free-ranging individuals into the area. The 1990 community still shows a low degree of similarity with that of 1988, possibly due to the increase in densities as a consequence of the heavy migration of animals to the Reserve. By 1991 the degree of similarity with the 1988 community increased to 61.92. The community seems to be in the process of returning to its original community structure (fig. 15.5).

TOTAL DENSITY

Estimates of total primate density show a general trend of density increase from 1989 to 1990 and then a decrease in 1991 (table 15.3). These changes are consistent with the assumption that the animals inside the forests of the reservoir moved to both study sites between August 1989 and May 1990 and then dispersed to adjacent forest between November 1990 and May 1991. Even though 1989 data for Jusante do not exist, I suspect that the total 1989 primate density for Jusante was, like the Reserve, similar to its 1991 estimate.

The sharper density decrease at Jusante is most likely related to lower capacity of the forest to support high primate densities. This is perhaps most obvious for *Cebus*, whose density remained at high levels in the Reserve but decreased drastically at Jusante (table 15.3). Because mean group size for *Cebus* and *Saimiri* at Jusante was much higher in 1990 than in 1991, the number of individuals per km^2 shows a more abrupt decline in density than the group density.

CONCLUSIONS AND IMPLICATIONS FOR FUTURE DAMS

The cost-benefit ratio of rescue operations need to be reevaluated before future rescue operations are initiated. The rationale behind rescue operations has always been to save animals from unquestionable death. However, animal mortality at hydroelectric dam sites is related more to loss of habitat than to drowning. Several large mammals have relatively good swimming capabilities (Child 1968), birds may fly away, and amphibians and some reptiles may survive by swimming to dry land. On the other hand, habitat loss, measured by square kilometers flooded, is irreversible. Kariba reservoir flooded an area of 5,462 km^2 (ironically enough, its rescue operation captured the least total number of animals), Brokopondo 1,683 km^2, Tucuruí 2,430 km^2, Balbina 2,600 km^2, and Samuel, the smallest of all, 560 km^2.

Furthermore, animals captured during rescue operations are usually released on the nearest piece of dry land, without any concern for the animal community inhabiting the area of release. My results on primate density changes at the Samuel release site demonstrated not only that they can move from the flooded area on their own but also that the site's carrying capacity will determine if they will stay on the release area or not. Of the 3,729 mammals rescued from the Samuel reservoir, only 2,374 were released (1,352 of these were primates). The movement of animals after the flooding confirmed that primates can move on their own (provided that the filling of the reservoir does not cover the tree tops).

Rescue operations have become a public relations strategy used by power companies to appease public opinion. The only way to make power companies change their policies is to inform the public and to give power companies better options for rescue/conservation programs. Some suggestions for future rescue programs are listed below:

1. Rescue operations should be confined to (a) species known to be endangered and/or vulnerable, which should be rescued when stranded; (b) species that are unable to escape their flooded environment; sometimes females with infants are trapped on small islands unable to swim with their young; (c) species that could be used for research; and (d) species that could be used for reestablishing depleted populations elsewhere; some areas in the Amazon have been heavily hunted, and large species are sometimes rare.
2. Animals should only be released on sites that have been studied previously and that have abnormally low densities, such as heavily hunted sites.
3. The practice, established at Balbina, of donating rescued animals to research in-

stitutions should be continued. If rescued animals are used to supply research institutions, there will be a decrease in the number of wild animals removed from other sites for research purposes.

4. Protected areas should be created to compensate for the habitat lost due to the construction of reservoirs.

5. Companies should invest in professional staff to design, conduct, and supervise conservation program at dam sites.

16

Niche Partitioning Among Gray Brocket Deer, Pampas Deer, and Cattle in the Pantanal of Brazil

LAURENZ PINDER

The global introduction of domestic livestock into tropical savannas and temperate prairie ecosystems has raised debates about the conservation and management implications of these introductions on wildlife, in particular on native ungulates. In addition to the obvious impact caused by the introduction of extraneous pests and diseases on native ungulate populations, there is evidence that dietary overlap with livestock may derive positive or negative implications to the coexisting herbivores. For instance, seasonal grazing by livestock may improve the nutritive quality of autumn and winter browse for wild ungulates (Alpe, Kingery, and Mosley 1999). On the other hand, vegetation modification and overgrazing of shared scarce food items potentially causes a decrease of food availability for native species (Murray and Illius 2000; Puig et al. 2001).

Despite the fact that more and more savannas and prairies are encroached and replaced by agriculture and introduced pastures, extensive areas of South America still harbor a rich diversity and abundance of wildlife in coexistence with centuries-old extensive ranching on native pastures. In seasonally flooded plains, such as the Venezuelan Llanos and the Brazilian Pantanal, this long-lasting coexistence has been possible because of the low densities of livestock and the ecosystem's inaptitude for other agricultural economic activities. In these vast areas intensification of cattle operations and flood pulse/hydrologic alterations are the most threatening potential hazards to native ungulates.

Few studies have verified potential resource competition caused by livestock introduction into South American grasslands (Jackson and Giuletti 1988; Larghero 2001; Puig et al. 2001). The existing studies indicate that trophic resource competition would intensify with the decrease in food diversity and patchiness that result from pasture management or dry years. Such studies are highly significant, and better ranching management practices need to be developed to allow wildlife and livestock to coexist as ranching operations within native grasslands intensify. A baseline for the development of better management practices is an understanding on how species segregate ecologically, i.e., how species share available resources.

Segregation along one or more niche dimensions facilitates partitioning of resources, and, thereby, ecological separation of species (MacArthur 1972). Among sympatric ungulates, coexistence is presumably correlated with digestive anatomy and mouth dimensions (Hofmann 1973; Owen-Smith 1989; Gordon 1989; Illius and Gordon 1991; Gross, Alkon, and Demment 1996). Many studies of ungulates have indicated that coexistence is facilitated principally by dietary differences (plant growth stage and parts eaten) and to a lesser extent by the plant species per se or by spatial and temporal differences in habitat use (Dunbar 1978; Jarman and Sinclair 1979; Hansen, Mugambi, and Baun 1985; Murray and Brown 1993). Yet, simulation modeling, incorporating bioenergetic requirements, suggests that partitioning is achieved primarily through habitat segregation and plant parts favored (Owen-Smith 1989). Thus, morphological differences in mouth dimensions and energy requirements lead distinct species of ungulates to achieve optimal foraging performance in those habitats where selected vegetation structures are abundant (Owen-Smith 1989; Gordon 1989; Perez and Gordon 1999; Perez, Gordon, and Nores 2001).

In this paper I examine mechanisms for segregation among gray brocket deer (*Mazama gouazoubira*), pampas deer (*Ozotoceros bezoarticus*), and cattle at Caiman, a 500-km^2 ranch in southeastern Pantanal, Brazil. I tested the hypotheses that these ruminant species should be segregated by at least one niche dimension and that each species should select those habitats in which their preferred foods were most abundant. The results of this study are particularly important to the conservation of these cervids in the Pantanal, as studies in Argentina suggested that cattle compete with pampas deer (Jackson and Giuletti 1988). Most of the remaining habitat for pampas deer today is restricted to cattle ranches (Pinto 1994; Por 1995).

METHODS

The study area (19°57′ S, 56°25′ W) is located in a wetland savanna. Mean annual rainfall is 1,773 mm, with monthly means ranging from < 30 mm in August to > 320 mm in January. Flooding occurs in March at the end of the rainy season.

Five fresh fecal samples for each ruminant were collected monthly from November 1991 through October 1992. Samples were obtained opportunistically in the study area during behavioral sampling of different individuals of the deer and cattle populations. Pellets were collected and preserved in ethanol just after being dropped. Additionally, brocket deer pellets were collected from latrines early in the morning, at the edge of forest patches commonly used for rumination or resting. As individuals were observed throughout the ranch area, it is assumed that the fecal samples were representative of the deer population in the study area. Samples were pooled by season: rainy (November to January), flood (March to May) and dry (July to September). Fragments of plant species were identified microhistologically (Sparks and Malechek 1968).

The proportion of different plant species in the diet was estimated using frequency counts of 100 microscopic fields per sample (Sparks and Malechek 1968; Johnson 1982). Trophic diversity based on plant species present in the diet was expressed by the Shannon-Wiener Index and niche breadth by Levins' Index, as standardized by Hurlbert (Krebs 1989). Similarity of diet between paired ungulate species was calculated with the Percentage Similarity Index (Renkonen 1938) and niche overlap among ruminant species with Horn's Index of Overlap (Horn 1966; Ricklefs and Lau 1980).

Availability of plant species was estimated by surveying the vegetation, using a modification of the point-sampling method (Levy and Madden 1933; Croker and Tiver 1948; Mantovani and Martins 1990). Representative tracts of ten plant formations were selected for sampling in the study area: marsh ponds, moist basins, short grass, medium grass, medium/tall grass, short grass/tall grass, tall grass, scrub, forest edge, and improved pasture (native pasture seeded with introduced grasses). Plant formations are described by Pinder (1997) and Pinder and Rosso (1998). One hundred sampling points were collected using five 25-m^2 quadrats in each plant formation, totaling 1,000 sampling points per season, rainy, flood, and dry. The preference for consumption of a plant species by deer and cattle was calculated by subtracting the proportion of the species in the diet from the proportion of the species available in the environment (Strauss 1979).

Two monthly surveys were conducted in the early morning and late afternoon along a 30-km trail connecting the SE and NW ends of the ranch to estimate the density and habitat use of cervids and cattle (Routledge and Fyfe 1992). The maximum sighting distances were 150 m for pampas deer and 140 m for gray brocket deer. Density of cattle was calculated based on the average cattle population in the ranch in 1991/1992. It is assumed that deer or cattle neither avoided nor were attracted by the trail. Additionally, one adult male of each cervid species was radio-collared to provide an unbiased sample of habitat use during daylight hours. Data on habitat use for all individuals were analyzed only for animals observed foraging or traveling. Radio-collared individuals were sampled at 30-minute intervals to avoid autocorrelation of data (Swihart and Slade 1985).

The availability of the different habitats was estimated by randomly selecting five coordinates within the ranch as start points for five 500-m-long transects. The direction of each transect was also randomly selected. Plant formations were registered at 50-m intervals along the transects, totaling 500 points distributed randomly within the floodplain (Marcum and Loftsgaarden 1980). Use versus availability of habitats was tested with the Chi-square goodness-of-fit-test and the related multiple comparisons of the Bonferroni z-test (Neu, Byers, and Peek 1974; Byers, Steinhorst, and Krausman 1984).

Friedman's test was used to verify consistency in use of food categories, i.e., grasses, forbs, browse, and fruits, among the ungulates, and the STP test to verify which food categories were consumed in greater proportions by each species (Sokal and Rohlf 1969). The Kruskal-Wallis test was employed in multiseasonal comparisons of food categories within species (Sokal and Rohlf 1969). Differences between consumption of food categories among paired seasons were determined by the Mann-Withney U-test (Sokal and Rohlf 1969).

RESULTS

DENSITY AND BIOMASS OF CATTLE AND CERVIDS

Cattle had the highest ecological density among the three species of ruminants ($36/km^2$) and almost the totality of the biomass (8,640 kg/ km^2). Cattle density increased on higher elevations during the peak of flooding when part of the foraging area was covered by 10 cm of water. Pampas deer and gray brocket deer occurred in low densities (0.68 and $1.08/km^2$, respectively). The estimated cervid population for the ranch was 205 (95% CI = 121–359) pampas deer and 412 (95% CI = 229–653) gray brocket deer. The sympatric marsh deer (*Blastocerus dichotomus*) was rare (0.05 individuals/km^2), and used swampland avoided by the other species. The red brocket deer (*Mazama americana*) did not occur in the study area.

DIETARY SIMILARITIES

Fecal analysis demonstrated that cattle, pampas deer, and gray brocket deer consumed the same food categories but in different proportions (Friedman, P < 0.01). The cumulative number of species consumed by each ruminant throughout the year indicated that samples were adequate to perform the comparative analysis between and among species (fig. 16.1). Cattle concentrated their foraging on grasses, whereas pampas deer and brocket deer selected forbs and browse respectively (STP, P < 0.01) (fig. 16.2).

Similarities in consumption of food categories between cattle and the cervids were not significant but were high between pampas and brocket deer. Cattle selected grasses and sedges year round, whereas pampas deer preferred forbs and browse and gray brocket deer browse and forbs. The largest overlap in diet between cattle and pampas deer occurred during the rainy season when about 20% of the pampas deer diet was newly grown grasses. The burst of pasture growth makes competition among pampas deer and cattle unlikely. In fact, cattle may be beneficial to pampas deer, as grazing by cattle increases the abundance of sprouting grass, which is more nutritious and palatable for pampas deer. Additionally, cattle reduce the height of the pasture and probably make forbs within the pasture more accessible for pampas deer. The consumption of different plant species within the same food categories resulted in an overall low diet similarity between pampas and gray brocket deer (table 16.1a). Similarities in diet were higher between the two deer species during the flood season when pampas deer increased their

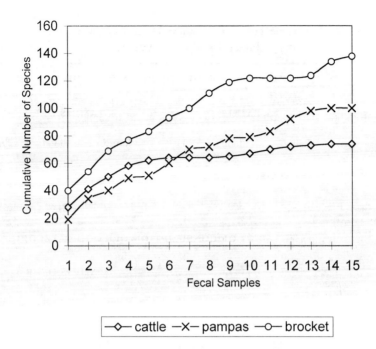

FIGURE 16.1 Relationship between cumulative number of plant species in fecal samples, as determined by microhistological analysis, and number of cattle, pampas deer, and gray brocket deer fecal samples, Pantanal Study Area (Caiman Ranch), 1991–1992.

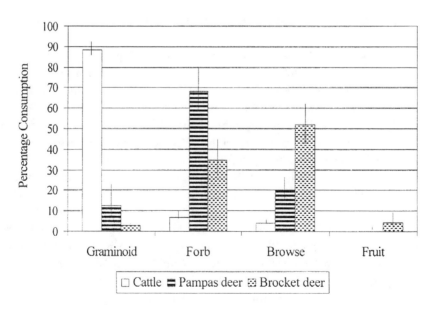

FIGURE 16.2 Mean percentage consumption of food categories by cattle, pampas deer, and gray brocket deer, Pantanal Study Area, 1991–1992. Vertical bars indicate 95% confidence intervals.

TABLE 16.1 Percentage Similarity Indices for Plant Species and Food Categories Consumed Seasonally by Cattle, Pampas Deer, and Gray Brocket Deer, Pantanal Study Area, 1991–92

(A)	SIMILARITY OF CONSUMPTION OF FOOD SPECIES					
	CATTLE			**PAMPAS DEER**		
	Rainy	Flood	Dry	Rainy	Flood	Dry
Cattle	—	—	—	—	—	—
Pampas deer	33.44	10.85	9.49	—	—	—
Brocket deer	7.42	8.17	6.23	17.17	24.67	26.04

(B)	SIMILARITY OF CONSUMPTION OF FOOD CATEGORIES					
	CATTLE			**PAMPAS DEER**		
	Rainy	Flood	Dry	Rainy	Flood	Dry
Cattle	—	—	—	—	—	—
Pampas deer	39.22	16.67	13.79	—	—	—
Brocket deer	14.30	11.06	13.55	55.62	81.49	65.48

consumption of browse, and gray brocket deer increased their consumption of forbs (table 16.1b).

However, only two plant species were consumed by both pampas and gray brocket deer in quantities exceeding 6% within the same season: *Melochia villosa* (Sterculiaceae forb) in the flood season (12.9% and 8.3%, respectively) and *Sida santamarensis* (Malvaceae forb) in the dry season (6.5% and 8.5%, respectively) (table 16.2). No plant species that contributed more than 6% in the diet of one cervid also contributed more than 6% in the diet of the other during the rainy season.

Gray brocket deer consumed species mostly associated with forest edge or scrub during the rainy and flood seasons: *S. santamarensis* (forb), *Melochia pyramidata* (shrub), *M. villosa* (forb), *Chomelia obtusa* (shrub), and *Vernonia scabra* (shrub). *Euphorbia thymifolia* (shrub) was the only species abundantly consumed by gray brocket deer that was associated with open vegetation (short grass). During the dry season, however, gray brocket deer consumed a greater proportion of forbs associated with tall grass and scrub: *Richardia grandiflora*, *Wedelia brachycarpa*, and *C. castaneifolia* (table 16.2). The majority of the forage consumed by pampas deer, especially during the dry season, was associated with moist soils, e.g., *Melochia simplex*, *Caperonia castaneifolia*, and *Hydrolea spinosa*.

TABLE 16.2 Seasonal Consumption of Plant Species (> 6% of the diet) by Cattle, Pampas Deer, and Gray Brocket Deer, Pantanal Study Area, 1991–92

FAMILY	SPECIES	CONSUMPTION BY CATTLE (%)			
		Rainy	Flood	Dry	Mean
Poaceae	Axonopus purpusiii	10.10	8.06	20.60	12.92
	Hymenachne amplexicaulis	6.96	5.43	2.86	5.08
	Mesosetum chaseae	15.20	33.40	4.84	17.81
	Paspalum pontanalis	5.16	5.73	6.24	5.71

FAMILY	SPECIES	CONSUMPTION BY PAMPAS DEER (%)			
		Rainy	Flood	Dry	Mean
Poaceae	Mesosetum chaseae	16.16	0.77	0.32	5.75
Euphorbiaceae	Caperonia castaneifolia	2.68	0.00	6.05	2.91
Hydrophyllaceae	Hydrolea spinosa	1.81	0.62	9.96	4.13
Malvaceae	Sida santamarensis	0.52	0.07	6.47	2.35
Onagraceae	Ludwigia longifolia	8.56	24.70	0.00	11.08
Pontederiaceae	Eichhornia azurae	0.00	16.97	0.00	5.66
Sterculiaceae	Melochia simplex	17.39	0.72	43.78	20.63
	Melochia villosa	9.61	12.90	3.39	8.63

FAMILY	SPECIES	CONSUMPTION BY BROCKET DEER (%)			
		Rainy	Flood	Dry	Mean
Compositae	Vernonia scabra	6.81	16.16	1.02	8.00
	Wedelia brachycarpa	0.08	2.05	13.58	5.24
Euphorbiaceae	Caperonia castaneifolia	1.00	0.09	8.37	3.15
	Euphorbia thymifolia	0.04	7.21	0.00	2.57
Malvaceae	Sida santamarensis	12.21	11.52	8.48	10.74
Rubiaceae	Chomelia obtusa	3.57	10.49	0.42	4.83
	Richardia grandiflora	3.16	0.82	17.46	7.15
Sterculiaceae	Bytneria dentata	14.17	2.35	0.67	5.73
	Melochia pyramidata	19.29	1.57	5.00	8.62
	Melochia villosa	2.87	8.33	5.09	5.43

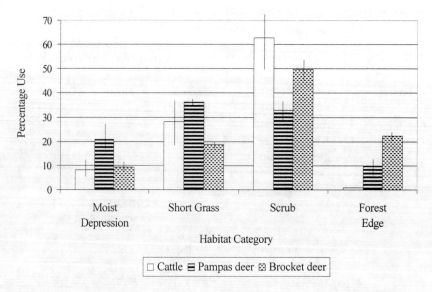

FIGURE 16.3 Mean percentage use of habitats by cattle, pampas deer, and gray brocket deer, Pantanal Study Area, 1991–1992. Vertical bars indicate standard deviations.

HABITAT USE

Pampas deer used open vegetation (moist depression and short grass) most frequently, whereas gray brocket deer selected habitats that were less affected by the flood and that had a higher density of shrubs (scrub and forest edge, fig. 16.3). Cattle had somewhat intermediary preferences, selecting grasslands and scrub and avoiding moist depression and forest edge.

Because climatic changes were extreme in the Pantanal (flooding versus drought), ungulates exhibited distinct habitat selection among seasons (table 16.3). Cattle selected short grass during the rainy season but used scrub almost exclusively during the flood season when water covered the short grass. During the dry season, observations of short grass habitat showed that graminoids were rapidly depleted there, and cattle once again selected scrub, where they found higher biomass of forage—medium and tall grasses growing in these plant formations along with shrubs and vines are eaten by cattle. Pampas deer were habitat-limited only during the flood season when inundation almost totally covered the vegetation in moist depressions. Gray brocket deer increased the use of open habitats only in the dry season when new growth in their most utilized habitats, scrub and forest edge, was reduced. During the dry season many trees shed their leaves as protection against excessive evapotranspiration. The only sources of green and nutritious leaves are the moist soils in depressions existing amidst the grasslands that become flooded during a few weeks of the year.

As a result of the shifts in habitat selection, the similarity in habitat use among

TABLE 16.3 Seasonal Habitat Selection by Cattle, Pampas Deer, and Gray Brocket Deer, Pantanal Study Area, 1991–92

		HABITAT SELECTION[a]			
SEASON	SPECIES	MD[b]	SG[c]	SC[d]	FE[e]
Rainy	Cattle	Avoided	Selected	NS	Avoided
Rainy	Pampas deer	NS	Selected	Avoided	Avoided
Rainy	Brocket deer	Avoided	Avoided	Selected	NS
Flood	Cattle	Avoided	Avoided	Selected	Avoided
Flood	Pampas deer	Avoided	NS	NS	NS
Flood	Brocket deer	Avoided	Avoided	Selected	NS
Dry	Cattle	Avoided	Avoided	Selected	Avoided
Dry	Pampas deer	NS	Selected	NS	Avoided
Dry	Brocket deer	Avoided	NS	NS	Selected

[a]Bonferroni z-test, $P < 0.05$. Avoided is habitat used less than expected based on its availability; selected, habitat used more than expected based on its availability; NS, no selection, habitat used in proportion to its availability.
[b]Moist depression.
[c]Short grass.
[d]Scrub.
[e]Forest edge.

TABLE 16.4 Percentage Similarity Indices for Habitats Used Seasonally by Cattle, Pampas Deer, and Gray Brocket Deer, Pantanal Study Area, 1991–1992

	SIMILARITY OF USE OF HABITATS					
	CATTLE			PAMPAS DEER		
	Rainy	Flood	Dry	Rainy	Flood	Dry
Cattle	—	—	—	—	—	—
Pampas deer	75.91	53.92	68.03	—	—	—
Brocket deer	65.40	69.92	77.11	61.70	78.00	72.30

the three ungulates differed from season to season. The greatest habitat overlap between cattle and pampas deer occurred during the rainy season, whereas the greatest overlap between cattle and gray brocket deer occurred during the dry season. In contrast, the greatest overlap between pampas and gray brocket deer took place during the flood season (table 16.4).

NICHE OVERLAP AND PARTITIONING

Overlap indices for the three ungulates indicated that they were segregated more on the basis of plant species consumed than either food categories consumed or habitats used. Dietary overlap among the three ungulates collectively, considering both plant species and food categories, was inversely related to the overlap in habitat use (table 16.5). Overlap in diet was greatest and overlap in habitat use was least during the rainy season when new growth was abundant.

CATTLE AND DEER RESPONSES TO ENVIRONMENTAL CHANGES

Cattle and cervids exhibited the greatest diet diversities and niche breadths during the rainy season, which coincided with the season of surplus forage. Climate in the Pantanal affected cattle only during the flood season when their diet was less diverse. The inclusion of some browse species in the diet during the dry season allowed cattle to maintain a diet diversity comparable to the rainy season (table 16.6).

Gray brocket deer maintained diet diversity and niche breadth throughout the

TABLE 16.5 Niche Overlap Indices for Cattle, Pampas Deer, and Gray Brocket Deer by Season, Pantanal Study Area, 1991–92

	NICHE OVERLAP INDEX		
SEASON	Plant Species	Food Categories	Habitat
Rainy	0.83	1.50	2.14
Flood	0.73	1.32	2.16
Dry	0.66	1.34	2.20

TABLE 16.6 Seasonal Diversity and Niche Breadth for Cattle, Pampas Deer, and Gray Brocket Deer by Season, Pantanal Study Area, 1991–92

	CATTLE		PAMPAS DEER		BROCKET DEER	
	H'[a]	Ba[b]	H'	Ba	H'	Ba
Rainy	3.15	0.20	3.08	0.11	3.24	0.11
Flood	2.79	0.09	2.82	0.07	3.20	0.10
Dry	3.31	0.18	2.39	0.04	3.20	0.09

[a]Shannon-Wiener Diversity Index
[b]Levins' Standardized Niche Breadth

TABLE 16.7 Percentages of Food for Cattle, Pampas Deer, and Gray Brocket Deer by Season, Pantanal Study Area, 1991–92

	CATTLE			PAMPAS DEER			GRAY BROCKET DEER		
	Rainy (%)	Flood (%)	Dry (%)	Rainy (%)	Flood (%)	Dry (%)	Rainy (%)	Flood (%)	Dry (%)
G[a]	87.04	90.52	87.62	26.25[f]	7.18	3.79	1.32	3.04	3.55
F[b]	10.00	7.00	4.41	41.02[f]	47.02	81.13	22.71[f]	34.12	47.13
B[c]	2.96	2.48	5.60	31.84[f]	45.62	14.98	60.32	55.71	40.41
Ft[d]	0.00	0.00	0.00	0.05	0.00	0.10	8.42	2.76	1.71
U[e]	0.00	0.00	2.50	0.84	0.08	0.00	7.23	4.37	7.20

[a]Graminoids
[b]Forbs
[c]Browse
[d]Fruits
[e]Unidentified
[f]Percentage consumption is significantly different among seasons (Kruskal-Wallis test, $P < 0.05$).

year by increasing the contribution of grassland forbs in their diet during the flood and dry seasons. As browse matured and became less digestible, and presumably richer in secondary compounds, gray brocket deer foraged away from the forest edge more frequently.

Pampas deer were highly affected by seasonal changes in food availability as evidenced by a narrowing of their diet diversity and niche breadth from the rainy season to the dry season. To illustrate, pampas deer foraged on graminoids, forbs, and browse during the rainy season, principally on hydrophytic forbs and browse during the flood season, and mainly on forbs during the dry season (table 16.7). Foraging observations during the dry season indicated that pampas deer traveled from one moist depression to the next to forage on forbs. Therefore the dry season was the period when pampas deer diet was most specialized.

DISCUSSION

NICHE PARTITIONING

Ungulates in the Pantanal partitioned resources along all three niche dimensions (plant species, plant categories, and habitat type). These ruminants consumed different plant species, many of which occurred mainly on their preferred habitats, i.e., where those species were more abundant (Pinder 1997). Despite the fact that diet overlap decreased as habitat overlap increased among the three ungulates, there was no evidence that current competition was responsible for such segregation. Competition theory would predict that, when resources are limiting, the cervids should diverge in feeding characteristics because of competitive effects (Smith et al. 1978).

In contrast, the dyad pampas deer/gray brocket deer simultaneously increased habitat and diet similarities from rainy to dry season in a way comparable to that observed for some ungulates in the African savanna (Jarman 1971; Dunbar 1978). In those studies, ungulate species with low diet overlap during the rainy season increased overlap in the dry season, but species with generally high diet overlap in the rainy season decreased habitat overlap in the dry season when food availability was presumably reduced. Dunbar (1978) suggested that the simultaneous increase in diet and habitat similarities between those ungulates with low diet overlap in the rainy season indicated that there was no interspecific competition. Their niche overlap was probably too low to require any further reduction during the dry season, which appears to be the case with pampas deer and gray brocket deer in the Pantanal. Thus, the feeding habits of these ungulates supported the existence of a positive relationship between habitat and diet, i.e., as deer species increased use of the same habitat, diet overlap increased also (Gordon 1989; Owen-Smith 1989; Pinder 1997).

Distinct nutritional requirements of each ruminant and their respective gut capabilities for processing plant material were more likely responsible for the segregation observed. The gut capacity of herbivores determines, in part, their capability for digestion of fiber (Demment and Van Soest 1985; Hofmann 1988). Large herbivores, such as cattle, are able to survive on high-fiber diets because their larger digestive systems allows a lower passage rate necessary for the digestion of cell walls (Parra 1978; Van Soest 1982; Illius and Gordon 1991). Large herbivores would not be expected to feed on a low-fiber diet because low-fiber foods are rare, and thus the amount of energy expended in searching and consuming these foods is prohibitive in comparison to the total daily energy requirements of large herbivores (Parra 1978; Demment and Van Soest 1985; Murray and Brown 1993).

In contrast, a small ruminant such as the gray brocket deer requires less total energy but more energy per unit of biomass. A faster energetic return from digestion of plant material can be reached by the fermentation and digestion of cell contents instead of cell walls (Van Soest 1982). Consequently, small ruminants need a low-fiber diet, which can be digested directly by vertebrate enzymes or fermented rapidly by microbes. In contrast to graminoids, low-fiber tissue is found in leaves of trees and shrubs, which constituted the main diet of gray brocket deer in the Pantanal. Pampas deer, which weigh twice as much as gray brocket deer, presumably have a larger gut capacity but require less energy relative to their body size compared to gray brocket deer. This larger gut capacity allows for a slower passage rate and better utilization of lower quality food than gray brocket deer.

Owen-Smith's hypothesis (1989), however, does not completely explain habitat segregation among pampas and gray brocket deer. If bioenergetic requirements were the only factor influencing habitat selection in the Pantanal, pampas deer and gray brocket deer presumably could exhibit a larger overlap in their diet. Other factors, such as predation, past and/or present, may have influenced the evolution of habitat choice by each cervid. Predation is a powerful force limiting the distribution of many animal populations to particular habitats. Dingos (*Canis dingo*), for example, can greatly depress kangaroo populations (Caughley et al. 1980). Many other vertebrate predators also can restrict prey species to particular habitats (Kettlewell 1955; Wilbur, Morin, and Harris 1983; Robinson 1985; Main, Weckerly, and Bleich 1996).

Thus different strategies of inherited behavior of predator avoidance adopted by pampas and gray brocket deer may have resulted in the observed habitat segregation. Pampas deer avoid predation by running through the open grassland, whereas gray brocket deer seek refuge within thickets. This differential pattern of habitat use by gray brocket deer and pampas deer at least partially explains the minimal overlap in their proportionate consumption of shared plant species.

Historically, interspecific competition as an explanation for niche partitioning may have been overemphasized (Cody and Diamond 1975). There is little evidence of current competition among vertebrate herbivore species because most of the observed partitioning may reflect past competition (Schroder and Rosenzweig 1975). In the few cases in which interspecific competition among herbivores has been demonstrated, it involved either the recent introduction of an exotic species into an indigenous herbivore community or the introduction of a herbivore into a habitat different from that in which it exists normally (Fox 1989; Wray 1994; Putman 1996). If niche partitioning among herbivores is manifest, it is more likely maintained by optimal foraging (Owen-Smith 1989) rather than by interspecific competition.

In the case of the introduction of cattle into the Pantanal, there is no evidence that cattle are directly competing with native deer for food. However, bad management and overstocking of cattle in native habitats cause habitat alterations that indirectly may affect food and shelter availability for deer and other herbivores. For example, the habit of cattle sheltering within forest patches is causing an evident opening of the habitat and a decrease of tree and shrub species in the forests. As cattle trample and feed on the saplings and seedlings, little regeneration is occurring in these forests. Similarly, overgrazing on the grasslands may cause an increase of weed species, which in turn may decrease the availability of forbs and other species useful for deer. Finally, high abundance of cattle may increase the chances of interspecific disease infection.

CLASSIFICATION OF THE CERVID-FEEDING STRATEGIES

Ruminants have been classified as concentrate selectors, grazers, roughage eaters, and intermediate types as determined by their diet and grastrointestinal anatomy (Hofmann 1988). Concentrate selectors, most commonly referred to as browsers, include cervids whose body size ranges from large, e.g., moose (*Alces alces*), to small, e.g., pudu (*Pudu puda*). Grazers are principally bovid species that also range in size from large, e.g., gaur (*Bos gaurus*), to small, e.g., sheep. Intermediate feeding types include representatives of cervids and bovids, which may shift diets between browse and grass depending on seasonal conditions (Hofmann 1988). Bodmer (1990) expanded this classification by including frugivory as a feeding strategy distinct from browsing. Frugivorous ruminants occur in tropical forests of South America, Africa, and Asia where fruits are abundant but where high quality browse for terrestrial herbivores is rare. Thus, ruminants can be classified along a continuum that ranges from frugivores to browsers to grazers.

If we accept this classification, the gray brocket deer would be classified in the frugivore-browser segment of the continuum, whereas the pampas deer would occupy the browser-grazer segment. Both cervids are predominantly browsers and concentrate selectors as inferred from their diet in the Pantanal. However, their digestive anatomy and possibly the absence of many competitors allow them to consume a greater proportion of other food types. This versatility is reflected in the diversity of habitats occupied currently by gray brocket deer and pampas deer. The former range from tropical forests to the savanna and the latter from the savanna to the temperate grasslands of South America (Avila-Pires 1959; Redford and Eisenberg 1992; Pinder 1997). This versatility also may have been responsible for their survival through the several climatic changes that eliminated a number of herbivorous species during the Pleistocene era (Stehli and Webb 1985).

ACKNOWLEDGMENTS

Research funds were provided by Conservation International, National Geographic Society, Tropical Conservation and Development Program (University of Florida), Wildlife Conservation Society, World Wildlife Fund-U.S.A., and Conselho Nacional de Desenvolvimento e Pesquisa (Brazil). Authorization to capture wildlife was issued by Instituto Brasileiro do Meio Ambiente (Brazil). We thank Arnildo Pott, Euzelita A. S. Pinder, Jane Kraus, Jay Harrison, Roberto Klabin, Sergio Rosso, and Valle Pott for their support in several aspects of this research. George W. Tanner, John F. Eisenberg, and Richard E. Bodmer made valuable comments on this paper.

17

Ecology and Conservation of the Jaguar (*Panthera onca*) in Iguaçu National Park, Brazil

PETER G. CRAWSHAW, JR., JAN K. MÄHLER, CIBELE INDRUSIAK,
SANDRA M. C. CAVALCANTI, MARIA RENATA P. LEITE-PITMAN,
AND KIRSTEN M. SILVIUS

Jaguars (*Panthera onca*) now occupy less than 50% of their historic range (33% in Central America and 62% in South America; Swank and Teer 1989). With the exception of occasional dispersing animals, they have been extirpated from the southern United States, northern Mexico, coastal northern and western South America, and southern Argentina, among other regions (Sanderson et al. 2002). Athough their greatest stronghold is in the continuous forest of the Brazilian Amazon, jaguar populations also persist in highly fragmented and threatened regions, such as the Atlantic forest, Central American dry and moist forest and pine savannas, northern South American dry forests, and South American savannas and savanna-parklands (*cerrado*). Sanderson et al. (2002) regard these populations as having low probability of long-term survival. These populations, however, are important to biodiversity conservation in general and jaguar conservation in particular because they contribute to species-level genetic diversity, help to maintain functioning ecosystems in isolated protected areas, and, perhaps most importantly, interact with humans and thus enter into the political and cultural consciousness of the countries in which they occur.

Iguaçu National Park in southwestern Brazil exemplifies an isolated fragment of natural habitat surrounded by intensive human activities. Created in 1939 to protect the world famous Iguaçu Falls, the park currently holds about 80% of the remaining subtropical forest that once covered Paraná state east of the coastal Serra do Mar (Poupard et al. 1981). Today, only about 6% of the state is covered by forest. At the time of the study, the park sustained populations of jaguars, pumas (*Puma concolor*), ocelots (*Leopardus pardalis*), and large ungulates such as collared peccaries (*Tayassu tajacu*), tapirs (*Tapirus terrestris*), and white-lipped peccaries (*Tayassu pecari*).

Jaguar studies in Central and South America (Rabinowitz and Nottingham 1986;

Emmons 1987; Ludlow and Sunquist 1987; Crawshaw and Quigley 1989; Konecny 1989; Sunquist, Sunquist, and Daneke 1989; Crawshaw and Quigley 1991; Emmons 1991; Quigley and Crawshaw 1992; Sunquist 1992) have usually been conducted in large tracts of continuous habitats. By studying jaguars in Iguaçu, we can look at the impact of fragmentation on individuals and populations of a large predator. From April 1990 through December 1994, we collected information on the ecology of jaguars in Iguaçu National Park. In 1991 the project was extended to neighboring Iguazu National Park, located in Argentina and separated from the principal study site by the Iguaçu river. Our objectives were to evaluate the conservation status of jaguars (and ocelots; Crawshaw 1995) in the park in order to determine whether genetic interchange through dispersal occurs between the (sub) populations of the two parks and to describe the dispersal patterns (sensu Shields 1987) of subadults.

STUDY AREA

Iguaçu National Park (INP) comprises 175,000 ha on the western border of Paraná state, southern Brazil (25°05' to 25°41' S and 53°40' to 54°38' W), along the international boundary with Argentina and Paraguay (fig. 17.1). The climate in the region is temperate subtropical, with mean monthly temperatures ranging from 25.7°C in

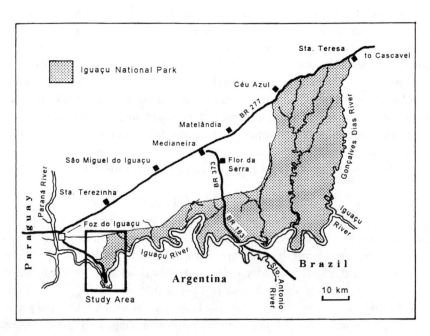

FIGURE 17.1 Location of Iguaçu National Park and Parque Nacional Iguazu at the border between Brazil and Argentina.

February to 14.6°C in July (Crespo 1982). Mean annual rainfall during the study period (1990–1994) was 1,700 mm, with one peak in May (180 mm) and one in October (240 mm). July and August are the driest months of the year. The vegetation in the park consists of a rather uniform forest cover, with some regenerating patches on the western half, where settlers had cleared or selectively logged the original forest. Four hundred families living in the park were removed and compensated in the late 1970s and early 1980s.

The region's rich soils sustain thriving agribusinesses and small farms, and human population density has increased to about 616 inhabitants/km^2 from 31 inhabitants/km^2 in the mid-70s (Poupard et al. 1981). On the Brazilian side the park is completely bounded by soybean, wheat, and rice plantations and by cattle pastures; on the Argentinean side it borders the 55,500-ha Parque Nacional Iguazu. Together, the two parks harbor the last large tract of subtropical rainforest that once covered much of the highlands on the west side of the Serra do Mar in southern Brazil. Although contiguous, the two parks are administratively separate.

The shape of the INP resembles that of an "L" turned 90° counterclockwise, with an average width (N/S) of only 5 km in the western half (fig. 17.1). The eastern portion, with approximately 50 km on the N-S axis and 23 km on the E-W axis (roughly 1,100 km^2), has the most intact habitat but the highest poaching levels. There is no buffer zone or effective fencing between the park and the private properties along the 136 km of dry boundary of the northern perimeter. Peccaries frequently enter the agricultural areas to feed on crops, while jaguar and puma prey on domestic animals (poultry, dogs, sheep, and cattle). An unpaved road runs along the north border of the western half for less than 60 km. The remaining 304 km of the perimeter is bounded by rivers. Park boundaries are difficult to patrol, and in the same way that wildlife enter agricultural areas, poachers (both subsistence hunters and wealthier landowners who hunt for sport) easily enter the park and place significant pressure on ungulate populations.

The study area encompassed about 80 km^2 of the westernmost part of the park, near the falls (fig. 17.2). Access to the falls is provided by a federal highway (BR-469), which ends at the tip of the peninsula that cuts into Argentina. This area is visited by close to one million people a year (mean of 802,587 during the study period 1990–1994), with peaks in the months of January, February, and July. During some national holidays the number of daily visitors may exceed 10,000. Most of the trapping and ground monitoring was conducted along the 8.8-km Poço Preto road (PPR) (fig. 17.2), which had been closed to the public since 1986. Three other trails also were used: Represa (TR; 1.2 km), Bananeiras (TB; 1.5 km), and Macuco (TM; 4.5 km), the latter of which is used for jungle-and-boat tours by a concessionaire.

METHODS

Jaguars were captured using custom-made wood or iron-bar box-traps, measuring 210 x 80 x 80 cm and baited with live chickens housed at the back of the trap with

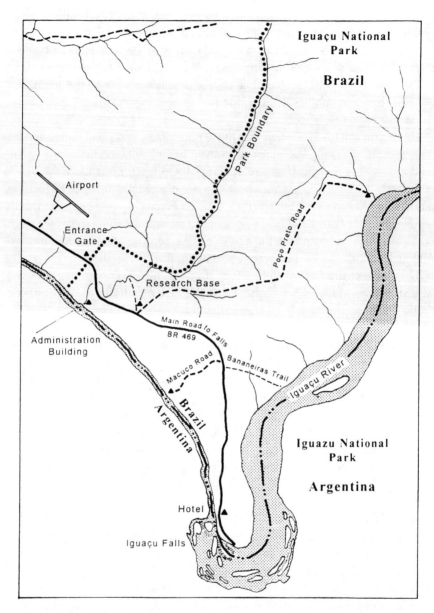

FIGURE 17.2 Map of Iguaçu National Park study area showing trails used for trapping and monitoring.

wire mesh. Once trapped (or, on three occasions, treed by trained dogs), animals were chemically restrained using a projectile dart shot with a CO_2 pistol (Telinject U.S.A., Saugus, CA 91350) or powder rifle (Capchur, Palmer Chemical and Equipment, Douglasville, GA 30133). Animals were anesthetized with either ketamine hydrochloride (Parke-Davis, Detroit, MI) or Zoletil (same as Telazol, or CI-744;

Virbac do Brasil, São Paulo, SP 04021), examined for general body condition, measured, weighed, ear-notched and/or tattooed, photographed, and fitted with radio-collars (150–152 MHz; Telonics, Mesa, AZ 85204). Relative age was estimated mainly on the basis of presence of milk or permanent dentition and tooth color and wear (juvenile, subadult, adult) in combination with other physical characteristics, such as weight, size, sign of previous reproduction (for females), and texture and color of the skin of the heel pads (Crawshaw 1992). Anesthetized animals were kept under observation until ambulatory.

Radio-equipped animals were located from a vehicle with a mounted omnidirectional antenna, or on foot with a directional antenna. Once a signal was heard, locations were obtained through triangulation, using the hand-held directional antenna. Given the limited ground range of the equipment, about 300 m in dense vegetation, a light aircraft was used at approximately 14-day intervals to obtain an unbiased sample of locations. The flyby method described by Mech (1983) was used to define animal locations from aircraft. Accuracy tests performed by project personnel indicated that locations could be described within a circle of 50 m radius (0.8 ha).

Locations were plotted on a 1:25,000 map of the study area, divided into 1.5-ha cells by a transparent grid overlay. Grid coordinates (vertical and horizontal) were assigned to each location. In the latter period of the study (subsequent to 1992), geographical coordinates were also obtained with a GPS (Global Positioning System) receiver (Transpak, Trimble, and Pronav 100, Garmin).

Home ranges were estimated with the Home-Range (Ackerman, Samuel, and Leban 1991) and the Mcpaal software packages(Microcomputer Program for the Analysis of Animal Locations; Stüwe 1985; National Zoo, Washington, D.C.). Comparisons between individuals in this study were made using the Minimum Convex Polygon, corrected upon visual analysis of plotted locations (Harris et al. 1990). Outliers, including locations of dispersing subadult individuals, were omitted from calculations.

Fecal samples were routinely collected along roads and trails, dried in large cardboard boxes with permanently lighted lamps, and stored until analysis. Contents were separated into the various components in running water over a wire mesh sieve and identified to the lowest taxon possible, using a local reference collection of hair, bones, reptile scales, and fruit. Some items were identified at the Capão da Imbuia Museum, in Curitiba, Paraná state. Whenever possible, the predator species that left the scat was identified at the collection site based on tracks and/or other circumstantial evidence. When such identification was not possible, hairs ingested in autogrooming were identified by comparison of cuticle and medullar patterns with hairs of known samples. Hairs were selected from samples by macroscopic characteristics, clarified with xylene, and compared with known samples at 400x magnification.

To correct for differences in prey size in the diet, we multiplied the mean weight of each food item by the number of times it was found in the sample of scats and by the mean number of individuals per scat (as indicated by the number of body parts found), thus obtaining an estimate of biomass consumed for each taxa (B_{est}). When available, we used weights from local animals; if unavailable, weights from the literature were used. The relative importance of each prey item was then expressed as the percentage of that item in relation to the combined weight of all items.

RESULTS

Seven jaguars were captured and radio-collared, including two adult males, one adult female, three subadult males, and one subadult female (table 17.1). One of the adult males (M48) was captured in Iguazu National Park, Argentina; all other animals were captured in INP, Brazil. Five of the jaguars in INP were probably related. Judging from sightings and spoor prior to capture and from subsequent monitoring after radio-collaring, the adult female F17 was presumed to be the mother of M13 and F21 (littermates) and M32 (from a subsequent litter). As the resident adult, M33 was probably the father of both litters. There was evidence of the presence of other jaguars in the trapped area—one female and subadult were observed but not captured, and one captured subadult (M34) must also have been the offspring of this unknown female.

HOME RANGES AND MOVEMENTS

A total of 236 locations (81% aerial) were obtained on the study animals, 142 (60%) for males and 94 (40%) for females. Adult and subadult animals were monitored about equally, with 49% and 51% of locations, respectively. Overall, the mean interval between locations was 10.1 ± 15.2 days (range: 0–102 days).

TABLE 17.1 Home-range Estimates for Seven Jaguars in Iguaçu and Iguazu National Parks

AN#	N	MCP (KM^2)
F17	69	70.0
F21	25	8.8
M13[a]	10	22.6
M13[b]	11	138.6
M32[a]	9	5.4
M32[b]	6	104.0
M33	17	25.7
M34	21	16.4
M48	15	86.5

Note: Uses the MCP (Minimum Convex Polygon) method. AN# is the animal's ID number, and N the number of radio locations.
[a]Natal area
[b]Adult home range (incomplete; *see* Dispersal)

Home-range estimates varied considerably for the study animals (Crawshaw 1995; table 17.1), ranging from 8.8 km² (F21) to 138 km² (M13). Undoubtedly, some of this variation can be accounted for by differences between sex and age classes. However, cumulative area curves (Odum and Kunzler 1955) indicate that the ranges of most individuals are underestimated because of small number of locations. The only exception is F17, which used a total area of 70.0 km² during 14.5 months of monitoring and for which an asymptote was reached at fifty locations.

The movements of most of the radio-tagged jaguars were influenced by human activities. Adult male M48 was translocated and released in an area 8 km from his first capture site because of his habit of preying on domestic dogs. Two months later, he was back in his original area and was wounded with a shotgun by a local resident as he was attacking a dog by the house. He was recaptured thirty-nine days later in a trap baited with the remains of another dog he had killed the previous night in an indigenous village, and permanently removed to a zoo after treatment of the wound.

The movements of M33, a male well past his prime, were likely influenced by an apparent acquired dependency on the live baits in our traps (eight recaptures). This dependency may have resulted from a decrease in his hunting ability, since his condition deteriorated steadily between captures. Similarly, the movements of F21, a subadult female, were influenced by a dependency on the garbage dump at the hotel by the falls and then by a short translocation (circa 5 km) following capture. This dependency was likely developed because of an infection on one of the upper canines, that prevented her from normal hunting. She was treated during capture and had apparently resumed normal feeding three months later when she was killed by a poacher inside the park, as her stomach contained the remains of a collared peccary. Her remains, together with the transmitter, were found buried in the forest. Initially captured as subadults, M13 and M32 were in the process of establishing their adult home ranges when they were also killed by poachers in the park.

DISPERSAL

Two subadult males, M13 and M32, were monitored during dispersal. After a few long-range movements (> 5 km) outside of his natal area, M13 (at an estimated age of eighteen months) crossed the Iguaçu river into Iguazu National Park, in Argentina, on September 24, 1991. He remained in Argentina until November 9, and was located five times during the interval, with a mean linear distance between locations (MLDBL) of 2.4 ± 2.7 km. On November 10 he was back in INP, where he remained until November 13, with a MLDBL of 2.0 ± 1.4 (0.6–3.8; N = 4) km. On November 21 he was found again in Argentina, where he traversed the whole peninsula and on December 1 was found in Paraguay, having crossed the Paraná river at a point where it is > 400 m wide. He was confined to a small island of forest surrounded by farmland. On December 17 he was back once again in his natal area in INP, a linear distance of 16.2 km from his previous location. On January 18, 1992, he was located 33 km northeast of where he was located on January 9.

When he was recaptured on March 13 to change his collar, he weighed 88 kg. He had several open cuts, likely incurred during intraspecific fights, and his upper right canine was broken off almost at the base. Between January 18 and May 16, the mean distance between thirteen locations was 12.9 ± 11.1 km (range 1.0–33.0 km). On May 26 we located the collar at a house by the Iguaçu river, on the opposite side from the Park—M13 had been killed by the owner and his fifteen-year-old son while they were poaching white-lipped peccaries in INP.

During his first nine locations following capture on November 17, 1992, M32 covered an area of 5.4 km², presumably still in his natal area. After January 21, 1993, the signal of his transmitter could not be found within the study area. After gradually increasing the area searched during flights, he was found on February 22 on the eastern sector of the park, 64 km east of his last location. Assuming he was born in September 1991, his age then was sixteen months. His dispersal may have been precipitated by the removal of his mother, F17, from the park in October.

His next ten locations encompassed an area of 308 km², with a mean linear distance between locations of 9.0 ± 3.6 km (range 4.4–16.7 km). His movements for the next ten locations decreased to a mean of 3.4 ± 1.2 km (range 1.8–4.9 km) and were confined to an area of 17 km². In his next ten locations he increased the area used, as a result of two long-range movements (17.5 and 10.2 km) to an area where he was later seen twice (during recapture attempts) with an adult female. Therefore he was already becoming an established adult when he was killed by poachers in June 1994.

DENSITY

Despite heavy human disturbance within and around the study area, jaguar sign (mainly tracks) was frequently encountered along roads and trails. Six jaguars were captured and radio-tagged in the 80 km² in a four-year period. Radiotelemetry data on the collared animals and indirect evidence indicated that at least one other adult male and two adult females (uncollared) used the entire area. On April 19, 1990, an uncollared adult male was photographed by an infrared-sensor remote camera. He was never captured and could have been either a resident or a transient. The estimated minimum density of adult animals therefore was 3.7 jaguar/100 km². Assuming the adult females had, on average, 1.5 young per litter in alternate years, the total estimate for the area would be six animals in the 80 km², or 7.5 jaguar/100 km². If applied to the 1,750 km² of INP, the total density would be approximately 64 adults, or 134 animals, including all ages. However, given that both dispersing subadult males were able to establish home ranges within the park, as well as the high jaguar mortality due to humans, the actual density was certainly much lower.

DIET

A total of seventy-three jaguar scats was found in INP. Mammals comprised 80% of the 106 items recorded, followed by birds (8.5%), reptiles (6.6%), fruit (3.8%), and invertebrates (1.9%) (table 17.2). Considering only the mammals, peccaries were the prey taken most often (45% of all items), followed by opossum, armadillo, and deer, to mention only those occurring above 10% in the sample. However, if the different food items are corrected by weight, the order of relative importance of these species changes considerably. Peccaries become even more prominent (77%), followed by deer (14%), coati (2.2%), armadillo (2.0%), agouti (1.1%) and opossum (1.0%). Therefore these species alone account for over 97% of the diet derived from the jaguar scat sample. It is noteworthy that these same species are among the most abundant in the study area, judging from our sighting data. However, the propor-

TABLE 17.2 Food Items Found in Seventy-three Jaguar Scats in Iguaçu National Park, Brazil

PREY ITEM	WGT (KG)	N	#IND	ITEMS (%)	SCATS (%)	B_{est} (KG)
Peccary (*Tayassu* sp.)	30.0	38	1	35.8	52.0	1,140
Opossum (*Didelphis aurita*)	1.50	10	1	9.4	13.7	15
Armadillo (*Dasypus novemcinctus*)	3.30	9	1	8.5	12.3	30
Deer (*Mazama* sp.)	22.5[a]	9	1	8.5	12.3	203
Bird (Cracidae, Tinamidae)	0.65	9	1	8.5	12.3	5.9
Lizard (*Tupinambis teguixin*)	1.60	6	1	5.7	8.2	9.6
Coati (*Nasua nasua*)	5.50	6	1	5.7	8.2	33.0
Agouti (*Dasyprocta azarae*)	3.2	5	1	4.7	6.8	16.0
Unidentified fruit	—	4	—	3.8	5.5	—
Squirrel (*Sciurus aestuans*)	0.80	2	1	1.9	2.7	1.6
Invertebrates	—	2	1	1.9	2.7	—
Rabbit (*Sylvilagus brasiliensis*)	0.8	1	1	0.9	1.4	0.8
Paca (*Agouti paca*)	10	1	1	0.9	1.4	10.0
Capuchin monkey (*Cebus apella*)	3.0	1	1	0.9	1.4	3.0
Tayra (*Eira barbara*)	5.5	1	1	0.9	1.4	5.5
Margay (*Felis wiedii*)	3.0	1	1	0.9	1.4	3.0
Unidentified snake	1.00	1	1	0.9	1.4	1.0
Total		106		99.7	145.1[b]	1,478

Note: Wgt equals mean live weight of prey item; N, number of scats in which the item was present; #Ind, number of individuals per scat; items, percentage of total number of items found; scats, percentage of scats containing that food item; and B_{est}, estimated total weight of that prey item.
[a] Average of the weights of the two species (*M. americana* and *M. nana*).
[b] Does not add to 100% because of more than 1 item per scat.

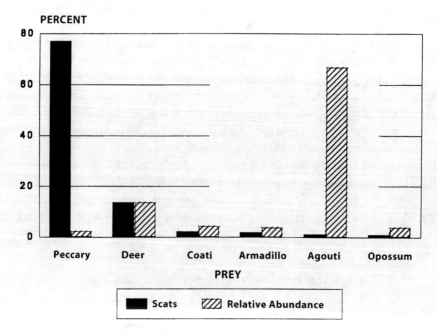

FIGURE 17.3 Percentage of corrected prey taken by jaguar in Iguaçu National Park, compared with relative abundance derived from sighting data.

tion in which they are taken indicates different levels of selectivity by the jaguar (fig. 17.3). Peccaries were taken much more often than the relative abundance estimate would predict, whereas agoutis were taken much less. Deer were the only species taken in equal proportion to their relative abundance.

MORTALITY AND INTERACTIONS WITH HUMANS

We lost all seven radio-tagged jaguars in a three-year period (Crawshaw 2002). Three were poached inside the park (M13, M32, and F21), two had to be removed from the park because of livestock depredation habits (F17 and M48; the latter had already been wounded), and two disappeared prematurely and were suspected of having been killed (M33 and M34). In addition, three other uncollared jaguars were known to have been killed during the study period, one for livestock depredation outside the Argentine park, one shot by a poacher, and one cub killed by a vehicle at the INP boundary.

Complaints of livestock depredation by jaguars were common during the study period, especially in the last two years. Incidents included predation of twenty-two sheep in a ten-day period, and of thirty-six calves and cows in a three- to-four-month period (Crawshaw 2002). Most people attributed the increase of instances to an increase in the population of jaguars in the park. However, the fact that two subadult males (M13 and M32) could establish adult home ranges within the INP indicates that there were still open spaces in the population (perhaps because of the high mortality of adults).

Other hypotheses can be postulated. Prior to the project, there was little response from park and wildlife authorities to predation complaints because of a lack of technical information and expertise on the subject. Therefore there was an unspoken policy between ranchers of resolving the situation by eliminating the problem animal. With knowledge of the project and of the critical status of the species, some of the more conscientious ranchers tended to report losses rather than to eliminate the animal. The widely publicized arrest of two poachers that killed radio-equipped jaguars also must have acted as a deterrent, with a widespread belief that all jaguars in the park were being radio-monitored.

DISCUSSION

Although conservation programs for large carnivores—quintessential charismatic megafauna—are successful in attracting funding (Weber and Rabinowitz 1996; Sanderson et al. 2002,), conservation successes on the ground are more difficult to achieve for several reasons. First, carnivore conservation projects must work with or change the negative attitudes of local peoples toward animals that pose real or perceived threats to livestock and human lives. Ranchers often consider all large carnivores as potential threats to their stock and shoot them on sight (Hoogensteijn 1993; Kellert et al. 1996; McNammee 1997).

Second, it is increasingly difficult for conservation units to provide the huge land areas required to maintain viable populations of top predators. When protected areas are surrounded by ranchlands, animals moving in and out of the area or living near its boundaries are susceptible to being killed by humans. They also suffer high mortality rates on roads because of collisions with vehicles.

Third, adequate prey bases must be maintained in and around protected areas to sustain carnivore populations. Carnivore management therefore involves management of ungulate and other larger herbivore populations, which are often the target of subsistence hunters and poachers. These issues are common to carnivore conservation worldwide, e.g., wolves in North America, tigers in Asia, and jaguars in South and Central America.

Despite its island nature and the extreme loss of habitat in the region, it is not habitat loss per se that is affecting jaguar populations in Iguaçu National Park. Rather, mortality related to the high human densities that accompany fragmentation have the greatest impact on jaguars. If such mortality could be controlled, viable jaguar populations could still persist. This viability presupposes, however, the maintenance of an abundant prey base. Polisar (2000) demonstrated a strong negative correlation between the availability of natural prey and the level of predation by pumas and jaguars on cattle in the Venezuelan Llanos.

In INP loss of white-lipped peccaries because of poaching and harassment by humans may pose one of the greatest threats to jaguar viability. Peccaries, especially white-lipped, were prominent in the diet of jaguars at Iguaçu. Assuming a jaguar needs about 5% of its body weight (4.0 kg) as the mean daily food intake (Emmons 1987), it would require approximately 1,460 kg per year. Using the percentages of the main food types in the scat sample, it would require the equivalent of thirty-eight peccaries, nine deer, six coatis, ten armadillos, five agoutis, and ten opossums to fulfill one individual jaguar's annual food requirements. The total figure for the entire park jaguar population (approximately seventy animals) would be 102,200 kg of required prey, of which 81,000 kg would presumably be provided by peccaries alone.

Our density estimate for white-lipped peccaries in the study area was about seventy individuals (in one herd that apparently split in subgroups occasionally) in the 80 km^2, or 0.9 individual/km^2. However, we knew of the recent killing of at least eight members of this herd when they invaded crops at a neighboring property. Fragoso (1994) found home-range sizes of 21.8 and 109.6 km^2 at Maracá Island Ecological Reserve, Amazonia, with densities ranging from 1.2 to 8.1 peccaries/km^2. Assuming a conservative estimate of 1.5 peccaries/km^2 for the 1,750-km^2 park, the resulting estimate would be 2,625 animals, or a total biomass of 78,750 kg (weight estimated as 30 kg, as opposed to 35 kg of adults, to account for young and subadult animals).

Collared peccary densities ranged from 1.0 to 3.3/km^2 in Maracá (Fragoso 1994). For Manu National Park in the Peruvian Amazon, Emmons (1987) estimated a density of 5.6 individuals/km^2 for the two species combined. Based on our sighting data in Iguaçu, collared peccaries were either scarcer than white-lipped or less conspicuous or both (36% for collared peccaries and 64% for white-lipped; N = 33 group sightings). Mean number of individuals per sighting was two for collared and twelve for white-lipped. Using this sighting ratio to estimate collared peccary density, as compared to white-lipped, the total estimate for the INP would be roughly 1,575 animals, or a biomass of 34,650 kg (mean weight of 22 kg). Therefore the estimated combined adult biomass of the two species would be approximately 113,400 kg for the entire park, or about 65 kg/km^2. To this figure should be added the potential yearly productivity, estimated at about 50,750 kg, for a total of 164,150 kg. The 81,000 kg required by the total jaguar population would thus comprise roughly half of the available peccary biomass. By itself, this level of predation would already be unsustainable, even in the short run. However, to this figure should be added competition from poachers, who also hunt selectively for peccaries (Becker 1981; Jorgenson and Redford 1993; Townsend 1995a) and the control exerted by neighboring farmers when peccaries raid their crops.

At the time of the study we predicted that, given the pronounced selection for white-lipped peccaries by jaguars (Crawshaw and Quigley 1984; Aranda 1994; Fragoso 1994; this study) and even allowing for some error in our estimates for this species in INP, white-lipped peccaries would not withstand current predation levels (both from jaguar and from humans) and would become extinct in the park within the foreseeable future. In this event jaguars, as an adaptable species, would likely gradually switch to smaller, abundant prey, as was probably the case in Belize (Rabinowitz and Nottingham 1986). Despte their initial abunddance, white-lipped peccaries did indeed disappear from the park in 1995 (Azevedo and Conforti 1998). Continued monitoring of the cat population is needed to determine the effect of this extinction on jaguar populations.

MANAGEMENT AND CONSERVATION

When one first looks at INP from the air and realizes its condition as an island, thoughts of inbreeding depression and genetic deterioration (Ralls and Ballou 1983; Ralls, Harvey, and Lyles 1986; Lacy 1992) come to mind, especially applied to species such as the jaguar. However, data presented herein suggest that inbreeding may not be a serious consideration in jaguar conservation in Iguaçu, as long as a corridor with undisturbed areas is maintained in Argentina. Dispersing subadults of both jaguar and ocelot (Crawhaw 1995) can, and do, eventually cross between countries and ensure gene flow between populations.

Some 12,000 km^2 of subtropical forest remain in the Argentinean province of Misiones (J. C. Chebez pers. comm.). A total of 4,140 km^2 (34.5%) of this area is protected to some degree, forming an almost continuous corridor, about 200 km in length, linking INP to Turvo State Park in the state of Rio Grande do Sul, also in Brazil (Ministerio de Ecologia y Recursos Naturales Renovables 1993). Turvo State Park protects the last known population of jaguar in that state and represents the southern limit of the species in Brazil (Crawshaw and Pilla 1994). Jaguar populations in the province of Misiones, Argentina, may well represent the southern limit of the species in South America.

Even though an alternative, stable prey base and the habitat for jaguar may be preserved, problems with their coexistence with humans will likely tend to escalate if present conditions persist. Direct competition from poachers taking its main prey (peccaries, deer, and paca) will tend to increase the frequency of livestock depredation around INP. Jorgenson and Redford (1993) showed extensive overlap in the prey taken by big cats and humans in several areas in the Neotropics that resulted in the decline of populations of the former. Results from Townsend (1995a) for lowland Bolivia also show the negative impact of human hunting on the same species of prey taken by jaguar in INP.

The large areas occupied by individuals, the narrow shape of INP, and the lack of effective deterrents will always lead jaguars to the periphery of the park in search of food. It is difficult to deter cats from feeding on livestock once they become habituated to this food source. This habituation is especially critical for females with cubs, who tend to pass this acquired behavior to their offspring (Ewer 1973; Crawshaw and Quigley 1984). The considerable economic impact that livestock depredation imposes is amplified for low-income landowners, and it is unreasonable to expect these small farmers and ranchers to accept jaguar predation without retaliation. Indeed, jaguar killing for livestock depredation control is presently one of the main mortality factors for the species throughout its remaining range (Rabinowitz 1986; Crawshaw and Quigley 1991; Quigley and Crawshaw 1992; Hoogesteijn, Hoogesteijn, and Mondolfi 1993). In this study two adult males from Argentina were shot because of livestock depredation. In addition to animals killed by ranch owners, an unknown number of jaguars will continue to be killed by poachers hunting for other species in the park. Such was the case with the two radio-tagged subadult males, M13 and M32.

The picture gleaned from some hard-won facts on the ecology of jaguar in the present study is not a very encouraging one. There were problems with almost every individual that was radio-monitored: cats that were shot because they killed animals outside the park or because they came too close to a poacher who was killing their prey inside the park; animals that became dependent on food scraps from humans; and animals that had to be removed because of acquired bad habits toward humans or their property. However, given the sheer size of INP and the fact that the habitat and some very productive prey species will remain, jaguars may just have a chance to endure there. Still, for jaguar survival in the park, three conditions have to be quickly met:

1. The park needs a completed and maintained road around its dry perimeter.
2. This road will help with the next requirment: more efficient control of poaching. Poacher control also demands the efforts of park police and a fence (combining conventional wire netting and electric) that would not only serve as a deterrent for poachers entering the park but also decrease the chances of animals (both carnivores and herbivores) from searching for food outside the park.
3. The final requirement is a permanent corridor with undisturbed habitat in the province of Misiones, Argentina.

Reinforced by this project and a similar study in Turvo State Park (Crawshaw and Pilla 1994), close ties between IBAMA, Argentine environmental authorities, and some NGOs (such as Fundación Vida Silvestre, Argentina; WWF-U.S.; WWF-Brazil; and the Sociedade de Preservação da Vida Silvestre, or SPVS) are beginning to develop. These organizations are working toward (a) the maintenance of a corridor of subtropical forest in Misiones, linking Iguaçu and the Parque Estadual do Turvo, in Brazil (WWF-U.S. 1991) and (b) the developmnet of a Trinational Corridor (Argentina, Brazil, and Paraguay) that will eventually link Morro do Diabo State Park in São Paulo state, as the limit to the north, with Parque Estadual do Turvo in Rio Grande do Sul state, as the southern limit (M. Di Bitetti pers. comm.). A further small step in crosscountry collaboration may allow for a metapopulation approach (Gilpin 1987; Guerrant 1992; Pádua 1993; Forys 1995) for the management and conservation of jaguar populations in the area.

As for the wolf recovery program in the United States (Mech 1995), the maintenance of a permanent local program of environmental education is essential to jaguar conservation efforts. This program should emphasize the importance of

Iguaçu not only for the conservation of natural resources, including the large predators, but also to the local economy in terms of tourism. Through this program close contact should be kept with neighboring landowners in order to help solve problems when they arise. One of the objectives of the recently created National Predator Center (IBAMA 1994) is to promote some form of compensation for losses inflicted by predators. This compensation may be accomplished by cuts in federal, state, or municipal taxes or by direct reimbursement using funds raised by NGOs. A similar attempt is underway in northwestern Argentina (Perovic 1993), and there is a law project in the parliament of Misiones province, Argentina, to implement a mechanism of compensation through provincial tax cuts. As a consequence of the reported jaguar depredation problems in the vicinity of Iguazu National Park and other areas, a management-conservation action plan is beginning to be implemented in Misiones (Chebez 1995).

Ultimately, the fate of the jaguar in INP and, for that matter, of any large predators constrained within relatively small, isolated parks and reserves, remains in the hands of the people who live around these areas and who coexist with these species. However, it is the responsibility of managing agencies to resolve local conflicts that are inevitable in the interface between the natural and the man-modified worlds. Only through the integration of applied research, implementation of management recommendations derived from these findings, involvement of NGOs, and participation of local communities through environmental education programs will these species have a chance to survive.

18

A Long-Term Study of White-Lipped Peccary (*Tayassu pecari*) Population Fluctuations in Northern Amazonia

ANTHROPOGENIC VS. "NATURAL" CAUSES

JOSÉ M. V. FRAGOSO

White-lipped peccaries (white-lips—*Tayassu pecari*) are among the largest of Neotropical forest mammals, reaching weights of about 50 kg (Sowls 1984; Fragoso 1998a). They can form groups with more than 100 animals, and anecdotal reports exist of groups with 1,000 and even 2,000 individuals (Mayer and Brandt 1982; Mayer and Wetzel 1987; Fragoso 1994). Occasionally, herds and even entire populations have disappeared from areas where they were usually found, leading some researchers to hypothesize that white-lips are migratory, probably in response to variations in food supply (e.g., Kiltie 1980; Kiltie and Terborgh 1983; Sowls 1984; Bodmer 1990; Vickers 1991). This hypothesis has been supported by the stories of several Amerindian groups, who tell that local disappearances of white-lipped peccaries are normal and that the herds return after a given time (e.g., Vickers 1991).

Further support for the migratory hypothesis was provided by Kiltie (1980) and Kiltie and Terborgh (1983), who using a mathematical model based on step length and number of animals, demonstrated that a herd could cover an area of more than 200 km². On the basis of this model, they hypothesized that herds may be nomadic and lack fixed home ranges. Some researchers have interpreted the results of this model as support for the hypothesis of white-lipped peccary migrations (e.g., Vickers 1991), which has come to be considered a fact by many biologists working with white-lips (e.g., Mayer and Brandt 1982; Sowls 1984; Bodmer 1990; March 1993). During fifteen years (1988–2003) of studying white lips in the northern Amazon region of Brazil, however, I have not found support for this hypothesis (Fragoso 1994, 1998a, 1998b, 1999, unpublished data), and have proposed that local disappearances are in fact in situ population declines caused by mortality and/or reproductive failure.

This article describes the disappearance of white-lipped peccaries from the northern portion of Roraima state, Brazil, an area of primary rainforest unimpacted by peoples of western descent, including colonists. The area is part of continuous rainforest that extends for about 1,500 km to the west, perhaps the same amount to the south, and hundreds of kilometers to the north of the study region (see below). To the east it borders the naturally occurring cerrado and savanna biome of the Guiana shield region. The Yanomami peoples, one of the least Westernized of native South American groups, inhabit much of the study area (Ricardo 1996). I consider three alternative hypotheses to explain the disappearance of white-lips in this area: (a) out-migration; (b) overhunting by humans; and (c) local in situ population decline resulting from an epidemic caused by an introduced pathogen.

METHODS AND STUDY AREA

Maracá Island Ecological Reserve is located at 3°25′ N latitude and 61°40′ W longitude, on the northern margin of the Amazon basin in the state of Roraima, Brazil. The riverine island was created by the bifurcation of the Uraricoera River, a tributary of the Rio Branco in the Amazon watershed. The 110,000 ha of rainforest, isolated savannas, and wetlands that make up the reserve are protected by IBAMA (Brazilian Institute of the Environment and Natural Resources) as a site for research and forest conservation. The river provides an incomplete barrier between Maracá and the surrounding forests. In the dry season it is bridged by stepping-stone islands, and white-lips and other animals are frequently sighted swimming the river at all times of year (J. Hemming pers. comm.; J. Thompson pers. com.).

The Yanomami Indigenous Reserve starts almost at the margin of Maracá and extends westward for more than 300 km. It covers 9,400,000 ha of rainforest, montane forest, transitional forest, and some savannas and woodlands. Elevations range from 100 m above sea level in the Amazonian lowlands to 2,000 m in the Parima and Pacaraima mountains that divide the Amazon and Orinoco watersheds. To date the flora and fauna of these two study sites have not been affected by Western peoples, with the exception of the recent invasion by artisanal gold miners (*garimpeiros*) of some parts of the Yanomami reserve (MacMillan 1995; Milliken and Ratter 1998).

POPULATION CENSUSES AND COUNTS ON THE EASTERN END OF MARACÁ ISLAND

Data on abundance and density of white-lipped peccaries on the eastern end of the island were collected during three separate periods. From June 1 to June 28, 1988, density transects were walked on a permanent trail system covering approximately 60 km (Fragoso 1998b). For each encounter the distance and angle of the herd to the transect line were noted along with the number of animals in the herd, and the information was used to calculate density using the King and Webb method (Schemitz 1980). Between June 1988 and June 1989, all individuals and herds encountered on the entire trail system during 227 days of walks were counted. All transect walks began between 7:00 and 8:30 A.M. and continued until 5:00 or 6:00 P.M., with a break between 11:00 or 11:30 A.M. to 1:00 or 2:00 P.M. On some days transect walks occurred only in the morning or afternoon. From January 1991 to December 1992, these transects were repeated in the same manner in the same area on 220 days.

RIVER SEARCHES

To assess the status of white-lipped peccaries in the central and western portions of the island, I traveled along 200 km of the north channel (Santa Rosa Channel) of the Uraricoera River and its tributary the Uraricaá during ten days in September 1991. During the trip I interviewed gold miners and Xiriana-Yanomami and opened four 5-km-long trails oriented toward the center of the island along which I searched for white-lips and their sign (tracks, rooted areas, feces, and hair). The transect start points were spaced to maximize coverage of the island: two near the center of the island (separated by approximately 5 km) and another two also separated by about 5 km near the westernmost point of the island. All interviews were informal and included the following questions: (a) when was the last time you or someone you know saw white-lipped peccaries or their sign, (b) when was the last time you or someone you know killed a white-lipped peccary, and (c) how many individuals were killed during hunts?

RADIO TELEMETRY

From January 1992 through January 1993, I radio-tracked seven white-lipped peccaries belonging to the only two herds using the eastern portion of Maracá island (region around the research station) and described their seasonal movement patterns and home range use (Fragoso 1994, 1998a, 1999). From June 1995 to May 1997, I again radio-tracked the two herds, which were still the only ones occupying this region of Maracá (Fragoso unpublished data).

INTERVIEWS IN THE MACUXI INDIGENOUS AREA NEAR MARACÁ ISLAND

To the southeast and northeast of Maracá, the Macuxi indigenous people live in widely spread communities. To assess the status of white-lipped peccary populations immediately east of the island, I interviewed four Macuxi Amerindian hunters from a community (Boqueirão) located approximately 20 km southeast of the Maracá study area. Between 1991 and 1992 I repeated the interviews with fifty hunters from Boqueirão and Mangueira, a neighboring Macuxi community located approximately 15 km to the southwest of Maracá. These were the nearest villages in these directions from Maracá.

INTERVIEWS AND SEARCHES IN THE YANOMAMI AREA

To assess the status of white-lipped peccaries in the Brazilian Yanomami area, I spent twenty-one days hunting with Yanomami hunters in April 1993, using the Paapiu area approximately 180 km southwest of Maracá as a base. I also interviewed representatives of thirty-eight Yanomami communities located throughout the reserve, including Yanomami people, indigenous health workers, Western teachers living in indigenous communities, and FUNAI (National Indian Foundation) personnel. These interviews were repeated with representatives of between thirty-five to fifty communities during the Yanomami Peoples Assemblies (which brings together Yanomami representatives from almost all villages) in 1996, 1997, and 1998. I collected information from at least one individual from each community represented at the assemblies. When one village was represented by more than one person, community members conferred before responding to my questions and then provided one community answer.

All interviews were informal and included the following questions: (a) when was the last time you or someone you know saw white-lipped peccaries or their sign and (b) when was the last time you or someone you know killed a white-lipped peccary? Because most Yanomami do not count beyond the number two (the exceptions are a few individuals with Western educations), I approximated the date of the last hunt by estimating the age of a child in the community who was said to have been born in the same year as the hunt.

DOMESTIC LIVESTOCK AND DISEASE

To assess the health of domestic livestock in Roraima state, I reviewed files kept by the Roraima Department of Agriculture in 1996. Complete yearly records were available dating to the early 1970s. In Roraima almost all domestic livestock, including cattle, sheep, goats, pigs, horses, chickens, turkeys, guinea fowl, and ducks are free ranging.

RESULTS

EASTERN END OF MARACÁ ISLAND

Transect data showed that white-lipped peccaries were extremely abundant at Maracá in June of 1988, reaching densities of 139 to 542 individuals per km^2 (Fragoso 1998b). Herds were encountered 478 times during the 227 search days between June 1988 and June 1989; however, within this time period no peccaries were sighted after March 1989.

During the January 1991 to December 1992 study period, herds were encountered only thirteen times despite a search intensity similar to that in the 1988–1989 study period (220 search days: 103 days in 1991 and 117 in 1992). Thus abundance and density fell sharply between the first and the second study periods; herd encounter rates indicate that the decrease occurred in or shortly before March 1989. Radio telemetry data collected from 1991 to 1992 indicated that the study area supported between 1.4 and 8.3 individuals per km^2.

RIVER SEARCHERS AND RADIO TELEMETRY

No white-lipped peccaries or their sign were found on the four 5-km-long transects cut in the central and western part of the island. Neither were animals or their sign found in the forest along the northern river channel toward the Yanomami Reserve. Fifteen gold miners interviewed during the river trip stated that they had not seen or killed white-lipped peccaries since 1989.

Radio telemetry data indicated that white-lipped peccaries have fixed home ranges, regularly use the same feeding sites (that is, they return to the same sites in subsequent years), and do not change their home range from season to season (Fragoso 1994, 1998a, 1999). The continuation of the radio telemetry study in 1995–1997 indicated that two herds continued using the same areas (Fragoso unpublished data). Herds were smaller in 1991–1992 (39 and 130 individuals per herd) than in 1995–1997 (70 and 200 individuals). Home-range size for the small and large herds increased from 21 and 109 km^2, respectively, during the 1991–1992 study to approximately 200 km^2 in 1997 for the herd with 200 individuals. The increase in home-range size over time as herd size increased and the fact that the large herd always had a larger home range than the smaller herd indicate that home-range size is related to herd size. Radio-collared individuals belonging to these herds were sighted up until November 2002 (J. M.V. Fragoso pers. obs.; G. de Oliviera pers. comm.).

The important point here is that from 1991 to 2002 the area continued supporting only the same two herds. During both tracking periods home ranges were spatially stable (no seasonal or yearly disjunctions in use of areas) within and among years. In other words, herds showed no sign of migratory behavior. They did not maintain two spatially distinct home ranges between seasons or years or between El Niño and non-El Niño years (the El Niño years of 1992 and 1997 caused marked declines in rainfall in the area, according to the climate records of the meteorological office of the Boa Vista, Roraima International Airport).

MACUXI INDIGENOUS AREA NEAR MARACÁ

If the white-lipped peccary population on Maracá and its surroundings had declined, then there should also have been a reduction in the number of peccaries killed by humans. In 1988 the four Macuxi hunters interviewed near Maracá each killed between three and five peccaries per month (Fragoso 1998b). In the year between December 1990 and December 1991, however, the fifty hunters interviewed in the same region jointly killed only seven white-lips.

YANOMAMI AREA

No data are available on white-lipped peccary harvests in the Yanomami area prior to 1990. Still, all the Yanomami interviewed, as well as persons associated with the Yanomami, reported that white-lipped peccaries were common prior to 1989. The anthropologist Bruce Albert who worked extensively in this region supported this report. He stated that during his Ph.D. fieldwork white-lips contributed up to 70% of the meat consumed by some Yanomami communities.

In 1993, during twenty-one days of hunting with Yanomami from five communities in the Paapiu area, no white-lips or their sign were encountered. Interviews with persons representing thirty-eight communities scattered throughout the Yanomami Reserve in 1996 and 1997 indicated that most of them had not seen white-lipped peccaries since 1989 or 1990 and some of them not since 1987.

At the 1998 Yanomami Assembly Davi Yanomami reported the return of one herd of white-lips to Ballalawu, a community located on the Parima mountain range that divides Venezuela (Orinoco drainage) from Brazil (Amazon drainage). In 2000 Marcos Wesley da Silva, the education coordinator for the nongovernmental organization Commission for the Creation of the Yanomami Reserve reported that from 1999 to 2000 hunters from Ballalawu had killed white-lips on two separate hunts spaced apart by approximately three months. Indigenous hunters from the communities of Auaris located in the Pacaraima mountains, (approximately 250 km to the northwest of Ballalawu) also noted the reappearance of white-lips in 2000 and reported killing multiple animals. In the Ballalawu case the interviewees reported that, when following the sign left by white-lips, they concluded that the herd(s) had come up over the mountains from the Orinoco drainage (Venezuela) and was (were) moving into the valleys and lowlands of the Amazon drainage. The Auaris informants did not provide information on the directional movement of herds; however, Auaris sits near the top of Pacaraima mountain range, which forms the northern barrier separating the Orinoco and Amazon drainages.

In 2001 white-lips were again killed at all the aforementioned indigenous communities and at Demeni (far southern region of the Yanomami reserve) and close to the communities of Catrimani (near the southeastern extreme of the Yanomami reserve). Essentially, these communities form an outer ring along the western and southeastern boundaries of the Yanomami reserve. In these regions the reserve borders (and forms part of) the greatest extent of contiguous forest of the Amazon basin. This forest extends for about 1,500 km up and over the Pacaraima and Parima mountain ranges to the Andes Mountains. To the south it extends over mountains of 1,000 to 1,500 m above sea level for about 800 km until encountering the main channel of the Amazon River.

DISCUSSION

At Maracá Island Ecological Reserve and in the surrounding Macuxi and Yanoma-mi indigenous areas, white-lipped peccary populations either disappeared or de-creased drastically to the point that even expert indigenous hunters could not find them. On Maracá the disappearance or decrease occurred approximately in March 1989. The sudden lack of white-lip sightings on the island and the reduction in white-lip harvests by indigenous peoples suggest that the phenomenon occurred throughout the contiguous forest and adjacent cerrado-savanna biome (with is-lands of forest) of the northern region of Amazonas and Roraima states in Brazil.

Here I consider three hypotheses that could explain the disappearance of white-lipped peccaries. The first hypothesis is that the white-lips migrated out of the study area. Most of the data collected do not support this hypothesis. The area over which disappearances took place is too extensive, and the vegetation, topography, and rainfall patterns too varied for the entire region to have become uninhabitable by white-lip herds searching for food. Rainfall ranges from 1,500 mm of rain per year in the east to over 2,000 mm in the west of the disappearance range (Barbosa 1997). This variation results in a diversity of biomes and plant associations, ranging from evergreen lowland rainforest to shrub-dominated areas, cerrado, savannas, and montane forest. Tree species identity also changes markedly from east to west: on Maracá alone, for example, 70% of the species that occur on the eastern end of the island are absent on the western end (Milliken and Ratter 1989).

If white-lipped peccary herds migrated seasonally or nonseasonally in search of food, they should have stopped once they encountered new plant communities or biomes with their potential for new and perhaps abundant food resources. At Maracá fruit availability does not seem to have varied greatly between the 1988–1989 and 1991–1992 sample periods (Moskovits 1985; Nunes 1992; Fragoso 1994). Even if fruit availability declined in some habitats due to changes in rainfall patterns (such as those caused by the El Niño Southern Oscillation, ENSO), many white-lip foods become superabundant in ponds and streams as they dry (many of these dry out only during El Niño years) (Fragoso 1999, pers. obs.).

Thus the hypothesis that the white-lips migrated is not supported because (a) fruit and seed availability did not vary between two critical time periods, (b) the white-lips should have appeared in some portion of the study area (110,000 ha on Maracá, 9 million ha in the Yanomami area), and (c) after a season or two, migrat-ing animals should have returned to their original territory as seasons changed. Furthermore, if migrations were linked to major changes in rainfall patterns rather than seasonal changes, the white-lips should have returned when rainfall patterns normalized. Note, however, that white-lips did not return to the Yanomami area af-ter the El Niño event of 1991–1992, and on Maracá they appeared to be unaffected by that El Niño event and those of 1997–1998 and 2002–2003 (Fragoso unpublished data). The herd with thirty-nine individuals in 1991 now contained ninety-seven an-imals in January 2003, and in March 2003 it split into two herds, one with twenty-seven and the other with seventy individuals (J. M. V. Fragoso and K. M. Silvius pers. obs.).

The radio-tracking data provides additional evidence against the migration hypothesis: during the years that two herds were tracked visually and by radio, they used their large home ranges in a regular fashion and gave no sign of migratory behavior (Fragoso 1994, 1998a, 1998b, 1999, unpublished data). They did not leave the area even when food availability was low in some portions of their home range. Their home ranges provided herds with a sufficient food in the appropriate spatiotemporal pattern (Fragoso 1999) to support population increases from 1991 to 1997 and possibly until 2002.

The only other published study that used radio telemetry to evaluate white-lip herd movements tracked thirty-six individuals in Corcavado, Costa Rica, from 1996 to 1998 and also found that herds did not migrate (Carrillo, Saenz, and Fuller 2002). White-lips did not exhibit migratory behavior at Corcavado even when rainfall decreased dramatically from one year to next because of an El Niño event. The decreased rainfall correlated with a drop in fruit production in the area; however, as in Fragoso's 1999 study the Corcavado white-lips responded by expanding home-range areas to include more habitats rather than by migrating (Carrillo, Saenz, and Fuller 2002).

The second hypothesis, that white-lipped peccary populations were reduced by overhunting, is also difficult to support. White-lips disappeared in 1989 from many areas, whereas the influx of gold hunters into the Yanomami area began in 1983 and reached its peak in 1987 (MacMillan 1995). One would expect overhunting to impact the populations much more rapidly, given that approximately 40,000 gold miners were living and hunting in the area. Furthermore, ten of the Yanomami communities that reported losing white-lips in their hunting areas (Yanomami hunting areas are somewhat larger than the home range of a white-lipped peccary herd; Good 1989) were not invaded by gold miners. If hunting by invaders had caused the loss of white-lips, these noninvaded areas should not have been affected. Additionally, while some illegal hunting by gold miners may have occurred along the northern margin of Maracá, white-lips disappeared from the eastern portion of the island at the same time as in the Yanomami Reserve. This part of Maracá was heavily patrolled by over 100 biological workers of the Maracá Research Project of the Royal Geographical Society and the National Institute for Research in the Amazon (INPA) from the beginning of 1987 to 1989 (Milliken and Ratter 1998).

The third and final hypothesis is that mortality from an epidemic caused a decline in white-lipped peccary populations at the regional level. The following observations support this hypothesis:

1. White-lip populations in noninvaded areas of the Yanomami reserve disappeared.
2. The factor that affected the white-lip populations must have been something that can affect individuals dispersed over an enormous area. Such a widespread agent

is a characteristic of epidemics in other mammal populations (Crosby 1986; Young 1994).

3. A researcher working with primates on Maracá at the time of the disappearance found the bodies of five white-lipped peccaries at one site (A. Nunes pers. comm.), and her description of the carcasses indicates that they were not killed by humans or other predators.

4. In 1989 there was an epidemic in free-ranging domestic pigs on ranches adjacent to Maracá and the Yanomami reserve. The disease, undocumented and unidenti-fied by veterinarians, killed over 50% of newborn piglets and caused a higher than normal mortality in adult pigs (J. Alves, local rancher, pers. comm.).

5. In 1989 there was a documented outbreak of foot and mouth disease in cattle on these same ranches (J. Alves pers. comm.). Further, the largest outbreak of foot and mouth disease documented for Roraima state occurred in 1989 (records maintained by the Roraima Department of Agriculture).

6. Both the pig and cattle disease outbreaks occurred at the same time that white-lips disappeared on Maracá and in the Yanomami area.

7. When gold miners entered the Yanomami area, they brought with them domestic pigs as a source of food (Fragoso pers. obs.).

8. White-lipped peccaries live in herds of more than 100 individuals, the number of animals necessary to maintain an epidemic in a reinfection cycle (A. Dobson pers. comm.).

9. Where they cooccur, free-ranging or feral domestic pigs and white-lips share use of wetland areas.

10. Because white-lipped peccaries are not territorial, they do not maintain exclusive use home ranges, and herds occasionally come together (Fragoso 1998a). This so-cioecology would allow the rapid dissemination of a disease among herds and across the entire population.

Disappearance patterns similar to those described here have been reported for white-lipped peccaries in other areas. For example, white-lips disappeared from Manu National Park, Peru, from 1981 to 1992 (L. Emmons and J. Terborgh pers. comm.), from the Siono-Secoya area of Ecuador from 1975 to 1985 (Vickers 1991), and from the Yuqui area in Bolivia in 1985 (Stearman 1990). In all these cases re-searchers working in the areas interpreted the disappearances as migrations. It should be noted that in all cases white-lips remained absent from the areas for ap-proximately ten years. Such long periodicity is not normal in migratory ungulates, but it is typical of mammal populations that exhibit boom and bust cycles (Krebs and Myers 1974). Note the reappearance of white-lips in the Yanomami area from 1997 to 2000 after having been absent for about eight to eleven years.

I suggest that what has been described as white-lipped peccary migrations may in most cases have been in situ population declines. Such population cycles are prob-ably a normal characteristic of this species' life history. I propose, however, that the characteristics that make white-lips susceptible to endemic diseases also make them susceptible to exotic diseases to which they have no resistance and which can cause more marked population declines with a longer recovery time than that of cycles caused by endemic diseases. The relationship between white-lips and domestic livestock may thus be similar to that between native Amerindian populations and the humans that migrated, together with their diseases, from the Old World.

Most of this report was first presented at the wildlife meeting in Iquitos, Peru, in 1995 and included data collected up until December 1992. As described above, I continued working at Maracá and in the Yanomami area from 1994 to the present (2003), tracking the same herds on the eastern end of Maracá in 1995–1997 and visiting Maracá and/or the Yanomami region yearly from 1998 to 2004 for at least one month each year. At least one herd was still using what appears to be the same home range in January 2004 (Fragoso pers. obs.). This constancy means that the same herds have remained in the same general area for at least thirteen years. During this time period herd size has increased, and the largest herd has expanded its home range to the north and west, but both herds still use the same areas they used in 1990–1992 and have not been absent from the study site during the entire period. In addition, the increase and splitting of the smaller herd in 2003 suggests that on Maracá populations are at or nearly at their high points and may once again enter a decline. If true, herds splintering off from the main groups should begin dispersal movements out of their natal home ranges, as discussed below.

On the basis of my extended observations, especially the increase in herd size and the apparent movement of white-lipped peccaries from the Orinoco basin into the Amazon basin, I propose the following scenario for the large scale spatiotemporal ecological dynamics of white-lip populations:

1. White-lip herd disappearances are the result of epidemic outbreaks of disease (either endemic or exotic) that kill off most individuals.

2. In areas where there is hunting by humans, this activity may extirpate white-lips that survive the epidemic.

3. Where there is/was no hunting, populations decline, but enough individuals survive to allow the slow repopulation of an area. At Maracá this has been a fifteen-year period from 1989 to the present, but abundance/density levels have yet to attain the levels observed in 1988–1989.

4. Where hunting may have extirpated populations, initial reappearance is slower (eight to twelve years in the Yanomami area), and this reappearance occurs through a process of recolonization by herds dispersing from very distant areas (hundreds of kilometers) that travel across major geographical barriers (e.g., mountain ranges).

White-lip abundances in the Yanomami reserve are only now approximating the levels observed on Maracá in 1990–1992. Of interest is how long it will take for white-lip populations in the Yanomami area to return to pre-1989 levels, when they formed up to 70% of the meat consumed by many Yanomami communities.

If the above scenario is correct, then white-lip populations fluctuate synchronously across very large spatial and temporal scales (perhaps decadal time scales). I predict that white-lip populations in both the Yanomami reserve and Maracá will continue increasing and should eventually attain the very high densities observed prior to 1989. At high densities herds will fission and subherds disperse away from the very large resident parent herds. Subherd dispersal will move individuals away from their natal home range toward areas unoccupied by other herds. This type of movement is best described as a population-level dispersal event and not as a migration. Note that this process may already have started at Maracá. There is evidence that herds have begun dispersing across the rivers, as they did prior to 1990. From December 2002 to March 2003, a herd was observed on numerous occasions on Nova Olinda island (G. de Oliviera pers. comm.), an approximately 1000-ha island that lies between Maracá and the mainland. White-lips did not use this island from 1991 to 1997.

Dispersal of this type is probably responsible for the recolonization of the Yanomami reserve by white-lips. This type of dispersal may also explain the cross-river movement of white-lips observed from Maracá in 1987–1988 during the period of peak population densities (J. Hemming, J. Thompson, and G. de Oliviera pers. comm.) and records of white-lips crossing the savannas of northern Roraima (J. Alves and G. de Oliviera pers. comm.). These dispersal events may also be what indigenous peoples refer to when they say that the "white-lips eventually return." There are many valid definitions of migrations (see Baker 1978), but all necessitate returning to the start point. Dispersion means movement away from a source, most likely without return. At the metapopulation level white-lip populations appear to be linked across scales of thousands of square kilometers. Maintaining white-lip metapopulation dynamics at this scale will continue to be a major challenge as the Western colonization zone penetrates deeper into the Amazon forest.

PART IV

───

Hunting Impacts—Biological Basis
and Rationale for Sustainability

19

Evaluating the Sustainability
of Hunting in the Neotropics

RICHARD E. BODMER AND JOHN G. ROBINSON

Rural people throughout the Neotropics hunt for subsistence food and to sell meat and hides in urban markets, actitivities that pose one of the greatest threats for tropical vertebrates and that create one of the most important conservation issues for developing countries (Robinson and Redford 1991; Robinson and Bennett 2000b). Many species are impacted more by hunting than by deforestation (Bodmer 1995b). Ensuring that wildlife hunting is sustainable is important both for the long-term benefits people receive from wildlife and for the conservation of species and ecosystems (Swanson and Barbier 1992; Freese 1997b). However, setting up more sustainable hunting is a complex process that must integrate the socioeconomics of rural people, the biology of species, institutional capacities, and national and global economic pressures. One of the fundamental aspects of sustainable wildlife use is the biological capacity of species to be used sustainably: if species are overhunted then there is no scope for sustainable use.

Wildlife management in the Neotropics takes a variety of forms, including community-based strategies, landowner strategies, and sport hunting programs. In tropical forests most wildlife hunting is done either by local indigenous or nonindigenous people or by small-scale timber operations (Robinson and Bennett 2000b). Managing this wildlife use requires information on the sustainability of hunting.

In most cases the first step in evaluating sustainability of hunting is to determine if current hunting appears sustainable or is obviously not sustainable. Most of the simple population models that have been used to evaluate the sustainability of hunting in the Neotropics have evaluated current hunting (Ojasti 1991; Fitzgerald, Chani, and Donadío 1991; Vickers 1991; Bodmer 1994; Alvard 1998; Hill 2000; Jorgenson 2000; Leeuwenberg and Robinson 2000; Mena et al. 2000; Peres 2000; Townsend 2000, among others). These studies have used a variety of models to evaluate sustainable use, including effort models, production models, age models, harvest models, and source-sink models (Robinson and Bodmer 1999). It is important to note that these models are not appropriate for predicting the outcome of increases in harvests or in developing harvest strategies in areas that are currently not hunted. Although model development has proceeded rapidly in the last decade, we are not yet at the stage where we can set harvesting levels based on the biology of species in specific areas.

This article will focus on models that have been used to evaluate sustainability of hunting in the Neotropics. These models are practical field-based approaches that can be used in rural and wilderness settings to assess hunting at a specific time and location. Wildlife management in the Neotropics will only be successful if field-based people, whether they be biologists or local hunters, have relatively straight-forward techniques that they can use to evaluate the sustainability of hunting. The strength of these models lies in the field data that is employed. Unlike more theoretical population models, the models described in this article have been developed in a way that allows people to input population parameters and hunting pressure collected from specific field sites.

Simple population models can indicate whether species are overhunted. There are some important guidelines for using these models and interpreting their outcomes (Robinson and Bodmer 1999). First, the confidence in assessing the sustainability of harvest is greatly enhanced by using a combination of models that use independent variables (Robinson and Bodmer 1999; Fragoso, Silvius, and Prada 2000). If the results of the different models point to the same conclusion, then the confidence of the conclusions is greatly augmented. If three different models all suggest that a certain species is overhunted or, conversely, that it appears to be hunted sustainably, then one has greater confidence in the conclusion.

Second, given the assumptions of models and the error margins of data, specific results should be considered as approximate values and the actual numerical results of models should not be used for management recommendations. Specific results should not be used to fine tune actual harvests. For example, if a model shows that a certain species has 20% of production harvested and the model uses 40% as the limit of sustainability, the model should not be used to make recommendations to increase the harvest to 40%. Rather it should be used to suggest that current levels of hunting appear to be sustainable.

Third, it is important to understand clearly the strengths and weaknesses of each model. Each model has assumptions, and these assumptions must be clearly grasped. In many cases we do not know how valid some of these assumptions are or whether they apply equally to different game species. As more information on tropical wildlife populations become available, these assumptions can be revised and perhaps become variables in the models rather than background unknowns.

Finally, the strength of the models lies in their ability to evaluate current hunting. The models presented in this article should not be used to model population projections, to increase harvests, or to initiate harvests in areas where hunting does not currently occur.

DESIGNING THE STUDY

To evaluate sustainability of wildlife hunting, it is necessary to have a clear understanding of the physical, biological, and temporal boundaries of the assessment (Robinson and Bodmer 1999). Physical space may include a nature or extractive reserve, a project area, or a state or region. In some cases the hunting of a single species might be evaluated, while in other groups of species might be evaluated. In terms of time the evaluation can be carried out over the short-term (three to five years) or over the long term (thirty to fifty years).

One must also define exactly what is meant by sustainable use. Some definitions are very general. For example, hunting is sustainable if a species population is healthy and stable under harvest. Other definitions might be specific to certain quantitative models. For instance, a harvest model might define a short-lived species to be sustainably hunted if less than 40% of its production is harvested.

One must be clear about the null hypothesis. In general it is much easier to demonstrate that hunting is not sustainable than to demonstrate that it is. If the null is accepted (or there is no evidence of overhunting), then hunting is apparently sustainable. If the null hypothesis is rejected, then the population is clearly overhunted. To actually prove that hunting is sustainable, one would need to measure all of the biological and socioeconomic variables that might influence sustainability. Such a feat is obviously not possible, and once you have shown sustainability, some variables will change, and the system would again be unsustainable.

Finally, one must determine which models can be used to accept or reject the null hypothesis. There are a variety of models to evaluate the sustainability of hunting, some of which will be discussed in this article. Each model has its strengths, weaknesses, and assumptions, and the different models require different types of data. In turn, the models determine what needs to be measured, what data needs to be collected, and the study design that needs to be set up.

The two approaches most commonly used to evaluate the sustainability of hunting of Neotropical wildlife are the comparative design and monitoring (Robinson and Redford 1994). The comparative design usually compares variables between nonhunted, slightly hunted, and heavily hunted sites. This design is useful if sustainability of hunting needs to be determined in a relatively short period of time.

The comparative design assumes that the sites being compared vary only in hunting pressure. This assumption is often not met, and studies need to be carefully designed to come as close as possible to meeting it or, failing this, to account for deviations in habitat quality between sites. Hunting pressure should be the variable that differs most between sites, with all other variables being as constant as possible. Thus sites being compared should have the same habitat. Adjacent sites with similar habitat have usually been used for comparisons. However, it is difficult to convincingly show the consistency of tropical habitats between sites, and the habitat variables measured in a study might not be the ones that are most important for the animals. Additionally, hunting pressure must be measured in order to discriminate between nonhunted, slightly hunted, and heavily hunted sites.

Monitoring implies a long-term commitment to a hunted site. The sustainability of hunting is evaluated by observing changes over time. Monitoring is a very important technique and is often used to evaluate hunting impact, especially for studies looking at animal exports under CITES regulations. As in the case of the comparative design, there are important assumptions about monitoring that must be understood in setting up and later analyzing the results of the study.

Once again, hunting pressure is the variable being measured, and all other variables should be as constant as possible. Hunting pressure must be monitored over time to document changes. Habitat at the site should be constant over time. Changes in the quality of the habitat can alter the food or nesting sites of species and override the impact of hunting. However, habitats do not remain constant over time, especially with respect to food availability. It is therefore important to monitor habitat quality and to account for it in the final analysis of hunting data. However, few studies monitor environmental changes in parallel with hunting changes, and those that do have not attempted to incorporate these variables into the models used to assess sustainability of hunting. Habitat constancy thus remains an important untested assumption in all models.

THE MODELS

Models that have been used to evaluate the sustainable use of tropical wildlife and that we will discuss in this article include:

1. abundance, densities, or standing biomass comparisons;
2. stock-recruitment models;
3. effort models;
4. age structure models;
5. harvest models;
6. unified harvest models;
7. production models;
8. source-sink models.

Data derived from research conducted in the Tamshiyacu-Tahuayo Community Reserve and the Pacaya-Samiria National Reserve will be used to illustrate the usefulness of the models. The Tamshiyacu-Tahuayo Community Reserve is an upland forest site in the northeastern Peruvian Amazon and is situated in the forests that divide the valley of the Amazon from the valley of the Yavari. The Pacaya-Samiria National Reserve is a flooded forest protected area in the confluence between the Marañon and Ucayali rivers (fig. 19.1). Data are available from nonhunted and

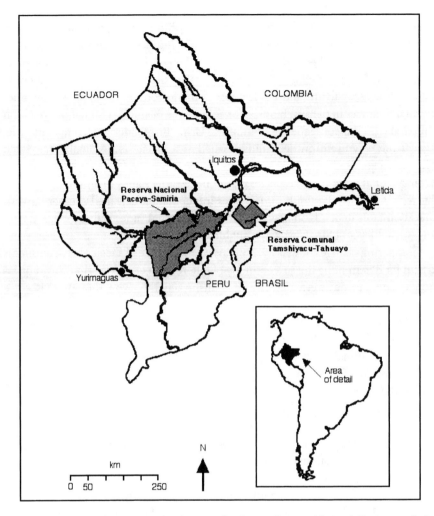

FIGURE 19.1 Map of Loreto, Peru, showing the Pacaya-Samiria National Reserve and the Tamshiyacu-Tahuayo Community Reserve.

hunted sites in both protected areas. The sites have been described in Bodmer (2000).

ABUNDANCE, DENSITIES, OR BIOMASS COMPARISONS

Changes in abundance, density, or biomass of species can be monitored in a site over time and such measurements are often used to determine whether a species is sustainably harvested. Impact of hunting can be indexed by the degree of decline. Harvests are usually classed as nonsustainable if an animal population continually decreases over time.

Changes in abundance, density, or biomass of species can also be compared between nonhunted, slightly hunted, and heavily hunted sites, and these comparisons have been used to evaluate the impact of hunting (Robinson and Redford 1994; Bodmer, Eisenberg, and Redford 1997; Peres 2000). Indices of abundance, such as tracks or other signs, have been used as proxies for direct measures of abundance (Fragoso, Silvius, and Prada 2000; Naranjo 2002). This approach assumes that a differences in density, biomass, or abundance of animal populations between different sites is a consequence of hunting. Using this model in a comparative design is not very useful in determining whether a species is sustainably harvested (Robinson and Redford 1994). The major concern is that differences in the density, biomass, or abundance of a species do not necessarily signify overhunting because harvests will generally result in decreases in population density. Whether harvests are sustainable depends on how the rate of recruitment varies with population density (Caughley and Sinclair 1994; Caughley 1997).

Comparing changes in density, biomass or abundance of species between sites is useful for determining the susceptibility of different species to overhunting. For example, comparisons were made in the Tamshiyacu-Tahuayo Community Reserve between the persistently hunted Blanco site and the slightly hunted Yavari-Miri site. The two sites had similar habitats, they were only 40 km apart, and they had continuous forests with no major rivers dividing the area. Differences in abundances of mammals were correlated to different life history characteristics including intrinsic rate of increase, longevity, and generation time (Bodmer, Eisenberg, and Redford 1997).

The comparison clearly showed that mammals with higher intrinsic rates of increase, shorter longevity, and shorter generation times had smaller or negative differences in their abundance between the slightly hunted and persistently hunted sites. In contrast, mammals with smaller intrinsic rates of increase, longer life spans, and longer generation times had greater differences in abundance between the slightly hunted and persistently hunted sites. This analysis indicates that mammals with higher intrinsic rates of increase, shorter life spans, and shorter generation times are less susceptible to overhunting than mammals with smaller intrinsic rates of increase, longer life spans, and longer generation times.

STOCK-RECRUITMENT MODEL

A variant of the density comparisons is the stock-recruitment model, which is based on density-dependent population models that use maximum sustained yield estimates (MSY) and carrying capacity (K). Most species of tropical wildlife that are hunted are K-selected species and should therefore have density-dependent recruitment (Caughley 1977). In turn, sustainable harvests of tropical wildlife populations will depend on relationships between rate of recruitment and population size. The stock-recruitment model predicts the riskiness of harvests for different populations sizes (McCullough 1987). The greatest base population is at carrying capacity (K) and the smallest at extirpation (o). A sustainable harvest can be realized at any base population size, but there is only one point at which the sustained harvest is at the maximum, or MSY (Caughley 1977).

A species population in a hunted area can be compared to a predicted K and MSY. This is accomplished by comparing the density of the hunted population (N) to an estimated K as N/K. MSY is also denoted as a proportion of K. In turn, the hunted population is positioned in relation to MSY, and this position is used to evaluate the riskiness of hunting in the sense that populations hunted at MSY or below are at greater risk of overhunting than populations that are hunted above MSY (fig. 19.2).

Harvesting species at the MSY is a risky management strategy and should be avoided. If attempts are made to manage a population at the theoretical MSY and small misjudgments occur, this slight overhunting would result in a decreased base population the following year. If this overhunting goes unnoticed and the population is again harvested in the same numbers, the effect of overhunting would be even more dramatic and would quickly lead to extirpation (McCullough 1987) (fig. 19.2).

Similarly, harvesting species with small base populations (to the left of MSY) is a risky management strategy and should be avoided. Again, if small misjudgements occur in calculating the sustainable harvest, this slight overhunting would result in a decreased base population the following year and would quickly lead to overexploitation and extirpation (McCullough 1987).

Harvesting species with large base populations (to the right of MSY), on the oth-

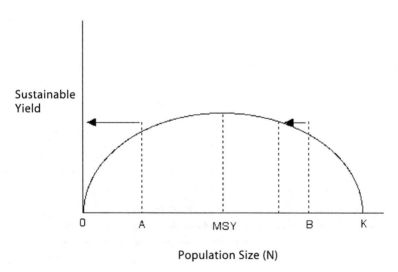

FIGURE 19.2 Representation of the stock-recruitment model, showing K and MSY. Overhunting at point A would drive the population to extirpation, whereas overhunting at point B would lead to a sustainable harvest at a lower population size.

er hand, is a safe management strategy that can be used for long-term sustainable use of a species (fig. 19.2). If small misjudgments occur with harvesting with a large base population, this slight overhunting would result in a decreased base population the following year. If this overhunting goes unnoticed and the population is again harvested in these same numbers, the population would stabilize at this new sustained harvest level and would not decrease further. Harvesting a species with a large base populations is a safer management strategy, since it is less likely to result in extirpation (McCullough 1987).

The stock-recruitment model used here does not actually evaluate the sustainability of current hunting. Rather, it is a powerful way to examine the potential for long-term sustainability. If animals are hunted in a risky manner, then there is less potential for long-term sustainability. In turn, if animals are hunted safely, then there is a much better potential for long-term sustainability.

In the stock-recruitment model a safe harvest is one that occurs to the right of the MSY point. MSY is species specific and is predicted to be at 50% of K for very short-lived species, 60% of K for short-lived species, and 80% of K for long-lived species. These differences derive mainly from variance in reproduction and the way in which this variance changes as the species approaches K in accordance with density-dependent interactions (Kirkwood, Beddington, and Rossouw 1994). Very short-lived species have the greatest variance in reproduction and show density dependent changes in reproduction as a normal distribution as their densities progress from low numbers to K. Short-lived species show slightly skewed changes in their reproduction with maximum production occurring at slightly greater population levels, usually at 60% of K. Long-lived species show little density-dependent responses to reproduction until their populations are actually quite large (Kirkwood, Beddington, and Rossouw 1994). Therefore the MSY is further to the right and is predicted to be at 80% of K.

The stock recruitment model was used to evaluate the riskiness of hunting of mammals in the Peruvian Amazon. For example, ungulates, rodents, and primates were studied in the Pacaya-Samiria National Reserve, in both hunted subsistence sites within the reserve and nonhunted fully protected areas (Bodmer 2000). Densities of these mammals in nonhunted areas were used to estimate K. MSY was set at 60% of K for peccaries, deer, and large rodents and 80% of K for lowland tapir (*Tapirus terrestris*) and primates. Lowland tapir was the only species for which hunting was risky. The tapir population in the subsistence zone was at 60% of the estimated K, a value that is below the predicted MSY of 80%. White-lipped peccary (*Tayassu pecari*), red brocket deer (*Mazama americana*), agouti (*Dasyprocta fuliginosa*), woolly monkey (*Lagothrix lagothricha*), and white capuchin (*Cebus albifrons*) populations in the subsistence use zone were all above the predicted MSY's (table 19.1). Collared peccaries (*Tayassu tajacu*), brown capuchin (*Cebus apella*), howler monkey (*Alouatta seniculus*), and monk saki monkey (*Cacajao calvus*) populations in the subsistence use zone were actually above the estimated

TABLE 19.1 Examples of the Stock-recruitment Analysis in the
Pacaya-Samiria National Reserve

SPECIES	MSY	N/K	STRATEGY
White-Lipped peccary	60%	84% ± 50%	safe
Collared peccary	60%	171% ± 28%	safe
Red brocket deer	60%	86% ± 28%	safe
Lowland tapir	80%	60% ± 80%	risky
Agouti	60%	83% ± 16%	safe
Woolly monkey	80%	92% ± 20%	safe
Brown capuchin	80%	130% ± 10%	safe
White capuchin	80%	81% ± 40%	safe
Howler monkey	80%	168% ± 15%	safe
Monk saki monkey	80%	136% ± 19%	safe

Note: MSY is maximum sustainable yield; N, density in the subsistence use zone; and K, carrying capacity. K was estimated from densities at the fully protected site. MSY is given as the percent of K. Strategies were either risky or safe.

K and thus were more abundant in the hunted area than the nonhunted area. This finding would obviously signify a safe hunting strategy.

Some concerns with this method are that estimating K from nonhunted populations represents an equilibrium population, not really K. An equilibrium population is the observed size of a population within the ecosystem and might be an underestimate of the real K (Caughley and Sinclair 1994). This danger is especially true for predator-limited species in which prey densities are held below K by predator mortality. An underestimate of K would lead to an underestimate of MSY and a misrepresentation in the relationship between N and the actual MSY.

Standard error in density estimations and hunting pressure calculations can be incorporated in the results. The riskiness of hunting can then be evaluated along a range of MSY estimates and a range of sizes for hunted populations.

The stock-recruitment model in this form is an important conservation strategy, since species will only be harvested sustainably in the long-term if their base populations are large. As previously mentioned, faster-reproducing species, such as peccaries, deer, and large rodents, are less vulnerable to overhunting. These faster-reproducing species have a predicted MSY at 60% of K. Thus base populations of these species must be above 60% of K to be considered sustainably harvested over the long term. Slow reproducing species such as tapir and primates are more vulnerable to overhunting. Because these species have less variance in their reproduction, they have a predicted MSY at 80% of K. Thus the base population of these species must be above 80% of K to be considered sustainably harvested over the long term and means that they almost have to be at K to be considered for harvesting.

EFFORT MODELS

Effort models examine relationships between hunting effort and hunting yield and commonly use catch per unit effort (CPUE). These models usually require extensive information about the daily activities of hunters to measure effort. Therefore most of the yield/effort models of Neotropical hunting have been conducted by anthropologists (Vickers 1991). Differences in the catch (or harvest) per unit effort are assumed to reflect differences in actual density or abundance. A decrease in the catch per unit effort suggests overuse (a decreasing population); a constant catch per unit effort, a stable population; and an increase in catch per unit effort, an increasing population. Catch per unit effort analysis can either use a comparative design that looks at nonhunted, slightly hunted, or heavily hunted areas, or it can be used to monitor an area over time.

Hunting registers were used to obtain hunting offtakes and effort (time spent hunting) to develop catch per unit effort relationships in the Tamshiyacu-Tahuayo Community Reserve (Puertas 1999; Puertas and Bodmer this volume). Catch per unit effort was tested by examining a hunted site during the high- and low-water seasons. During high water, access to hunted sites by canoe is relatively easy. During low water access is difficult. Abundance and CPUE of large mammals did not change between the two seasons (Puertas and Bodmer this volume), whereas hunting pressure and effort were considerably different between the seasons, being much greater during high water. Thus, catch per unit effort successfully reflected abundances of animal populations, and changes in catch per unit effort would therefore reflect changes in abundance and give an indication of overhunting.

Catch per unit effort relies on several conditions and assumptions. First, effort must be measurable, even though the choice of measures is somewhat arbitrary. In the Peruvian Amazon, effort was recorded as the number of days a person spent hunting (hunter-days) (Puertas and Bodmer this volume). Second, effort must be constant. If hunters change from bow and arrow to guns, the measure of off-take per effort would not be constant, and the comparison would be suspect (Robinson and Bodmer 1999). The activities of hunters during hunting trips might not always be constant, so assumptions are be made about the average activity of hunters during hunting trips. Third, the catch, or hunting pressure, must be accurately recorded. If animals are omitted from the catch, this will alter the catch per unit effort calculation and render the analysis suspect.

AGE MODELS

Age models examine the age structure of wildlife populations to see if changes in demography indicate overuse. There are two types of analysis that depend on whether hunting is selective or random (Caughley 1977). Type 1 models are used when hunting is selective and hunters harvest only certain age classes, such as the larger or older individuals of a species. If older or larger individuals are selectively hunted, the age distribution of the population will be skewed toward younger or smaller animals. A stable age distribution in a selectively hunted population indicates that the animals are probably not overhunted. In contrast, if the age distribution continues to decline toward younger or smaller animals, then the population is likely overhunted.

Age distributions from hunted samples can be used to evaluate whether animals are overhunted in a selectively hunted situation. Samples collected from individuals in a selectively hunted population will not reflect the age distribution of the population but will reflect the age distribution of the harvest. If the age distribution of the hunted sample decreases, becoming more biased toward younger ages, this finding would reflect a shortage in the preferred age or size classes and suggest overhunting.

Type 2 models apply when hunting is random with respect to age classes. Random hunting occurs when hunters have no choice of the individuals they are selecting and hunt individuals randomly with regard to age. This commonly happens with snaring or trapping. If hunting is truly random, then samples from randomly hunted populations should reflect the age distribution of the population, and changes in the age distribution of the hunted samples should directly reflect changes in the demography of the population.

However, the interpretation of overhunting from changes in the age distribution is problematic. Changes in age distribution in randomly hunted populations might be caused by (a) recruitment rates increasing with density declines, resulting in proportionally more younger animals; (b) behavioral shifts (i.e., more wariness) in certain age classes, making them less vulnerable (so hunting actually becomes selective); (c) a reduction of natural predators in hunted areas, which in turn results in less infant or young mortality and skews the population toward older animals; or (d) immigration of young individuals into overhunted sites from source areas.

Age distribution can be compared between slightly or nonhunted sites and heavily hunted sites, or it can be monitored in the same site over time. Nonhunted sites can only be included if there is the possibility of obtaining age data of animals from nonhunted samples (i.e., live trapping or capture).

Age distributions are instantaneous samples of individuals. Survival and mortality relationships can only be inferred if one assumes a stable age distribution (Caughley and Sinclair 1994). This stable age distribution assumes that there are no variations in the age distribution between cohorts, an unlikely situation in free-ranging animals. In addition, survival relations can only be inferred from a randomly hunted sample (Caughley and Sinclair 1994).

A variety of techniques are used to age wild animals. The most common technique for mammals is tooth wear. In the Tamshiyacu-Tahuayo reserve skulls from hunted animals were used to determine age through tooth wear and in turn to evaluate the age distributions of populations (Gottdenker and Bodmer 1998). Results are not easy to interpret in terms of overhunting. For example, age models would predict that randomly hunted populations of collared peccaries and white-lipped peccaries are different between persistently hunted and slightly hunted populations. However, within-site, interyear variability in age distribution is significant and overrides differences between hunting sites, masking any effect of hunting (fig. 19.3). On the other hand, obviously overhunted species, such as lowland tapir, show skewed age distributions between slightly hunted and heavily hunted sites, with age distributions being younger in the overhunted sites (fig. 19.4).

Age models have conditions and assumptions that must be considered. First, studies must determine whether a population is randomly or selectively hunted. Second, interpreting what causes changes in age distribution can be problematic in randomly hunted species. Third, it is difficult to determine whether a species is overhunted from the degree of change in age distribution. Indeed, it is possible for a randomly hunted population to show little variance in age distribution even if the population is being overhunted. This observation is especially true for species, such as primates, that have little variation in density-dependent reproduction and do not migrate between areas.

Our understanding of how hunting affects age structure of tropical forest species remains in its infancy. Evaluating sustainable use by understanding age structure has enormous potential because data collection agrees with the activities of local hunters. Hunters can easily collect skulls from animals they hunt with only a minimum of extra labor, thus creating large skull collections (Bodmer and Puertas 2000). Still, it is not advisable to make management decisions using only age data on randomly hunted populations with our current lack of understanding of the relationship between age distribution and sustainability of hunting.

HARVEST MODELS

Animal populations can theoretically be sustainably harvested at any population level, except at carrying capacity (K) and extirpation (0) (Caughley 1977). Thus one way to evaluate the sustainability of hunting would be to know the actual production at the population size being harvested. The harvest can then be compared to production in order to obtain a measure of the percent of production harvested at the time hunting is being assessed, as well as determining whether this percent is within sustainable limits. This analysis is known as the harvest model (Bodmer 1994).

The harvest model uses production estimates derived from reproductive productivity and population density. Reproductive productivity is determined from data on reproductive activity of females and uses information on (a) litter size and (b) gross reproductive productivity (number of young/number of females). Population density is determined from field censuses of wildlife species. Animal densities are then multiplied by reproductive productivity to give an estimate of production, measured as individuals produced/km^2 as:

$$P = (0.5D)(Y^*g),$$

where Y is gross reproductive productivity, g is the average number of gestations per

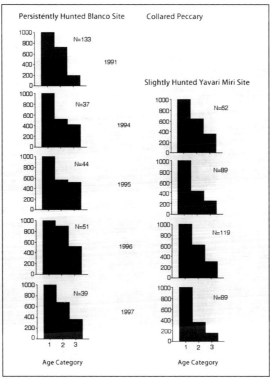

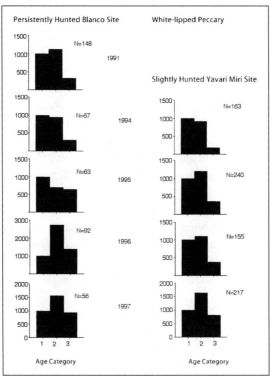

FIGURE 19.3 Annual age distribution of collared peccary and white-lipped peccary in the persistently and slightly hunted areas of the Tamshiyacu-Tahuayo Community Reserve.

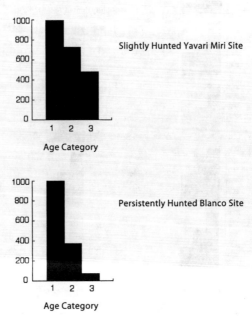

Lowland Tapir Age Distributions

FIGURE 19.4 Age distribution of lowland tapir in the persistently and slightly hunted areas of the Tamshiyacu-Tahuayo Community Reserve.

year, and *D* is the population density (discounted by 50% under the assumption that the population sex ratio is 1:1). Of course sex ratios are often not 1:1, and animal populations often vary in sex ratio for numerous reasons. Our current understanding of sex ratios in Neotropical mammals is still very incomplete. Thus we usually assume a 1:1 sex ratio. Hunted sex ratios probably do not reflect actual sex ratios, and with our current understanding, it is better to assume a 1:1 sex ratio than use hunted sex ratios.

The sustainability of hunting is determined by comparing harvest with production. Harvest data and catchment areas give an estimate of hunting pressure (individuals harvested/km²). The calculation of production assumes no prereproductive or adult mortality. Therefore an estimate of population growth rate must be incorporated in order to estimate the proportion of production that can be harvested sustainably.

Robinson and Redford (1991) suggest that the average lifespan of a species can be used as an index of population growth. Animals with longer lifespans have slow population growth and in turn a smaller percent of production should be harvested to maintain a sustainable hunt (Kirkwood, Beddington, and Rossouw 1994). Likewise, animals with shorter lifespans have a faster population growth and in turn a larger percent of production can be harvested to maintain a sustainable hunt.

Robinson and Redford (1991) propose that hunters can take 60% of the production of very short-lived animals (those whose age of last reproduction is less than five years) to maintain a sustainable hunt, 40% of the production of short-lived animals (those whose age of last reproduction is between five and ten years), and 20% of the production of long-lived animals (those whose age of last reproduction is greater than ten years).

An example using the subsistence zone of the Pacaya-Samiria National Reserve will illustrate the harvest model. Examination of the reproductive condition of female red brocket deer shot by hunters revealed that the gross productivity (total number of young examined/total number of females examined) was 0.44. Collared peccaries have an average of 2.0 gestations per year (two birthing periods) (Gottdenker and Bodmer 1998). This number resulted in an annual reproductive productivity of 0.88 young/female-year. The reproductive productivity was multiplied by one-half of density of red brocket deer (0.6 ind./km^2) since it was assumed that one-half of the population of red brocket deer were females. The product, annual production (0.5 ind. produced/km^2), was then divided into the annual hunting pressure of 0.06 red brocket deer hunted per km^2. This calculation yielded the percentage of production taken by hunters, which in this case was 12% of red brocket deer production. This figure is below the 40% maximum for a short-lived species, suggesting that harvests of red brocket deer in the subsistence zone probably are sustainable. As mentioned above the 40% maximum accounts for mortality due to factors other than hunting.

Sensitivity analyses can be used to see whether the error in estimating the different variables, such as density, hunting pressure, and reproduction, will influence the results. For example, the harvest model used to evaluate sustainability of hunting in the Pacaya-Samiria National Reserve incorporated standard error calculations. Red brocket deer in the subsistence use zone had error margins that did not exceeded the 40% limit, providing greater confidence in the conclusion (table 19.2). Over 100% of lowland tapir production was harvested in the subsistence use zone, clearly showing that this species is overhunted and that the base population was being depleted (table 19.2).

The harvest model is a useful way to evaluate the sustainability of hunting in an area because it uses information on production and harvests from the field sites. However, the model is a closed population model and does not take into account immigration or emigration of animals from adjacent areas. For example, the lowland tapir population is clearly overhunted according to the harvest model, and the harvest model predicts its extirpation. However, tapirs still occur in the subsistence use zone of the reserve. This presence suggests that individuals are immigrating to the area from adjacent nonhunted forests.

THE UNIFIED-HARVEST MODEL

The unified harvest model combines the stock-recruitment and harvest models into a unified analysis that evaluates both the sustainability of current hunting and

TABLE 19.2 Percent of Production Taken by Hunters in the Subsistence Zone of the Pacaya-Samiria National Reserve

| | PERCENTAGE OF PRODUCTION TAKEN BY HUNTERS | | |
SPECIES	Subsistence Zone	Proportion of Production Analysis That Can Be Harvested Sustainably	Sustainability
White-Lipped peccary	19 ± 17	40	Appears sustainable
Collared peccary	4.7 ± 2.2	40	Appears sustainable
Red brocket deer	12% ± 6	40	Appears sustainable
Lowland tapir	>100	20	Overhunted
Agouti	2.5 ± 0.6	40	Appears sustainable
Brown capuchin monkey	3.0 ± 0.5	20	Appears sustainable
White capuchin monkey	5.0 ± 2.4	20	Appears sustainable
Woolly monkey	15.0 ± 4.0	20	Appears sustainable
Howler monkey	36 ± 9	20	Overhunted
Monk saki monkey	3.3 ± 0.7	20	Appears sustainable

the potential for long-term sustainable use. The unified harvest model uses a modified population growth curve. As with stock-recruitment curves, the horizontal axis is the population size from extirpation (o) to carrying capacity (K) and the vertical axis is the sustainable limit of exploitation expressed as sustainable yield (SY). The SY mirrors the growth of the population dN/dt and has a maximum point of growth or a maximum sustainable yield (MSY).

The major difference between the unified harvest model and the population growth curve is that the vertical axis in the unified harvest model uses the percentage of production harvested as a measure of SY rather than population growth. Thus the harvest model can be used to evaluate the sustainability of offtake, and the line (known as the SY line) is in fact the 20%, 40%, or 60% limit to the percentage of production that can be harvested.

For example, a maximum of 40% of collared peccary production can be harvested sustainably according to the harvest model. Therefore the SY line in the unified harvest model is the 40% limit. If the harvest of collared peccary exceeds the 40% limit, then it is deemed unsustainable. If, however, the harvest is lower than the 40% limit, then the harvest appears to be sustainable. Thus sustainable harvests can occur at any collared peccary population size, as long as the harvest is less than 40% of production. In the case of the lowland tapir, the maximum level of harvest is 20% of production and in turn, the SY line represents the 20% limit for lowland tapir (table 19.3).

The unified harvest model also analyzes the riskiness of harvests in terms of the potential for long-term sustainability by incorporating the stock-recruitment analy-

TABLE 19.3 The Limits of Sustainability of the Unified Harvest Model

LIFE HISTORY STRATEGY	MAXIMUM % OF PRODUCTION HARVESTABLE	ESTIMATED MSY AS A % OF K
Short-lived	60%	50%
Medium-lived	40%	60%
Long-lived	20%	80%

Note: The limits reflect life history strategies of species and are of two types: (1) limits of the maximum percent of production that can be harvested before a species is overhunted, and (2) the estimated MSY (maximum sustainable yield) of species that is used to determine the proximity of harvested populations to MSY and in turn the riskiness of hunting. K is carrying capacity.

sis. This procedure is done by determining the proximity of the current harvest to carrying capacity (K) and to the estimated maximum sustained yield (MSY). The unified harvest model is used to evaluate whether a harvest level is risky or safe depending on the population size relative to the predicted MSY.

The unified harvest model can then combine the percentage of production of a harvested population with its position relative to MSY in order to give both a measure of the current sustainability and the long-term riskiness of the harvest. This result can be very useful since sustainability can be represented by a single line, which denotes both the percent of production harvested in relation to the SY line and the position of the harvested population relative to the species estimated MSY.

The unified harvest model was used to evaluate the sustainability of hunting in the subsistence use zone of the Pacaya-Samiria National Reserve by combining results from the stock-recruitment and harvest models. In the case of the white-lipped peccary, 19% of production was harvested, which is below the 40% limit, and the harvested population was at 83% of K, well above the estimated MSY at 60% of K (fig. 19.5A). Thus, harvests of white-lipped peccary appeared sustainable and the harvested population was being safely hunted in terms of its long-term sustainability. Similarly, in the case of the red brocket deer, 12% of production was harvested, well below the 40% limit, and the harvested population was at 86% of K (fig. 19.5B). Thus harvests of red brocket deer also appeared sustainable. The case of the woolly monkey was similar to the white-lipped peccary and red brocket deer, but the SY line represents 20% of production harvested, and the MSY is set at 80% of the estimated K (fig. 19.5C).

In the case of the lowland tapir, well over 100% of production was harvested, above the sustainable limit of 20% and obviously not sustainable. Similarly, the tapir population was harvested at 60% of K, which was below the predicted MSY of 80% K. Thus lowland tapir were both hunted at unsustainable levels and at risk in terms of long-term sustainability (fig. 19.5D).

Sensitivity analysis can be incorporated into the model on both the SY and riski-

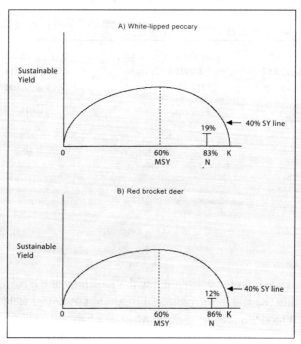

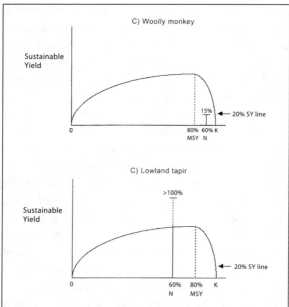

FIGURE 19.5 Diagram of the unified harvest model. This example evaluates the sustainability of hunting in the subsistence zone of the Pacaya-Samiria National Reserve. The height of the solid vertical line represents the percent of production harvested, whereas the position of the vertical line represents the proximity of the harvested population to K and MSY. The SY line is the estimated limit of sustainable harvests, which for white-lipped peccary (A) and red brocket deer (B) is 40% of production, and for woolly monkey (C) and tapir (D) is 20% of production.

ness calculations. The possibility of misjudging the sustainability of hunting can be evaluated by looking at the error margins of density, hunting pressure, and reproductive variables.

The unified harvest model reflects the conservation requirements of species by setting SY limits and MSY levels in accordance to species vulnerability to overhunting. Species susceptibility to overhunting is correlated to life history characteristics, including reproduction, longevity, and generation time. Species with greater vulnerability to overhunting have higher bars in the unified harvest model, as with tapir whose SY limit is set at 20% of production and whose predicted MSY is set at 80% of K. In contrast, species that are less vulnerable to overhunting have lower bars, as with collared peccary whose SY limit is set at 40% of production and whose predicted MSY is set at 60% of K (table 19.3).

The unified harvest model combines both the stock-recruitment analysis and the harvest model; therefore it relies on the same assumptions and carries the same concerns and potential weaknesses as these models. Some concerns with the unified harvest model are that estimating K from nonhunted populations might be an underestimatation of the real K. The SY calculation in the unified model assumes a closed population and does not take into account immigration or emigration of animals from adjacent areas.

PRODUCTION MODEL

The most commonly used model for evaluating sustainability of hunting in tropical wildlife is Robinson and Redford's population-growth model (Robinson and Redford 1991, 1994; Robinson 2000), termed here the production model. The sustainability of hunting can be evaluated by calculating maximum possible production of a species and comparing this figure to actual harvests in the absence of detailed information about species density and reproduction at a specific site. The model evaluates whether an actual harvest is unsustainable, but it can not evaluate whether an actual harvest is sustainable.

The production model assumes that populations of wildlife are density dependent, with maximum production at 0.6 K. As with the other models, K is estimated from nonhunted, undisturbed populations. Maximum production (P_{max}) is calculated by multiplying the density at maximum production (estimated as 0.6 K) by the finite rate of population increase (λ_{max}) and subtracting it from the previous year's density (also estimated at 0.6 K), using:

$$P_{max} = (0.6K * \lambda_{max}) - 0.6K,$$

where λ_{max} is the exponential of r_{max}, being the intrinsic rate of population increase (highest rate of population increase when a species is not limited by food, space, resource competition, or predation). The intrinsic rate of population increase can be calculated using Cole's 1954 equation:

$$1 = e^{-r_{max}} + be^{-r_{max(a)}} - be^{-r_{max(w+1)}},$$

where a is the species-specific age of first reproduction, w is the age of last reproduction, and b is the annual birth rate of female offspring. These reproductive parameters are available in the literature from captive and wild individuals for commonly hunted species. Cole's equation is actually a measure of maximum reproductive productivity.

As with the harvest model, the production model assumes no prereproductive or adult mortality. Hence, the average lifespan of a species is used as an index of an animal's population growth. As with the harvest model, the harvest limits are set at 60% of production for very short-lived species, 40% of production for short-lived species, and 20% of production for long-lived species. This modification of λ_{max} by a factor of f_{rr} of 0.6, 0.4, or 0.2 means that the effective rate of population growth, λ_{rr} is

$$\lambda_{rr} = 1 = (\lambda_{max}^{-1}) f_{rr}$$

and the maximum possible production available to hunters is:

$$P_{rr} = (\lambda_{rr}^{-1}) 0.6K$$

(Slade, Gomulkiewicz, and Alexander 1998). λ_{rr} has been used to show maximum possible harvests for a number of species (Robinson and Bodmer 1999).

Collared peccary harvests in the Tamshiyacu-Tahuayo Community Reserve were evaluated using the production model. The maximum production rate of collared peccary is 1.99 (Robinson and Redford 1991). The density of collared peccary at the nonhunted site was 1.6 individuals/km^2, yielding a maximum sustainable harvest ($0.4P_{max}$) of 0.38 ind/km^2, which was above the actual harvest of 0.33 ind./km^2 at the hunted Blanco site. In contrast, the maximum sustainable harvest of lowland tapir was well below the observed harvest at the hunted site. The density of lowland tapir at the nonhunted site was 0.21 individuals/km^2, yielding a maximum sustainable harvest ($0.2P_{max}$) of 0.0055 ind/km^2, which was well below the actual harvest of 0.07 ind./km^2 at the hunted Blanco site.

The production model does not tell us much about collared peccary hunting in the Blanco site, only that it is not obviously overhunted. It does not tell us whether the hunting is sustainable or not because the model does not use density data from the Blanco site. In contrast, the model clearly implies that lowland tapir harvests are not sustainable at the Blanco site.

The production model can help determine if a species is overhunted when harvests exceed maximum levels, but it cannot determine whether a harvest is sustainable if levels are below the maximum. One concern of the model is that it includes many parameters without using data from the actual site of harvesting. This failing can result in estimations that are not appropriate for a particular site.

SOURCE-SINK MODELS

The models described above are useful indicators of the sustainable use of populations, but they do not embody the complexities of natural ecosystems and the unpredictable fluctuations of wildlife populations. Some of the models, such as abundance comparisons, stock-recruitment model, age structure analysis, and effort models, are open models that intrinsically incorporate immigration and emigration from neighboring areas. Others, such as the harvest model and production model, are closed population models that do not take into account immigration or emigration.

One promising way to include concerns of complexity, unpredictable fluctuations, and animal movements into management strategies for tropical wildlife is to adjoin nonhunted source populations to hunted (or sink) areas (McCullough 1996; Townsend 1996b; Hill et al. 1997; Fragoso, Silvius, and Prada 2000; Novaro, Redford, and Bodmer 2000). Both intact habitats with continuous animal populations or fragmented habitats with a metapopulation structure can be used with a source-sink strategy (Novaro, Redford, and Bodmer 2000).

Source-sink models require information on the status of source and sink populations and movements of animals between sources and sinks. Analyses of source-sink systems in the tropics have usually used information on the populations of wildlife in source and sink areas but have generally not known the degree of movement between sources and sinks (Novaro, Redford, and Bodmer 2000).

For example, the unified harvest model can incorporate source and sink areas by estimating the percentage of production harvested and the riskiness of harvests in heavily hunted sinks, slightly hunted sources, and nonhunted sources. In nonhunted sources the percentage of production harvested is zero. It is then possible to combine source and sink areas in order to reach an approximation of the percentage of production harvested and the riskiness of the harvest throughout the entire source-sink area.

It is important also to appreciate the importance of landscape features and the spatial relationship between source and sink populations. In the example above, it is assumed that animals will disperse from the source to the sink. This case might be true for the fully protected area of the Pacaya-Samiria National Reserve since it is situated in the center of the reserve and surrounded by the subsistence use zone. In other areas, however, this might not be the case and consideration must be given to the spatial assumptions.

The Tamshiyacu-Tahuayo Community Reserve and its surroundings can demonstrate how source-sink analysis can be combined with the unified harvest model. The Tamshiyacu-Tahuayo Community Reserve was divided into three hunting zones: (a) a persistently hunted area of 1,700 km^2, (b) a slightly hunted areas totaling 4,000 km^2, and (c) a nonhunted area totaling 5,300 km^2. The nonhunted and slightly hunted areas were potential source populations for the persistently hunted area. The size of hunting zones were estimated from data on harvests and catchment area collected from hunters over an eight-year period (Novaro, Redford, and Bodmer 2000)

The effectiveness of the source-sink strategy was examined for lowland tapir, peccary, and deer populations. The unified harvest model showed that, in the persistently hunted Blanco site, 140% of lowland tapir production was hunted; the harvest was ruled risky. This region is clearly a sink area for lowland tapir. The slightly hunted site had an estimated 16% of lowland tapir production hunted, a figure that is below the 20% limit; the hunting was deemed safe. Thus the slightly hunted sites can be considered part of the source area. The nonhunted sites had 0% of production hunted, and the slightly hunted plus nonhunted site together made up the aggregate source area.

Hunters were taking an estimated 8% of the lowland tapir production from this aggregate source area, which is within sustainable levels. Within the entire source-sink area, including the persistently hunted, slightly hunted, and nonhunted sites, hunters were taking an estimated 18% of lowland tapir production. This figure suggests that hunting of lowland tapir in the entire source-sink area appears to be sustainable and that the sustainability of hunting in the persistently hunted area depends largely on immigration (or replenishment) rates from adjoining source areas.

But are lowland tapir actually moving between the hunting zones? Continued persistence of tapir in the Blanco site suggests that recruitment via immigration from the source area is important. In addition, tapir populations in the Blanco site are considerably younger than tapir population in the slightly hunted area, suggesting that younger animals might be moving from the source to the sink (see age model above).

The effectiveness of the source-sink strategy was also examined with peccary and deer populations (table 19.4). The risky harvesting levels of white-lipped peccary hunting in the Blanco site and the proximity of collared peccary and brocket deer

TABLE 19.4 Results of the Harvest Model for Ungulates in Source and Sink Areas in and Around the Tamshiyacu-Tahuayo Community Reserve

	PERCENTAGE OF PRODUCTION HARVESTED		
SPECIES	Sink	Source	Source-Sink
Lowland tapir	140%	8%	18%
Collared peccary	31%	3.3%	6%
White-lipped peccary	11%	1.5%	2%
Red brocket deer	38%	2.1%	9%

harvests to the sustainable limits suggests that these animals might be at risk of over-hunting during some years. However, if a management strategy includes the slight-ly hunted and nonhunted source areas, this risky strategy is more acceptable be-cause source areas could replenish overhunting in the persistently hunted site.

METHODS

All of the models described above require the input of data collected from field studies. Thus the level of accuracy of the models depends largely on the data col-lected from the field. Weak data sets will yield inaccuracies, and the results of the models will be suspect. It is important to acknowledge the importance of collecting reliable data with sufficient sample sizes. All of the models require information on hunting pressure, and the comparative and monitoring designs themselves require a knowledge of hunting pressure.

Hunting pressure is usually collected by involving hunters in the study, and many studies have involved local hunters and their families in data collection (Vickers 1991; Bodmer 1994; Alvard 1998; Jorgenson 2000; Townsend 2000; among others). This participatory approach has several advantages over nonparticipatory methods, which do not involve local people in the design and implementation of data collection:

1. The participatory approach permits researchers to collect information on hunting pressure.
2. It allows researchers and hunters to work together and better understand each other's needs.
3. It sets the stage for local involvement in future management of wildlife resources.
4. It teaches hunters how to collect data so that in the future they will be directly in-volved with analyzing the sustainability of their own hunting.
5. Hunters can easily collect animal parts such as skulls, reproductive tracts, kidney fat, and genetic material, among other biological material.

These participatory methods are useful for collecting catch per unit effort data. They are also useful for collecting skull samples that can be used for age structure analysis and reproductive samples that are needed to determine gross reproductive productivity.

Many of the models require information on density (stock-recruitment model, harvest model, production model, and unified harvest model), while others can use measures of abundance (abundance comparisons). In the Neotropics most studies have used line transect censuses to estimate large mammal densities. These census data have usually been analyzed using the computer program DISTANCE (Buckland et al. 1993). The major drawback to this method is that a large sample size of direct observations is required, whereas small sample sizes are usually ob-tained in the field. Large distances using numerous transects are necessary to ob-tain sample sizes sufficient to estimate densities. For example, in the Tamshiyacu-

Tahuayo Reserve there have been about 1,000 km of censuses conducted in each of the sampled areas (nonhunted, slightly hunted, and heavily hunted) for a total of around 3,000 km censused. In the Pacaya-Samiria National Reserve there has been around 2,300 km censused. Many of the rarer species have not been sighted sufficiently to estimate density, even with these efforts of censusing.

DISCUSSION

Over the past decade many studies have begun to evaluate the sustainable use of wildlife (Robinson and Bennett 2000b). The most common model that is used is the production model developed by Robinson and Redford (1991). However, other models outlined in this article are also regularly used, and when information is available, these models may be preferable.

One of the major findings is that almost all of the sites examined had some species that were overhunted, while others were used more sustainably (Robinson and Bennett 2000b). In the Neotropics it was usually the rodents and faster-reproducing ungulates, such as collared and white-lipped peccaries and deer, that were hunted more sustainably, and the primates and slower-reproducing ungulates that were overhunted.

If hunting is to be sustainable, then overhunting should be stopped or reduced. In almost all areas studied a reduction in overhunting would entail a decrease in the hunting of certain species. Some authors have suggested that this reduction in harvests can be supplemented by increasing the harvests of sustainably used species (Alvard 1998). However, this is a very risky strategy that should not be used at the current time because of our incomplete understanding of tropical wildlife, the simplicity of the models used, and the limited accuracy of evaluating the sustainability of hunting.

The models described in this article are useful tools with which to evaluate whether measured hunting appears to be sustainable, but they are not sufficiently precise to determine exact hunting quotas or to recommend increases in harvests. The models can suggest whether current hunting pressures appear sustainable and whether this hunting can be continued at the measured levels. The models are also useful at revealing overhunting and in turn if hunting levels should be reduced. These models are not precise enough to suggest increases in hunting or to initiate hunting in nonhunted areas. The models are not population projections; rather, they are analysis of population size, harvest rates, and demography.

Evaluating the sustainability of hunting is only the first step toward converting unsustainable hunting to more sustainable hunting. It is important to evaluate sustainability as part of a process in order to manage hunting in a more sustainable manner. In many areas this assessment can be best done by including hunters in the evaluation through participatory approaches. Hunters, either local or sport, are often the people who use the wildlife most and who are the most interested in the continued well-being of the wildlife populations. They are the ones who visit and spend time in remote areas where wildlife abounds. Hunters are also often willing to contribute actively toward wildlife management. Thus, if hunters have a better understanding about wildlife management, they will be better suited to implement wildlife management actions.

Involving hunters and their families in the collection of data has some very important ramifications for conservation. First, such involvement produces a common ground in order to discuss wildlife issues among wildlife extension personnel, researchers, and the hunters. But more importantly, it persuades the hunters to become involved in analyzing the impact of their own hunting. They can then better understand the consequences of hunting, which in turn helps them think about ways of managing harvests in a way that agrees with their own realities. This self-monitoring process will be essential to the long-term implementation of sustainable wildlife harvests, and it frequently begins with participatory research techniques.

ACKNOWLEDGMENTS

We are indebted to the tremendous support provided by the communities of the Reserva Comunal Tamshiyacu-Tahuayo and the Pacaya-Samiria National Reserve, who participated with many wildlife projects; to Pablo Puertas, Rolando Aquino, César Reyes, Miguel Antunez Correa, and José G. Gil who helped with the fieldwork; and to Julio Curinuqui and Gilberto Asipali for their dedicated field assistance. The following organizations provided logistical and financial support for the projects: Pacaya-Samiria National Reserve-INRENA, the Wildlife Conservation Society; the Chicago Zoological Society, University of Florida's Programs in Tropical Conservation, Instituto Nacional de Recursos Naturales-Peru, the Universidad Nacional de la Amazonía Peruana, the Asociación para la Conservación de la Amazonía, and the Rainforest Conservation Fund.

20

Hunting Sustainability of Ungulate Populations in the Lacandon Forest, Mexico

EDUARDO J. NARANJO, JORGE E. BOLAÑOS, MICHELLE M. GUERRA, AND RICHARD E. BODMER

Wildlife has been and continues to be an important resource for the subsistence of rural people worldwide, providing food, hides, tools, medicine, income, and many other benefits (Redford and Robinson 1991; Shaw 1991; Freese 1998; Robinson and Bennett 2000b). There are many documented cases of continuous use of vertebrate species by native people in the tropics. Three good examples are the hunting of white-tailed deer (*Odocoileus virginianus*) by Mayan Indians of the Yucatan Peninsula, Mexico (Mandujano and Rico-Gray 1991; Jorgenson 1995), the use of duikers (*Cephalophus* spp.) by Mbuti tribes in the Ituri Forest of Central Africa (Hart 2000), and the harvest of Celebes pigs (*Sus celebensis*) by the Wana people of Central Sulawesi, Indonesia (Alvard 2000). However, the persistence of these harvested ungulate populations does not necessarily mean that local hunting systems have been sustainable in the past or that they are sustainable today (Robinson and Bodmer 1999).

On the basis of the arguments of Caughley and Sinclair (1994) and Prescott-Allen (1996), we define sustainable hunting as occurring when the number of animals taken does not exceed their production rates for a given period and when their long-term viability is not impaired. In this study we rely on five currently available models to evaluate the sustainability of hunting of ungulate populations in the Lacandon tropical rain forest of Mexico. Ungulates (Mammalia: Artiodactyla and Perissodactyla) are usually ranked as one of the most important groups of game mammals in Latin America because of their high yield, the good taste of their meat, and the usefulness of their skins (Ojasti and Dallmeier 2000).

Five ungulate species were included in this study: (a) Baird's tapir (*Tapirus bairdii*), which is endangered throughout its range (INE 2000; IUCN 2000); (b) white-lipped peccary (*Tayassu pecari*), whose populations and range seem to be declining in Mexico; (c) collared peccary (*Tayassu tajacu*); (d) white-tailed deer; and

(e) red brocket deer (*Mazama americana*). The last three species are still locally abundant and constitute an important source of protein for subsistence hunters in southeastern Mexico and Central America (March et al. 1996; Reid 1997; Bolaños 2000; Escamilla et al. 2000; Naranjo 2002). The objectives of this study were to evaluate the hunting sustainability of ungulate populations around Montes Azules Biosphere Reserve (MABR) in the Lacandon Forest of Chiapas, Mexico; and to determine if hunting sustainability of ungulates is affected by spatial scale in the study area.

MODELS OF HUNTING SUSTAINABILITY

Hunting sustainability in tropical forests has been assessed using a diversity of methods and models. The simplest methods consist of evaluations of the effects of hunting on game populations. These effects have been estimated by comparing hunting effort, densities, and age structure of game populations in unhunted and hunted areas (Robinson and Bodmer 1999). For hunting effort the measurement of catch per unit effort (i.e., prey killed/100 man hours) allows one to infer whether the abundance of game populations has changed over time as a result of hunting (Puertas 1998; Puertas and Bodmer this volume). An evident decline in catch per unit effort between two periods may suggest that the population has been over-hunted. For densities the assumption is that hunting provokes a decline in population density. Therefore, habitats being equal, hunted areas would maintain lower densities than unhunted areas (Robinson and Bennett 2000b). For age structure one may expect that hunted areas would sustain populations with higher proportions of young individuals than unhunted areas because of the selectivity of hunters for the largest (and older) animals (Bodmer et al. 1997a).

These three methods may be helpful in assessing the effects of hunting on game populations. However, they also have potential biases: The differences in population densities or age structures between hunted and unhunted areas may not be due to hunting pressure but to such natural factors as variations in soils, primary productivity, availability of water, and demographic stochasticity, among many others. Similarly, a decay of catch per unit effort through time could be a result of changes in the livelihood of local hunters more than an effect of population decline.

The two most widely used models of hunting sustainability rely on comparisons of production rates and actual harvest rates of game populations (Robinson and Redford 1994; Robinson and Bodmer 1999). Robinson and Redford's 1991 production model works with estimations of maximum production rates (n animals produced/km^2/year; P_{max}) that are compared to actual harvest rates (n animals taken/km^2/year). This model is particularly useful in the absence of data on densities and actual production rates of hunted populations; it allows the detection of overharvest but not of sustainable hunting (Robinson and Bodmer 1999).

The harvest model proposed by Bodmer (1994) uses calculations of actual density, reproductive productivity (number of young/female/km²/yr), and harvest rates of hunted populations. This model can be used to estimate whether hunting is sustainable or not at a given site (Robinson and Bodmer 1999).

Both the production and the harvest models assume that production, maximum potential harvests, and growth rates of game populations are density-dependent. Therefore, sustainable harvest rates should not exceed 20%, 40%, or 60% of the production rates of long-lived species (i.e., tapir), short-lived species (i.e., peccaries and deer), and very short-lived species (i.e., spiny rats), respectively (Robinson and Bodmer 1999). Likewise, both models assume that Maximum Sustained Yield (MSY) of Neotropical rain-forest mammals may be achieved at about 0.6 K in species with relatively high reproductive productivity (e.g., deer and peccaries), but the same MSY is reached at up to 0.8 K in slowly reproducing animals (e.g., tapirs; Bodmer and Robinson this volume).

Evaluation of sustainability of hunting systems has frequently involved the model of Maximum Sustained Yield (Bennett and Robinson 2000a). MSY is conceived as the maximum possible number of animals harvested without driving the population into decline (Caughley 1977; Eltringham 1984). Theoretically, MSY is achieved when the harvest rate equals the population's recruitment rate by reproduction and it may occur at about 50 to 60 % of carrying capacity (K) for density-dependent populations in a given habitat (Caughley 1977; Riney 1982; McCullough 1987). However, MSY has suffered criticism as a management goal because under its guidance many wild populations seemed to have been depleted (Larkin 1977; Freese 1998). One of the problems of using MSY as a goal is that sustainable yields eventually may be obtained at densities well below 0.5 or 0.6 K (Caughley 1977; Caughley and Sinclair 1994). This implies that it is not possible to obtain the MSY of a wild population in an unhunted area because its density is already at carrying capacity (Eltringham 1984). In such cases there is a great risk of driving the population to local extinction when harvests are mistakenly higher than they should be (Caughley and Sinclair 1994).

Two more models that have recently been applied to evaluate hunting sustainability in tropical forests are the stock-recruitment model and the unified harvest model (Bodmer 2001). The stock-recruitment model (logistic model) can help determine the riskiness of harvests (McCullough 1987). It has been used to evaluate the status of a harvested population by analyzing the distance between its actual size (N) and K (Bodmer et al. 1997a). The model assumes that MSY is reached at about 0.5 K (Caughley 1977).

The unified harvest model proposed by Bodmer and Robinson (this volume) integrates the information needed to evaluate hunting sustainability through both the harvest and the stock-recruitment models: productivity, harvest rates, and density at hunted and unhunted sites. In addition to its usefulness in assessments of hunting sustainability, the model can be applied to predict the potential of populations for sustainable use (Bodmer and Robinson this volume).

METHODS

STUDY AREA

The Lacandon Forest of Mexico comprises the southwestern sector of the Greater Maya Forest. The area is located in the northeastern portion of the state of Chiapas (from 16°05 to 17°15′ N, and from 90°30′ to 91°30′ W) and is delimited by the Guatemalan border on the east, north, and south and by the Chiapas Highlands on the west. Average monthly temperatures range from 24°C to 26°C with maximum and minimum values in May (28°C) and in January (18°C), respectively. Mean annual rainfall is 2,500 to 3,500 mm, with roughly 80% of the rains falling between June and November. The area was originally covered by over a million hectares of rain forest, of which about half remain today. Among the protected areas extant in the Lacandon Forest, Montes Azules Biosphere Reserve (MABR) is the largest with over 3,300 km^2, harboring some of the largest Mexican populations of precious hardwood trees and large vertebrate species, harvested by both Indian and mestizo residents (Vasquez and Ramos 1992; Medellín 1994; March et al. 1996).

ABUNDANCE AND DENSITY

From May 1998 through December 2000, two of the authors (EJN and JEB), assisted by two trained biologists and three local hunters, recorded individuals, tracks, and fecal groups of large- and medium-sized species along 1,908 km of line transects established at two slightly hunted (n = 8 transects) and two persistently hunted (n = 7 transects) sites of the Lacandon Forest. Slightly hunted sites were located within MABR, while persistently hunted sites were both within MABR and in community lands contiguous to the protected area (table 20.1; fig. 20.1). Line transects

TABLE 20.1 Hunting Intensity and Distance Traveled by the Authors in Four Sampling Sites of the Lacandon Forest, Mexico (1998–2000)

SITE	HUNTING SIGNS[*] /100 KM (N)	HUNTING INTENSITY	DISTANCE TRAVELED (KM)
MABR-Chajul	0.42 (2)	Light	478.2
MABR-Playón de la Gloria	0.55 (5)	Light	905.3
MABR-Reforma Agraria	3.10 (11)	Persistent	355.6
Lacanjá-Bonampak	5.33 (9)	Persistent	169.0
Total			1,908.1

[*]Hunting signs included direct sightings of hunters, used cartridges, and animal carcasses with clear signs of shots or cuts made by humans. N = number of hunting signs found at each site.

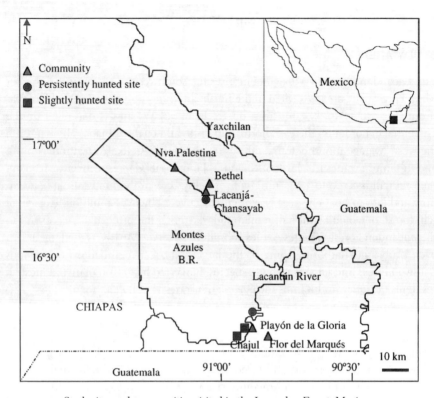

FIGURE 20.1 Study sites and communities visited in the Lacandon Forest, Mexico.

were 1 to 5 km long, and each of them was walked at least six times during the study. We walked the transects quietly and slowly (1 to 1.5 km/h), searching for individuals and other ungulate signs during the first and the last hours of daylight.

Perpendicular distances between the centerlines of the transects and ungulate individuals or groups seen were recorded. Peccary group size was assessed whenever possible. We only counted fresh tracks (clean, neat, and very contrasting to surrounding soil) observed within a 1-m strip along every transect. We erased all tracks found after every count. Encounter rates (number of individuals/100 km), indices of abundance (number of tracks or fecal groups/km; Conroy 1996), and density (number of individuals or groups/km^2) of each ungulate species were calculated for all sites, years, and seasons using the DISTANCE computer software (Buckland et al. 1993; Thomas et al. 1998).

HUNTING RATES

From September 1999 through August 2000, two of the authors (MMG and EJN), assisted by local hunters, obtained systematic records of hunting in five communities adjacent to MABR: (a) Bethel and (b) Lacanjá-Chansayab, populated by La-

TABLE 20.2 Population Size, Number of Hunters, and Number of Interviews Carried Out in Five Communities of the Lacandon Forest, Mexico (1999–2000)

	BETHEL	LACANJÁ-CHANSAYAB	NUEVA PALESTINA	FLOR DEL MARQUÉS	PLAYÓN DE LA GLORIA
Ethnic group	Lacandon	Lacandon	Tzeltal	Mestizo	Mestizo
Population	210	350	15,000	200	300
Hunters	30	50	850*	25	35
Interviews	44	43	45	44	56
% of population interviewed	21.0	12.3	0.3	22.0	18.7
Catchment area (km²)	113.1	201.1	452.4	28.3	28.3
Number of species used	35	37	32	42	37

*Only the fifty most active hunters were monitored during the study.

candon Indians; (c) Nueva Palestina, inhabited by Tzeltal Indians; and (d) Playón de la Gloria and (e) Flor del Marqués, whose residents are mestizo (table 20.2; fig. 20.1) We conducted a total of 232 formal interviews (range = 40–56/community) through structured questionnaires to both men and women of age fifteen and older. With the help of hunters from each community, we further recorded the species, sex, approximate age category (young, juvenile, or adult), weight, site and date of capture, and hunting method used for each animal consumed during the study.

We were also assisted by the hunters from all communities in estimating their catchment areas through the calculation of the radius of hunting around the villages. We used data on catchment areas and hunting frequencies to calculate specific annual harvest rates in each community (individuals hunted/ km²; Robinson and Redford 1991; Robinson and Bodmer 1999).

HUNTING SUSTAINABILITY MODELS APPLIED

We applied five different methods to evaluate hunting sustainability of ungulate populations in the Lacandon Forest.

Density Comparisons To assess hunting effects on population density, estimated densities of each ungulate species were contrasted between slightly and persistently hunted sites, between seasons, and among localities and years of study using chi-square analysis and t-tests (Sokal and Rohlf 1995). The hypothesis tested was that densities are lower in persistently hunted than in slightly hunted sites of the study area.

Production Model We compared actual estimates of harvest rates (H) obtained from visual records of hunting at the five communities visited and also the theoretical maximum production rates (P_{max}) of each species (Robinson and Redford 1991). We calculated P_{max} through an equation combining maximum finite rates of increase (λ_{max}) and values of 60% of population density at carrying capacity (0.6 K):

(1) $$P_{max} = (\lambda_{max} - 1)\, 0.6D$$

where P_{max} is maximum production rate; λ_{max}, maximum finite rate of increase; and D, density.

λ_{max} was in turn derived from the iterative solution of Cole's equation (Cole 1954; Robinson and Redford 1986b):

(2) $$1 = e^{-r_{max}} + b\, e^{-r_{max}(a)} - b\, e^{-r_{max}(w+1)}$$

where e is base of natural logarithms (2.71); r_{max}, maximum intrinsic rate of increase; b, fecundity rate; a, age at first reproduction; and w, age at last reproduction. We assumed that hunting was not sustainable if harvest rates exceeded 20% of P_{max} estimated for tapirs or 40% of P_{max} estimated for deer and peccaries (Robinson and Redford 1991, 1994).

Harvest Model We used data on annual production rates (n individuals produced/km^2) calculated from population densities in persistently hunted areas of the Lacandon Forest, coupled with annual fecundity rates (n female young produced/female) estimated by Bodmer (1994) and Bodmer, Eisenberg, and Redford (1997) in the Peruvian Amazon (Equation 3):

(3) $$P = (Y * G)\, 0.5D$$

where P is production rate (number of individuals produced/km^2/year); Y, annual number of young born per female; G, annual number of gestations; and P, population density at hunted sites (number of individuals/km^2).

Since sex ratios were not significantly different from the expected 1:1 for all species in the study area (Naranjo 2002), we used 50% of total density (the density of females) at hunted sites for calculations of production. We considered that hunting was sustainable if a harvest rate (H) was well below 40% of production (P) of deer and peccaries and if H was well below 20% of P of tapirs (Bodmer 1994; Robinson and Bodmer 1999).

Stock-Recruitment Model We used this model to evaluate the status of ungulate populations at persistently hunted sites through the analysis of the distance between actual population size (N) and its environmental carrying capacity (K) (McCullough 1987; Robinson and Bodmer 1999). We assumed that (a) carrying capacity (K) was represented by population density of each species at slightly/unhunted sites within MABR (Bodmer et al. 1997a); and (b) maximum sustained yield (MSY) is achieved at about 0.6 K in deer and peccary populations and at roughly 0.8 K in tapir populations (McCullough 1987; Bodmer et al. 1997a; Robinson and Bennett 2000b). We considered that a population was in safe condition

for sustainable hunting if N was at or over 60% of K in deer and peccary populations or was equal to or greater than 80% of K in tapir populations (Robinson and
Redford 1994; Bodmer et al. 1997a).

Unified Harvest Model This model constitutes an integration of the harvest and
the stock-recruitment models, combining data on actual productivity, harvest rates,
and population densities in slightly and persistently hunted sites (Naranjo 2002;
Bodmer and Robinson this volume). We used the information on population densities, harvest rates, and reproductive productivity mentioned above to construct a
graph that displays a vertical bar representing the status of hunted populations with
respect to their K (x-axis), and with respect to their corresponding MSY (y-axis).
Following the criteria applied in the harvest and the stock-recruitment models, we
assumed that a population was being sustainably harvested and in safe condition if
its vertical bar was well under its corresponding curve representing MSY (the harvest did not exceed production), and on the right side of the graph (N approached
to K; Bodmer 2001).

RESULTS

ABUNDANCE AND DENSITY

We observed 411 individuals and 1,153 tracks of tapirs, peccaries, and deer during
the study (tables 20.3 and 20.4).). We saw tracks, but not individuals, of white-tailed

TABLE 20.3 Comparative Encounter Rates (ER) and Densities of Ungulate Populations in
Slightly Hunted and Persistently Hunted Sites of the Lacandon Forest, Mexico (1998–2000)

	SLIGHTLY HUNTED SITES			PERSISTENTLY HUNTED SITES		
SPECIES	N (ind.)	ER[a] (ind/100 km)	Density (ind/km^2 ± SE)	N (ind.)	ER[b] (ind/100 km)	Density (ind/km^2 ± SE)
Tapirus bairdii	14	1.07	0.24 ± 0.09	3	0.50	0.05 ± 0.04
Tayassu pecari	211	16.15	7.93 ± 5.95	52	8.65	1.08 ± 0.87
Tayassu tajacu	87	6.66	1.53 ± 0.39	25	4.16	1.15 ± 0.47
Mazama americana	13	0.99	0.20 ± 0.07	6	1.00	0.33 ± 0.19
Odocoileus virginianus	P[c]	—	—	P	—	—
Total	325	x̄ = 6.22	x̄ = 2.48 ± 1.63	86	x̄ = 3.58	x̄ = 0.65 ± 0.39

[a]Distance traveled = 1306.7 km.
[b]Distance traveled = 601.4 km.
[c]Present but not seen during transect sampling.

TABLE 20.4 Frequency of Ungulate Tracks in Slightly Hunted and Persistently Hunted Sites of the Lacandon Forest, Mexico (1998–2000)

SPECIES	SLIGHTLY HUNTED SITES		PERSISTENTLY HUNTED SITES		OVERALL	
	Tracks	Tracks/ 100 km[a]	Tracks	Tracks/ 100 km[b]	Tracks	Tracks/ 100 km[c]
Tapirus bairdii	307	23.49	90	6.89	397	20.81
Tayassu pecari	72	5.51	41	3.14	113	5.92
Tayassu tajacu	312	23.88	119	9.11	431	22.59
Mazama americana	145	11.10	64	4.90	209	10.95
Odocoileus virginianus	1	0.08	2	0.15	3	0.16
Total	837	$\bar{x} = 12.81$	316	$\bar{x} = 4.84$	1,153	$\bar{x} = 12.09$

[a]Distance traveled = 1306.7 km
[b]Distance traveled = 601.4 km
[c]Distance traveled = 1908.1 km

deer during transect samplings; they were eventually sighted in the study area. Average group sizes of the four ungulate species seen were 20.2 for white-lipped peccary (5–60 ind., n = 13 groups); 2.3 for collared peccary (1–15 ind., n = 49 groups); 1.1 for red brocket deer (1–3 ind., n = 18 groups); and 1.1 for Baird's tapir (1–2 ind., n = 17 groups). The rank order of group densities was collared peccary > red brocket deer > tapir > white-lipped peccary. Average densities of the four species were 1.34 (0.6 individuals/km^2) and 0.38 (0.17 groups/km^2; table 20.3).

Seventy-nine percent of all deer, peccaries, and tapirs sighted during the study were found at slightly hunted sites within MABR, easily explained by a greater sampling effort in these areas compared to persistently hunted sites (table 20.3). Our sampling effort was not the same in both kinds of sites because forest patches with similar structure and composition (and consequently the length of transects) were smaller outside MABR. To reduce this bias in the contrast of ungulate abundance, we used population density (number of individuals/km^2). In this comparison we found a significantly higher density of ungulates at slightly hunted sites than at persistently hunted sites ($\chi^2 = 177.2$; df = 3; P < 0.0001), which suggests that heavy hunting pressure probably has had an effect on local ungulate populations.

However, such an effect did not appear to be the same for all species. Individual densities of the white-lipped peccary and Baird's tapir were almost seven and five times greater at slightly hunted than at persistently hunted sites, respectively. In contrast, the densities of the collared peccary were similar at both sites, while the red brocket deer was slightly more abundant at persistently hunted sites (table 20.3).

These results suggest that Baird's tapirs and white-lipped peccaries are more vulnerable to hunting pressure than collared peccaries and red brocket deer in the Lacandon Forest. Bodmer, Eisenberg, and Redford (1997) observed a similar pattern in the Peruvian Amazon, where the abundance of peccaries was similar, but the abundance of tapirs was very different between hunted and unhunted areas.

We performed an additional analysis of ungulate abundance using data on track frequencies recorded in the Lacandon Forest (table 20.4). In this analysis, we detected a higher relative abundance of collared peccary and tapir tracks compared to white-lipped peccary and red brocket deer tracks (Kruskal-Wallis; $H = 159.9$; $df = 4$; $P < 0.0001$). The higher abundance of collared peccary tracks may be explained by their actual higher density and relative tolerance to human disturbance compared to other ungulates in the study area (Fragoso, Silvius, and Villa-Lobos 2000). We found considerably more ungulate tracks per 100 km in slightly hunted than in persistently hunted sites ($\chi^2 = 9.48$; $df = 3$; $P = 0.044$; table 20.4). This result supports the hypothesis that hunting has reduced the abundance of ungulate populations in the study area.

HUNTING SUSTAINABILITY

We obtained an overall annual harvest rate of 0.77 ungulates per km^2 (table 20.5). Collared peccaries were more frequently harvested than the rest of the ungulate species in the Lacandon Forest ($P < 0.05$) (table 20.6). Collared peccaries accounted for 55% of overall harvest rates, followed by red brocket deer (28%), white-tailed deer (11%), white-lipped peccaries (5%), and Baird's tapirs (1%). In contrast, red brocket deer ranked first in biomass extracted (1469 kg, or 27%) to total ungulate biomass extracted in the five communities (table 20.5), followed by white-tailed deer (21%), collared peccaries (21%), Baird's tapirs (19%), and white-lipped peccaries (12%).

TABLE 20.5 Numbers of Hunters, Ungulates Taken, Biomass Harvested, and Catchment Areas of Three Ethnic Groups in the Lacandon Forest, Mexico (1999–2000)

VARIABLE	LACANDON n	LACANDON %	TZELTAL n	TZELTAL %	MESTIZO n	MESTIZO %	ALL COMBINED n	ALL COMBINED %
Number of hunters monitored	80	42.1	50	26.3	60	31.6	190	100
Number of ungulates hunted	93	48.2	53	27.5	47	24.3	193	100
Ungulate biomass extracted (kg)	2,308	42.5	2,219	40.8	906	16.7	5,433	100
Catchment area (km^2)	314	38.2	452	54.9	57	6.9	823	100
Overall ungulate harvest rate (individuals/10 km^2/year)	1.7	21.9	0.9	12.3	5.1	65.8	7.7	100

TABLE 20.6 Evaluation of Hunting Sustainability of Ungulate Populations in the Lacandon Forest Through the Production and the Harvest Models

SPECIES	DENSITY (IND/ KM^2)	P_{MAX}[a] (IND/ KM^2)	P[b] (IND/ KM^2)	H[c] (IND/ KM^2)	P_{MAX}	P	MFP[d]	SUST?
Tapirus bairdii	0.05	0.007	0.007	0.003	0.40[e]	0.44[e]	0.2	No
Tayassu pecari	1.08	0.853	0.508	0.013	0.02	0.03	0.4	Yes
Tayassu tajacu	1.15	1.718	0.874	0.140	0.08	0.16	0.4	Yes
Mazama americana	0.33	0.097	0.109	0.072	0.74[e]	0.66[e]	0.4	No
Odocoileus virginianus	0.29[f]	0.187	0.218	0.027	0.15	0.12	0.4	Yes
Total	2.61	2.862	1.716	0.255				

[a]Maximum production rates based on r_{max}.
[b]Production rates based on actual densities estimated in the Lacandon Forest and reproductive data from R. E. Bodmer (pers. comm.).
[c]Harvest rates obtained from visual hunting records in five communities.
[d]Maximum fraction of production that can be sustainably harvested (Robinson and Redford 1991).
[e]Unsustainable hunting under Robinson and Redford's criteria (1991).
[f]Estimated from track frequency using data for red brocket deer.

The results, pooling data from all communities visited (table 20.6), obtained through the production model (Robinson and Redford 1991) suggest that the harvests of Baird's tapir and red brocket deer were unsustainable on a "regional" scale. Following Robinson and Redford's criteria, an unsustainable harvest consists of taking more than 20% of the maximum annual production (0.2 of P_{max}) of long-lived mammals, such as tapirs, and over 40% of P_{max} of short-lived mammals, such as deer and peccaries. The average fraction of Baird's tapir P_{max} harvested in the study area was 40%, which denotes an overhunting of this species. Nonetheless, this over-harvest was actually concentrated in the Tzeltal community of Nueva Palestina, where 105% of P_{max} was taken. In contrast, the Lacandon communities of Bethel and Lacanjá-Chansayab harvested only 15% of tapir P_{max}, while mestizos from Playón de la Gloria and Flor del Marqués did not hunt tapirs (table 20.7). An analogous situation was observed for the red brocket deer, which was overharvested by Lacandon and mestizo hunters, who took over 100% of P_{max}, but not by Tzeltal hunters, who took only 9% of P_{max} (table 20.7).

Under this model the harvests of the remaining three ungulate species (both peccaries and the white-tailed deer) were not high enough to be regarded as unsustainable. However, this model does not allow for the verification of the hypothesis that such harvests are actually sustainable (Robinson and Bodmer 1999).

The harvest model (Bodmer 1994) was helpful in confirming a similar pattern of unsustainable offtake at a landscape scale in the Lacandon Forest: Hunters took an estimated 44% and 66% of production (P) of tapirs and red brocket deer, respectively (table 20.8). The overharvest of tapirs was again located in Nueva Palestina,

TABLE 20.7 Hunting Sustainability of Ungulates Taken by Three Ethnic Groups in the Lacandon Forest Through the Production Model

SPECIES	DENSITY (IND/ KM^2)	P_{MAX}[a] (IND/ KM^2)	LACANDON H[b]	LACANDON H/P$_{max}$	TZELTAL H	TZELTAL H/P$_{max}$	MESTIZO H	MESTIZO H/P$_{max}$
Tapirus bairdii	0.05	0.007	0.001	0.15	0.007	1.05[c]	0	0
Tayassu pecari	1.08	0.853	0.020	0.02	0.009	0.01	0.009	0.01
Tayassu tajacu	1.15	1.718	0.029	0.02	0.038	0.02	0.354	0.21
Mazama americana	0.33	0.097	0.111	1.14[c]	0.009	0.09	0.097	1.00[c]
Odocoileus virginianus	0.29[d]	0.187	0.007	0.04	0.031	0.17	0.044	0.24
Total	2.61	2.862	0.168		0.094		0.504	

[a]Maximum production rates based on r_{max}.
[b]Harvest rates obtained from visual hunting records in five communities.
[c]Unsustainable harvest under Robinson and Redford's criteria (1991).
[d]Estimated from track frequency using data for red brocket deer.

TABLE 20.8 Hunting Sustainability of Ungulates Taken by Three Ethnic Groups in the Lacandon Forest Through the Harvest Model

SPECIES	DENSITY (IND/ KM^2)	P[a] (IND/ KM^2)	LACANDON H[b]	LACANDON H/P	TZELTAL H	TZELTAL H/P	MESTIZO H	MESTIZO H/P
Tapirus bairdii	0.05	0.007	0.001	0.15	0.007	1.04[c]	0	0
Tayassu pecari	1.08	0.508	0.020	0.04	0.009	0.02	0.009	0.02
Tayassu tajacu	1.15	0.874	0.029	0.03	0.038	0.04	0.354	0.41[c]
Mazama americana	0.33	0.109	0.111	1.02[c]	0.009	0.08	0.097	0.89[c]
Odocoileus virginianus	0.29[d]	0.218	0.007	0.03	0.031	0.14	0.044	0.20
Total	2.61	1.716	0.168		0.094		0.504	

[a]Production rates based on actual densities estimated in the Lacandon Forest and reproductive data from R.E. Bodmer (pers. comm.).
[b]Harvest rates obtained from visual hunting records in five communities.
[c]Unsustainable harvest under Robinson and Redford's criteria (1991).
[d]Estimated from track frequency using data for red brocket deer.

where Tzeltal hunters extracted an estimated 104% of P. In the same way Lacandon and mestizo hunters took an estimated 102% and 89% of P of red brocket deer in their respective communities. A noteworthy difference in the evaluation of sustainability through the harvest model compared to the production model was the detection of unsustainable hunting of collared peccaries in mestizo communities. Mestizo hunters obtained about 41% of collared peccaries' P, barely exceeding the limits of sustainability proposed by Robinson and Redford (1991) for this species.

Since the harvest model works with actual data on population densities, production rates, and harvest rates of local populations, it allows the inference that offtakes may be sustainable when they are well below the maximum fractions of P (0.2, 0.4, or 0.6, depending on species longevity) that can be taken without driving the population into decline (Robinson and Redford 1991). Consequently, this model suggests that the hunting of collared peccaries, white-lipped peccaries, and white-tailed deer appears sustainable, while tapirs and red brocket deer are being overhunted on a landscape scale in the Lacandon Forest. On a local scale Tzeltal hunters seem to be overhunting tapirs, Lacandon hunters are probably overharvesting red brocket deer, and mestizo hunting rates of collared peccaries and red brocket deer are beyond the limits of sustainability.

The stock-recruitment model has been used to assess the status of populations and to predict their potential for sustainable harvests (Caughley 1977; McCullough 1987; Bodmer et al. 1997a). The data on densities of ungulates at unhunted, slightly hunted, and persistently hunted sites in the Lacandon Forest indicate that the populations of collared peccary and red brocket deer are in safe condition, while tapir and white-lipped peccary populations seem to be in risky condition (table 20.9; fig. 20.2).

At 75% of its carrying capacity (0.75 K), the hunted collared peccary population is well above its respective point of MSY (0.6 K; Robinson and Redford 1991), implying that this species probably has a good potential to support sustainable harvests at a landscape scale. The red brocket deer is an interesting case because its density was higher at persistently hunted sites than at slightly hunted sites. Thus it is likely that the hunted population of this species (N) is at its carrying capacity (1.0 K) and so it is in safe condition to allow sustainable hunting. On the other hand, the hunted populations of Baird's tapir and white-lipped peccary apparently have been negatively affected by hunting at the study area, since they were at only 0.21 K and 0.14 K, respectively (table 20.9; fig. 20.2).

Bodmer (2001) has recently proposed the unified harvest model, so its use is still incipient. This model provides an integrated and graphical view of the results ob-

TABLE 20.9 Status of Ungulate Populations in Persistently Hunted Sites of the Lacandon Forest, Mexico

SPECIES	K (IND/KM^2)	D$_{PH}$ (IND/KM^2)	D$_{PH}$/K	STATUS
Tapirus bairdii	0.24	0.05	0.21	Risky
Tayassu pecari	7.93	1.08	0.14	Risky
Tayassu tajacu	1.53	1.15	0.75	Safe
Mazama americana	0.20	0.33	1.65	Safe?
Odocoileus virginianus	?	0.29	?	?

Note: Density estimates from persistently hunted sites (D$_{ph}$) were compared to density estimates from unhunted sites (assumed as carrying capacity K).

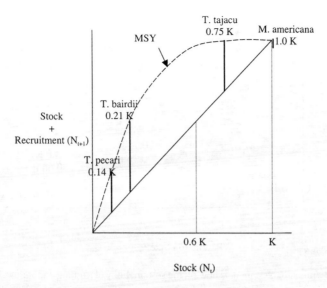

FIGURE 20.2 Stock-recruitment model comparing the status of persistently hunted populations of ungulates with respect to their carrying capacity (K) in the Lacandon Forest, Mexico.

tained through the harvest and the stock-recruitment models, showing that the hunting systems of ungulates are not producing equal effects on different species in the Lacandon Forest (figs. 20.3 and 20.4). The apparently sustainable harvest and safe condition of the hunted collared peccary population, as well as the unsustainable hunting and risky condition of the tapir population estimated through the other models was confirmed with the unified harvest model (figs. 20.3 and 20.4).

Both the production and the harvest models indicated that the hunted white-lipped peccary population was sustainably harvested on both local and landscape scales in the Lacandon Forest. The unified harvest model provided a different view: hunting of white-lipped peccaries in fact seems to be within the theoretical sustainability limits but this result is actually a consequence of the species' very low density in persistently hunted areas. This finding concurs with the information provided by interviewed hunters, as well as with our own visual records of hunting in the communities visited during the study.

The unified harvest model offered a clear view of this situation: the white-lipped peccary population was at a very low density and it was harvested at a low rate (fig. 20.3). The analysis of hunting sustainability for the red brocket deer population is particularly interesting by means of this new model. The red brocket deer had a higher density at persistently hunted than at slightly hunted but still was apparently overharvested (fig. 20.3). An explanation of this contradictory result may be based on the hypothesis that the red brocket deer population maintains a relatively high density at persistently hunted sites via immigration of individuals from slightly hunted areas of MABR (a source-sink dynamic; Pulliam 1988).

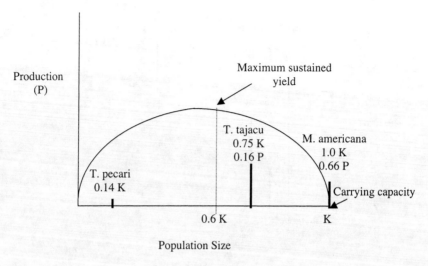

FIGURE 20.3 Unified harvest model showing the sustainability of hunting and status of red brocket deer, collared peccary, and white-lipped peccary populations at persistently hunted sites of the Lacandon Forest, Mexico. Bar position on the x-axis indicates population status with respect to carrying capacity (K). The height of the bar denotes the harvested fraction of population production (P) in relation to the maximum sustained yield curve.

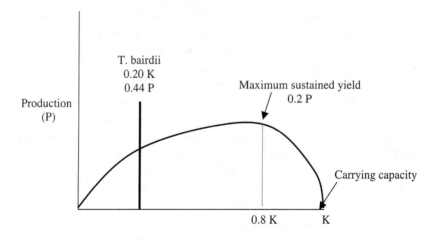

FIGURE 20.4 Unified harvest model showing the sustainability of hunting and status of Baird's tapir populations at persistently hunted sites of the Lacandon Forest, Mexico. Bar position on the x-axis indicates population status with respect to carrying capacity (K). The height of the bar denotes the harvested fraction of population production (P) in relation to the maximum sustained yield curve.

Finally, given the lack of reliable data on white-tailed deer densities in the study area, we cannot draw conclusions about the sustainability of its hunting. However, considering the harvest rates obtained for this species, as well as the information provided by local hunters, it is likely that its populations benefit by habitat transformation outside existing protected areas in the Lacandon Forest.

DISCUSSION

POPULATION DENSITY AND ABUNDANCE

The abundance and density of ungulate populations at slightly hunted sites of Montes Azules Biosphere Reserve (MABR) are within the range observed in other Neotropical rain forests. Baird's tapir encounter rates estimated in MABR were similar to those obtained by Cruz (2001) and Naranjo and Cruz (1998) in La Sepultura Biosphere Reserve, Mexico. However, the density of tapirs was lower in MABR than in Corcovado National Park (CNP), Costa Rica (Naranjo 1995; Foerster 1998), and Barro Colorado Island (BCI), Panama (Glanz 1982).

We attribute these differences largely to a lower hunting pressure in the latter two areas compared to MABR and a substantial difference of size between MABR (> 3,000 km²) and the other two areas (< 500 km²). Furthermore, Baird's tapir abundance was higher in slightly hunted sites of the Lacandon Forest than in an unhunted area of Chiquibul Reserve, Belize (Fragoso 1991), and northeastern Honduras (Flesher 1999), where there was relatively high hunting pressure on this species.

Population densities of collared peccaries, white-lipped peccaries, and red brocket deer estimated in slightly hunted areas of the Lacandon Forest were not very different than the densities obtained in several Central and South American localities (Glanz 1982; Bodmer et al. 1997a; Fragoso 1998a). This finding suggests that local populations of the three species are in good condition at unhunted and slightly hunted sites of our study area. However, the situation seems notably different for tapirs and white-lipped peccaries at persistently hunted sites of the Lacandon Forest. The clear differences in the density of these two species between slightly and persistently hunted sites may be a combined effect of overhunting and habitat transformation. Human density and activity has dramatically increased in the study area during the last twenty-five years (INEGI 2002), and the concomitant need for land, timber, and food has caused forest fragmentation and overexploitation of many wildlife populations outside extant protected areas (Naranjo 2002).

Both the tapir and the white-lipped peccary are vulnerable to heavy hunting pressure in different ways: The first has a very low reproductive productivity (Eisenberg 1989), and its populations often cannot recover from an intense or even a moderate harvest rate (Bodmer 1995b). In addition, because of its habitat requirements, the tapir is sensitive to habitat fragmentation and other effects of human activity (e.g., noise, odors, dogs, and cattle; Matola, Cuarón, and Rubio-Torgler 1997; Naranjo and Cruz 1998).

The white-lipped peccary is similarly sensitive to habitat fragmentation because of its extensive home range and its feeding strategies (Fragoso 1998a). Moreover, hunters capitalize on its social behavior and very large herd size to decimate their populations in the study area (Naranjo 2002). The impact of these factors was evident in many communities adjacent to MABR (e.g., Nueva Palestina, Bethel, and Playón de la Gloria), where both tapirs and white-lipped peccaries appear to be close to local extinction outside the protected area.

Estimations of wildlife population densities and abundances in tropical rain forests are almost always complicated by factors such as the natural rarity of the species, the restricted visibility within the forest, and the high sensitivity of these mammals to human noises and odors. In spite of these difficulties, the use of density and abundance comparisons between areas with different hunting pressure should not be discarded as a useful technique in evaluating the effects of hunting on populations of Neotropical ungulates, especially if they are combined with other kinds of information such as production and harvest rates, that allow us to run models of hunting sustainability.

HUNTING SUSTAINABILITY

Evaluation of hunting sustainability of ungulate populations in the Lacandon Forest benefited from the simultaneous application of several models. The production and the harvest models gave similar results on a landscape scale, but there were some differences on a local scale. Both models suggested unsustainable hunting of tapirs and red brocket deer in the study area. However, these species were not overharvested in all communities. In most cases the harvest/production ratios (H/P) obtained through the harvest model tended to be higher than H/P ratios estimated with the production model, primarily because the first is based on actual data of population productivity and the second uses theoretical calculations of P_{max}. This trend made it possible to detect unsustainable offtake of collared peccaries by mestizo hunters through the harvest model only.

It was remarkable that the production model did not indicate overhunting of white-lipped peccaries, while the harvest model in fact suggested sustainable hunting of this species in the study area. However, synergic effects of habitat fragmentation, heavy hunting pressure, and extensive home ranges of herds seem to have driven this mammal to local extinction in most of the rain-forest patches remaining around MABR. In this sense both the stock-recruitment model and the unified harvest model provided a reasonable answer to this apparent incongruity: Hunting of white-lipped peccaries is within sustainability limits largely because the group density of this species is very low at persistently hunted sites. Hence, hunters of the Lacandon Forest have a smaller chance of finding a white-lipped peccary herd than any other ungulate group when they go out to search for prey. This fact was particularly evident in mestizo communities, in which the interviewees themselves explained the extremely low harvest rate of this species because of its scarcity in the small catchment areas within their communities. In addition, most of the oldest hunters interviewed during the study concurred in the perception that white-lipped peccary populations have severely declined or even disappeared near their villages in the last two decades.

All these observations lead to the conclusion that current hunting on the remaining white-lipped peccary populations outside MABR should be lowered to facilitate their recovery through local reproduction and immigration from slightly hunted areas of MABR. Forest management to improve connectivity between large forest fragments in community lands would also help to facilitate migration of white-lipped peccaries from MABR.

There were some similarities but also some differences between the results of the evaluations of hunting sustainability of tapirs and white-lipped peccaries in the study area. The analysis of interviews with residents of the Lacandon Forest revealed that hunting pressure on tapirs has been relatively low in the last three years. In fact, this mammal does not appear within the most frequently hunted species in the study area (Naranjo 2002). However, tapirs were overhunted in Nueva Palestina, where Tzeltal hunters took a little more than 100% of P. Meanwhile, Lacandon hunters extracted only 15% of P, and mestizo hunters did not take tapirs at all.

The causes of these variations may be related to the geographical, cultural, and socioeconomic contexts of hunters and their communities. Tzeltal hunters of Nueva Palestina (n = 850) outnumbered by far the Lacandon and mestizo hunters combined (n = 140) and used a larger catchment area than the latter two. These differences imply that Tzeltal hunters had a higher probability of finding a tapir in their home ranges than Lacandon and Tzeltal hunters. On the other hand, as noticed by March (1987), most Lacandon hunters interviewed in this study said that they did not like to hunt tapirs because they are too heavy and too bulky to be carried back to their homes. Consequently, it would be a waste of time and effort to hunt one of these large mammals. The reason why mestizo hunters did not harvest tapirs during the study was simple: the species is rarely found in their *Ejidos*.

As in the case of white-lipped peccaries, the stock-recruitment and the unfiied harvest models were helpful in recognizing that the tapir population has been depleted in the most persistently hunted sites, maintaining extremely low densities (around 0.05 ind/km^2) in a few localities outside the protected areas of the Lacandon Forest. Under these circumstances a few animals hunted per year can constitute an unsustainable offtake, as was observed on both local (Nueva Palestina) and landscape scales.

Red brocket deer seemed to be regionally and locally overhunted by two ethnic groups (Lacandon and mestizo). Nevertheless, its populations at persistently hunted sites were at its carrying capacity (1.0 K; figs. 20.2 and 20.3). The production, harvest, and stock-recruitment models provided partial arguments for understanding the whole situation of red brocket deer hunting in the study area. In contrast, the unified harvest model was the only one that showed the complete scenario at once: The persistently hunted population of red brocket deer is being overharvested, but at the same time it is in safe condition compared to the slightly hunted population (fig. 20.3).

This result suggests that the red brocket deer population may be functioning in a source-sink system in the area, where MABR would be the source of individuals and the surrounding communal lands would be the sinks. A similar case was observed by Bodmer (2000), who found that a source-sink system seemed to be maintaining a constant overhunting of lowland tapirs recorded in persistently hunted areas of the Peruvian Amazon. In this condition the overharvest appears less risky for red brocket deer than for tapirs and white-lipped peccaries in the Lacandon Forest. However, a decrease of current harvest rates of red brocket deer in Lacandon and mestizo communities should benefit both the hunters and their prey.

CONCLUSION

Using several models to evaluate hunting sustainability was a first step toward understanding the current status of ungulate populations harvested in the Lacandon Forest. It must be recognized, however, that the results of models applied in this study do not constitute exact measurements of reality but rather offer an overview of the general trends of the hunting systems present in the study area.

Estimates of hunting sustainability may be affected by scale factors (Wiens 1989; Novaro, Redford, and Bodmer 2000). Novaro, Redford, and Bodmer (2000) found that most evaluations of hunting sustainability did not consider the potential effects of animal migration from large protected areas to hunted areas (a source-sink system; Pulliam 1988; Pulliam and Danielson 1991). In consequence, local hunting systems that apparently are unsustainable may actually be sustainable at a much larger, regional scale, as shown recently by Novaro (1997); Bodmer (2000); Fragoso, Silvius, and Villa-Lobos (2000); and Hill and Padwe (2000). Migration of animals between slightly hunted and persistently hunted areas of the Lacandon Forest has not been measured, but the evidence gathered over the three years of this study suggests that the movement of individuals from large protected areas (i.e., Montes Azules and Lacantún Biosphere Reserves) into nonprotected community lands is an important variable in the functioning of regional ungulate hunting systems. However, much more field evidence is needed to support the hypothesis of source-sink dynamics in this and other hunting systems (Clutton-Brock 1997; Novaro, Redford, and Bodmer 2000).

A common recommendation derived from this kind of evaluation is that hunting should be controlled where it seems to be unsustainable (Robinson and Redford 1991; Bodmer et al. 1997a, 1997b; Robinson and Bennett 2000b). The process of steering unsustainable hunting systems toward sustainability is complex. Comprehensive and detailed assessments of the status of game populations and their harvest rates are certainly very important but not sufficient to predict and encourage sustainability. A realistic strategy to promote hunting sustainability must not only address the biology of game species but also the cultural and socioeconomic context of local people along with the needs and interests of local hunters (Western, Wright, and Strum 1994; Bennett and Robinson 2000a; Bodmer 2000). Residents of the Lacandon Forest will not participate in programs of wildlife conservation and sustainable use unless they are convinced that they can share the benefits of hunting regulation in their own lands. Research and education can help to boost this process.

ACKNOWLEDGMENTS

We thank Allan Burns, John Eisenberg, José Fragoso, F. Wayne King, Kirsten Silvius, Melvin Sunquist, and George Tanner for their suggestions to improve the manuscript. Carlos Muench, Rausel Sarmiento, Isidro López, Romeo and Caralampio Jiménez, Antonio Navarro-Chankín, and Celedonio Chan assisted us in the field. We are grateful to the residents of Playón de la Gloria, Flor del Marqués, Reforma Agraria, Nueva Palestina, Bethel, and Lacanjá-Chansayab for their hospitality during the fieldwork. Funds were provided by Mexico's National Commission for Biodiversity, Mexico's National Council of Science and Technology, the U.S. Fish and Wildlife Service, The Compton Foundation, the University of Florida's Program for Studies in Tropical Conservation, the U. S. Man and Biosphere Program, and Idea Wild. El Colegio de la Frontera Sur facilitated infrastructure, vehicles, and logistical support. The Dirección General de Vida Silvestre of the Instituto Nacional de Ecología and the staff of Montes Azules Biosphere Reserve kindly gave the permits to carry out this project. Conservation International (Chiapas chapter), the Department of Wildlife Ecology and Conservation of the University of Florida, and the Universidad de Ciencias y Artes de Chiapas (UNICACH) provided material support in different ways.

21

Human Use and Conservation of Economically Important Birds in Seasonally Flooded Forests of the Northeastern Peruvian Amazon

JOSÉ A. GONZÁLEZ

Wildlife plays a key role for people inhabiting the Amazonian rain forest (Terborgh, Emmons, and Freese 1986; Redford and Robinson 1991; Vickers 1991). Subsistence hunting has been very significant for the economy of the Amazon region and, in particular, for the well-being of thousands of rural families (Dourojeanni 1972). Indeed, in many parts of the northeastern Peruvian Amazon, wildlife provides most of the animal protein consumed by local households (Pierret and Dourojeanni 1967; Ríos, Dourojeanni, and Tovar 1973; Bodmer et al. 1994). Although mammals are always the most important prey for susbistence hunters, birds may comprise a significant amount of the total biomass intake (Ayres et al. 1991; Vickers 1991; Zent 1997). Birds are especially important in terms of the number of animals taken and may account for up to 27% of captures (Ojasti 1993). Bird eggs are also an important source of food for local people in many areas of the Peruvian Amazon (González 1999a). Several Amazonian ethnic groups also hunt birds for ornamental, medicinal, or magical purposes (Redford and Robinson 1991).

Finally, birds such as parrots and macaws are popular as pets and are heavily targeted by the pet market (Thomsen and Brautigam 1991). The resultant international trade in psittacines is of great concern to conservation biologists (WCI 1992; Wright et al. 2001). Until 1973 more than a hundred species were exported regularly from the Peruvian Amazon, with psittacines and other ornamental birds comprising more than 40% of this trade (Dourojeanni 1972). Nowadays, despite being banned by national laws, the harvesting and trade of parrots and macaws is still a common practice in many parts of the region (González 1999b).

Notwithstanding the importance of subsistence hunting and the pet-bird trade in the Neotropics for both human welfare and wildlife conservation, most of the field data required to develop sustainable management programs is still lacking (Peres 1997). Some recent studies have focused on assessing the sustainability of ungulate harvests in the Peruvian Amazon (Bodmer et al. 1994, 1999), but there are still few reports that deal with the sustainability of bird hunting (Begazo 1997). Even though the sustainability and implications of the bird trade have been extensively analyzed in recent years (Beissinger and Bucher 1992; Thomsen and Mulliken 1992; WCI 1992), figures that quantify illegal trade at the national level and its impact on the populations of the exploited species are still lacking (Beissinger 1994). In this paper I document the patterns of harvesting of wild birds in the Ucayali-Puinahua floodplain (northeastern Peruvian Amazon), evaluate the relative importance of each species in social and economic terms, and finally assess the effects of harvesting on bird populations as a first step in developing sound management strategies for bird conservation.

STUDY AREA AND METHODS

Field work was conducted in the southern part of the Pacaya-Samiria National Reserve (PSNR), a protected area located in the Department of Loreto between the Marañón and Ucayali/Puinahua rivers (from 04° 26′ 36″ to 06° 08′ 01″ S and from 73° 26′ 59″ to 75° 34′ 33″ W). The reserve covers an area of 2,150,770 ha, being one of the largest conservation units in Latin America (fig. 21.1).

The climate is tropical with a mean annual temperature around 27°C and a mean annual rainfall over 2,900 mm, most falling between October and May, when much of the land becomes flooded. Rivers reach their highest levels from

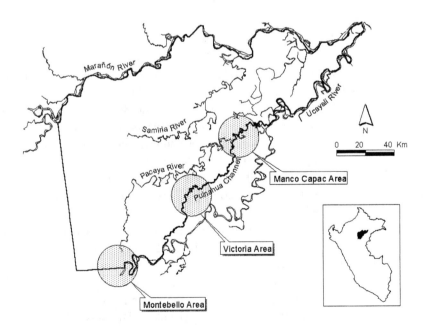

FIGURE 21.1 Location of the study sites at Pacaya-Samiria National Reserve.

March to May and their lowest in August and September (Soini 1995). The flood-plain ecosystem consists of varied landforms and aquatic surfaces sculptured by the river. These features include natural levees (*restingas*), mud and sand bars (*bar-reales* and *playas*), backswamps (*tahuampas*), palm swamps (*aguajales*), lakes (*co-chas* and *tipishcas*), side channels (*caños*), and rivers (Hiraoka 1995). Most of the study area is occupied by palm swamps (37%) and seasonally flooded forests (34%). A detailed description of vegetation communities present at PSNR is provided by Rodríguez, Rodríguez, and Vásquez (1995).

Within the reserve and the surrounding areas, humans live in 173 rural settle-ments, most of which (89%) are small villages with fewer than 500 inhabitants lo-cated on the borders of the Marañón and Ucayali/Puinahua rivers. The last popu-lation census recorded 32,241 persons living in the southern part of the reserve (Ucayali/Puinahua basin), 27% inside the boundaries of the protected area, and 73% in the buffer zone (Rodríguez, Rodríguez, and Vásquez 1995). The inhabi-tants of the study area include people of mixed origins (*mestizos*), as well as detrib-alized natives from the ethnic groups Cocama-Cocamilla and Shipibo-Conibo.

Major economic activities include fishing, agriculture, extraction of forest prod-ucts, and game hunting, with the relative importance of each of these activities varying in different parts of the reserve. Although the population is mainly engaged in subsistence production, the growing need to integrate into the market economy has forced some inhabitants to distance themselves from the traditional systems of production, threatening the ecological sustainability of the area as well as the hu-man population's opportunities for survival and development (Junglevagt for Ama-zonas 1995).

ESTIMATES OF BIRD USE

Between January and April 1997 I conducted interviews with local families in a to-tal of 194 households in seventeen communities located in the southern part of the PSNR (fig. 21.1). This sample represents 23.3% of the households in the visited vil-lages and 2.9% of the population in the whole study area. Since 1992 the selected communities have been part of two large development projects financed by US-AID/TNC/FPCN (Project Employment and Natural Resources Sustainability No. 527-0341) and AIF-WWF/DK (Juglevagt for Amazonas: Programa Integral de De-sarrollo y Conservación Pacaya Samiria). I was always accompanied by well-known managers of these projects in order to increase trust between myself and local peo-ple. However, as interviews and questionnaires may bias the estimation of the total number of animals harvested (Wright 1978), I included several crossquestions to as-sess the credibility of the answers. Six households (3.1%) were excluded from the analyses because of incoherent responses or hostile atmospheres. Several especially designed booklets, containing pictures of the most commonly taken game birds (Begazo 1997), were distributed among twenty-six selected hunters to keep a record of the number of birds they harvested monthly. Even so, the calculations of hunting pressure should be considered as estimates because of biases inherent in the interview and participatory methods of data collection (Silva and Strahl 1991; Townsend this volume).

ESTIMATES OF BIRD ABUNDANCE AND DENSITY

I also estimated the abundance and density of the most important game birds in three heavily hunted areas (close to villages: Urarinas, Montebello, and Victoria) and three areas of low/absent hunting pressure inside the protected area (close to ranger stations: Alfaro, Santa Cruz, and El Dorado). A total of 143.5 km of transect lines (82.4 inside the reserve, 61.1 outside) and 67.8 km of streams (40.8 inside and 27.1 outside) were surveyed following the methodology proposed by Strahl and Silva (1997). Nineteen, twelve, and ten transects were surveyed at each of the protected sites, respectively, and nine, eleven, and eleven at each of the hunted sites near villages. Transects were walked along streams and preexisting trails and were variable in length, ranging from 1.1 to 8.8 km. Each transect was surveyed only once between January and May 1998, between 7:00 and 11:00 A.M. or between 4:00 and 6:00 P.M. Both visual and auditory sightings were recorded, but only visual sightings were used for abundance estimates. The perpendicular distance between the bird and the transect was measured using steps. Birds sighted on the trail or over the stream were assigned a distance of 1 m. Results are expressed as number of birds per 10 km surveyed and for terrestrial birds as number of birds per km^2. Density was calculated using the King method (Overton 1971).

Differences in bird abundance are not likely to be attributable to habitat type since the censused areas were very similar. I used the harvest model proposed by Robinson and Redford (1991) to evaluate the sustainability of subsistence hunting of terrestrial birds in the study area, employing reproductive and demographic data recorded during the study, as well as unpublished information provided by A. Begazo. Conservative parameters were used for estimating the maximum production and the size of the catchment area.

ESTIMATES OF EGG-HARVESTING

During the heron nesting seasons of 1997–1998 (March to May), I monitored the total number of eggs harvested from two heronries located close to the villages of Padre López and Nueva Cajamarca. I visited the heronries with two groups of egg-collectors to assess the impact of egg-harvesting on nesting success.

ESTIMATES OF PARROT AND MACAW HARVESTING

Between 1996 and 1999, during the nesting season of parrots and macaws (February to April), I monitored the total number of nestlings collected in two Mauritia palm swamps (total area: 3,890 ha) located close to the village of Victoria, where most of

the parrot trade took place. During the nesting seasons of 1998 and 1999, I also estimated nest density and chick productivity of the seven most commonly harvested
species in this area.

Density of parrot and macaw nests was estimated in variable-sized plots, ranging
from 20 to 47 ha, randomly placed along two transect lines that crossed the most
frequently used Mauritia palm swamps. Eight plots were set in 1998 (total: 268 ha)
and nine in 1999 (total: 278 ha). Each plot was surveyed during two consecutive
days by the researcher and two experienced poachers looking for parrot and macaw
nests. Nest density estimates were done before the harvesting season began. However, at one of the swamps in 1999, it is possible that some undetected harvesting
had occurred in two of the plots before our arrival in the area. Productivity (fledg-
lings/nest) was calculated with the information provided by poachers and by means
of the nests opened during the harvesting process. Only those nests containing fully fledged young were considered. The total area of harvest was mapped using aerial photographs and GPS data gathered during our visits. Sustainability was assessed by comparing the recorded annual harvesting rates with the estimated
annual production of nestlings in the whole area.

RESULTS

BIRDS AS A SOURCE OF FOOD

At least forty-one bird species were hunted for food in the southern part of PSNR
and its surroundings during 1996 (table 21.1); the use of six additional bird species
was recorded during our visits to the study site in 1997–1998. However, only eight
species were hunted regularly by more than 25% of the households, with these
species accounting for 75% of the total number of birds harvested in the study area
(table 21.2). Undulated tinamous (*Crypturellus undulatus*), anhingas (*Anhinga anhinga*), razor-billed curassows (*Mitu tuberosa*), muscovy ducks (*Cairina moschata*),
and olivaceous cormorants (*Phalacrocorax olivaceus*) were the most-frequently
hunted species. The razor-billed curassow and the muscovy duck, because of their
larger size, were the most important species in terms of biomass. The white-eyed
parakeet (*Aratinga leucophthalmus*) and the dusky-headed parakeet (*Aratinga weddellii*), locally considered as agricultural pests, were also killed in large numbers using nets in corn and rice fields but were not considered in our analyses because
they were not eaten.

Subsistence hunting showed a marked seasonality in the study area related to the
scarcity of fish and agricultural products during the flood season, which forced local people to dedicate more time and effort to hunting activities. Furthermore,
hunting is much easier when waters reach their highest levels because wildlife is
concentrated in the few small areas that remain unflooded (restingas).

In general, subsistence hunters took significantly more birds during the flood
season (November to May) than during the dry season (June to October) (Mann-

TABLE 21.1 Birds Consumed in Seventeen Rural Settlements of the
Ucayali/Puinahua Floodplain During 1996

LATIN NAME	COMMON NAME	HOUSEHOLDS		BIRDS	
		N	%	N	%
Mitu tuberosa	Razor-billed curassow	129	66.49	382	5.75
Cairina moschata	Muscovy duck	107	55.15	481	7.24
Anhinga anhinga	Anhinga	106	54.63	484	7.28
Crypturellus undulatus	Undulated tinamou	89	45.87	740	11.14
Pipile cumanensis	Blue-throated piping-guan	71	6.59	269	4.05
Phalacrocorax olivaceus	Olivaceous cormorant	68	35.05	657	9.89
Ardea cocoi	Cocoi heron	61	31.44	266	4.00
Tinamus major	Great tinamou	55	28.35	181	2.72
Aratinga leucophthalmus	White-eyed parakeet	54	27.83	832	12.52
Ortalis guttata	Speckled chachalaca	35	18.04	110	1.65
Ara spp.	Macaws	27	13.91	177	2.66
Columbidae[a]	Pigeons	26	13.40	139	2.09
Penelope jacquacu	Spix's guan	26	13.40	95	1.43
Ara severa	Chestnut-fronted macaw	25	12.88	266	4.00
Aratinga weddellii	Dusky-headed parakeet	23	11.85	933	14.04
Amazona spp.	Parrots	23	11.85	138	2.08
Aramus guarauna	Limpkin	22	11.34	54	0.81
Aramides cajanea	Gray-necked wood-rail	21	10.82	107	1.61
Crypturellus cinereus	Cinereous tinamou	15	7.73	52	0.78
Tigrisoma lineatum	Rufescent tiger-heron	11	5.67	31	0.47
Psophia leucoptera	Pale-winged trumpeter	8	4.12	21	0.32
Ramphastos spp.	Toucans	7	3.60	13	0.19
Pteroglossus spp.	Aracaris	7	3.60	7	0.11
Ceryle torquata	Ringed kingfisher	5	2.57	51	0.77
Anhima cornuta	Horned screamer	5	2.57	8	0.12
Cochlearius cochlearius	Boat-billed heron	4	2.06	40	0.60
Agamia agami	Agami heron	4	2.06	17	0.26
Icteridae[b]	Oropendolas	3	1.54	18	0.27
Odontophorus gujanensis	Marbled wood-quail	3	1.54	8	0.12
Crax globulosa	Wattled curassow	2	1.03	4	0.06
Graydidascalus brachyurus	Short-tailed parrot	1	0.51	20	0.30
Butorides striatus	Striated heron	1	0.51	15	0.23
Gymnomystax mexicanus	Oriole blackbird	1	0.51	15	0.23
Busarellus nigricollis	Black-collared hawk	1	0.51	3	0.05
Rosthramus sociabilis	Snail kite	1	0.51	2	0.03
Buteo magnirostris	Roadside hawk	1	0.51	2	0.03
Nothocrax urumutum	Nocturnal curassow	1	0.51	2	0.03

TABLE 21.1 (*Continued*)

LATIN NAME	COMMON NAME	HOUSEHOLDS N	HOUSEHOLDS %	BIRDS N	BIRDS %
Mesembrinibis cayennensis	Green ibis	1	0.51	2	0.03
Casmerodius albus	Great egret	1	0.51	2	0.03
Jabiru mycteria	Jabiru	1	0.51	1	0.01

Note: Sample size was 194 households. The number and percentage of households that consumed each species and the total number of individuals of each species consumed during the year are presented in the table. During our visits to the study area in 1997 and 1998, we recorded the use for food of six additional bird species: *Tinamus tao, Tinamus guttatus, Crypturellus soui, Porphyrula martinica, Jacana jacana,* and *Opisthocomus hoazin.*
[a]Includes *Leptotila rufaxilla* and *Columba* spp.
[b]Includes *Psarocolius angustifrons* and *Cacicus cela.*

TABLE 21.2 Estimate of the Number and Biomass of the Most Important Game Birds Consumed During 1996 in Seventeen Rural Settlements of the Ucayali-Puinahua Floodplain

	NUMBER $\bar{x}$	NUMBER C.I. 95%	BIOMASS (KG)[a] $\bar{x}$	BIOMASS (KG)[a] C.I. 95%
Mitu tuberosa	976	801–1,168	2,942	2,415–3,521
Cairina moschata	926	726–1,151	2,199	1,724–2,733
Anhinga anhinga	992	784–1,226	1,339	1,058–1,655
Crypturellus undulatus	1,059	792–1,368	601	449–777
Pipile cumanensis	559	417–709	712	531–903
Tinamus major	375	267–492	408	290–535
Phalacrocorax olivaceus	759	542–1,009	1,376	983–1,830
Ardea cocoi	475	342–625	1,077	775–1,417

Note: Only the eight species that were consumed by more than 25% of the 834 households (6,622 inhabitants) during 1996 are represented (see table 21.1 for the complete list of birds consumed). The $\bar{x}$ is mean, and C.I. the confidence interval.
[a]Masses (males and females averaged) reported for these species in Ayres et al. (1991), Ojasti (1993), and Peres (1997) were averaged and used in the calculations.

Whitney U-test, $U = 0$, $P < 0.05$). However, this seasonal pattern was not statistically significant in the case of waterbirds alone (Mann-Whitney U-test, $U = 14.5$, $P > 0.05$). The average number of terrestrial birds harvested monthly per hunter was 7.8 ± 2.5 (mean $\pm$ sd.) during the flood season and only 2.2 ± 1.7 in the dry season. In the case of waterbirds only 1.4 ± 1.2 and 1.1 ± 0.9 birds were harvested monthly per hunter in the flood and dry seasons, respectively.

All the species, except the olivaceous cormorant, were more abundant in the surveys conducted inside the protected area than in heavily hunted areas outside the reserve (table 21.3). However, the abundances (number of birds/10 km) of the most commonly hunted birds did not show significant differences between heavily hunted areas and areas where hunting pressure is low or even absent, except for the great tinamou (Kolmogorov-Smirnov tests, $P > 0.05$; table 21.3). The harvest model of Robinson and Redford (1991), applied to the most commonly hunted terrestrial birds, suggests that only the razor-billed curassow is being hunted at the maximum sustainable harvest rate (20% of the production), whereas current harvest rates of the other species are under the estimated maximum sustainable level (table 21.4).

BIRDS AS A SOURCE OF EGGS

Bird eggs are frequently consumed in the study site, especially in villages located close to breeding colonies. During 1996 I recorded the use of eggs of twenty-two species, of which the great egret (*Casmerodius albus*), cocoi heron (*Ardea cocoi*), boat-billed heron (*Cochlearius cochlearius*), and agami heron (*Agamia agami*) were the most commonly harvested. Eggs of these species were consumed in 11% of the households monitored (table 21.5). The greater ani (*Crotophaga major*), striated heron (*Butorides striatus*), great tinamou (*Tinamus major*), hoatzin (*Opisthocomus hoazin*), and horned screamer (*Anhima cornuta*) were also important sources of eggs for local people.

The use of heron eggs was monitored in the villages of Padre López and Nueva Cajamarca, where most of the egg harvesting took place. The collection of eggs is practiced almost every year by some households from these villages during the

TABLE 21.3 Abundance (Number of Birds/10 Km; x ± SD) of the Eight Most Commonly Hunted Bird Species in Heavily Hunted Areas (Outside PSNR) and Lightly Hunted Areas (Inside PSNR)

		HEAVILY HUNTED	LIGHTLY HUNTED
Razor-billed curassow	*Mitu tuberosa*	1.77 ± 1.34	1.80 ± 0.97
Muscovy duck	*Cairina moschata*	1.43 ± 2.48	4.44 ± 6.35
Anhinga	*Anhinga anhinga*	2.84 ± 2.73	3.26 ± 3.89
Undulated tinamou	*Crypturellus undulatus*	2.67 ± 1.92	3.87 ± 2.17
Blue-throated pipin-guan	*Pipile cumanensis*	1.01 ± 0.97	2.32 ± 1.02
Great tinamou	*Tinamus major*	1.06 ± 1.12	5.23 ± 2.82
Olivaceous cormorant	*Phalacrocorax olivaceus*	25.40 ± 42.95	14.37 ± 13.25
Cocoi heron	*Ardea cocoi*	7.86 ± 8.95	10.70 ± 1.34

TABLE 21.4 Maximum Production and Maximum Sustainable Yield for the Most Important Terrestrial Game Birds

	DENSITY[a] (#/KM^2)	MAXIMUM PRODUCTION (#/KM^2)	MAXIMUM SUSTAINABLE HARVEST (#/KM^2)	CURRENT HARVEST (#/KM^2)	% OF PRODUCTION HARVESTED
Mitu tuberosa	5.28	1.88	0.38	0.38	20.05
Pipile cumanensis	4.43	1.97	0.39	0.21	10.52
Penelope jacquacu	2.69	1.20	0.24	0.07	6.13
Ortalis guttata	3.17	1.82	0.36	0.08	4.39
Tinamus major	25.28	39.06	7.81	0.14	0.36
Crypturellus undulatus	25.80	46.41	9.28	0.41	0.89
Crypturellus cinereus	1.35	2.43	0.49	0.04	1.57
Psophia leucoptera	2.03	0.63	0.13	0.01	2.02

Note: Estimates done using a conservative harvest model, compared with current harvest rates in the study area.

[a] The observed densities in areas of low/absent hunting pressure were used as a conservative estimation of densities at carrying capacity, except for *P. leucoptera* and *P. jacquacu* for which the maximum observed density was used.

TABLE 21.5 Bird Eggs Consumed in 101 Households of the Study Area in 1996

LATIN NAME	COMMON NAME	NO. OF EGGS	NO. OF HOUSEHOLDS
A. agami/C. Cochlearius	Agami/Boat-Billed Herons[a]	489	7
C. albus/A. cocoi	Great egret/cocoi heron[a]	360	4
Crotophaga major	Greater ani	167	9
Butorides striatus	Striated heron	94	4
Tinamus major	Great tinamou	86	7
Opisthocomus hoazin	Hoatzin	84	9
Anhima cornuta	Horned screamer	68	6
Crypturellus undulatus	Undulated tinamou	37	6
Pitangus sulphuratus	Great kiskadee	36	1
Cairina moschata	Muscovy duck	32	2
Other birds[b]		59	4

[a]Eggs of *Agamia agami/Cochlearius cochlearius* and *Casmerodius albus/Ardea cocoi* are collected, consumed, and sold together, so they were placed in the same category.
[b]Includes *Phaetusa simplex, Mitu tuberosa, Crypturellus cinereus, Leptotila rufaxilla, Crotophaga ani, Aramus guarauna, Aramides cajanea, Mesembrinibis cayennensis,* and *Ortalis guttata*

month of April in two large mixed-species heronries (an activity that is locally known as *garceada*).

The exploitation of eggs in the heronry at Padre López began in 1986. Since then, the heronry has moved three times from its position but has always remained close to its previous location. Local people report that up to six species of waterbirds bred in the heronry during the first years of exploitation (great egret, cocoi heron, boat-billed heron, agami heron, olivaceous cormorant, and anhinga), but only great egrets and cocoi herons were present during our study.

Eleven households of the village (33.3%) have participated at least once in the harvesting of heron eggs. The number of eggs taken during the study period ranged from 5,400 in 1996 to only a dozen in 1998 (table 21.6). Some of the eggs were consumed by the collectors and their families, but most were sold in the neighboring villages of Victoria and Obreros (ca. US$ 1 per dozen). The occasional hunting of chicks and adults from this colony was also reported by local people but did not occur during our study period.

At Nueva Cajamarca (inhabited by natives from the ethnic group Shipibo), nine households (52.9%) have participated in the collection of heron eggs in recent years. The number of eggs taken ranged from 780 in 1996 to 210 in 1998 (table 21.6). Most of these eggs were consumed by the collectors and their families. When harvesting eggs, Shipibo Indians practiced a traditional management technique of leaving at least one egg in each nest in the belief that this guarantees the permanence of the heronry in the same site the following year.

TABLE 21.6 Number of Heron Eggs Harvested in the Heronries at Padre López and Nueva Cajamarca Between 1996 and 1998

		HARVESTING GROUPS[a]	EGGS/TRIP/GROUP $\bar{X} \pm SD$	TOTAL EGGS HARVESTED
	1996	7	77.9 ± 69.1	780
Nueva Cajamarca	1997	6	55.6 ± 30.7	445
	1998	4	52.5 ± 61.0	210
	1996	3	1,800.0 ± 1,039.2	5,400
Padre López	1997	3	700.0 ± 264.5	2,100
	1998	1	12	12

Note: $\bar{x}$ = mean.
[a]Each harvesting group was usually formed by two to three people.

MAGICAL, MEDICINAL, AND OTHER USES OF BIRDS

Several bird species were sometimes used in the study area for medicinal, magical, ornamental, or domestic purposes (table 21.7). The use of the crimson-crested woodpecker's beak (grated and macerated in alcohol) to strengthen virility and the use of macaw feathers as ornaments were the most frequently recorded practices. However, none of these uses is very widespread in the region since only three of the households (1.5%) stated that they hunt birds regularly for medicinal, ornamental, or other related purposes.

BIRDS AS PETS

At least thirty-three species of birds were kept as pets by local households in the study area. The most popular were the canary-winged parakeet (*Brotogeris versicolorus*), cobalt-winged parakeet (*B. cyanoptera*), tui parakeet (*B. sanctithomae*), orange-winged parrot (*Amazona amazonica*), festive parrot (*A. festiva*), yellow-crowned parrot (*A. ochrocephala*), and blue-and-yellow macaw (*Ara ararauna*).

There is also an important trade of these birds to neighboring large cities. Globally, 77.2% of the nestlings harvested by local collectors in 1996 were sold to middlemen, who brought them to the big markets of Pucallpa, Iquitos, or Lima. Of the households in the study area, 26.3% sold parakeets (*Brotogeris* spp.) in 1996, 18.5% sold parrots (*Amazona* spp.), 8.8% sold macaws (*Ara* spp.), and 4.1% sold other bird species. A total of 934 birds of fourteen species were sold during 1996 in the 194 households visited (table 21.8). Although parakeets were the most frequently sold pets, the orange-winged parrot, festive parrot, and blue-and-yellow macaw were the most important species in terms of gross profit.

Most of the harvesting and trade of parrots and macaws in the study area took place in the village of Victoria and its surroundings because of the near presence of

TABLE 21.7 Wild Bird Species Used for Ornamental, Medicinal, Magical, or Domestic Purposes in the Study Area

LATIN NAME	COMMON NAME	TYPE OF USE	PARTS USED
Eurypyga helias	Sunbittern	Magical (pusanga)[a]	Wing bone
Daptrius americanus	Red-throated caracara	Magical (pusanga)	Eye fluid
Campephilus melanoleucos	Crimson-crested woodpecker	Medicinal (virility)	Beak
Opisthocomus hoazin	Hoatzin	Medicinal (cough)	Meat
		Medicinal (virility)	Coccyx
Coragyps atratus	Black vulture	Medicinal (epilepsy)	Heart/Blood
Ara spp.	Macaws	Ornamental (adornment)	Feathers
Mitu tuberosa	Razor-billed curassow	Domestic (feather duster)	Feathers
Ramphastos spp.	Toucans	Magical (pusanga)	Tongue
Trogon spp.	Trogons	Medicinal (depilatory)	Fat
		Magical (pusanga)	Heart/brain
Herpetotheres cachinnans	Laughing falcon	Medicinal (snake bites)	Fat
Nyctidromus albicollis	Pauraque	Medicinal (birth)	Eggs

[a]Pusanga is a potion or an amulet used to attract the love of another person.

TABLE 21.8 Birds Traded as Pets in 194 Households of the Study Area During 1996

SPECIES	NO. OF SELLERS	NO. OF BIRDS SOLD		TOTAL INCOME[a]	
		N	%	S/.	%
Brotogeris versicolorus	45	497	53.2	507	9.6
Brotogeris cyanoptera	14	73	7.8	85	1.6
Brotogeris sanctithomae	20	151	16.2	382	7.2
Ara ararauna	10	40	4.3	981	18.5
Ara macao	6	12	1.3	343	6.5
Ara chloroptera	5	8	0.8	307	5.8
Amazona festiva	20	52	5.6	1,013	19.1
Amazona amazonica	11	52	5.6	883	16.6
Amazona ochrocephala	11	30	3.2	658	12.4
Amazona farinosa	4	6	0.6	90	1.7
Other species[b]	5	13	1.4	52	1.0

[a]Using the average prices paid by middlemen to local collectors (US$ 1 = S/. 3.5).
[b]Includes *Aratinga weddellii*, *Aratinga leucophthalmus*, *Graydidascalus brachyurus*, and *Mitu tuberosa*.

a large area of Mauritia palm swamps in which these species concentrate in large numbers to breed every year. We estimated that 79% of the macaws and 58% of the parrots traded to towns came from this part of the reserve. The harvesting of parrot and macaw nestlings is a major economic activity (locally called *loreada*) practiced by many people from Victoria every year between the months of February and April.

Between 1996 and 1999 I monitored the harvesting of parrots and macaws in two large Mauritia palm swamps (1,230 and 2,660 ha), located close to the village. Three species of parrots and four species of macaws were collected by local poachers (*loreros*) in these sites, the orange-winged parrot (61.1% of the captures) and the blue-and-yellow macaw (25.9%) being the most commonly harvested. The total number of nestlings taken during the four-year study period was 1,718, ranging from 680 birds harvested in 1996 to 166 in 1998. The number of households that took part in the loreada ranged from forty-two (45.2%) in 1996 to eighteen (19.3%) in 1998. The reasons for these changes probably relate to the presence of the researcher in the harvesting area and to the increasing number of birds confiscated by regional authorities, causing a reduction in the demand for nestlings (González 1999b).

Two methods were generally used to collect nestlings in the study site: cutting down the nesting tree (for species like macaws that nest very high) or hacking open the nest cavities in order to remove the chicks. Both methods are very destructive because nests become useless and the next generation is completely removed. Mortality during the harvesting process is another matter of great concern, especially when the collectors cut down the nesting trees. Overall, 229 of 1,142 nestlings died during the harvesting process (20.1%) in the study site. Mean mortality rates varied between parrots (3.2%) and macaws (29.2%). Figures were particularly high for the blue-and-yellow macaw (48.4% of the nestlings died during the harvest).

The red-bellied macaw, orange-winged parrot, and blue-and-yellow macaw were the most abundant species breeding in the studied swamps (mean of 19.1, 14.8, and 6.4 nests/100 ha, respectively). The nestling production (number of fledglings/successful nest) ranged from 1.35 fledglings/nest for the scarlet macaw to 2.33 fledglings/nest in the festive parrot. Overall productivity in the swamps was estimated with these data and compared with the average annual harvest rate for each species (table 21.9). Following the categories of Robinson and Redford (1991), parrots and macaws should be considered as long-lived species, for which a maximum sustainable harvest rate of 20% of the production can be assumed (Robinson and Redford 1991). In this case our data suggest that some species, like the red-bellied or the chestnut-fronted macaws, which have little demand, are being harvested under the maximum sustainable level, while other species, like the scarlet macaw, the blue-and-yellow macaw, and the orange-winged parrot are being overharvested and may be seriously threatened in the long term.

TABLE 21.9 Estimate of the Number of Nests, Total Number of Nestlings Produced, and Current Number of Birds Harvested Annually in the Mauritia Palm Swamps Located Close to the Village of Victoria

	ESTIMATED NO. NESTS[a]	PRODUCTION (NO. NESTLINGS)[a]	MAXIMUM HARVEST RATE (20%)	CURRENT ANNUAL HARVEST[b]
A. ararauna	249	379	75.8	111.5
A. macao	35	47	9.4	15.0
A. severa	89	169	33.8	7.8
A. manilata	742	1,485	297.0	10.5
A. amazonica	576	1,014	202.8	262.5
A. festiva	78	181	36.2	16.8
A. ochrocephala	17	31	6.2	3.5

[a]Average values recorded during the 1998 and 1999 nesting seasons.
[b]Average values recorded in the village of Victoria from 1996 to 1999.

DISCUSSION

HUNTING OF BIRDS FOR FOOD

The PSNR and its buffer zone provide local people with most of the resources they need to survive, including medicines, building materials, and food. Hunting in the study area is practiced mainly for subsistence. Game is used to satisfy the food needs of the hunter and his family, although sometimes the surplus may be sold in the village (Rodríguez, Rodríguez, and Vásquez 1995). The commercialization of game meat to large cities is banned. Peruvian laws (D.S. No. 934-73-AG/DGFF and D.S. No. 158-77-AG/DGFF) only permit subsistence hunting of some species of the Cracidae, Columbidae, and Tinamidae families. Despite the laws, our data shows that the range of birds used by local people in the study area is much wider and includes some species that may be endangered at the national level (CDC-UNALM 1993).

Soini et al. (1996) reported that people living in the PSNR or its surroundings use more than sixty animal species for subsistence hunting, including thirty mammals, twenty-five birds, and five reptiles. In this study we recorded the use of more than forty bird species, cracids, tinamous, and waterbirds being the most commonly hunted groups.

Cracids are traditionally considered the most important birds for subsistence hunting in tropical forests, and they are always present in the diet of all Amazonian rural settlements (Pierret and Dourojeanni 1967; Vickers 1991; Ojasti 1993). However, management of cracid populations is difficult because of their low capacity to recover from losses caused by hunting. As a consequence, cracids do not tolerate high harvesting rates, and their populations usually decline under continuous hunting (Silva and Strahl 1991; Vickers 1991). Some authors have pointed out that the best way to conserve these species is to give them total protection, at least until we have precise demographic information to determine the optimum harvesting rates (Dobson 1997). Others, however, argue that most of the cracid populations might tolerate some level of extraction, especially if hunting is in extensive areas surrounded by unhunted buffer populations (Silva and Strahl 1991; Begazo 1997).

Tinamous are the second most important group of game birds, the undulated tinamou being the most commonly hunted species in the study area. Tinamous are easily located by their sounds, which are imitated by hunters to attract the birds. Many undulated tinamous are also captured using a slip knot trap (*tuclla*) close to villages and agricultural fields.

Among waterbirds, the most important species in terms of biomass consumed is the muscovy duck. The hunting of this species is frequently associated with rice fields, where ducks concentrate to feed during the months of October and November when waters begin to rise and shallowly inundate these areas. Soini et al. (1996) reported a sharp decline suffered by whistling ducks (*Dendrocygna autumnalis* and *D. bicolor*) in the study area, suggesting that this decline may be related to the extensive use of pesticides in the rice fields where these birds forage. The effect of pesticides on muscovy ducks is unknown as there are no long-term data on this species. Experimental studies to assess this point are urgently needed. Olivaceous cormorants and anhingas are usually hunted with shotguns during fishing activities. Because of their foraging habits, these species are also frequently trapped in fishing nets.

The comparative method of assessing hunting sustainability, in which abundances or densities of species are compared between hunted and unhunted sites, depends on many problematic and often untested assumptions, such as similarity of habitat and constancy of all variables except hunting pressure (Robinson and Redford 1994; Bodmer and Robinson this volume). However, when long-term monitoring data are not available and an assessment must be made of hunting sustainability, the comparative method can be used as a good initial diagnosis of the effect of hunting on animal populations (Silva and Strahl 1991; Bodmer et al. 1994; Peres 1997).

My data suggest that most of the game birds are not being overharvested in the southern part of PSNR, although sample size may be too small and variances too high to extract definitive conclusions. The harvest model of Robinson and Redford (1991) also indicates that none of the terrestrial game birds are being overhunted in the study area, although the current harvest rate of razor-billed curassows is at the maximum sustainable level.

Vickers (1991) pointed out that, when human population density is low and rural settlements are dispersed, subsistence hunting can be practiced on a sustainable basis for most of the species used by Amazonian people. Furthermore, the effects of hunting may be reduced if the hunting occurs over extensive areas surrounded by unhunted buffer populations (Begazo 1997; Bodmer et al. 1997a).

Several reasons may contribute to the sustainability of bird hunting in the study site. First, in the southern part of PSNR hunting is only a secondary activity compared with fishing and agriculture, which are the main sources of food for local people (González 1999c). Second, hunting pressure in not continuous throughout the year and shows a clear seasonal pattern, increasing during the months of maximum flooding (February to April) but decreasing sharply during the rest of the year. Finally, because of the presence of a huge protected area, it is likely that unhunted populations from inside PSNR serve as a source of new individuals to repopulate overhunted areas, maintaining relatively stable populations of game birds in spite of hunting.

HARVESTING OF BIRD EGGS

Bird eggs are an important source of food in some areas (Cott 1954; Redford and Robinson 1991). Flamingo eggs are eaten in certain areas of the Andes of Bolivia and Chile (Campos 1986), collection of seabird eggs is a traditional practice in the Caribbean islands that dates back several centuries (Haynes 1987), and harvesting eggs by indigenous peoples from beach-nesting birds in the Peruvian Amazon is common (Redford and Robinson 1991). However, there are very few reports on the harvesting of eggs from mixed-species colonies of wading birds (heronries), and the impact of this harvest has rarely been evaluated (Feare 1976).

Thomas (1987) reports on the consumption of maguari stork (*Ciconia maguari*) nestlings in the llanos of Venezuela, and Luthin (1987) reports that wood stork (*Mycteria americana*) nestlings are consumed in Central America. Soini et al. (1996) observed the harvesting of cocoi heron nestlings from heronries in a small village in northeastern Peru. My data suggest that collection of heron eggs is a common activity, traditionally practiced by several native communities in some parts of the Peruvian Amazon.

While the harvesting of eggs of noncolonial nesting birds in the study area can be considered as occasional and probably has little effect on wild bird populations, the harvesting of eggs in heronries may have severe consequences because any negative impact to a breeding colony may affect a large proportion of the local population. Human activities are a major factor in the disturbance of colonial waterbirds (Ellison and Cleary 1978; Tremblay and Ellison 1979; Frederick and Collopy 1989). In some cases birds may abandon the site because of frequent human disturbance; in other cases the colony may persist but with lower reproductive success (Parnell et al. 1988).

The impact of human disturbance on local heronries is analyzed in González (1999a). All the evidence suggests that heronries in the study area are highly sensitive to human disturbance during early stages of nesting and that this disturbance may represent a major threat to local wading bird populations. In 1998 a group of hunters entering the colony at Padre López during egg laying caused the total abandonment of the heronry. Egg harvesting during early stages of nesting probably caused the abandonment of the colony at Nueva Cajamarca by agami herons.

Management of large water bird colonies for egg harvesting has proven problematic, and outright protection of the colonies may be the only way to prevent depletion (Feare 1976; Haynes 1987). Subsistence egg harvesting practiced on a small scale by people of Nueva Cajamarca for their own consumption seems to be less detrimental to herons than commercial egg harvesting practiced in Padre López. However, because of the great sensitivity of breeding herons to human disturbance during early stages of nesting, it is my opinion that heronries in the study area cannot be properly managed for long-term harvesting of eggs on a sustainable basis.

HARVESTING OF NESTLINGS FOR THE PET TRADE

The tradition of keeping wild animals as pets is quite common among people inhabiting the Amazon region (Redford and Robinson 1991). There is also a thriving export trade in wild animals for pets. No less than 150 animal species have been regularly exported as pets from the Amazon region, parrots and primates being the most important groups (Dourojeanni 1972).

In Peru a total of 1,958,000 animals were legally exported from Iquitos between 1965 and 1973 (before a national law banned the trade of Amazonian wildlife); 39% of these animals were psittacines and 4.9% other birds (Dourojeanni 1985). Prior to the enactment of the Wild Bird Conservation Act in 1992, documented U.S. imports of live birds since 1900 totaled nearly thirty million birds (WCI 1992). Although significant, these figures represent only a small fraction of the total number of birds removed from the wild because they do not include smuggled birds, birds dying during the capture and holding process prior to export, and birds sold in domestic pet markets (Iñigo-Elías and Ramos 1991; Thomsen and Mulliken 1992). Although there is a strong tradition in Neotropical countries of keeping birds in captivity, we lack detailed studies on the magnitude of the domestic cage-bird trade (Thomsen and Brautigam 1991)

My data show that, despite being banned by national laws since 1973, the harvesting and domestic trade of psittacines is still a common practice in the Peruvian Amazon. It is a matter for concern that large numbers of long-lived species (*Amazona* spp. and *Ara* spp.) are harvested every year in some parts of the PSNR. Because of their low reproductive rates, large, long-lived species cannot sustain high levels of exploitation (Munn 1988; Thomsen and Brautigam 1991; WCI 1992). The comparison of estimated production and current harvest rates in the village of Victoria showed that at least three species are being harvested over sustainable levels.

Several authors have explored in recent years the feasibility of sustainable harvest of wild parrot populations (Munn et al. 1991; Thomsen and Brautigam 1991;

Beissinger and Bucher 1992). As discussed by Beissinger and Bucher (1992), the sustainable harvest of certain species of parrots is biologically possible and could contribute to both habitat protection and the local economy. The social, political, and economic feasibility of sustainable parrot harvests, however, is still in question. As there are no documented examples of any sustainable harvesting project for the pet trade, Snyder, James, and Beissinger (1992) and WCI (1992) strongly recommended the implementation of pilot sustainable management projects, designed to test how different management techniques benefit the local communities and how effectively these techniques can be controlled. Because of the amount of information available and the willingness of collectors to participate, the palm swamps located close the village of Victoria offer a good opportunity to develop one such experimental project, intended to evaluate the biological, social, and economic sustainability of the harvesting of wild parrot and macaw nestlings.

ACKNOWLEDGMENTS

I thank all the staff of ProNaturaleza, Junglevagt for Amazonas, and Centro de Datos para la Conservación for assistance in the field and for providing accommodation and logistic support. I am especially grateful to my local guides F. Cachique, L. Isamani, A. Pacaya, A. Navarro, M. Navarro, J. Seopa, E. Seopa, B. Pacaya, M. Ihuaraqui, E. Tangoa, W. Severiano, and S. Vela and the rangers M. Vásquez, R. Cárdenas, R. Armas, A. Valles, and J. Rodríguez. I also thank the management staff of PSNR and INRENA for giving me all the facilities to develop the field work. P. Soini, J. Alvarez, A. Begazo, P. Vásquez, and K. Silvius gave valuable comments on previous drafts of the manuscript. This research was supported by grants from Fondo Lende-Simmons and Agencia Española de Cooperación Internacional.

22

Patterns of Use and Hunting of Turtles in the Mamirauá Sustainable Development Reserve, Amazonas, Brazil

AUGUSTO FACHÍN-TERÁN, RICHARD C. VOGT,
AND JOHN B. THORBJARNARSON

Turtles have been, and continue to be, one of the principal sources of protein from the wild for indigenous and riverine populations in Amazonia. Pressure on the resource increased with the arrival of the first European colonizers, who exploited almost all species of Amazonian quelonians (Ayres and Best 1979). The most heavily exploited species was *Podocnemis expansa*, sought after for its size, its eggs, and the quality of its meat.

Several authors have reported on the exploitation of female turtles and their eggs, especially those in the genus *Podocnemis* (Bates 1863; Smith 1979a; Fachín-Terán 1994; Fachín-Terán, Chumbe, and Taleixo 1996; Rebêlo and Lugli 1996; Landeo 1997). Turtles are more vulnerable during their annual reproductive period than at other times, and the protection of eggs and nesting areas is considered a high priority. Turtle exploitation levels have been quantified and recorded through research projects of short duration (Bates 1863; Smith 1979a; Moll 1986; Polisar 1995; Fachín-Terán, Chumbe, and Taleixo 1996; Thorbjarnarson, Perez, and Escalona 1997; Landeo 1997), but few studies have monitored the capture of Amazonian turtles (Rebêlo and Lugli 1996; Fachín-Terán, Vogt, and Thorbjarnarson 2000). In combination with data on harvest levels, long-term research projects describing the characteristics of turtle populations will enable us to evaluate the biological impact of current harvests.

Historically, populations of three species of *Podocnemis* that occur in the Mamirauá Sustainable Development Reserve (Reserva de Desenvolvimento Sustentável Mamirauá, or RDSM) were abundant. Bates (1863) reports that from this section of the Solimões River, in the neighborhood of Ega (present-day Tefé), and from the Madeira River, approximately forty-eight million *P. expansa* eggs were collected annually between 1848 and 1859 for exportation to Pará.

Pressure on the resource continues to this day, bringing *P. expansa* to the brink of extinction in this part of the Amazon. Interviews with the oldest inhabitants of the area indicate that populations of two other species in the genus, *Podocnemis unifilis* and *Podocnemis sextuberculata*, have also diminished drastically because of the continuous hunting pressure to which they have been subjected. In a recent assessment by the IUCN Freshwater Turtle and Tortoise Specialist Group, *P. unifilis*, *P. sextuberculata*, and *Geochelone denticulata* were placed on Appendix I of CITES as endangered species and *P. expansa* on Appendix II as a species at low risk but dependent on conservation (IUCN 1996).

Podocnemis expansa was placed on Appendix II rather than Appendix I because for the past twenty years it has been under an intensive conservation program by the Brazilian government. The program includes nesting beach protection and the release of over two million hatchlings per year into the wild. However, there is no scientific proof that this program is working, and in fact populations are diminishing within some of the protected reserves, notably the Rio Trombetas Biological Reserve. Throughout the remainder of its range in other countries, *P. expansa* populations have been drastically reduced (e.g., Peru; Soini 1997) and remain at high risk of local extinction.

The present study examined turtle hunting patterns in the RDSM. These patterns include species, number and size of individuals extracted, hunting methods and season, and habitats where turtles are most frequently captured.

METHODS

The study was carried out in the Jarauá sector of the RDSM. The reserve covers 1,124,000 ha between the Japurá, Solimões, and Auti-Paraná rivers, near the city of Tefé, Amazonas state, Brazil (03° 08′ S, 64° 45′ W, and 2° 36′ S, 67° 13′ W). Crampton et al. (this volume) give a detailed description of the reserve.

Information on species, number, sex, weight, method, habitat, and use of turtles was gathered in the communities of São Raimundo de Jarauá (2° 51′ S, 64° 55′ W), Nova Colômbia (2° 54′ S, 64° 54′ W), Novo Pirapucu (2° 53′ S, 64° 51′ W), and Manacabi (2° 50′ S, 54° 52′ W) through both interviews and direct observations. These communities were selected because they are located near turtle-nesting beaches and near lakes designated for preservation, personal use, or commercialization on the Japurá river and the Jarauá Paraná. Fifty families were interviewed on two occasions in these four communities, the first time between September 22 and October 12, 1996, and the second between November 17–18, 1997. All communities had few households and were therefore completely sampled. Data on turtle captures were collected from September 1996 to April 1998.

The consumption of turtles during the study period was identified based on the presence of ectodermal shields (Thorbjarnarson, Perez, and Escalona 1993). Species were identified using external shell characteristics. Carapace length was measured in a straight line at the point of greatest separation between the anterior and posterior edges (Medem 1976). Sex was determined by size, head color, carapace length, plastron shape, invagination of the anal plate, precloacal length, and thickness of the tail (Ponce 1979; Pritchard and Trebbau 1984) and in some cases by asking the interviewee if he had noted the gonads of the turtle before it was consumed. Turtles were weighed with spring scales.

Várzea ecosystems comprise a diversity of aquatic features and habitat types, many of which appear only seasonally. Often there are no direct, concise translations for the names of these unique habitats, and we use the local Portuguese language terms in the text. Here we give a brief description of these habitat types and also of terms that refer to changes in the hydrological cycle:

Enchente: rising water phase in the annual hydrological cycle

Vazante: dropping water phase in the annual hydrological cycle

Repiquete: temporary rises in the water level (oscillating water levels) that usually precede the main enchente. In the Central Amazon these oscillations usually occur around December and January; from February to April the water usually rises steadily.

Remanso: an eddy that occurs in little curved inlets along the edge of main river channels. These curved inlets are usually caused when a chunk of forest falls into the river (*terra caida*); the water flows into the new inlet and forms an eddy.

Restingas: levees in the floodplain

Poças: ponds or static pools of water in the forest or sometimes on beaches, created by rainwater or when water is stranded in the floodplain when water levels drop

Paraná: side branch of a main river channel that winds its way through the várzea floodplain. It is always connected at both ends to a whitewater river.

Canos: channels that drain lakes in the várzea

Lagos: floodplain lakes

Enseada: outer curve of a meander or curve in a river where erosive processes are at their strongest. Remansos often form along the enseadas.

Ressaca: an inlet or branch to any water body (usually lake or channel) that dead-ends

RESULTS

COMMUNITIES STUDIED

São Raimundo de Jarauá is the largest and most important community in the Jarauá sector. It comprises eighteen houses with twenty families, who engage primarily in commercial fishing and who also practice subsistence farming. Turtles are captured in the Japurá river and in the paranás, ressacas, canos, and floodplain lakes of the Paraná do Jarauá Hydrological System. The high number of *P. sextuberculata* (n = 386) and *P. unifilis* (n = 177) registered for this community reflects the knowledge that community dwellers have of turtle behavior and of the areas where turtles occur.

Novo Pirapucu comprises nine houses and ten families. The community engages in both subsistence and commercial fishing activities and in subsistence agriculture. Its location near a *P. sextuberculata* nesting beach on the Japurá river explains the high consumption of females of this species (n = 37) by community members. During the 1996 nesting season, community members agreed to preserve 50% of the area of the beach in order to allow turtle researchers from the Mamirauá project to study the species' reproductive biology (protection of the beach continued through 2002). This protection reduced the rate of capture of reproductive females in the area. In 2003 this community offered to preserve 75% of the beach.

The community of Manacabi comprises nine families in nine houses and relies primarily on subsistence agriculture and subsistence fishing. Small turtle nesting beaches emerge during the dry season in the paraná that provides access to the community. In 1996 four *P. sextuberculata* and one *P. unifilis* females were captured there.

Nova Colômbia, with ten houses and eleven families, relies primarily on agriculture and less intensively on subsistence fishing. Of the eleven *G. denticulata* registered for this community, eight were captured in the restingas of the Paraná do Jarauá and three in the restingas of Nova Colômbia.

Through interviews and direct observations in the field, we were able to locate the carapaces of dead turtles. Members of the communities eat four species of quelonians (table 22.1), in the following order of importance: *P. sextuberculata* (66.6%, n = 447), *P. unifilis* (30.0%, n = 201), *G. denticulata* (2.8%, n = 19), and *Chelus fimbriatus* (0.6%, n = 4). Of 671 quelonians captured by community members, 655 (97.6%) were consumed, ten (1.5%) were sold, and six small individuals (0.9%) were kept to be raised in captivity.

TABLE 22.1 Species, Sex, and Number of Quelonians Consumed in Four Communities of the Jarauá Sector of the RDSM

SITE	Podocnemis sextuberculata			Podocnemis unifilis			Geochelone denticulata		Chelus fimbriatus		TOTAL
	M	F	N/D	M	F	N/D	M	F	M	F	
São Raimundo de Jarauá	210	130	46	84	85	8	3	3	1	1	571
Novo Pirapucu	12	37	0	0	10	0	0	0	0	0	59
Manacabi	4	6	0	0	7	0	2	0	0	1	20
Nova Colômbia	0	2	0	3	4	0	6	5	0	1	21
Total	226	175	46	87	106	8	11	8	1	3	671

Note: M is male; F, female; and N/D, not determined.

SPECIES AND SIZE

Two members of the Pelomedusidae family (*Podocnemis sextuberculata* and *P. unifilis*), one of the Chelidae (*Chelus fimbriatus*), and one of the Testudinidae (*Geochelone denticulata*) were recorded. There was variation in size and number of animals captured in each species. *Podocnemis sextuberculata* and *P. unifilis* showed sexual dimorphism in size, males being smaller than females (table 22.2).

CAPTURE METHODS AND SEASON

Of the 447 *P. sextuberculata* registered in the study, 363 were captured with gill nets in different aquatic habitats of the Paraná do Jarauá, and 45 females at nesting beaches. Of the 301 *P. unifilis* registered, 51 were captured with gill nets and 74 females were captured by probing in the mud of shallow lakes with a wooden pole. The eleven *G. denticulata* were captured by hand. One male and two female *C. fimbriatus* were captured with gill nets and one female with a harpoon (table 22.3).

Podocnemis sextuberculata is captured primarily during the dry season and when water levels begin to rise (start of the enchente) (fig. 22.1). Almost all size classes are captured during this period (fig. 22.2). During the nesting season in August, September, and October, individuals are captured by hand when they emerge to lay eggs on the beaches. Gill nets are used in the paranás and ressacas during temporary oscillations in the water levels in October and November and at the start of the flood season from December/January through March.

The size of the mesh influences the size classes of turtles captured with nets. Mid-sized males and females are most frequently caught with gill nets. This size

TABLE 22.2 Statistical Summary of Measurements of Quelonians Captured in the Jarauá Sector of the RDSM

| SPECIES | SEX | CARAPACE LENGTH (CM) | | | | WEIGHT (KG) | | | |
		X	S.D.	Range	N	X	S.D.	Range	N
P. sextuberculata	M	20.7	1.9	11.1–24.4	162	0.9	0.2	0.17–1.4	160
	F	20.9	4.7	12.4–31.2	90	1.1	0.7	0.2–3.8	78
P. unifilis	M	21.6	5.8	7.6–31.0	68	1.5	0.8	0.06–3.3	46
	F	32.9	10.2	8.7–46.0	74	4.4	3.5	0.105–14	16
G. denticulata	M	44.4	3.6	39.3–50.3	9			9	1
	F	42.3	3.9	36.6–46.5	9	8.9	1.0	7.55–10	4
C. fimbriatus	M			29.2	1			3.3	1
	F			40.5–43.0	2			13	1

Note: M is male; F, female; X, mean; S.D., standard deviation; and N, sample size.

TABLE 22.3 Methods Used to Capture Four Quelonian Species in the Jarauá Sector of the RSDM

METHODS	Podocnemis sextuberculata			Podocnemis unifilis			Geochelone denticulata		Chelus fimbriatus		TOTAL
	M	F	N/D	M	F	N/D	M	F	M	F	
Gill net	217	106	40	31	20	1	0	0	1	2	418
By hand	0	45	0	2	31	1	11	8	0	0	98
Jaticá	1	15	0	0	3	0	0	0	0	0	19
Wooden pole	0	0	0	46	23	5	0	0	0	0	74
Harpoon	4	7	0	2	16	0	0	0	0	1	30
Diving	0	2	1	0	0	0	0	0	0	0	3
Drag nets	3	0	0	0	0	0	0	0	0	0	3
Unknown	1	0	5	6	13	1	0	0	0	0	26
Total	226	175	46	87	106	8	11	8	1	3	671

Note: M is male; F, female; N/D, not determined.

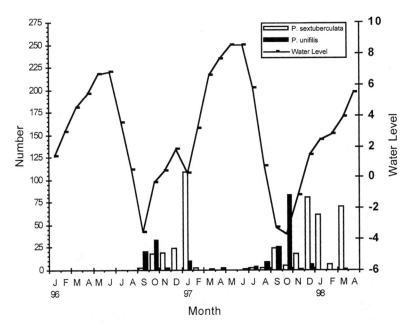

FIGURE 22.1 Capture of two species of *Podocnemis* in the Jarauá sector of the RDSM (Mamirauá Sustainable Development Reserve) with respect to water levels.

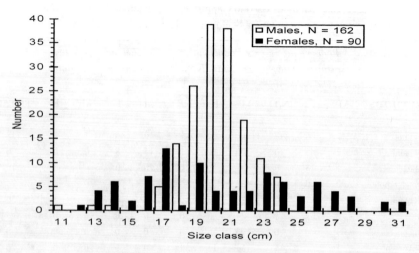

FIGURE 22.2 Size-Class distribution for *P. sextuberculata* captured in the Jarauá sector of the RDSM (Mamirauá Sustainable Development Reserve).

bias in net captures, which also occurs with the drag nets used in the remansos, has led to the popular local belief that the population contains many males and few females. Studies using Trammel nets indicated that gill nets used by fishermen are not efficient enough to capture the larger females, which are found in the deeper sections of the water channel where the current is weaker. As a result, a large portion of the adult population escapes capture.

During the dry season the *P. unifilis* population is concentrated in the canos and poças of the ressacas and lakes. In the 1996 dry season, adult males and females were the most common age class captured (fig. 22.3). Because of the concentration of the population in these habitats during the dry season, however, the potential exists for fishermen to capture animals of all sizes and of both sexes, as they did in October of 1997 (fig. 22.4).

Podocnemis unifilis is captured primarily with wooden poles (see below), with gill nets, by hand, and with harpoons. Capture with gill nets is occasional and occurs primarily when fishermen are seeking tambaqui fish (*Colossoma macropomum*). These nets have a stretched mesh size of 22 cm and do not capture hatchlings and juveniles.

Following the nesting season, *P. unifilis* females remain in canos and small pools with abundant macrophytes, burying themselves to a depth of about 20 cm in the mud until the water level rises again. Inhabitants of São Raimundo de Jarauá know this behavior, as well as the sites where *P. unifilis* can be found in the dry season from September to October. They have developed a searching technique that involves the use of a three-meter long wooden pole. They locate the buried turtle by the characteristic sound produced when the stick impacts on its carapace, then

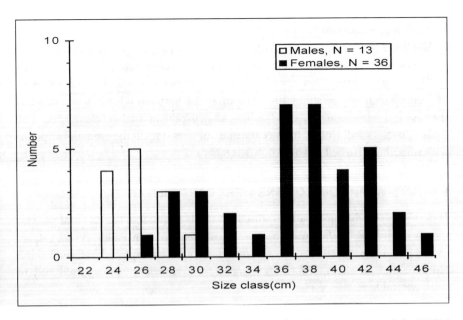

FIGURE 22.3 Size-Class distribution for *P. unifilis* captured in the Jarauá sector of the RDSM (Mamirauá Sustainable Development Reserve) from September 1996 through April 1997.

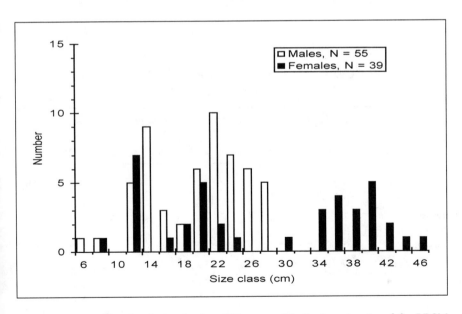

FIGURE 22.4 Size-Class distribution for *P. unifilis* captured in the Jarauá sector of the RDSM (Mamirauá Sustainable Development Reserve) from May 1997 through April 1998.

capture it by hand. Using this method, one community member captured twelve females in one day.

Additionally, individuals of *P. unifilis* swimming just under the surface of the water are identified by the size and shape of air bubbles visible at the surface and then captured with harpoon, *jaticá*, or by diving. A harpoon consists of a long pole with an iron tip secured to a strong line. The tip of the harpoon releases when it enters the prey but remains attached to the line, allowing the animal to be captured. The strike causes a small wound in the carapace but does not kill the animal. Harpoons with smooth, unbarbed tips are known as jaticás.

HABITATS WHERE QUELONIANS WERE CAPTURED

Podocnemis sextuberculata was captured primarily in ressacas (54.8%) and paranás (28.6%), while *P. unifilis* was captured most frequently in canos (37.8%), lakes (20.4%), and ressacas (18.9%). *Chelus fimbriatus* was also captured in three habitat types: paranás, ressacas, and canos. *Geochelone denticulata* was captured only in restingas (table 22.4).

DISCUSSION

HUNTING OF TURTLES

The three most sought-after genera of quelonians in Amazonia are *Podocnemis, Peltocephalus,* and *Kinosternon.* The genus *Podocnemis* includes six species, all ex-

TABLE 22.4 Habitats Where Quelonians Were Captured in the Jarauá Sector of the RDSM

HABITAT	P. sextuberculata		P. unifilis		Chelus fimbriatus		G. denticulata	
	N	%	N	%	N	%	N	%
Unknown			15	7.5			2	10.5
Ressaca	245	54.8	38	18.9	1	25.0		
Paraná	128	28.6	7	3.5	2	50.0		
River-Beach	43	9.6	12	6.0				
River-Remanso	15	3.4						
River-Enseada	7	1.6	8	4.0				
Lake	9	2.0	41	20.4				
Cano			76	37.8	1	25.0		
Restinga							17	89.5
Island			4	2.0				
Total	447	100.0	201	100.0	4	100.0	19	100.0

ploited to varying degrees depending on local preference, but *P. expansa* and *P. unifilis* are the most favored for the preparation of several dishes considered to be delicacies (Alho 1986).

The capture of quelonians in the RDSM primarily provides food for local consumption. When *P. sextuberculata* and *P. unifilis* are captured in large quantities by professional fishermen outside of RDSM, however, it is always for commercial (illegal) purposes. These two species are also eaten in appreciable quantities in Itacoatiara and other cities near Manaus, the capital of Amazonas State, with *P. unifilis* the most frequently consumed species (Smith 1979a; Santos 1996).

In the RDSM unsustainable exploitation of the Amazonian turtle (*P. expansa*) has virtually eliminated the species from the area. Quelonian hunting therefore focuses primarily on two species, *P. sextuberculata* and *P. unifilis*, which make up 96.6% of captures. Similarly, quelonian extraction in Jaú National Park focuses on three species: *Peltocephalus dumerilianus*, *Podocnemis unifilis*, and *P. erythrocephala*, which represent 95% of captures (Rebêlo and Lugli 1996). In Cinaruco-Capanaparo National Park, Venezuela, 100% of captures are of three species, *P. unifilis*, *P. vogli*, and *P. expansa* (Thorbjanarson, Perez, and Escalona 1997). In all three locations, populations of *P. expansa* are small, with isolated females nesting only when river levels are low. According to Rebêlo (1985), the 261 quelonians confiscated in Manaus and on the Purus, Negro, and Uatumâ rivers included four commercial species: *P. expansa*, *P. unifilis*, *P. sextuberculata*, and *Peltocephalus dumerilianus*. Of these four *P. sextuberculata* was the species most frequently sold on the Purus river (50%) and *P. unifilis* the most frequently sold in Manaus (63%).

SIZES OF CAPTURED TURTLES

All size classes of *P. sextuberculata* are affected by fishing because fishermen use a mesh size 10 or 15 cm in length. Furthermore, the entire animal is cooked, so consumption is independent of the size of the animal. On the other hand, capture of *P. unifilis* focuses primarily on adults, with females being the most affected. A similar pattern was observed by Thorbjarnarson, Perez, and Escalona (1993) on the Capanaparo River, Venezuela, where adult males and females are the most frequently captured size classes.

The smallest reproductive female *P. sextuberculata* recorded in this study, captured while egg laying at a beach on the Paraná do Manacabi, measured 26 cm. Vanzolini (1977) examined eleven *P. sextuberculata* from three rivers in the Brazilian Amazon and found that the smallest female containing eggs measured 27.1 cm. If 26 cm is the minimum reproductive size for the species, then 19% (n = 17) of the ninety females for which we obtained a carapace length were sexually mature. The proportion of adult females in samples is probably higher, given that we were unable to measure carapace length in 49% (n = 95) of the 175 captured females.

The mean size of *P. unifilis* females captured within the RDSM was smaller ($\bar{x}$ = 32.9 ± 10.2 cm [mean ± sd.], n = 74) than that reported by Smith (1979) for Itacoatiara, Amazonas, Brazil, ($\bar{x}$ = 35.5 ± 5.5 cm, n = 15) or by Thorbjarnarson, Perez, and Escalona (1993) for the Capanaparo River in Venezuela ($\bar{x}$ = 33.1 ± 3.3 cm, n = 109). The largest sizes documented by any study for captured females are those reported by Fachín-Terán, Chumbe, and Taleixo (1996) for Pacaya-Samiria National Reserve in Peru ($\bar{x}$ = 41.0 ± 3.0 cm, n = 145).

Of ten individuals of *P. unifilis* examined by Vanzolini (1977) in Brazil, the smallest female with eggs measured 33 cm. If this is the minimum reproductive size for the species, then 64% (n = 47) of the seventy-four females captured during this study for which carapace length was measured were above this size. This proportion may be somewhat larger because we could not measure carapace length for 30% (n = 32) of the 106 females captured. This number is still a smaller proportion than that reported by Thorbjarnarson, Perez, and Escalona (1993), who found that 82.6% of 109 females measured on the Capanaparo River in Venezuela were of reproductive age. An even higher value was obtained by Fachín-Terán, Chumbe, and Taleixo (1996) for Pacaya-Samiria in Peru, where 95% of females eaten in the communities that border the reserve were of reproductive age. These differences are due to the fact that animals of all sizes were captured at RDSM, while in the other studies, which report only large individuals of reproductive size, the turtles were apparently captured at the nesting beaches.

Mean carapace lengths of *Geochelone denticulata*, both for males ($\bar{x}$ = 44.4 ± 3.65 cm, n = 9) and females ($\bar{x}$ = 42.3 ± 3.97 cm, n = 9), were larger than those reported by Fachín-Terán, Chumbe, and Taleixo (1996) for Pacaya-Samiria in Peru, where mean carapace length for ten males was 36.5 ± 5.6 cm and for nine females 32.4 ± 11.3 cm. Capture of *G. denticulata* in the RDSM is occasional and can occur at any time of year.

CAPTURE METHODS AND SEASON

Podocnemis sextuberculata is primarily captured with gill nets. Fishermen near Itacoatiara use the same method, leaving their nets up overnight and capturing turtles as well as the targeted fish (Smith 1979a). In the dry season reproductive females are captured at night on the beaches of the Japurá and Solimões rivers in the RDSM. The same occurs near Itacoatiara (Smith 1979a).

Podocnemis unifilis is captured during the dry season in canos, lakes and ressacas, primarily using the wooden pole method and gill nets. A different pattern was reported for this species in the Itacoatiara area, where it is captured year round with *espinhel* (long line with multiple baited hooks) and with harpoons. Most captures occur during the season of rising water levels (May and June) when *P. unifilis* moves into the flooded forest to feed on flowers and fruits (Smith 1979a). The long line was also the method most frequently used to capture turtles on the Capanaparo river in Venezuela (Thorbjarnarson, Perez, and Escalona 1997). Harpoons are also used on the Orinoco river, Venezuela, to capture *P. expansa* during the rainy season (Ojasti 1971) and in Belize to capture *Dermatemys mawei* while it floats on the surface of the water (Moll 1986).

Podocnemis unifilis is also captured by hand at night when the females emerge to lay eggs on islands and lake margins (Smith 1979a). In the RDSM capture by hand is also accomplished at night during the egg-laying season when females emerge on the beaches of the Japurá and Solimões rivers.

Data gathered from twenty-one *extrativistas* (subsistence forest dwellers) from the lower and middle Jaú river, who capture primarily *Peltocephalus dumerilianus*, *Podocnemis unifilis*, and *P. erythrocephala*, suggest that 64% of stocked turtles were attracted with a bait of fish and captured with jaticá, 20% were caught with several types of baited hooks (float line, rod and line, and long line), and 15% were captured on land by ambushing (*viração*). *Geochelone denticulata* was captured during incidental encounters (Rebêlo et al. 1996). Polisar (1995) reported three techniques used to capture *Dermatemys mawei* in Belize: harpoon, net, and diving. Free diving is the most efficient of these techniques. When well organized, diving can lead to the almost complete removal of turtles from an area.

Although *Geochelone denticulata* is a preferred diet item, this species is captured only occasionally because, not nesting communally, they are usually hard to find. Santos (1996), however, reported an unusual incident whereby in one flood season one person captured about sixty animals at a small settlement in the Barroso sector of the RDSM. The water level that year was extremely high and the normally terrestrial tortoise was easy to spot floating in the flooded forest.

Chelus fimbriatus is only occasionally captured and is of less importance in the local diet than are *Podocnemis* turtles (Smith 1979a). Fachín-Terán, Chumbe, and Taleixo (1996) observed the same ranking in Pacaya-Samiria National Reserve in Peru, and this finding was confirmed for the RDSM in this study, in which *C. fimbriatus* was the least captured species. *Chelus* is uncommon in RDSM. Additionally, because of its strong musky odor, most people do not find it an attractive food item.

HABITATS WHERE QUELONIANS WERE CAPTURED

In the RDSM *P. unifilis* uses several microhabitats for reproduction but relies on the beaches of the Japurá River less than on reproductive sites on the margins of its lakes, ressacas, and canos. The majority of the population remains in these habitats until the next flood, making them vulnerable to predation by humans. In other portions of its range, such as Pacaya-Samiria in Peru and the Guaporé and Trombetas Biological Stations in Brazil, the species nests primarily on sand beaches that emerge along river edges, and for this reason females are captured more frequently than males (Fachín-Terán 1992; Soini and Coppula 1995; Soini and Soini 1995; Fachín-Terán, Chumbe, and Taleixo 1996).

Podocnemis unifilis uses lakes, ressacas, and the flooded forest during the enchente, while in the dry season the majority of the population remains buried in the mud of canos and pools that form in the ressacas. Such estivation behavior has not previously been reported for the species. One of us (RCV) observed this same behavior in December of 1990 when he collected forty-three specimens of *P. unifilis* at the Trombetas Biological Reserve. Estivation thus appears to be common but less predictable and probably less documented in areas with short dry seasons (Gibbons, Greene, and Congdon 1990). According to Vaillant and Grandidier (1910) and Tronc and Vuillemin (1973) (cited in Kuchling 1988), most *Erymnochelys* bury themselves in the mud during the dry season, even though the habitats they occupy are not completely dry.

CONSERVATION AND MANAGEMENT

Following the decline of *P. expansa* populations in Peru and Brazil, more pressure was placed on *P. unifilis*, *P. sextuberculata*, and *Peltocephalus dumerilianus* (Fachín-Terán, Chumbe, and Taleixo 1996; Vogt and Soini in press). The progressive substitution of large, valuable species by smaller species was confirmed in the RDSM. Here, *P. unifilis* and *P. sextuberculata* are more frequently consumed, especially the latter species, which is captured in large quantities by professional fishermen at different locations of the reserve and sold in the cities of Tefé and Alvarães. This trend in the exploitation of Amazonian turtles is reminiscent of that observed in the whaling industry, leading Mittermeier (1975) to characterize the aquatic quelonians of Amazonia as suffering from the whaling syndrome.

Factors contributing to the decrease in turtle populations in the reserve include lack of protection of nesting sites; capture of adult females and overharvesting of their eggs; artisanal and commercial fisheries in paranás, ressacas, canos, and lakes by both community dwellers and those commercial fishermen who carry out their activities within the area of the reserve; the commercial demand in urban centers; lack of control of illegal trade in urban centers; and loss of nests due to the repiquete. However, the most predatory hunting method, which causes the greatest harm to the population and which poses the greatest threat to reproductive animals, is the use of drag nets by professional fishermen and some community dwellers who capture large numbers of *P. sextuberculata* in river remansos. Using this method, one community member captured 130 individuals of the species in August of 1996 in the remanso of Praia Machado on the Solimões River. In 1997, in the same area and season, we only captured two males and two females during forty-eight hours of sampling. Similarly, Ojasti (1971) observed fishermen using drag nets to capture *P. expansa* in remansos of the Orinoco River.

To initiate the recuperation of quelonian populations in the reserve, a protection and management program accompanied by an environmental education program must be developed in the short term and with the agreement of the communities. Community leaders have already agreed during assemblies to forbid the use of gill nets, fence or encirclement nets (*redinhas*), and *arrastadeiras* (drag nets used along beaches) near turtle nesting sites. They also identified lakes and nesting sites that should be preserved in different locations of the reserve by prohibiting the capture of turtles, eggs, and hatchlings.

It is a priority to implement these agreements with the full participation of community members, who would thus help to protect the resource and would no longer represent a threat to the survival of these reptiles. According to Rebêlo and Lugli (1996), only the active participation of local inhabitants in the planning and implementation of a quelonian management plan will guarantee the success of the plan, as local peoples are both the problem and the long-term solution for large areas that require permanent inhabitants to protect them. Additionally, the Brazilian Agency for the Environment and Renewable Natural Resources (IBAMA) should invest greater effort in inspecting and monitoring turtle sellers and intermediaries in the trade.

To reduce the intentional capture of turtles, which is illegal in Brazil, it is necessary to apply our knowledge of the species' ecology (Alho 1985). There is a proven synchronization between the flood regime and nesting behavior. Therefore, in addition to protecting nesting and breeding sites, it is possible and essential to protect the migrating population and the migratory routes themselves as turtles move out of their aquatic habitats. This is the time when the population is most vulnerable to gill nets as they move through narrow river channels. In the river the reproductive population is vulnerable to drag nets because turtles are located in the remansos near the beaches. Hunting in these locations and seasons will negate any benefit derived from turtle protection in water bodies.

Captive breeding has been suggested as a way of minimizing illegal harvest of turtles in Amazonia (Alho 1985). Captive breeding may be a temporary salvation for some species that have reached the point were they are unable to survive in the wild. However, it is not recommend to spend money raising species in captivity when the funds are better spent on measures to prevent their extinction in the wild (Magnusson 1993). In várzea areas, where it is difficult carry out captive breeding for both socioeconomic and ecological reasons, other alternatives such as management in the wild should be implemented in agreement with local communities.

Podocnemis sextuberculata nesting sites in the Japurá and Solimões rivers should be identified and permanently protected, and the capture of the species should be temporarily forbidden. Some beaches should be completely protected, while others are managed by dividing them half and half into a protected section and a section that can be used for egg collection by local inhabitants. The lower beaches where nests are bound to fail should be considered as harvestable beaches. These actions should be coordinated between the personnel of the Mamirauá Project and the local communities. If natural predation and egg loss due to the repiquete and capture of juveniles are controlled, then high survivorships may be obtained.

Unlike *P. sextuberculata*, whose nests are concentrated on sandy beaches, *P. unifilis* nests on dispersed sites and uses a variety of substrates. Therefore it is crucial to identify *P. unifilis* nesting sites within the reserve and to protect them permanently from human interference during the dry season. Nest loss due to the repiquete should be minimized by translocation of eggs to sand beaches on the Japurá and Solimões rivers. Hatchlings should then be released at the site where eggs were collected.

Because of the risk of extinction faced by *P. expansa* in the reserve, all surviving individuals must be protected until the population recovers, a period that may occur in 80 to 120 years. Initially, the recovery of P. *expansa* populations can be aided by total protection of nesting beaches to prevent predation by humans and translocation of nests that are in danger of flooding by the repiquete. It may also be necessary to release into the reserve hatchlings of this species collected on nesting beaches farther upstream in the Solimões and Japurá rivers. The release of five to ten thousand PIT (Passive Integrated Transponder)-tagged hatchlings per year during ten years would serve as an experiment to determine the effectiveness of this technique.

Protective measures should be implemented year round. For now, we lack demographic data to determine whether areas such as the RDSM are sufficient to conserve quelonians in the Amazon basin (Santos 1996). The most basic information about the minimum area required to protect *Podocnemis* turtles is still unknown. In the Trombetas Biological Reserve all P. *expansa* nesting beaches are protected. However, once the reproductive season is over, the turtles move at least 65 km downstream to feeding sites, where they are often captured by fishermen (Moreira and Vogt 1990). In the RDSM the opposite situation exists: feeding areas are protected, but little effort is put into protecting nesting sites. Still, a new factor may soon alter this scenario. The Amanã Sustainable Development Reserve was created in 1997. Together with Jaú National Park and the RDSM, it makes up the Central Amazonian Ecological Corridor, which may enable the maintenance of genetically healthy populations with increasing recruitment rates.

This study shows that commercial sale of turtles is low in the communities. However, there is a substantial trade in P. *sextuberculata* in Tefé. Over 300 animals are sold every eight to ten days during the season of vazante and the start of the enchente. It is interesting to note that most animals sold are male P. *sextuberculata*. Capture for commerce takes place in the ressacas and remansos, primarily with gill nets and drag nets; the latter capture nearly all of the reproductive population. This finding was confirmed by our field observations, when in a remanso of the Piranhas beach, on the Solimões river, we captured 132 males and 19 females in forty-eight hours of sampling.

Individuals of P. *unifilis* from the Tefé, Japurá, and Juruá rivers are sold in Tefé. The few female P. *expansa* that emerge to lay eggs on the beaches of the Solimões and Japurá rivers are also captured for sale. Johns (1987) reported that local inhabitants of Tefé estimated that 300 P. *expansa* are sold annually. This number may be an underestimatation because some turtles are transported directly to Manaus. Santos (1996) recorded 400 P. *sextuberculata* unloaded at the Tefé market on one occasion. Smith (1979) estimated that 8,000 P. *unifilis* were captured annually in a 60-km radius around Itacoatiara; of these, about 6,000 were unloaded in Tefé and half of them were eventually sent to Manaus. Cooked *Podocnemis expansa* were still being served openly in restaurants in Tefé in September 2003 as part of the local noon luncheon buffet, along with other wild game (R. C. Vogt pers. obs.)

To preserve and make rational, sustainable use of this important resource, the following studies are needed in the short term: quantify and describe the *P. sextuberculata* and *P. unifilis* trade in Tefé and Alvarães; map the nesting beaches and their level of exploitation on the Japurá and Solimões rivers; and evaluate and monitor *P. unifilis* and *P. sextuberculata* populations in all of the RDSM. According to Calouro (1995), the effects of hunting on animal populations are not easily quantified because one must estimate both the hunting pressure to which the populations are exposed and their basic population parameters. In parts of the RDSM where turtles are captured by riverine communities and where commercial fisheries still take place, a long-term study is required to compare the population structure and densities in hunted and unhunted areas. This will allow us to determine whether populations decline because of exploitation and whether current hunting patterns are efficient. This information will serve as a base from which we can design strategies for the recovery and management of turtles in the RDSM, so that they may continue to act as a source of food for its inhabitants.

ACKNOWLEDGMENTS

We are indebted to Marcio Ayres and the Instituto de Desenvolvimento Sustentável Mamirauá and their staff for use of facilities and logistical support. A tremendous thanks to Marcio for his enthusiasm and interest in encouraging us to initiate and continue with our turtle work in IDSM. Miriam Elenit Lima de Fachín is thanked for her astute suggestions on the content of the paper and keeping Augusto content through the traumatic writing phase of this project. Other biologists, José Gerley, Juarez Pezzuti, Mercês Bezerra, Masidonio Pinho de Carvalho, and Ozimar Neves da Silva are alternately thanked and cursed for the help they gave us while collecting the data. This study was financed by the Comunidade Econômica Europea, Overseas Development Administration, and CNPq/MCT. El Pato of Renascer is thanked for keeping us sane during our stay in Tefé. Richard Vogt—Pit Bicha—it was very funny during the field work. Fachín-Terán received a doctoral scholarship from CAPES during his tenure in the Departamento de Ecologia do INPA. Will Crampton provided the definitions of várzea habitat types and water cycles on p. 364.

23

Fisheries, Fishing Effort, and Fish Consumption in the Pacaya-Samiria National Reserve and Its Area of Influence

SALVADOR TELLO

Fisheries in the Peruvian Amazon constitute an important source of animal protein and of income for *ribereño* people (local inhabitants who live alongside rivers in várzea ecosystems). Fish is the principal component of family diets, and the overall fisheries yield is approximately 80,000 tons per year (Bayley et al. 1992), representing a contribution of about US$ 80 million per year to the regional economy. In the Peruvian Amazon the largest volumes of fish originate in the lower reaches of the Ucayali and Marañon rivers, where the Reserva Nacional Pacaya-Samiria (RNPS) is located. Named for the two rivers that run through its territory, the RNPS has a surface area of 2,156,770 ha and covers portions of the provinces of Loreto and Requena in the Loreto region.

Despite the importance of fish in the region, to date we lack detailed information that will allow us to answer key questions about the factors that affect the fisheries. The current study contributes information necessary to improve planning and orientation of sustainable use of fish resources. It does not attempt to present all information pertinent to the fisheries linked to the reserve because that would require data on captures, effort, and consumption for communities located in the interior of the reserve and in the buffer zone. Although this study obtained data on overall fish consumption in several ribereño communities, more information is needed on fish consumption patterns and preferences.

METHODS

CATCH AND CATCH PER UNIT EFFORT

Information was obtained by means of direct interviews carried out between April 1994 and February 1995. Four experienced field assistants were hired and trained as data collectors in the cities of Iquitos, Requena, and Nauta. Additionally, the volume of fish exported from the RNPS and its buffer zone to Iquitos, Pucallpa, and Yurimaguas were estimated on the basis of information provided by the Ministry of Fisheries, whose technicians collect daily information on the provenance of fish landings at these cities.

To obtain information on amount of fish caught, fish landed at the ports were recorded daily according to their method of preservation, species, and source area. Weights of fresh salted and dry salted fish were converted to fresh weight using the factors 1.8 and 2.5, respectively. The number of trips, number of fishermen in the trip, number of days spent fishing, and number of casts (line or net) were recorded to estimate fishing effort for the commercial fishery fleet in Iquitos, Nauta y Requena. Data were collected daily by the field assistants as each vessel arrived in port when fishermen opened their coolers to sell their catch. This procedure enabled identification of the catch to species.

Information was also collected on a daily basis on cargo and passenger vessels that cover the routes to Iquitos, Pucallpa, Yurimaguas, Requena, and Nauta. These vessels mainly transport salted fish, and they carry a detailed registry of the amounts transported.

INTERVIEWS TO DETERMINE FISH CONSUMPTION

In addition to the sampling in the ports, in the cities of Nauta and Requena we carried out household interviews to determine the amount of fish consumed. In Nauta (8,548 inhabitants) interviews were carried out daily during the study period because of the population size, while in Requena (14,690 inhabitants) interviews were carried out ten days per month, with the days selected using a random numbers table. Data on fish consumption in Iquitos were already available from government sources.

Data collection was designed using street maps at scales of 1:500 and 1:2,500 for Nauta and Requena, respectively. Each household was numbered on the maps. Since each map consisted of three sheets that divided the cities into three zones, we considered these divisions as blocks for statistical analyses. These blocks correspond to zones located at differing distances from the river; in each zone households are also characterized by different socioeconomic indicators. In each block we randomly selected houses, allowing for replacement houses in case we were unable to complete interviews at the selected houses.

LOCATION OF FISHING AREAS

The origin of each catch was noted on the data sheets, allowing us to locate precisely fishing areas using a 1:400,000-scale map. We measured distances from the fishing areas to Iquitos on the Fotocarta Nacional (scale of 1:100,000) prepared by the Instituto Geográfico Nacional (National Geographic Institute). To classify fishing areas as within the reserve, within the buffer zone, or outside the buffer zone, we used as a reference the RNPS Master Plan (Plan Maestro de la RNPS; COREPASA 1986), which defines the buffer zone as a 5-km wide strip around the protected area. Data on capture and fishing effort where analyzed with FoxPro 2.6 (Microsoft Corporation 1989-1994), while data on fish consumption were analyzed with StatPac (Statistical Analysis Package, version 5.2; Walconick 1985).

RESULTS

CHARACTERISTICS OF THE FISHERY

Three classes of fishery can be distinguished in the study area, differentiated by the final destination of the catch. Subsistence fishing is carried out by ribereños as a daily activity for self-sustenance. They use canoes or small boats, along with nets, primarily gill nets, of various mesh sizes and lengths. During seasons when fish are plentiful, the uneaten excess catch is salted, dried, and sold to an intermediary, who accumulates fish to supply markets in larger cities such as Iquitos, Pucallpa, and Yurimaguas. Dry salted fish is carried by both cargo and passenger vessels that cover these routes.

Local commercial fishing is carried out daily by fishermen from medium-sized cities, such as Requena and Nauta, to supply these urban centers with fresh fish. Fishermen have moderate-sized boats powered by 9 to 16 HP engines known as *peque-peque*; these are stationary motors adapted with a long extension for navigation in shallow areas. They use *honderas*—medium sized nets with a stretched mesh size of two inches—and frequently depart on fishing trips in the afternoon, returning at sunrise on the following day.

Regional commercial fishing supplies fresh fish to large cities such as Iquitos, Pucallpa, and Yurimaguas. The fishing fleet is comprised of vessels of varying size and design which frequently fish at large distances from the port of origin and cross regional boundaries. Vessels use very large hondera nets of two-inch stretched mesh size and are equipped with iceboxes, allowing them to remain out of port for an average of 30 days.

STATISTICS ON FISH LANDED

A total of fifty species of fish were sold in the markets of Iquitos, Pucallpa, Yurimaguas, Requena, and Nauta (table 23.1). The number of species is actually larger because often a single name is used for more than one species of fish. Five species essentially sustain the fisheries, making up 80% of the catch. The *bo-quichico* (*Prochilodus nigricans*) alone makes up 40% of the catch (table 23.2).

In the Peruvian Amazon the *paiche* (*Arapaima gigas*) can be legally sold from April to September, while its sale is prohibited from October to March. Despite this restriction, fishermen find ways to land their catch in places of difficult access

TABLE 23.1 List of Fish Captured by the Commercial and Subsistence Fleets
in the Study Area

COMMON NAME	LATIN NAME	FAMILY	ORDER
Arahuana	*Osteoglossum bicirrhosum*	Osteoglossidae	Osteoglossiformes
Paiche	*Arapaima gigas*	Arapaimidae	Osteoglossiformes
Pez torre	*Acestrorhynchus* sp.	Characidae	Characiformes
Sábalo cola roja	*Brycon erythropterum*	Characidae	Characiformes
Sábalo cola negra	*Brycon melanopterus*	Characidae	Characiformes
Palometa	*Mylossoma* sp.	Characidae	Characiformes
Gamitana	*Colossoma macropomum*	Characidae	Characiformes
Paco	*Piaractus brachypomus*	Characidae	Characiformes
Paña	*Serrasalmus* sp.	Characidae	Characiformes
Sardina	*Triportheus* sp.	Characidae	Characiformes
Huapeta	*Hydrolicus scomberoides*	Cynodontidae	Characiformes
Chambira	*Raphiodon vulpinus*	Hemiodontidae	Characiformes
Yulilla	*Hemiodus* sp.	Hemiodontidae	Characiformes
Shuyo	*Erythrinus* sp.	Erythrinidae	Characiformes
Fasaco	*Hoplias malabaricus*	Erythrinidae	Characiformes
Boquichico	*Prochilodus nigricans*	Prochilodontidae	Characiformes
Yaraquí	*Semaprochilodus* sp.	Prochilodontidae	Characiformes
Ractacara	*Curimata* sp.	Curimatidae	Characiformes
Chío chío	*Psectrogaster* sp.	Curimatidae	Characiformes
Llambina	*Potamorhina altamazonia*	Curimatidae	Characiformes
Yahuarachi	*Potamorhina latior*	Curimatidae	Characiformes
Lisa común	*Schizodon fasciatus*	Anostomidae	Characiformes
Lisa	*Leporinus* sp.	Anostomidae	Characiformes
Turushuqui	*Oxydoras niger*	Doradidae	Characiformes
Bocón	*Ageneiosus* sp.	Ageneiosidae	Siluriformes
Saltón	*Brachyplatystoma filamentosum*	Pimelodidae	Siluriformes
Dorado	*Brachyplatystoma flavicans*	Pimelodidae	Siluriformes
Manitoa	*Brachyplatystoma vaillanti*	Pimelodidae	Siluriformes
Zúngaro alianza	*Brachyplatystoma juruense*	Pimelodidae	Siluriformes
Mota	*Gallophysus macropterus*	Pimelodidae	Siluriformes
Achara	*Leiarus marmoratus*	Pimelodidae	Siluriformes
Cunchi	*Pimelodella* sp.	Pimelodidae	Siluriformes
Doncella	*Pseudoplatystoma fasciatum*	Pimelodidae	Siluriformes
Tigre zúngaro	*Pseudoplatystoma tigrinus*	Pimelodidae	Siluriformes
Zúngaro mama	*Paulicea lutkeni*	Pimelodidae	Siluriformes
Shiripira	*Sorubim lima*	Pimelodidae	Siluriformes
Tabla barba	*Goslynea platynema*	Pimelodidae	Siluriformes
Maparate	*Hypophthalmus* sp.	Hypophthalmidae	Siluriformes
Shirui	*Hoplosternum* sp.	Callichthydae	Siluriformes
Carachama	*Pterygoplichthys multiradistus*	Loricariidae	Siluriformes
Shitari	*Loricariichthys* sp.	Loricariidae	Siluriformes

TABLE 23.1 (*Continued*)

COMMON NAME	LATIN NAME	FAMILY	ORDER
Corvina	*Plagioscion* sp.	Sciaenidae	Perciformes
Bujurqui	*Satanoperca jurupari*	Cichlidae	Perciformes
Tucunaré	*Cichla monoculus*	Cichlidae	Perciformes
Acarahuazú	*Astronotus ocellatus*	Cichlidae	Perciformes
Añashúa	*Crenicichla* sp.	Cichlidae	Perciformes

TABLE 23.2 Fresh Fish Captured by the Commercial
Fishery and Landed in the Study Area in 1994

SPECIES	TONS	% OF TOTAL
Boquichico	1,036	43.7
Llambina	284	12.0
Palometa	278	11.7
Chio-Chio	166	7.0
Lisa	124	5.2
Other	480	20.4
Total	2,368	100.0

where there is no legal control. On the basis of our daily samples, we calculated that sixteen tons of paiche were landed in Requena alone in 1994. Of this amount, 62% was caught in the reserve and 38% in the buffer zone.

We estimated the total catch landed by the commercial fishery in 1994 in Iquitos, Yurimagua, Requena, and Nauta at approximately 19,000 tons of fresh fish (table 23.3). Of this tonnage about 27% came from the interior of the reserve, 13% from the buffer zone, and 60% from sites outside the protected area. These results allow us to state unequivocally that the RNPS is an important area for the regional fisheries.

FISHING AREAS

The Iquitos commercial fleet fishes in different rivers, depending on the abundance of fish. During this study vessels operated in the Amazonas, Ucayali, Marañon, Napo, and Nanay rivers (table 23.4). Close to 57% of fish landed in Iquitos came from the lower Amazonas, making it the most important source of fish that year.

The commercial fishing fleets of Requena and Nauta fish with more frequency

TABLE 23.3 Fish Catch Landed in 1994

LOCATION UNLOADED	SOURCE AREA			
	RNPS Reserve	Buffer Zone	Outside Reserve	TOTAL
Iquitos	1,388	252	2,231	3,871
Requena	485	367	87	939
Nauta	284	58	91	434
Pucallpa	1,633	712	3,841	6,216
Yurimaguas	1,201	1,052	5,024*	7,277
Total	5,021	2,441	11,274	18,737

Note: Catch is in tons and is categorized by source area and location of sale. Fresh salted and dry salted fish are converted to fresh weight using the factors 1.8 and 2.5, respectively.
*Includes fish transported from Iquitos.

TABLE 23.4 Most Important Fishing Zones in the Study Area, Frequented by the Commercial Fleets of Iquitos, Requena, and Nauta

FISHING AREA	APPROXIMATE DISTANCE (KM)*	CATCH IN TONS	% OF TOTAL
Ucayali river		1,390	80.4
Requena	262	400	23.8
Tipishca	262	198	11.5
Curahuaytillo	260	83	4.8
Pucate	280	70	4.1
Carocurahuayte	316	69	4.0
Yuracocha	172	64	3.7
Contamanillo	256	62	3.6
Huarmi Isla	300	60	3.5
Montebello	580	60	3.5
Machín Tipishca	580	52	3.0
Marañon river		335	19.4
Sarapanga	152	131	7.6
Nauta	136	62	3.6
Shiriyacu	160	24	1.4
San Pablo Tipishca	212	10	0.6

*Distance from Iquitos

in the Ucayali and Marañon rivers, with the most important sites located in the RNPS and its buffer zone (table 23.4; fig. 23.1). The largest catches came from the main river channels, where fishermen take advantage of fish migrations.

The Ucayali is the most important river basin with respect to fish production,

FIGURE 23.1 Fishing zones on the Ucayali and Marañon Rivers in the RNPS (Reserva Nacional Pacaya-Samiria) study area.

providing 80% of captures recorded for the entire study area. At least 70% of the dry salted fish that were landed in Pucallpa originated in the Puinahua channel, a tributary of the Ucayali, which itself is the main route for the stocking and transporting of fish extracted from the reserve.

ESTIMATING CATCH PER UNIT EFFORT

Because there is a lot of variation in vessel type and fishing methods among and within the different fisheries, we calculated the coefficient of variation to select the most representative measure of effort among all the measures recorded (number of net casts, days spent fishing, number of crew members, total catch, etc.). For Iquitos number of fishermen was the best measure of effort because the index of kg/fisherman had the lowest coefficient of variation of all possible indices. Number of casts (number of times a net is set or thrown) was not considered as a reliable measure of effort because fishermen do not remember with any precision the numbers of times they used a net. Unfortunately, there is no system by which this information is registered during daily fishing tasks. The number of trips carried out also does not represent an adequate measure of effort because of the huge variability in the storage capacity of the different vessels that make up the Iquitos fleet.

For Requena and Nauta the choice of an index was more easily made because the boats, fishing methods, number of fishermen, and days spent fishing show little variation within the fleet. For this reason we chose number of trips as the unit of measure and kg/trip as the index of catch per unit effort.

On the basis of the selected indices of catch per unit effort, the greatest fish abundance for the Iquitos fleet occurred in May and June, and the most important fishing sites were in the lower Amazonas river (table 23.5). For Nauta and Requena the largest catches occurred in August and September, coinciding with the highest indices of catch per unit effort (table 23.5). The largest catches occurred during the migrations when fish move out of the flooded area with the receding water levels and become concentrated in the main river channels.

FISH CONSUMPTION AND FISHERIES YIELD

Using as a reference data on fish consumption provided by the Instituto Nacional de Estadísticas (National Institute of Statistics; INEI 1972, 1993), we estimated the supply and demand of fish inIquitos. The INEI gives a per capita consumption of 20.4 kg of fish per year, which multiplied by the population of that time (225,000 inhabitants), gives a demand of 4,590 tons of fresh fish per year. According to our study, an average of 354 tons of fish are landed each month in Iquitos, representing a total of 4,250 tons per year. Commercial fishing thus satisfies 92% of the fish demand, with the remaining 8% supplied by subsistence fishing carried out near Iquitos, an activity not measured in this study.

To analyze supply and demand in the RNPS and its buffer zone, we estimated the per capita consumption of fish in Requena and Nauta based on our interview

TABLE 23.5 Mean Catch Per Unit Effort (CPUE) Per Boat During 1994 for Commercial Fisheries

| | **MONTHS** | | | | | | | | | |
SITE	Apr.	May	June	July	Aug.	Sept.	Oct.	Nov.	Dec.	Jan
Iquitos[a]	540.0	556.0	578.0	536.0	524.0	537.0	639.0	542.0	504.0	522.5
Requena[b]	86.5	114.2	179.0	205.6	138.9	236.0	116.5	72.8	91.7	56.7
Nauta[b]	161.2	117.6	183.0	218.0	281.0	165.6	118.3	141.0	98.4	76.5

[a]CPUE = Kg/fisherman.
[b]CPUE = Kg/trip.

TABLE 23.6 Per Capita Fish Consumption in Requena and Nauta

	INDICE	
SITE	Gr/person/day	Kg/person/year
Requena	215	78.6
Nauta	327	119.5

data (table 23.6). For rural areas we used the value of 55.8 kg/person per year estimated by Tello (1993) and based on data collected by IIAP (Instituto de Investigaciones de la Amazonía Peruana) in nearly fifty communities in the RNPS. The estimated fish consumption in Requena and Nauta was 1,152 and 1,029 tons, respectively, while for the rural areas it was 2,905 tons, for a total of 5,000 tons.

Thus the total amount of fish landed by the commercial and subsistence fishery fleets at Iquitos, Pucallpa, Yurimaguas, Requena, and Nauta, together with the dry salted fish carried in passenger and freight vessels to the different cities, is nearly 7,800 tons, of which 66% originates in the RNPS and 34% in its buffer zone. If we add to this figure the amount consumed by local populations, we arrive at a total annual catch of more than 12,800 tons. On the basis of the above proportions, approximately 8,500 and 4,300 tons of fish per year are extracted from the RNPS and the buffer zone, respectively. To summarize, 12,800 tons of fish are caught in the RNPS and its zone of influence per year, 5,000 tons are consumed locally, and 7,800 tons are exported.

Using the percentages and amounts estimated by this study and combining them with fish consumption by local communities, we can estimate the total commercial catch for the area, including catches derived from areas outside of the RNPS and its influence zone, such as the Amazonas, Juanache, Tapiche, and other rivers. Figure 23.2 illustrates the dynamics of the fisheries and fish trade for the entire study area, including the amount of fish consumed in each community and the final destination of exported catches.

In his estimate of the yield of fisheries in the 520,000 km^2 of the Peruvian Amazon that lie under 270-m altitude above sea level, Hanek (1982) determined that 75% of the volume of the total catch is captured by subsistence fisheries and 25% by commercial fisheries. Using these proportions, if the approximately 19,000 tons of commercial catch estimated by the present study (fig. 23.2) represent 25% of the total catch, then subsistence fisheries contribute an additional 57,000 tons (75%), for a total of 76,000 tons caught in the study area. The fish yield calculated in this study (76,000 tons) is very similar to that of the 80,000 tons calculated by Bayley et al. (1992), who used as a reference for his calculations Hanek's 1982 value of 277 grams/day per capita consumption and a 3.1% growth rate of the population.

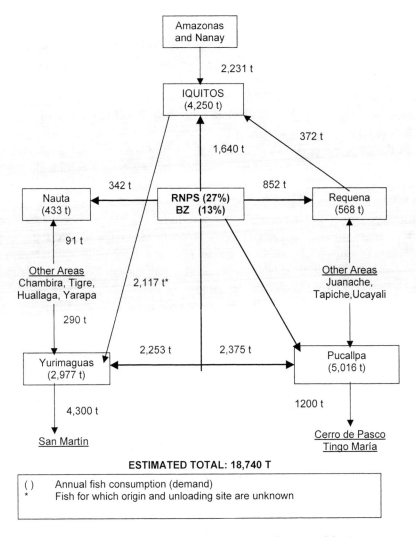

FIGURE 23.2 Diagram showing flux of catch landings. BZ = Buffer Zone of the Reserve.

DISCUSSION, CONCLUSIONS, AND RECOMMENDATIONS

In flooding river systems fish productivity, migrations, and populations dynamics are profoundly influenced by interactions between physical, chemical, and biological characteristics, including flooding regimes, extent of flooded area, water temperature, and pH. These factors are in turn influenced primarily by environmental and geographical issues, which themselves vary among sites and seasons. Therefore it is not an easy task to develop fisheries evaluation and management programs. Studies carried out in the last two decades indicate that the simplest approach is to obtain capture and effort data (Petrere 1978a,b; Welcomme 1985; Bayley and Pe-trere 1989; Montreuil and Tello 1990; Montreuil et al. 1997). These two parameters are used to calculate an index of catch per unit effort, which when recorded for a given length of time, allows an evaluation of the abundance of the resource (Rick-er 1975). One of the great limitations in the Peruvian Amazon is the lack of ade-quate data on catch per unit effort, making it difficult to compare our effort and production values with those of previous years. Also, it will be difficult to carry out future comparisons if this work is not continued.

In the catches landed in the study area, large species are being replaced with smaller species of high yield and low price because of increases in fishing intensity and changes in fishing methods. This selective process results from the preference by fishermen and consumers for large species in combination with the susceptibil-ity of large species to high mortality levels from fishing (Welcomme 1985). When fishing efforts intensify, larger species are overexploited and progressively replaced in biomass by smaller, shorter-lived species that are better able to withstand inten-sive extraction due to their high production/biomass indices (Regier and Hender-son 1973; Turner 1985; Lowe McConnell 1987).

Changes in species composition have occurred in many flooded river systems. Novoa (1989) noted a reduction in the mean size of catfish and in the proportion of captures made up of large catfish in the Orinoco. Similarly, Bayley and Petrere (1989) found evidence of the disappearance of such large species as *Colossoma macroponum* and *Arapaima gigas* in catches of the Manaus fishing fleet, in Brazil.

Despite these changes in catch composition and the size reduction noted for some species, I agree with Bayley (1992) in the assessment that, as long as there are no drastic changes in environmental conditions (caused by the construction of dams, for example), fish resources in the Peruvian Amazon will not fail, and the human population will continue to benefit from them. The continued decrease in fishing stocks is caused more by indirect human activities than by fishing activities. For example, deforestation leads to significant changes in aquatic systems, includ-ing increases in daily temperature fluctuations; higher turbidity because of sedi-ment outflow from the clear-cut area; decrease in areas available for dispersal, shel-ter, and reproduction; and abnormal changes in the water levels. All of these conditions negatively affect the development of fish populations.

Significant catches result when fishermen take advantage of fish migrations. Lo-cal fishermen indicate that the migrations that sustain the local commercial fish-eries in Nauta and Requena frequently originate inside the RNPS. This finding again highlighting the ecological importance of this area. There are few studies of migrations in the Peruvian Amazon. We know that there are two main types: one for reproduction and one for trophic purposes. Reproductive migrations take place in relatively short periods, and their importance for fish catches is minimal. On the other hand, migration in search of feeding areas can be lengthy—Goulding (1979) calculates that some species cover 450 kilometers—and are more important for fish supplies in the study area. Water level is the most important of several key environmental factors influencing the migratory behavior of tropical species. Fishermen believe that rainfall and lunar cycles also affect the timing of characin migrations (Goulding 1979). Information on these migrations, along with records of the hydrological regime and monthly precipitation at strategic sites on the Ucayali, Marañon, Pacaya, and Samiria rivers, would be of great value for the sustainable management of the fisheries in the study area.

ACKNOWLEDGMENTS

I would like to extend my most sincere thanks to Drs. Peter Bayley and Humberto Guerra, for their reviews and contributions to the document on which this paper is based. I would also like to thank the biologists Rosario del Aguila, Aurea García, Ronald Rodríguez, and Enrique Chalco, along with the Regional Fisheries Directorates of Loreto, Ucayali, and Yurimaguas, without whose participation and support it would not have been possible to carry out this study.

24

Implications of the Spatial Structure of Game Populations for the Sustainability of Hunting in the Neotropics

ANDRÉS J. NOVARO

Harvest theory has been built almost entirely on assumptions of uniformly distributed populations (Beddington and May 1977; Walters 1986; Caughley and Sinclair 1994). Most natural populations, however, are spatially structured, and this structure has profound effects on the dynamics of the populations and, consequently, on their responses to hunting (Kareiva 1990; Hanski and Gilpin 1997). Harvest theory therefore must be revised to incorporate spatial structure, and the resulting models need to account for the spatial heterogeneity of populations and their environments.

The spatial heterogeneity of hunted game populations in the Neotropics has begun to be analyzed explicitly only in recent years (Hill and Padwe 2000; Novaro, Redford, and Bodmer 2000). This lack of previous analyses may be because of the lack of comprehensive demographic data on game populations and in particular on the spatial variation of hunting in this region. The importance of spatial heterogeneity of hunting also has been recognized for wildlife populations in other regions where spatial data are scarce: African forests (Fimbel, Curran, and Usongo 2000; Hart 2000) and savannas (Owen-Smith 1988). Spatial heterogeneity has long been suggested, although not incorporated into harvest models, as a relevant factor in the dynamics of game in regions where data are more abundant (Pyrah 1984; Bergerud 1988; Knick 1990; Slough and Mowat 1996).

In this paper I first review the processes that can structure populations in ways that are relevant to the sustainability of game hunting. Then I discuss the approaches that are being used to account for the spatial structure of game when evaluating hunting sustainability and consider some of the limitations of these approaches. Finally, I suggest some topics for research, discuss difficulties of identifying game spatial structure, and present a model to incorporate the effects of one type of spatial structure on the population dynamics of game when demographic data are scarce.

SPATIALLY STRUCTURED POPULATIONS

Game populations can be spatially structured naturally or as a result of human activities. Nonhuman factors that structure populations range from naturally fragmented habitats to spatial heterogeneity in habitat quality that produces local differences in reproduction and/or mortality rates. These local differences in turn may be due to interspecific interactions (including availability of food, competitors, predators, and pathogens) and availability of refuge or breeding sites (Pulliam 1996; Tilman and Kareiva 1997). Human activities that can structure populations, on the other hand, include habitat transformation, degradation, and fragmentation (Hanski and Simberloff 1997) and hunting itself when it is spatially heterogeneous (Novaro, Redford, and Bodmer 2000).

Whether caused by natural or human-related processes, a wide range of patterns of spatial structure in populations has been described (Kareiva 1990; Hanski and Gilpin 1997; Tilman and Kareiva 1997). There is a continuum of probable patterns of spatial structure between totally structured and continuous populations. Spatial structure can involve discrete, more-or-less isolated subpopulations separated by an unsuitable matrix and connected by dispersal, or continuous populations without discontinuities in distribution but with spatially specific demographies (Hanski and Simberloff 1997). In the former case a series of subpopulations may function as a metapopulation of different types(e.g., classical or mainland-island) if dispersal allows for recolonization of habitat patches where subpopulations have gone extinct. If the rate of dispersal is high, the subpopulations are likely to function as a population with a patchy distribution (Harrison 1991). If dispersal is low and does not allow for recolonization, the structure is termed a nonequilibrium metapopulation, and the fate of each subpopulation is independent of the others.

Continuous populations can also be heterogeneous and have spatial structure if there are habitat-specific differences in mortality or reproduction. Both in the case of discrete and continuous populations, differences in mortality and reproduction between habitats or patches can determine that rates of local population increase (r; in the absence of immigration) may be larger or smaller than zero. Local populations with r's larger and smaller than zero have been termed sources and sinks, respectively (Pulliam 1988; Hanski and Simberloff 1997). In the absence of migration from nearby sources, sink populations will go extinct. Local populations that have low productivity and densities that will decline in the absence of migration but that do not necessarily go extinct are called pseudosinks (Watkinson and Sutherland 1995). Finally, as one of the many possible combinations, the type of spatial structure in which discrete metapopulation patches act as sources or sinks (depending on their quality) has been termed a source-sink metapopulation (Hanski and Simberloff 1997).

Hunting of a population that has any of the types of spatial structure mentioned above (or any combination of different types) has implications for its dynamics and its probability of persistence. In the rest of the paper I will refer mostly to metapopulations and source-sink populations because they are two models that have been considered in more detail. It is important to bear in mind, however, that most real game populations are likely to have spatial structures that are combinations of these or other types of structures.

EVALUATIONS OF THE SUSTAINABILITY OF HUNTING AND SPATIAL STRUCTURE

Different studies have used diverse methods to evaluate the sustainability of game hunting or to propose ways to increase the likelihood of sustainability. Some have assumed homogeneous populations, particularly in areas such as the Neotropics where demographic and spatial data are scarce, whereas others have begun to consider the spatial structure of these populations. In this section I will describe the sustainability analyses that have been done to date, assuming different types of spatial structure, including no structure, and considering some of the limitations.

Until the early 1990s the sustainability of hunting was evaluated mostly using sustainability indices. These indices (reviewed by Robinson and Redford 1994) included population trends of hunted species and comparisons of age structures and hunting yields across time and space. They allowed only a measure of relative levels of sustainability (Bodmer and Robinson this volume). By including comparisons among sites, however, it became evident that there were marked differences in the intensity of hunting across space and that these differences could have implications for the regional dynamics of game populations.

During the 1990s several evaluations of hunting sustainability were published using sustainability models. These models are used to estimate population production at a site with data on densities and reproductive rates (Bodmer 1994) or to estimate maximum population growth with data on reproduction (Robinson and Redford 1991) and on survival (Slade, Gomulkiewicz, and Alexander 1998). Estimations of population production or growth are then compared to harvest rates that are estimated at a specific site in order to determine if harvest is sustainable. These models have been applied for the most part assuming that all of the recruitment into game populations came from reproduction within the harvested area. The models have been useful in many cases and have been applied widely (Robinson and Bennett 2000c; earlier studies reviewed by Robinson and Bodmer 1999). These authors and others, however, have found inconsistencies between model predictions and actual population trends or harvest rates, and many of them have attributed these inconsistencies to potential dispersal of game from nearby sources into harvested areas (Alvard et al. 1997; Robinson and Bodmer 1999; Hill and Padwe 2000).

Another approach to estimating hunting sustainability has been to apply the models mentioned above while attempting to incorporate the potential effect of dispersal from adjacent and presumed sources of game. This approach estimates the current harvest for a combined area that includes the harvested area plus adjacent, potential sources that may be producing immigrants to repopulate the harvested area. The current harvest for the combined area is then compared to the maximum population growth (calculated with Robinson and Redford's method) in order to evaluate hunting sustainability. This approach was used by Townsend (1995a) and by Hill and Padwe (2000) to evaluate hunting by the Sirionó in Bolivia and the Aché in Paraguay, respectively. The approach uses the models developed for homogeneous populations but assumes that hunting creates a system of sources and sinks. It then tries to account for this heterogeneity by extrapolating the harvest pressure to the entire source-sink area

One of the limitations of this approach is that if dispersal of game between the sources and the hunted areas can not compensate for intense hunting levels, local extinction can occur. Other limitations are that the nature and location of source areas are assumed, but not known, and that source areas may be different among game species, depending on the species's different dispersal abilities.

During the last decade Joshi and Gadgil (1991) and McCullough (1996) began to consider more explicitly the effect of spatial structure of populations on the sustainability of hunting. Joshi and Gadgil described a system of protected and hunted areas created by indigenous people in India through trial and error. The researchers conducted simulations to evaluate the population implications of the presence of unhunted sources of game. McCullough, on the other hand, proposed the implementation of a similar trial-and-error system to designate a mosaic of refuge areas within a continuous population (spatial control). By monitoring the total harvest and by changing the number of hunted and refuge areas, it is possible to maximize the harvest without risking overexploitation.

More recently, Novaro, Redford, and Bodmer (2000) suggested that many of the hunted systems that have been studied in the Neotropics may be source-sink systems because hunted areas often are adjacent to large lightly hunted or unhunted areas. These authors proposed combining the methods of Joshi and Gadgil and McCullough, modified an equation given by Joshi and Gadgil to account for stochasticity, and showed its use to estimate the proportion of sink area that could be harvested at an unregulated rate without risking an overall population decline.

These methods do not require much field data or enforcement of quotas, and they are more conservative than quota systems because they maintain refugia that are free of hunting. One of the limitations of the modified method proposed by Novaro, Redford, and Bodmer is that it only allows estimation of the maximum size of sink area that can be completely overharvested. This area is very small for most species, so it is perhaps too conservative to be practical.

McCullough (1996) also evaluated the potential effects of harvest on metapopulations. He concluded that the likelihood for sustainability of hunting of metapopulations is low because hunting reduces the dispersal between subpopulations, increasing their likelihood of extinction. This conclusion is crucial because many game populations in the Neotropics and in other regions occur in increasingly fragmented habitats (Cullen et al. this volume). Game species that evolved in relatively continuous habitats may have been able to tolerate high harvest rates in continuous or naturally fragmented habitats in the past. These same species, however, may now experience rapid local extinction of subpopulations and eventual regional collapse in anthropogenically fragmented habitats in which the distances between patches and the nature of the matrix between patches drastically reduce or preclude dispersal. Unfortunately, it may not be enough to state that sustainability of hunting of game metapopulations is unlikely. Harvest of these game metapopulations will continue because enforcement in regions with poor social and economic conditions is weak or nonexistent and because social and economic pressures are not likely to decline. In the next section I offer some general recommendations to begin to evaluate the ecological implications of harvest in metapopulations.

Finally, in a simulation study of the effect of harvest in spatially structured populations, Lundberg and Jonzén (1999) analyzed the optimal harvesting strategies for a source-sink population, evaluating equilibrium densities, yield, and stability under different harvest options. Lundberg and Jonzén considered subpopulations that had intrinsic differences in their dynamics (population growth surplus in the source and dependence on immigration in the sink) regardless of the level of hunting. Conversely, in the studies mentioned previously (Hill and Padwe, 2000; review of earlier studies in Novaro, Redford, and Bodmer 2000), the sinks may be created by the heterogeneous distribution of hunting. Lundberg and Jonzén's main conclusion, as expected, was that the decision as to whether or not to harvest the source and/or the sink would greatly influence density, yield, and stability of the entire population, three factors that are crucial for sustainable harvest and conservation. The authors concluded that harvesting the sink always produces the highest yields and that harvesting is destabilizing (measured by the range of parameters that produces stable populations) under all conditions. Populations with intermediate growth rates that are harvested only in the sinks, however, are more resilient (measured by the time needed to return to initial conditions) than those in which the source is harvested and are even more resilient than unharvested populations.

Lundberg and Jonzén provide an equation with which to estimate optimal harvest rate in sinks. This equation, however, requires knowledge of dispersal rates between sources and sinks, among other variables, which are difficult to estimate (Novaro, Redford, and Bodmer 2000). Another limitation of Lunberg and Jonzén's analysis, which the authors point out, is the difficulty of identifying source and sink subpopulations that are caused by natural heterogeneity. I discuss this difficulty in more detail in the next and the last section.

EVALUATING THE EFFECT OF SPATIAL STRUCTURE

HARVEST OF METAPOPULATIONS

As indicated above, McCullough concluded that the likelihood of sustainable hunting in metapopulations is low, but increasing fragmentation of habitats in underdeveloped regions may determine that more and more game populations will occur as metapopulations. In areas where hunting of these populations is likely to continue, it is crucial whenever possible to carry out research that can produce information useful in predicting its effects and that can increase its chances of sustainability. The effects of harvest on naturally occurring metapopulations, on the other hand, also are unknown and equally important to understand. Naturally occurring metapopulations may have evolved in naturally fragmented habitats but are not necessarily more likely to withstand harvest because of their population characteristics (i.e., dependence on dispersal for recolonization) and probable alteration of the matrix between patches by anthropogenic changes. Important aspects that need to be researched are

1. the combined effects of fragmentation and hunting on game dispersal, which are likely to be synergistic;
2. the effects of harvest on the extinction probability at the patch and landscape levels;
3. the effects on connectivity (ease of game movement) of different degrees of development of the matrix between habitat fragments;
4. the community- and ecosystem-level effects of game harvest in metapopulations; that is, removal from or reduced dispersal of key species among habitat patches may have negative effects at the patch and landscape level;
5. the potential benefits of harvesting subpopulations of species that are likely to have detrimental effects on their habitat in certain patches (McCullough 1996).

Novaro, Redford, and Bodmer (2000) reviewed some of the empirical methods available for the study of game dispersal. Because of the difficulties of field studies on dispersal of medium-sized and large vertebrates, a combination of empirical data and simulations may be most useful, particularly for population-level processes. This combination may allow predictions and management recommendations more rapidly than from field studies alone and would permit avoiding experimental evaluations that are impractical. Long-term monitoring of harvest of metapopulations would enable scientists and managers to review and update structured population models and to correct management recommendations.

IDENTIFICATION OF GAME SOURCES IN SOURCE-SINK SYSTEMS

Harvest of source-sink populations is more likely to be sustainable than harvest of metapopulations. As for metapopulations, unfortunately, lack of field data and knowledge of processes and mechanisms involved in the harvest of source-sink populations also limit the ability to predict and make management recommendations for these populations.

One important limitation for research and management of source-sink populations that are hunted is the difficulty of identifying sources and sinks. Identification is fairly simple when there is relative homogeneity of habitat, and sources and sinks are produced by the spatial distribution of hunting. In this case, if hunting is intense, hunted areas are likely to be sinks. If sources and sinks occur naturally as a result of local differences in habitat quality, the identification of each type of sub-population is not trivial. Furthermore, the small size of some source areas can complicate their identification. Even in regionally stable populations, theoretical (Howe, Davis, and Mosca 1991) and empirical studies (reviewed by Pulliam 1996) have shown that sources can be relatively small in area as compared to sinks.

Population density is one of the variables most commonly studied for game populations. Density is often used as an indicator of the sustainability of hunting at a site and sometimes as an indicator of habitat quality. However, there are important limitations in the use of density as an indicator. In areas where hunting is spatially heterogeneous, for example, low density at a site may not necessarily mean that hunting is unsustainable and that the regional population is being overharvested. This lack of correlation was shown by Hill and Padwe (2000) in the Mbaracayú reserve of eastern Paraguay where the Aché hunt. Game densities are low in the vicinity of the Aché village where most hunting takes place, but harvest levels of most species have not declined through time, probably because of immigration from a large, lightly hunted portion of the reserve.

The problem with the use of density levels to identify the location of game sources is perhaps more disturbing. The density level at a site is a poor indicator of habitat quality and thus of the potential source or sink condition of the site, as has been shown empirically and theoretically (van Horne 1981; Watkinson and Sutherland 1995; Pulliam 1996; Lundberg and Jonzén 1999). Population densities in natural sink habitats can often be high because they receive large numbers of dispersers that are forced to leave source areas as a result, for example, of territoriality by source residents. Source populations, conversely, can be very productive and may be able to supply a constant surplus of individuals to sinks, but they may still have low density levels. In other words, low game density in an unhunted area does not mean that this area cannot act as a game source.

Considering the limitations of density as an indicator, it could be misleading to suggest to managers that game sources that need protection should be expected to have high densities (Bennett and Robinson 2000b; Robinson and Bennett 2000c). If only areas of high density are protected, managers may end up protecting sinks and losing small and more sparsely populated sources. The establishment of game reserves in sinks may be advantageous in terms of optimizing the harvest, particularly when dispersal from the sink to the source is low (Lundberg and Jonzén 1999), but it is unlikely to be the safest strategy from a conservation stand point. It is perhaps safer to suggest that source areas that need protection should be the most productive areas for the game species of interest.

Estimating game productivity or habitat quality for game at different sites is often more difficult than estimating game density. Nevertheless, by considering a small set of additional measurements, it may be possible to obtain preliminary indicators of productivity or habitat quality that can aid in the identification of game sources and thus complement estimates of population density. These measurements may be the ratio of adults to offspring or the availability of key food resources or breeding sites. Furthermore, it is important to keep in mind that the source or sink condition of a site can change through time because of natural or anthropogenic disturbances (Pulliam 1996). For this reason a field-monitoring system needs to be implemented in order to evaluate potential changes in the source or sink condition that is estimated initially.

A MODEL TO STUDY AND TO MONITOR HUNTED SOURCE-SINK POPULATIONS

It is necessary to develop new models to begin to understand the effects of hunting in source-sink systems and in order to make predictions that can be useful for management. In a recent study I evaluated the effect of changes in the proportion of source and sink habitat on the sustainability of harvest of culpeo foxes (*Pseudalopex culpaeus*) in Argentine Patagonia (Novaro 1997). Sources and sinks are distributed in a mosaic, with sources occurring in cattle ranches where there is no hunting and with sinks in sheep ranches where hunting is intense because culpeos prey on sheep. Periodically, some ranches switch from sheep to cattle ranching and vice versa so that the proportion and area of sources and sinks changes through time.

I used field data to develop a culpeo population model and simulated their dynamics in order to study the relationship between the rate of increase of the regional population and the proportion of the landscape in sources and sinks. Field data that were used included demographic data for source and sink populations (abundance, reproduction, and mortality), dispersal rates between sources and sinks, and the proportion of area in sources and sinks. In the case of the culpeos, a sensitivity analysis showed that their dynamics were strongly affected by adult mortality in the sources, suggesting that occasional killing in cattle ranches should be controlled to prevent a regional decline. The level of culpeo hunting in sinks was predicted to be sustainable for the current proportion of source area in the landscape (ca. 37%), but hunting would not be sustainable if this proportion fell below ca. 30%. The prediction about sustainability of current hunting was confirmed by field monitoring that showed a positive rate of increase of the regional population.

The conceptual model developed for culpeos may be useful for studying the relationship between the dynamics of other game populations and their spatial structure when it is possible to identify potential source and sink subpopulations. This relationship may be studied without estimating all the demographic variables mentioned above. Figure 24.1 shows the relationship between the observed rate of increase (r) of the regional population and the proportion of habitat in sources and sinks (sn). This relationship could be used to predict changes in the sustainability of hunting when there are changes in the proportion of area in sources or sinks. These changes could result from altered land use practices in sources and sinks, as

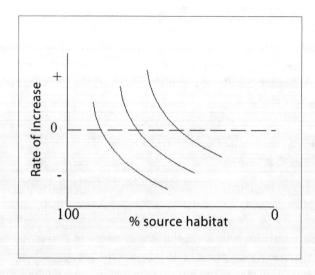

FIGURE 24.1 Relationship between the regional rate of increase and the proportion of source habitat in a source-sink population. Different lines correspond to different levels (increasing toward the right) of adult mortality in the sources.

was the case for culpeos, or from human encroachment into previously unhunted areas or abandonment of hunted areas.

The relationship between r and sn for the game populations of interest can be studied by obtaining a series of estimates of the rate of increase of the populations and the corresponding proportions of area in sources and sinks. Rates of increase can be estimated from population trend data that must involve estimates of absolute abundance (Eberhardt and Simmons 1992; Caughley and Sinclair 1994). Rates of increase of source and sink subpopulations should be averaged in order to estimate the regional rate of increase (for the entire mosaic of sources and sinks) for the game population, which is the most relevant demographic variable for sustainability. The estimation of game population trends over a number of years and under changing hunting patterns requires a large and long-term monitoring effort. It can, however, be conducted by local people (see Hill and Padwe 2000) and does not require detailed ecological studies that tend to be expensive and of short duration.

If a plot of the r-sn relationship is obtained, it can be used to predict whether hunting would be sustainable or not for a given proportion of source and sink area and a given direction of change between the source and sink condition. At least three conditions are necessary for the relationship between r and sn to remain relatively constant: source areas must remain free from hunting, habitat quality in sources must remain high, and the spatial arrangement of source and sink areas must remain relatively constant. First, as indicated, the r-sn relationship for culpeos, and probably for most game species, is relatively robust to changes in most demographic parameters except adult mortality in the source. Thus, even if low levels of hunting in the source occurred, the *r-sn* curve would shift markedly to the right (fig. 24.1), making hunting unsustainable even if the actual *sn* proportion remained constant. This marked shift is probably due to a density-dependent effect on dispersal from source areas. Second, a relatively constant habitat quality in sources means that their high productivity remains unchanged. Finally, dramatic changes in the spatial configuration of source and sink areas as a result, for instance, of large-scale habitat conversion or fragmentation would also affect the *r-sn* relationship, invalidating its application under the new conditions.

The model presented here provides a first approximation to an approach to explicitly evaluating the interaction between hunting, population dynamics and the spatial heterogeneity of some game populations. Other possible approaches were reviewed in the previous section (Joshi and Gadgil 1991; McCullough 1996; Lundberg and Jonzén 1999; Hill and Padwe 2000; Novaro, Redford, and Bodmer 2000). The application of one or the other approaches to hunting studies would depend on the amount of data available and, most importantly, on the spatial-structure model assumed to best represent the real game populations of interest. It is important to keep in mind that unexpected results will be encountered when studying the dynamics of or managing game populations because of stochastic events, failure of enforcement measures, and inconsistencies between model assumptions and real game populations. Long-term monitoring systems of game populations and their habitats are an essential tool needed to periodically adjust our assumptions and to improve our management recommendations.

25

Hunting and Wildlife Management
in French Guiana

CURRENT ASPECTS AND FUTURE PROSPECTS

CÉCILE RICHARD-HANSEN

AND ERIC HANSEN

French Guiana is a French overseas department, located between Brazil and Suriname. Human density is very low, averaging less than two people/km². However, the population is not evenly distributed, and most people are concentrated in the coastal area. Ninety-five percent of the territory is covered by evergreen tropical moist forest, representing over eight million hectares of almost intact and nonfragmented forest. Many different ethnic groups share the country: creoles, Bushinengue, Hmong, Chinese, people from metropolitan France, and six different Amerindian groups (Wayãpi, Wayana, Teko, Kali'na, Palikur, and Arawak). For most of the people—except perhaps the Chinese, whose economic focus is on restaurants and food shops—hunting is both a strong local tradition and a current practice. Subsistence hunters are frequent in remote isolated areas. In small rural villages and for people with low income, despite government aid, hunting for meat and selling the surplus represents a nonnegligible economic contribution. More-or-less organized commercial hunting also exists, as well as sport hunting near the main cities.

Starting approximately ten years ago, the Ministry of Environment took a new interest in French Guianan conservation problems. Following the recommendations of the Rio conference, a national park project was initiated, and in 1993 the National Game and Wildlife Service (Office National de la Chasse et de la Faune Sauvage, or ONCFS) began to work for the first time in the country. In France, as well as overseas, ONCFS is responsible for studying and monitoring wildlife and its habitats and for monitoring and regulating hunting. The environmental police fall under ONCFS's jurisdiction, and the agency is also charged with conducting applied studies of wildlife management.

In 1999 national funds were proposed for research projects aimed at studying human impact on the environment in tropical areas. The scientific group Silvolab, which comprises ten scientific and management organizations working on tropical forest issues, engaged in a two-year program called Hunting in French Guiana: Toward Sustainable Management. The main objective of this multidisciplinary program is to establish the necessary scientific, ecological, and sociological bases for the development of a sustainable use of wildlife in French Guiana. Although initial results are not yet published, new and complementary studies have been designed and are currently underway. This paper summarizes the status of wildlife management in the year 2002 in French Guiana, which is not well known in neighboring Neotropical countries, and then describes the current status and orientation of research in wildlife management.

PRESENT HUNTING LAWS IN FRENCH GUIANA

Although French Guiana is a French department, French hunting law does not apply here. French hunting law specifies that its decrees apply throughout the French territory (including the French West Indies, for example) with the notable exception of French Guiana. No specific reason is given for this omission, but one can imagine that the department was so distant, complex, and different from the national reality that, at the time the law was written, the problem was simply avoided.

As a consequence, in French Guiana there is currently no hunting season nor requirement for a hunting license—anyone can hunt anytime and anywhere, except in protected areas. Additionally, there are very few restrictions to hunting. A general review of the hunting regulations in the various Amazonian countries (Richard-Hansen 1998) has shown that French Guiana has one of the mildest restrictions on hunting practices.

The first regulation on hunting in French Guiana was set in 1975, but as a local decree it has relatively weak enforcement power. Later, in 1986, ministerial decrees enacted basic rules for wildlife protection; those for marine turtles, cetaceans, and sirenians were completed in 1991 and 1995. Also in 1995 another decree gave a more precise statement about the local trade.

At the present time, there are in French Guiana three main legal categories of wildlife: (a) fully protected species (table 25.1), for which any use is forbidden; (b) species for which hunting is allowed but trade in is forbidden; and (c) species for which both hunting and local trade in are allowed. Eight species of mammals can be traded locally: the two species of peccaries (*Tayassu tajacu* and *Tayassu pecari*); the tapir *Tapirus terrestris*; three species of rodents, paca (*Agouti paca*), red-rumped agouti (*Dasyprocta agouti*), and capybara (*Hydrochaeris hydrochaeris*); and two armadillos (*Dasypus novemcinctus* and *Dasypus kappleri*). Anyone concerned with conservation and sustainable use will certainly notice that trading tapir meat can hardly be justified from a biological point of view. Most likely such trade is permitted because of its importance to local people; political reasons might have weighed heavily in these decisions.

The same problem occurs for birds. The black curassow (*Crax alector*), trum-

TABLE 25.1 Fully Protected Animals in French Guiana

MAMMALS	BIRDS	REPTILES
Chironectes minimus	*Anhinga anhinga*	*Melanosuchus niger*
Cyclopes didactylus	*Phalacrocorax olivaceus*	*Chelus fimbriatus*
Tamandua tetradactyla	*Pelecanus occidentalis*	*Platemys platycephala*
Myrmecophaga tridactyla	*Fregata magnificens*	*Podocnemis cayanensis*
Priodontes maximus	*Phoenicopterus ruber*	*Corallus caninus*
Lutra enudris	*Cairina moschata*	*Dermochelys coriacea*
Pteronura brasiliensis	Ciconiformes: all spp.	*Caretta caretta*
Eira barbara	*Mesembrinis cayanensis*	*Lepidochelys olivacea*
Galictis vittata	*Eudocimus ruber*	*Lepidochelys kempii*
Speothos venaticus	*Ajaia ajaja*	*Eretmochelys imbricata*
Cerdocyon thous	Ardeidae: all spp.	*Chelonia mydas*
Procyon cancrivorus	Falconiformes: all spp.	
Herpailurus yagouaroundi	Strigiformes: all spp.	
Trichechus manatus	Lariformes: all spp.	
Ateles paniscus	*Opisthocomus hoazin*	
Chiropotes satanas	*Aburria pipile*	
Pithecia pithecia	*Ara ararauna*	
Aotus trivirgatus	*Ara macao*	
Odocoileus virginianus	*Ara chloroptera*	
Leopardus pardalis	*Rupicola rupicola*	
Leopardus tigrinus		
Leopardus wieidii		
Marine mammals: all spp.		

peter (*Psophia crepitans*), and guan (*Penelope marail*) may be among the most vulnerable to hunting pressure, yet owing to a strong hunting tradition, they are the three species that can be traded locally. Among the reptiles, the green iguana (*Iguana iguana*) is the only species authorized for commerce.

Concerning international trade, France is a party to CITES. Twenty-one species occurring in French Guiana are listed in Appendix I of CITES, and 170 in Appendix II. Moreover, it is illegal to export most species of the French Guianan fauna, even to metropolitan France.

BUSH MEAT TRADE

The local branch of the French government's Veterinary Services has monitored the bush meat trade for several years at the Cayenne Market, a central place for the local food trade. Estimates of tons of meat and individual small animals sold between 1986 and 1997 (figs. 25.1 and 25.2; adapted from Tyburn 1994) show that tapir

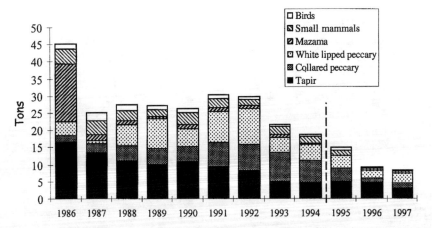

FIGURE 25.1 Tons of bush meat sold in the Cayenne market between 1986 and 1997. The dotted line indicates that data have been collected less systematically after this date. (Data from Tyburn 1994 and Veterinary Services)

and peccary represent the major part of the traded biomass. In 1986 the effect of the new regulation that declared brocket deer (*Mazama sp.*) illegal to trade is clearly seen (fig. 25.1).

The general tendency is for a decrease in the global amount of meat sold at the Cayenne market. However, after 1995 monitoring became less systematic (indicated by the dashed line on the graphs), and the data therefore less reliable. In reality, the presence of bush meat dealers in the market place has become less and less frequent and regular, and controls on the trade are more difficult. In fact, although we assume that a decrease in total amount of bush meat traded has occurred, the decline in meat sales is also related to the emergence of new commercial channels (Magnat 2000; C. Richard-Hansen pers. obs.). Hunters and dealers are more often selling their meat directly to restaurant owners or even to individual people and less often at the market. This situation makes control (site visits followed by application of sanctions if illegal meat is detected) by various police services much more difficult. Restaurant owners are required to declare their bush meat purchases on a specific registry, but this requirement is not very well respected and is also difficult to control.

IMPLEMENTATION

Legal control and police surveillance were very rare before 1993 because no specialized service was present. Beginning in 1993, French Game and Wildlife rangers have been present in French Guiana, but they are still in very small numbers. To deal with this situation, their main means of action is to enhance collaborations

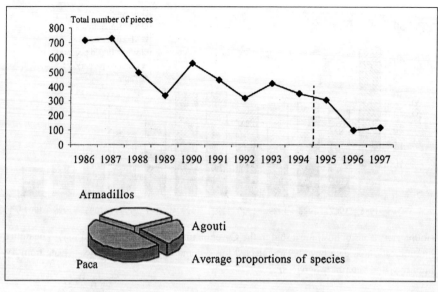

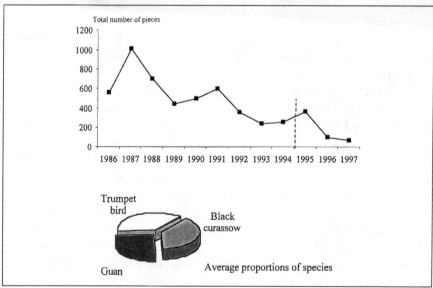

FIGURE 25.2 Number of pieces and mean proportions of total pieces made up by three species of small mammals and birds sold at the Cayenne Market between 1986 and 1997. The dotted line indicates that data have been collected less systematically after this date.   (Data from Tyburn 1994 and Veterinary Services)

with other police services, which have the ability to control but previously were not interested nor informed enough about wildlife to take action. Rangers of the French Game and Wildlife service regularly train customs and various policemen in particular aspects of wildlife conservation. Controls are then usually carried out in cooperation with one another.

Specific action has been undertaken for the control of restaurants, based on the assumption that the sales of bush meat to these establishments in the main cities may be the most important cause of wildlife harvest and may bring the greatest risk of overexploitation. Moreover, it is easier and more efficient to control the restaurant trade than to patrol wide forest areas. According to the 1995 decree, restaurants that wish to sell bush meat must request authorization, are required to respect trade and hunting bans (table 25.1), and must fill in a registry indicating all traded pieces. A massive information campaign was carried out, followed by regular restaurant inspections. Monitoring quantities is difficult because the registries are not always very well filled out, but flash inspections allow the detection of illegal species in restaurant freezers. In the main cities of Cayenne and Kourou, the situation has greatly changed (pers. obs.). It is now harder to find illegal species like caiman or brocket deer on the menu, although such listings were very frequent a few years ago when almost all wildlife species could be eaten in restaurants.

Hunting is also regulated in protected areas, but their isolated situation and the small number of agents makes control difficult. Therefore, coastal areas and sensitive species, such as marine turtles nesting on beaches, were given priority. As a further step in protection, the first national reserve was created in 1992. At present time, five national reserves exist: a marine reserve for nesting seabirds, such as terns and frigate birds; two in the interior, nonfragmented forested zone; and two in coastal areas, designed to protect marshes, aquatic habitats, and sea turtles. In the forest reserves, which are uninhabited, hunting is strictly forbidden, but the Kaw marshes and Amana coastal reserves have several different zones, in which hunting or fishing may be forbidden, restricted, or permitted for resident and nonresident people. Those areas are regularly patrolled by local workers and occasionally by national rangers.

In the rest of the country hunting controls involve only checking for protected species or illegal trade. As the country is large and very few means of communication exist, most inspections are made along the roads and rivers. Therefore the situation remains almost uncontrolled in small and remote villages.

PUBLIC INFORMATION

As enforcement actions are relatively recent, it was necessary to inform local people, tourists, and residents of the existing regulations because many bad habits had been acquired. The law texts were translated and explained, and brochures, booklets, posters, and a book (Hansen and Richard-Hansen 2000) were produced to describe the threats to wildlife and to detail which species are protected and which can be legally traded. Environmental education is difficult because of the low literacy rate, and follow-up studies to evaluate the success of the campaign have yet to be carried out. However, it does appear that more and more people in the cities, and especially in restaurants, are aware of the regulations.

SCIENTIFIC STUDIES

CONTEXT AND OBJECTIVES

Applied studies on wildlife management are a very recent concern in French Guiana, and research programs on hunting have been in place for a few years only. Still, detailed studies have previously been conducted on hunting practices of the Wayãpi, an Amerindian group living in southern French Guiana (Grenand 1996; Ouhoud-Renoux 1998).

As a first step in the Silvolab program, we carried out a global review of studies on hunting practices and impacts, ecological knowledge of main game species, sustainable use models, and hunting legislation in Amazonian countries (Richard-Hansen 1998; Richard-Hansen and Hansen 1998). The aim was to assess existing knowledge in order to orient the studies in French Guiana by taking into account and adapting previous experiments in neighboring countries.

Since 2000 study programs on hunting management have been in place. The coordination unit Silvolab was in charge of the first stage, which involved six scientific institutions, ONF (Office national des Forêts), IRD (Institut de la Recherche pour le Dévelopment), CIRAD (Centre de cooperation internationale en recherché agronomique pour le development), CNRS (Centre National de la Recherche Scientifique), ENS (Ecole Normale Supériore), and ENGREF (Ecole Nationale du Génie Rural des Eaux et Forêts), as well as ONCFS and the park project commission (Mission pour la création du Parc). The latter two institutions are now in charge of follow-up and complementary studies.

The context in which these studies are undertaken is a consideration of both the importance of local hunting traditions and the modern changes in human demography and hunting habits. Both traditions and new changes have a growing impact on wildlife populations, particularly in the densely inhabited coastal area. In fact, although local hunters are frightened by the possibility of new hunting regulations, more of them are coming to agree with the need to control the excesses because they feel that it has become difficult to hunt in proximate areas. However, they ask for locally adapted regulations, which take into account the social and faunal particularities of the country. The program is thus based on a bifocal approach combining biological and socioethnological studies.

In the long term the ecological part of the program primarily aims at, first, enhancing basic biological and ecological knowledge on the main hunted species: population structure and dynamics, reproductive rates and periods, relative species abundance in hunted and nonhunted areas, diet, and habitat use. Second, it aspires to develop and perfect simple ecological and hunting indices that allow monitoring of the status of populations and hunting impacts: kilometric index of abundance, fecundity and productivity indices for females, harvest yields, etc.

At the same time, a socioethnological study was conducted on hunting practices. The first stage of the study consisted of (a) characterizing hunting practices quantitatively and qualitatively, quantifying harvests, analyzing the social and ethnic context of hunting, and understanding the traditional representation and value of natural environment for the various ethnic groups, and (b) mapping hunting areas

METHODS

The above studies have been or are being established at selected focal study sites (fig. 25.3) in which both the social and ecological aspects of hunting are examined simultaneously. Additional nonhunted sites have been selected as reference areas for estimating animal densities (fig. 25.3). At focal study sites, a survey is made with hunters who have agreed to collaborate with the study. Local investigators daily visit the hunters and record the place, locality, type of weapon used, mode of transportation, number of hunters, and quantity of harvest for each hunting event.

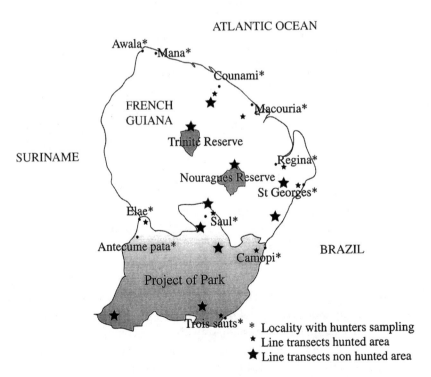

FIGURE 25.3 Main study sites for hunting and game species abundances in French Guiana.

Game harvest data obtained by hunter interviews provide the basic information needed about the overall number of animals harvested, their specific distribution, and the age-sex structure of the harvested population, as well as reproductive parameters of the main species and estimations of the hunting territory for people at each study site. Collecting jaws from hunted animals is undertaken at some study sites for more detailed analysis of age structure and survival.

Game densities and abundances are estimated in the hunted areas previously determined by the survey by using standardized line-transect and distance sampling methods. The same method is also conducted on the nonhunted sites,which are used as reference sites for the analyses of sustainability of hunting. Comparing animal densities in the different nonhunted areas also allows us to document and understand the ecological influences in addition to the human on animal densities. Although there are no strong ecological differences in French Guiana equivalent to those between várzea and terra firme forests in other parts of the Amazon, the carrying capacity may vary according to different animal species and forest types. Hunting impact cannot be clearly assessed if the basic ecological influence on animal densities is not known. For that reason habitat composition is to be described at each site at which animal counts are made.

The study is currently running in most of the coastal area sites. In the southern area establishment of a park has been proposed that would cover a large area occupied by a few indigenous populations (fig. 25.3). Five study sites have been selected in this area, which will allow a comparison of different sociocultural situations, integrating true situations of subsistence hunting. In contrast to the coast, the small size and isolated location of some villages will allow us to make a preliminary diagnosis of the sustainability of practices, applying models elaborated at other sites in Amazonia (Robinson and Redford 1991; Bodmer et al. 1997b; Bodmer and Penn 1997).

Sustainability of the current hunting practices will be estimated through the analysis of both the harvest quantities and the estimated production of various species in the hunted area (i.e., number of animals killed and number of births per square kilometer). The production will as much as possible be estimated locally, according to the reproductive parameters deducted from animals killed in the area. Then the Robinson and Redford 1991 model will be used to estimate the maximum sustainable harvest level, according to the lifespan of the species (20%, 40%, or 60% of the production as maximum sustainable use, for long-, short-, or very short-lived species, respectively).

The genetic structure of populations is also being studied by CNRS through the collection of tissue samples from hunted animals in order to analyze the genetic variability at local and regional scales. Concepts of metapopulation and source-sink systems may be underlying hypothesis with strong conservation implications for wildlife management. Finally, specific studies on the ecoethology of the main game species will then be initiated in order to conduct true management on the basis of ecological local knowledge.

PRACTICAL MANAGEMENT OBJECTIVES

In the park project ecological and hunting analyses will be the basis for a proposal for community management of wildlife resources. In the coastal area the situation is much more complex because several communities and hunter categories share all sites and resources. This sharing makes it very difficult to give people a sense of responsibility for management because there is no appropriation of the resource and its future.

However, we hope that the results of the study will help to integrate local needs, constraints, and realities in future management rules. French Guiana has the opportunity to convert its backwardness in the area of wildlife management in a favorable way. It can do so by integrating as a practical outcome, and at a very large scale, the results of the discussions and models that R. Bodmer, J. G. Robinson, and the five International Conferences on Wildlife Management and Conservation in Latin America and the Amazon have presented. It must do so without the constraints and advantages of an existing legal or protected-area system.

ACKNOWLEDGMENTS

Institutions collaborating in the program are Silvolab, overall project coordination; ONCFS, abundance index, ecological analysis of harvest impact, wildlife monitoring, and controls; Institut de la Recherche pour le Dévelopment, sociological and ethnological analysis and scientific coordination; Centre National de la Recherche Scientifique, genetic analysis; Centre de cooperation internationale en recherché agronomique pour le development, bush meat trade; Office national des Forêts, hunting survey and abundance index; Ecole Normale Supériore, abundance of game and nongame species of birds; and Direction Régionale de l'Environment/Guyane, Mission pour la création du Parc de la Guyane, hunting and ecological studies in the southern area.

Bibliography

Ab'Saber, A. N. 1977. Os domínios morfoclimáticos da América do Sul. *Geomorfologia* 52:1–23.

Ackerman, B. B., M. D. Samuel, and F. A. Leban. 1991. *User's Manual for Program HOME RANGE Version 2*. Moscow: Forestry, Wildlife, and Range Experiment Station, University of Idaho.

Acuña, C. de. 1942. *Nuevo descubrimiento del gran rio del Amazonas, por el P. Cristobal de Acuña, el cual fué por la provincia de Quito el año de 1639 (Publicacíon dirigida por Raúl Reyes).* Quito: Imprenta del Ministerio de Gobierno.

Akaike, H. 1973. Information theory and an extension of the maximum likelihood principle. In B. N. Petran and F. Csàaki, eds., *International Symposium on Information Theory*, 2d ed., pp. 267–288. Budapest: Akadèemiai Kiadi.

Albernaz, A. L. M. and J. M. Ayres. 1999. Selective logging along the middle Solimões River. In C. Padoch, J. M. Ayres, M. Pinedo-Vasquez, and A. Henderson, eds., *Várzea: Diversity, Conservation, and Development of Amazonia's Whitewater Floodplains*, pp. 135–154. New York: New York Botanical Garden Press.

Albert, B. and M. Gomez. 1997. *Saude Yanomami: Um manual etnolinguistico*. Belém, Brazil: Museu Paraense Emilio Goeldi.

Alexander, B. 1994. People of the Amazon fight to save the flooded forest. *Science* 265:606–607.

Alho, C. J. R. 1985. Conservation and management strategies for commonly exploited Amazonian turtles. *Biological Conservation* 32:291–298.

Alho, C. J. R. 1986. Uso potencial da fauna silvestre através de seu manejo. In *Anais do 1° Simpósio do Trópico Úmido*, pp. 359–369. Belém, Brazil.

Alho, C. J. R., Z. M. S. Campos, and H. C. Gonçalves. 1987: Ecologia da capivara (*Hydrochoerus hydrochaeris* — Rodentia) do Pantanal. I. Atividade, sazonalidade, uso de espaço e manejo. *Revista Brasileira de Biologia* 47:99–110.

Allen, A. C. and J. F. M. Valls. 1987. *Recursos forrageiros nativos do Pantanal Mato-Grossense*. Documentos 8. Brasília: EMBRAPA-CENARGEN.

Allendorf, F. W. and N. Ryman. 2002. The role of genetics in population viability analysis. In S. R. Bessinger and D. R. McCullough, eds., *Population Viability Analysis*, pp. 50–85. Chicago: University of Chicago Press.

Alonso, M. T. A. 1977. Vegetação. Região Sudeste. In *Geografia do Brasil*, pp. 91–118. Rio de Janeiro: Instituto Brasileiro de Geografia e Estatística.

Alpe, M. J., J. L. Kingery, and J. C. Mosley. 1999. Effects of summer sheep grazing on browse nutritive quality in autumn and winter. *Journal of Wildlife Management* 63(1):346–354.

Alvard, M. 1998. Indigenous hunting in the Neotropics: Conservation or optimal foraging? In T. M. Caro, ed., *Behavioral Ecology and Conservation Biology*, pp. 474–500. Oxford: Oxford University Press.

Alvard, M. 2000. The impact of traditional subsistence hunting and trapping on prey populations: Data from Wana horticulturalists of upland central Sulawesi, Indonesia. In J. G. Robinson and E. L. Bennett, eds, *Hunting for Sustainability in Tropical Forests*, pp. 214–230. New York: Columbia University Press.

Alvard, M., J. G. Robinson, K. H. Redford, and H. Kaplan. 1997. The sustainability of subsistence hunting in the Neotropics. *Conservation Biology* 11:977–982.

Alvard, M. S. 1994. Conservation by native peoples: Prey choice in a depleted habitat. *Human Nature* 5:127–154.

Amoroso, M. C. M. 1981. Alimentação em um bairro pobre de Manaus, Amazonas. *Acta Amazonica* (Suppl.) 11:23–26.

Anderson, E. N. 1996. *Ecologies of the Heart: Emotion, Belief, and the Environment.*New York: Oxford University Press.

Anderson, L. 1986. *The Economics of Fisheries Management.* Baltimore: Johns Hopkins University Press.

Andrade, P. C. M., A. Carvalho, M. E. Elias, A. Lavorenti, and S. L. G. Nogueira-Filho. 1996. Curva de crescimento de Capivaras (*Hydrochoerus hydrochaeris hydrochaeris*) em cativeiro. In *Anais 48ª Reunião Anual da Sociedade Brasileira para o Progresso da Ciência. Vol. II, p.5.* São Paulo.Sociedade Brasileira para o Progresso da Ciência.

Andrade, P. C. M., A. Lavorenti, and S. L. G. Nogueira-Filho. 1998. Efeitos do tamanho de área, da dieta e da idade inicial de confinamento sobre capivaras (*Hydrochoerus hydrochaeris hydrochaeris* L. 1766) em crescimento. *Revista Brasileira de Zootecnia* 27:292–299.

Anon. 2000. Serviço publico: Concurso para 12,3 mil vagas. *A Crítica, Manaus* September 22.

Aranda, M. 1994. Importancia de los pecaries (*Tayassu* spp.) en la alimentación del jaguar (*Panthera onca*). *Acta Zoologica Mexicana* (n.s.) 62:11–22.

Araujo-Lima, C. and M. Goulding. 1997. *So Fruitful a Fish: Ecology, Conservation, and Aquaculture of the Amazon's Tambaqui.* New York: Columbia University Press.

Araujo-Lima, C. A. R. M., B. Forsberg, R.Victoria, and L. Martinelli. 1986. Energy sources for detritivorous fishes in the Amazon. *Science* 234:1256–1258.

Arteaga, A., I. Cañizales, G. Hernández, M. C. Lamas, A. De Luca, M. Muñoz, A. Ochoa, A. E. Seijas, J. Thorbjarnarson, A. Velasco, S. Ellis, and U. Seal. 1997. *Reporte del Taller de Análisis de la viabilidad poblacional y del hábitat del Caimán del Orinoco* (Crocodylus intermedius). Apple Valley, Minn.: IUCN/SSC Conservation Breeding Specialist Group.

Auzel, P. and D. S. Wilkie. 2000. Wildlife use in northern Congo: Hunting in a commercial logging concession. In J. G. Robinson and E. L. Bennet, eds., *Hunting for Sustainability in Tropical Forests*, pp. 499–520. New York: Columbia University Press.

Avila-Pires, F. S. 1959. As formas sul-americanas do "veado-virá." *An. Acad. Bras. Ciên.* 31:547–556.

Ayala, G. 2002. *Monitoreo de* Tapirus terrestris *en el Izozog (Cerro Cortado) mediante el uso de telemetría como base para un plan de conservación.* M.S. thesis. La Paz: Instituto de Ecología, UMSA.

Ayala, J. and A. Noss. 2000. Censo por transectas en el Chaco boliviano: Limitaciones biológicas y sociales de la metodología. In E. Cabrera, C. Mercolli, and R. Resquin, eds., *Manejo de fauna silvestre en Amazonía y Latinoamérica*, pp. 29–36. Asunción, Paraguay: CITES Paraguay, Fundación Moises Bertoni, University of Florida.

Ayala, W. 2000. Participación de los comunarios izoceños en el manejo de fauna. In E. Cabrera, C. Mercolli, and R. Resquin, eds., *Manejo de fauna silvestre en Amazonía y Latinoamérica*, pp. 529–534. Asunción, Paraguay: CITES Paraguay, Fundación Moisés Bertoni, University of Florida.

Ayarzagüena, J. 1987. *Conservación del caimán del Orinoco* (Crocodylus intermedius) *en Venezuela. Parte I. Río Cojedes.* Caracas: FUDENA, WWF-U.S., Proyecto 6078.

Ayarzagüena, J. 1990. An update on the recovery program for the Orinoco crocodile. *Crocodile Specialist Group Newsletter* 9:16–18.

Ayres, J. M. 1986. *Uakaris and the Amazonian Flooded Forests.* Ph.D. dissertation. Cambridge: University of Cambridge.

Ayres, J. M. 1993. *As matas de várzea do Mamirauá: Médio Rio Solimões.* Brasilia: MCT-CNPq/ Sociedade Civil Mamirauá.

Ayres, J. M. and R. Best. 1979. Estratégias para a conservação da fauna amazônica. *Supl. Acta Amazônica* 9:81–101.

Ayres, J. M. and G. A. B. Fonseca. 1997. Corredores ecológicos das florestas neotropicais. *Megadiversitas: Boletim do Grupo de Trabalho em Biodiversidade (CNPq)* 1: 3.

Ayres, J. M., A.R. Alves, H.L. Queiroz, M. Marmontel, E. Moura, D. de M. Lima, A. Azevedo, M. Reis, P. Santos, R. de Silveira, and D. Masterson. 1999. Mamirauá: The conservation of biodiversity in an Amazonian flooded forest. In C. Padoch, J. M. Ayres, M. Pinedo-Vasquez, and A. Henderson, eds., *Várzea: Diversity, Conservation, and Development of Amazonia's Whitewater Floodplains*, pp. 203–216. New York: New York Botanical Garden Press.

Ayres, J. M., D. de Magalhaes Lima, E. de Souza Martins, and J. L. K. Barreiros. 1991. On the track of the road: Changes in subsistence hunting in a Brazilian Amazonian village. In J. G. Robinson and K. H. Redford, eds., *Neotropical Wildlife Use and Conservation*, pp. 82–92. Chicago: University of Chicago Press.

Azcarate-Bang, T. 1980, Sociobiologia y manejo del capibara (*Hydrochoerus hydrochaeris*). *Doñana-acta Venezuelana* 7: 6.

Azevedo, F. C. and V. A. Conforti. 1998. *Predation Dynamics of Wild Carnivores on Livestock Ranches Surrounding Iguaçu National Park: Evaluation, Impact, and Implementation of Preventive Methods.* Unpublished report. Curitiba, Brazil: Boticário Foundation.

Baitello, J. B., J. A. Pastore, O. T. Aguiar, F. C. Serio, and C. E. Silva. 1988. A vegetação do Parque Estadual do Morro do Diabo, município de Teodoro Sampaio, Estado de São Paulo. *Acta Botânica Brasileira* 1:221–230.

Baker, R. R. 1978. *The Evolutionary Ecology of Animal Migration.* New York: Holmes and Meier.

Barbarán, F.R. 1999. Comercialización de cueros de pecari (*Tayassu* spp.) en el Chaco semiárido de la Província de Salta, Argentina, período 1973–1997. In T. G. Fang, O. L. Montenegro, and R. E. Bodmer, eds., *Manejo y conservación de fauna silvestre en América Latina,* pp. 195–206 La Paz: Instituto de Ecología.

Barbier, E. B. 1992. Economics for the wilds. In T. M. Swanson and E. B. Barbier, eds., *Economics for the Wilds: Wildlife, Diversity, and Development,* pp. 15–33.Washington, D.C.: Island Press.

Barbosa R. I. 1997. Distribuição das chuvas em Roraima. In R.I. Barbosa, E. J. G. Ferreira, and E. G. Castellón, eds., *Homem, ambiente e ecologia no Estado de Roraima,* pp. 325–336. Manaus, Brazil: INPA.

Barret, C. B., K. Brandon, C. Gibson, and H. Gjersten. 2001. Conserving tropical biodiversity amid weak institutions. *Bioscience* 51(6):497–502.

Barrientos, J. and L. Maffei. 2000. Radio-telemetría en la hurina *Mazama gouazoubira* en el campamento Cerro Cortado, Izozog, Santa Cruz, Bolivia. In E. Cabrera, C. Mercolli, and R. Resquin, eds., *Manejo de fauna silvestre en Amazonía y Latinoamérica,* pp. 369–372. Asunción, Paraguay: :CITES Paraguay, Fundación Moises Bertoni, University of Florida.

Barrientos, J. and E. Cuéllar. 2003. Radiotelemetría de antas (*Tapirus terrestris*) en el Chaco seco, Izozog, Santa Cruz, Bolivia. In R. Polanco-Ochoa, ed., *Manejo de Fauna silvestre en Amazonía y Latinoamérica: Selección de trabajos V Congreso Internacional,* pp. 140–142. Bogotá: Fundación Natura.

Barthem, R. and M. Goulding. 1997. *The Catfish Connection: Ecology, Migration, and Conservation of Amazon Predators.* New York: Columbia University Press.

Barthem, R. B. 1999a. A pesca comercial no Médio Solimões e sua interação com a Reserva de Desenvolvimento Sustentável Mamirauá. In. H. Queiroz and W.G.R Crampton, eds., *Estratégias para manejo de recursos pesqueiros em Mamirauá,* pp. 72–107. Brasília: SCM-CNPq.

Barthem, R. B. 1999b. Várzea fisheries in the middle Rio Solimões. In C. Padoch, J. M. Ayres, M. Pinedo-Vasquez, and A. Henderson, eds., *Várzea: Diversity, Conservation, and Development of Amazonia's Whitewater Floodplains,* pp. 7–28. New York: New York Botanical Garden Press.

Barthem, R. B. and M. J. Petrere. 1992. Fisheries and population dynamics of *Brachyplatysoma vaillantii* (Pimelodidae) in the Amazon Estuary. *Abstract Bulletin World Fisheries Congress* 3:31–32.

Barthem, R. B., M. J. Petrere, V. Isaac, M. C. L. de Brito Ribeiro, D. G. McGrath, I. J. A.Vieira, and M. V. Barco. 1997. A pesca na Amazônia: Problemas e perspectivas para o seu manejo. In. C. Valladares-Pádua and R. E. Bodmer, eds., *Manejo e conservação de vida silvestre no Brasil,* pp. 173–185. Rio de Janeiro: MCT-CNPq/Sociedade Civil Mamirauá.

Bates, H. W. 1863. *The Naturalist on the River Amazons.* London: John Murray.

Batista, V. de S. 1998. *Distribuição, dinâmica da frota e dos recursos pesqueiros da Amazônia Central.* Ph.D dissertation. Manaus, Brazil: INPA-UA.

Bauman, D. E., B. A. Corl, L. H. Baumgard, and J. M. Griinari. 1998. Trans fatty acids, Conjugated Linoleic Acid and milk fat synthesis. In *Proceedings of Cornell Nutritional Conference,* pp. 95–103. Ithaca, N.Y.: Division of Nutritional Science, Cornell University.

Bayley, P. B. 1980. Accounting for effort when comparing tropical fisheries in lakes, river floodplains, and lagoons. *Limnol and Oceanogr.* 33:963–972.

Bayley, P. B. 1989. Aquatic environments in the Amazon basin, with an analysis of carbon sources, fish production, and yield. *Special Publications of the Canadian Journal of Fisheries and Aquatic Sciences* 106:399–408.

Bayley, P. B. and Petrere, M. J. 1989. Amazon fisheries: Assessment methods, current status, and management options. *Special Publications of the Canadian Journal of Fisheries and Aquatic Sciences* 106:385–398.

Bayley, P. B., F. Vázquez, P. Ghersi, P. Soini, and M. Pinedo. 1992. *Environmental Review of the Pacaya-Samiria National Reserve in Peru and Assessment of Project (527-0341).* Unpublished report. Washington D.C.: Nature Conservancy.

Becker, M. 1981. Aspectos da caça em algumas regiões de cerrado de Mato Grosso. *Brasil Florestal* 11:51–63.

Beckerman, S. 1994. Hunting and fishing in Amazonia: Hold the answers, what are the questions? In A. Roosevelt, ed., *Amazonian Indians from Prehistory to the Present: Anthropological Perspectives*, pp. 177–200. Tucson: University of Arizona Press.

Beddington, J. R. and R. M. May. 1977. Harvesting natural populations in a randomly fluctuating environment. *Science* 197:463–465.

Begazo, A. J. 1997. Use and conservation of the Cracidae in the Peruvian Amazon. In S. D. Strahl, S. Beaujon, D. M. Brooks, A. J. Begazo, G. Sedaghatkish, and F. Olmos, eds., *The Cracidae: Their Biology and Conservation*, pp. 449–459. Washington, D.C.: Hancock House.

Beissinger, S. R. 1994. La conservación de los Psitácidos del neotrópico: Retos para los biólogos, gerentes y gobierno. In G. Morales, I. Novo, D. Bigio, A. Lucy, and F. Rojas-Suárez, eds., *Biología y conservación de los Psitácidos de Venezuela*, pp. 141–147. Caracas: Gráficas Giavimar.

Beissinger, S. R. and E. H. Bucher. 1992. Sustainable harvest of parrots for conservation. In S. R. Beissinger and N. F. R. Snyder, eds., *New World Parrots in Crisis: Solutions from Conservation Biology*, pp. 73–115. Washington, D.C.: Smithsonian Institution Press,.

Beissinger, S. R. and D. R. McCullough, eds. 2002. *Population Viability Analysis*. Chicago: University of Chicago Press.

Bendayán, N. 1991. *Influencia socioeconómica de la fauna silvestre en Iquitos, Loreto*. Undergraduate thesis. Iquitos, Peru: Universidad Nacional de la Amazonia Peruana.

Beneria-Surkin, J. 1998. *Socioeconomic Study of Five Izoceño Communities*. Technical Paper #6. Santa Cruz, Bolivia: Kaa-Iya Project.

Bennett, E. L. and J. G. Robinson. 2000a. Hunting for sustainability: The start of a synthesis. In J. G. Robinson, and E. L. Bennett, eds., *Hunting for Sustainability in Tropical Forests*, pp. 499–519. New York: Columbia University Press.

Bennett, E. L. and J. G. Robinson. 2000b. *Hunting of Wildlife in Tropical Forests: Implications for Biodiversity and Forest Peoples*. Environment Department Papers. Washington, D.C.: World Bank.

Bergerud, A. T. 1988. Caribou, wolves, and man. *Trends in Ecology and Evolution* 3:68–73.

Berkes, F., ed. 1989. *Common Property Resources: Ecology and Community-Based Sustainable Development*. London: Belhaven Press.

Bodmer, R. E. 1989. Frugivory in Amazonian Artiodactyla: Evidence for the evolution of the ruminant stomach. *Journal of Zoology* 219:457–467.

Bodmer, R. E. 1990. Responses of ungulates to seasonal inundations in the Amazon floodplain. *J. Trop. Ecol.* 6:191–201.

Bodmer, R. E. 1994. Managing wildlife with local communities: The case of the Reserva Comunal Tamshiyacu-Tahuayo. In D. Western, M. Wright, and S. Strum, eds., *Natural Connections: Perspectives on Community Based Management*, pp.113–134. Washington, D.C.: Island Press.

Bodmer, R. E. 1995a. Managing Amazonian wildlife: Biological correlates of game choice by detribalized hunters. *Ecological Applications* 5:872–877.

Bodmer, R. E. 1995b. Priorities for the conservation of mammals in the Peruvian Amazon. *Oryx* 29:23–28.

Bodmer, R. E. 1999. Uso sustentable de los ungulados amazónicos: Implicaciones para las áreas protegidas comunales. In T. G. Fang., O. L. Montenegro, and R. E. Bodmer, eds., *Manejo y conservación de fauna silvestre en América Latina*, pp. 51–58. La Paz: Instituto de Ecología.

Bodmer, R. E. 2000. Integrating hunting and protected areas in the Amazon. In N. Dunstone and A. Entwistle, eds., *Future Priorties for the Conservation of Mammals: Has the Panda had its Day?*, pp. 277–290. Cambridge: Cambridge University Press.

Bodmer, R. E. 2001. Evaluating wildlife sustainable use in the Amazon: The unified harvest model. In P. Sánchez, A. Morales, and H. F. López, eds., *Proceedings of the Fifth International Congress on Wildlife Management in Amazonia and Latin America*, p. 55. Bogotá: Fundación Natura.

Bodmer, R. E., C. M. Allen, J. W. Penn, R. Aquino, and C. Reyes. 1999. *Evaluación del uso sostenible de la fauna silvestre en la Reserva Nacional Pacaya-Samiria, Perú*. Documentos de Trabajo América Verde No. 4b. Arlington, Va.: Nature Conservancy.

Bodmer, R. E., R. Aquino, P. E. Puertas, C. J. Reyes, T. G. Fang, and N. L. Gottdenker. 1997a. Manejo y uso sustentable de pecaríes en la Amazonía Peruana. Occasional Paper of the IUCN Species Survival Commission, No. 18. Quito: UICN-Sur.

Bodmer, R. E., N. Y. Bendayán, A. L. Moya, and T. G. Fang. 1990. Manejo de ungulados en la Amazonía Peruana: Análisis de la caza de subsistencia y la comercialización local, nacional e internacional. *Boletin de Lima* 70:49–56.

Bodmer, R. E., J. F. Eisenberg, and K. H. Redford. 1997. Hunting and the likelihood of extinction of Amazonian mammals. *Conservation Biology* 11:460–466.

Bodmer, R. E., T. Fang, and L. Moya. 1988a. Primates and ungulates: A comparison of susceptibility to hunting. *Primate Conservation* 9:79–83.

Bodmer, R .E., T. G. Fang, and L. Moya. 1988b. Ungulate management and conservation in the Peruvian Amazon. *Biological Conservation* 45:303–310.

Bodmer, R. E., T. G. Fang, L. Moya, and R. Gill. 1994. Managing wildlife to conserve Amazonian forests: Population biology and economic considerations of game hunting. *Biological Conservation* 67:29–35.

Bodmer, R. E. and J. W. J. Penn. 1997. Manejo da vida silvestre em comunidades na Amazônia. In C. Valladares-Padua and R. E. Bodmer, eds., *Manejo e conservação de vida silvestre no Brasil*, pp. 52–69. Brasilia, Brazil: CNPq/Sociedade Civil Mamirauá.

Bodmer, R. E., J. W. Penn, P. Puertas, L. Moya, and T. G. Fang. 1997b. Linking conservation and local people through sustainable use of natural resources: Community-Based management in the Peruvian Amazon. In C. H. Freese, ed., *Harvesting Wild Species: Implications for Biodiversity Conservation*, pp. 315–358. Baltimore: Johns Hopkins University Press.

Bodmer, R. E. and E. Pezo. 1999. Análisis econômico de la venta de carne de monte y exportación de peles en Loreto, Perú. In T. G. Fang., O. L. Montenegro, and R. E. Bodmer, eds., *Manejo y conservación de fauna silvestre en América Latina*, pp. 171–182. La Paz: Instituto de Ecología.

Bodmer, R. E. and E. Pezo Lozano. 2001. Rural development and sustainable wildlife use in Peru. *Conservation Biology* 15:1163–1170.

Bodmer, R. E. and P. Puertas. 2000. Community-Based comanagement of wildlife in the Peruvian Amazon. In J. G. Robinson and E. L. Bennet, eds., *Hunting for Sustainability in Tropical Forests*, pp. 395–409.New York: Columbia University Press.

Bolaños, J. E. 2000. *Densidad, abundancia relativa, distribución y uso local de los ungulados en la cuenca del río Lacantún, Chiapas, México*. Undergraduate thesis. Tuxtla Gutiérrez, Chiapas, México: Universidad de Ciencias y Artes de Chiapas.

Bonacic, C. 1996. *Sustainable Use of the vicuña* (Vicugna vicugna) *in Chile*. M.S. thesis. Reading, U. K.: University of Reading.

Borges, P.A., M. G. Dominguez-Bello, and E. A. Herrera. 1996. Digestive physiology of wild capybara. *Journal of Comparative Physiology* B. 166:55.

Bostock, T. 1998. *Mamirauá Sustainable Development Reserve: Fish Processing and Marketing Consultancy*. DFID Consultant report. Mamirauá Project, phase 2. Chatham: Natural Resource Institute.

Bourdy, G. 2001. *Plantas del Chaco: Guía Para el Docente*. Santa Cruz, Bolivia: Kaa-Iya Project.

Bourdy, G. 2002. *Plantas del Chaco II: Usos tradicionales Izoceños-Guaraní*. Santa Cruz, Bolivia: Kaa-Iya Project.

Brandon, K. 1996. Traditional peoples, nontraditional times: Social change and the implications for biodiversity conservation. In K. H. Redford and J. A. Mansour, eds.,*Traditional Peoples and Biodiversity Conservation in Large Tropical Landscapes*, pp. 219–236. Washington, D.C.: Nature Conservancy.

Brown, J. H. and A. Kodric-Brown. 1977. Turnover rates in insular biogeography: Effect of immigration on extinction. *Ecology* 58:445–449.

Bruno, N. and E. Cuéllar. 2000. Hábitos alimenticios de cinco Dasypodidos en el Chaco boliviano. In E. Cabrera, C. Mercolli, and R. Resquin, eds,. *Manejo de fauna silvestre en Amazonía y Latinoamérica*, pp. 401–412. Asunción, Paraguay: CITES Paraguay, Fundación Moises Bertoni, University of Florida.

Buckland, S. T., D. R Anderson, K. P. Burnham, and J. L. Laake. 1993. *DISTANCE Sampling: Estimating Abundance of Biological Populations*. London: Chapman and Hall.

Byers, C. R., R. K. Steinhorst, and P. R. Krausman .1984. Clarification of a technique for analysis of utilization-availability data. *Journal of Wildlife Management* 48:1050–1053.

Caballero, M. and A. Noss. 2000. Comparación de la dieta de urina (*Mazama gouazoubira*) entre la época lluviosa y seca en la zona del Izozog, Santa Cruz, Bolivia. In E. Cabrera, C. Mercolli, and R. Resquin, eds., *Manejo de fauna silvestre en Amazonía y Latinoamérica*, pp. 377–384. Asunción, Paraguay: CITES Paraguay, Fundación Moises Bertoni, University of Florida.

Cabrera, E., C. Mercolli, and R. Resquin, eds. 2000. *Manejo de fauna silvestre en Amazonía y Latinoamérica.* Asunción, Paraguay: CITES Paraguay, Fundación Moises Bertoni, University of Florida.

Calouro, A. 1995. *Caça de subsistência: sustentabilidade e padrões de uso entre seringueiros ribeirinhos e não ribeirinhos do Estado do Acre.* M.S. thesis. Brasilia: Universidade de Brasília, Instituto de Ciências Biológicas.

Calzadilla V., F. 1948. *Por los Llanos de Apure.* Caracas: Ediciones del Ministerio de Educación Nacional, Dirección de Cultura.

Campo, M. and E. Rodríguez. 1997. *Evaluación de la calidad del ambiente acuático del río Cojedes.* Unpublished report. Caracas: PROFAUNA/MARNR, Dirección de Manejo de Fauna Acuática.

Campos, L. C. 1986. *Distribution, Human Impact, and Conservation of Flamingos in the High Andes of Bolivia.* M.A. thesis. Gainesville: University of Florida.

Campos-Rozo, C., H. Rubio-Torgler, and A. Ulloa. 2001. Manejo local por los Embera del Chocó Colombiano. In R. Primack, R. Rozzi, P. Feisinger, R. Dirzo, and F. Massardo, eds., *Fundamentos de conservación biológica: Perspectivas Latinoamericanas,* pp. 599–601. México, D.F.: Fondo de la cultura económica.

Campos-Rozo, C., A. Ulloa, and H. Rubio-Torgler. 1996. *Manejo de fauna con comunidades rurales.* Bogotá: Instituto Colombiano de Antropología e Historia-Fundación Natura.

Caro, T. M. 1999. Conservation monitoring: Estimating mammal densities in woodland habits. *Animal Conservation* 2:305–315.

Carrillo E., J. C. Saenz, and T. K. Fuller. 2002. Movement and activities of white-lipped peccaries in Corcavado National Park, Costa Rica. *Conservation Biology* 108:317–324.

Carrilo, G. 1999. Valor nutritivo de cuatro variedades de carne de monte de mayor consumo en la ciudad de Iquitos. In *Proccedings of the Fourteenth International Congress on Wildlife Management in Amazonia and Latin America,* p. 43. Asuncion, Paraguay.

Castello, L. In press. A method to count *Arapaima gigas*: Fishermen, assessment, and management. *North American Journal of Fisheries Management.*

Castro, R. 1991. *Behavioral Ecology of Two Coexistent Tamarin Species* (Saguinus fuscicollis nigrifrons *and* Saguinus mystax mystax, *Callithrichidae, Primates) in Amazonian Peru.* Ph.D. dissertation. Saint Louis, Mo.: Washington University.

Cattan, P. and A. Glade. 1989. Management of the *Vicugna vicugna* in Chile: Use of a matrix model to assess harvest rates. *Biological Conservation* 49:131–140.

Caughley, G. 1977. *Analysis of Vertebrate Populations.* New York: Wiley.

Caughley, G., G. C. Grigg , J. Caughley J, and G. J. E. Hill. 1980. Does dingo predation control the densities of kangoroos and emus? *Austr. Wildl. Res.* 7:1–12.

Caughley, G. and A. R. E. Sinclair. 1994. *Wildlife Ecology and Management.* Boston: Blackwell Scientific.

Cavalcante-Filho, M. F., M. A. M. Carvalho, G. V. Machado, and E. Bevilacqua. 1998. Estudo comparativo sobre a morfologia do estômago do queixada (*Tayassu pecari*) e cateto (*Tayassu tajacu*). *Brazilian Journal of Morphological Sciences* 15:203–207.

CDC-UNALM. 1993. *Evaluación ecológica de la Reserva Nacional Pacaya-Samiria.*—La Molina, Lima, Peru: Centro de Datos para la Conservación, Universidad Nacional Agraria.

Cerdeira, R. G. P., M. L. Ruffino, V. J. Isaac. 1997. Consumo de pescado e outros alimentos nas comunidades ribeirinhas do Lago Grande de Monte Alegre. *Acta Amazônica* 27:213–227.

Cerri, C. 1995. O crepúsculo do gigante. *Revista Globo Rural* 10:28–32.

Chao, N. L., P. Petry, G. Prang, L. Sonneschien, and M. Tlusty, eds. 2001. *Conservation and Management of Ornamental Fish Resources of the Rio Negro Basin, Amazonia, Brazil—Projeto Piaba.* Manaus, Brazil: Editora da Universidade do Amazonas.

Chávez, V. 1999. Determinación de la estacionalidad de partos en urina (*Mazama gouazoubira*). Laboratory practical. Santa Cruz, Bolivia: UAGRM.

Chiarello, A. G. 2000. Density and population size of mammals in remnants of Brazilian Atlantic Forest. *Conservation Biology* 14 (6):1649–1657.

Child, G. 1968. *Behaviour of Large Mammals During the Formation of Lake Kariba.* Salisbury: National Museums of Rhodesia.

Cincotta, R. P., J. Wisnewksi, and R. Engelman. 2000. Human population in the biodiversity hotspots. *Nature* 404:990–992.

Clay, J. W. 1988. *Indigenous People and Tropical Forests: Models of Land Use and Management from Latin America.* Cultural Survival Report 27. Cambridge: Cultural Survival.

Clutton-Brock, T. H. 1997. Stability and instability in ungulate populations: An empirical analysis. *American Naturalist* 149:195–219.

Codazzi, A. 1940 [1841]. *Resumen de la geografía de Venezuela. Venezuela en 1841. Tomo I.* Caracas: Geografía Física, Biblioteca Venezolana de Cultura. Taller de Artes Gráficas Escuela Técnica Industrial.

Cody, M. L. and J. M. Diamond. 1975. *Ecology and Evolution of Communities.* Cambridge: Belknap Press.

COICA. 1989. To the community of concerned environmentalists. Lima: Coordinadora de las Organizaciones Indigenas de la Cuenca Amazonica.

Colding, J. and C. Folke. 1997. The relationship among threatened species, their protection, and taboos. *Conservation Ecology* 1:6.

Cole, L. 1954. The population consequences of life history phenomena. *Q. Rev. Biol.* 29:103–137.

Combès, I. 1999. *Arakae: Historia de las comunidades izoceñas.* Santa Cruz, Bolivia: Kaa-Iya Project.

Combès, I., J. Aguirre, A. Noss, J. Ventocilla, and A. M. Saavedra. 1997. *Educación ambiental en el Izozog.* Santa Cruz, Bolivia: Kaa-Iya Project.

Combès, I., N. Justiniano, I. Segundo, D. Vaca, and R. Vaca. 1998. *Kaa-Iya reta: los dueños del monte.* Technical paper #40. Santa Cruz, Bolivia: Kaa-Iya Project.

Comizzoli, P., J. Peiniau, C. Dutertre, P. Planquette, and A. Aumaitre. 1997. Digestive utilization of concentrated and fibrous diets by two peccary species (*Tayassu pecari, Tayassu tajacu*) raised in French Guyana. *Animal Feed Science Technology* 64:215–226.

Conroy, M. J. 1996. Abundance indices. In D. E. Wilson, F. R. Cole, J. D. Nichols, R. Rudran, and M. Foster, (eds.), *Measuring and Monitoring Biological Diversity: Standard Methods for Mammals,* pp. 179–192. Washington, D.C.: Smithsonian Institution Press.

Coomes, O. 1992. *Making a Living in the Amazon Rain Forest: Peasants, Land, and Economy in the Tahuayo River Basin of Northeastern Peru.* Ph.D. dissertation. Madison: University of Wisconsin.

COREPASA. 1986. *Plan maestro de la Reserva Nacional Pacaya-Samiria.* Loreto, Peru: Editorial e Imprenta DESA.

Costa, F. de A., C. de S. F. Senna, S. Pereira, and D. C. Kern. 1986. *Levantamento arqueológico na área da UHE Cachoeira Porteira: Relatório global.* Belém, Brazil: Museu Paraense E. Goeldi.

Costa, J. M. M. 1992. Impactos econômico-territoriais do atual padrão de ocupação da Amazônia. In A. M. Matos, ed., *Amazônia: Desenvolvimento ou retrocesso,* pp. 40–115. Belém, Brazil: CEJUP.

Costa, L. R. F., R. Barthem, and M. A.V. Correa. 1999. Manejo da pesca do tambaqui (*Colossoma macropomum*) nos lagos de várzea da Reserva de Desenvolvimento Sustentável. In. H. Queiroz and W.G. R Crampton, eds., *Estratégias para manejo de recursos pesqueiros em Mamirauá,* pp. 142–158. Brasília: SCM-CNPq.

Cott, H. B. 1954. The exploitation of wild birds for their eggs. *Ibis* 96:129–149.

Coutinho, J. M. de S. 1868. Sur les tortues de l'Amazone. *Bulletin de la Societé Impériale Zoologique d'Acclimatation* 5:147–166.

CPT. 1992. *Os ribeirinhos. Preservação dos lagos, defesa do meio ambiente e a pesca comercial.* Rio de Janeiro: Comissão Pastoral da Terra (regional AM e RR).

Crampton, W. G. R. 1999a. The impact of the ornamental fish trade on the discus *Symphysodon aequifasciatus:* A case study from the flood plain forests of Estação Ecológica Mamirauá. In C. Padoch, J. M. Ayres, M. Pinedo-Vasquez, and A. Henderson, eds., *Várzea: Diversity, Conservation, and Development of Amazonia's Whitewater Floodplains,* pp. 29–44. New York: New York Botanical Garden Press.

Crampton, W. G. R. 1999b. Os peixes da Reserva Mamirauá: Diversidade e história natural na planície alagável da Amazônia. In H. Queiroz and W.G.R Crampton, eds., *Estratégias para manejo de recursos pesqueiros em Mamirauá,* pp. 10–36. Brasília: SCM-CNPq.

Crampton, W. G. R. 1999c. Plano de manejo para o uso sustentável de peixes ornamentais na Reserva Mamirauá. In. H. Queiroz and W.G.R Crampton, eds., *Estratégias para manejo de recursos pesqueiros em Mamirauá,* pp. 159–176. Brasília: SCM-CNPq.

Crampton, W. G. R. 2001. Diversidad e Conservación de los Peces de la Cuenca Amazónica. In S. Paitán and M. C. Portillo, eds., *Amazonia: Orientaciones para el desarrollo sostenible,* pp. 121–140. Lima: UNAP.

Crampton, W. G. R. and J. P. Viana 1999. Conservação e diversificação econômica da pesca nas várzeas do alto Rio Solimões: Uma breve revisão e sugestões para um futuro sustentável. In. T. G. Fang, O. L. Montenegro, and R. E. Bodmer, eds., *Manejo y conservacion de fauna silvestre en America Latina*, pp. 209–226. La Paz: Instituto de Ecologiá.

Crawshaw, Jr., P. G. 1992. Recommendations for study design on research projects on neotropical felids. In *Felinos de Venezuela: Biología, Ecología e Conservación, pp.* :187–222. Caracas: FUDECI.

Crawshaw ,Jr., P. G. 1995. *Comparative Ecology of Ocelot (*Felis pardalis*) and Jaguar (*Panthera onca*) in a Protected Subtropical Forest in Brazil and Argentina*. Ph. D. dissertation. Gainesville: University of Florida.

Crawshaw, Jr., P. G. 2002. Mortalidad inducida por humanos y la conservación de jaguares: El Pantanal y el Parque Nacional Iguaçu en Brasil. In R. A. Medellín, C. Chetkiewicz, A. Rabinowitz, K. H. Redford, J. G. Robinson, E. Sanderson, and A. Taber, eds. *Los jagurares en el nuevo milenio: Un análisis de su estado, detección de prioridades y recomendaciones para la conservación*. México, D.F.: Universidad Nacional Autonoma de México/Wildlife Conservation Society.

Crawshaw, Jr., P. G., and J. Pilla. 1994. *Ecologia e conservação das populações de felinos do "Parque Florestal Estadual do Turvo."* Report. Rio Grande do Sul, Brasil: World Wildlife Fund-U.S.

Crawshaw, Jr., P. G., and H. B. Quigley. 1984. *A ecologia do jaguar ou Onça pintada (*Panthera onca*) no Pantanal Matogrossense*. Final report. Brasilia: Instituto Brasileiro de Desenvolvimento Florestal-IBDF.

Crawshaw, Jr., P. G., and H. B. Quigley. 1989. Notes on ocelot movement and activity in the Pantanal region, Brazil. *Biotropica* 21:377–379.

Crawshaw, Jr., P. G., and Quigley, H.B. 1991. Jaguar spacing, activity, and habitat use in a seasonally flooded environment in Brazil. *J. Zool. London* 223:357–370.

Crespo, J. A. 1982. Ecología de la comunidad de mamíferos del Parque Nacional Iguazú, Missiones. *Rev. Mus. Argentino Ciencias Naturales "Bernardino Rivadavia," Ecología* 3:45–162.

Crocker, R. L. and N. S. Tiver. 1948. Survey methods in grassland ecology. *J. Br. Grassld Soc.* 3:1–26.

Crosby, A.W. 1986. *Ecological Imperialism: The Biological Expansion of Europe, 900–1900*. New York: Cambridge University Press.

Cruz, E. 2001. *Hábitos de alimentación e impacto de la actividad humana sobre el tapir (*Tapirus bairdii*) en la Reserva de la Biosfera La Sepultura, Chiapas, México*. M.S. thesis, San Cristóbal de Las Casas, Chiapas, México: El Colegio de la Frontera Sur.

Cuéllar, E. 1999. *Hunting of Armadillos by Izoceño Indians of the Gran Chaco, Bolivia*. M. S. thesis. Canterbury: University of Kent.

Cuéllar, E. 2002. Census of the three-banded armadillo, *Tolypeutes matacus*, using dogs, Southern Chaco, Bolivia. *Mammalia* 66:448–451.

Cuéllar, E., A. Arambiza, R. Miserendino, L. González, F. Soria, G. Castro, N. Ramón, and A. Noss. 1998. *Manual de la Fauna Izoceña*. Santa Cruz, Bolivia: Kaa-Iya Project.

Cuéllar, E., R.S. Miserendino, and A. Noss. 1998. Introducción a los estudios biológicos en el Gran Chaco, Provincia Cordillera, Santa Cruz, Bolivia. *Ecología en Bolivia* 31:1–15.

Cuéllar, E. and A. Noss. 1997. Conteo de huellas en brechas barridas: Un índice de abundancia para mamíferos. *Ecología en Bolivia* 30:55–67.

Cuéllar, R.L. 2000a. Fenología de las plantas frutales importantes en la dieta de vertebrados frugívoros en el Chaco boliviano. In E. Cabrera, C. Mercolli, and R. Resquin, eds. *Manejo de fauna silvestre en Amazonía y Latinoamérica*, pp. 47–56. Asunción, Paraguay: CITES Paraguay, Fundación Moises Bertoni, University of Florida.

Cuéllar, R.L. 2000b. *(r)Para qué sirven los animales de monte en el Izozog?* Santa Cruz, Bolivia: Kaa-Iya Project.

Cuéllar, R.L. 2000c. Uso de los animales silvestres por pobladores izoceños. In E. Cabrera, C. Mercolli, and R. Resquín eds., *Manejo de fauna silvestre en Amazonia y Latinoamérica*, pp. 471–484. Asunción, Paraguay: CITES Paraguay, Fundación Moises Bertoni, University of Florida.

Cullen, Jr., L, R. E. Bodmer, and C. Valladares-Padua. 2000. Effects of hunting in habitat fragments of the Atlantic forests, Brazil. *Biological Conservation* 95:49–56.

Cullen, Jr., L, R. E. Bodmer, and C. Valladares-Padua. 2001. Ecological consequences of hunting in Atlantic forest patches, São Paulo, Brazil. *Oryx* 35(2), 137–144.

Cullen, Jr., L, M. Schmink, C. Valladares-Padua, and I. Morato. 2001. Agroforestry benefit zones: A tool for the conservation and management of Atlantic Forest fragments, São Paulo, Brazil *Natural Areas Journal* 21:345–355.

Deem, S. L., A. J. Noss, R. Villarroel, M. M. Uhart and W. B. Karesh. 2003. Examen serológico para agentes seleccionados de enfermedades infecciosas en la urina (*Mazama gouazoubira*) en estado silvestre y ganado del doméstico (*Bos taurus*) en el Gran Chaco, Bolivia. In R. Polanco-Ochoa, ed., Manejo de Faune silvestre en Amazonia y Latinoamérica: Selección de trabajos V Congreso Internacional, pp. 81–88. Bogotá: Fundación Natura.

Deem, S. L., A. J. Noss, R. L. Cuéllar, R. Villarroel, D. J. Forrester and M. J. Linn. 2002. Sarcoptic mange in free-ranging pampas foxes in the Gran Chaco, Bolivia. *Journal of Wildlife Diseases*, 38:625–628.

Demment, M.W. and P. J. Van Soest. 1985. A nutritional explanation for body-size patterns of ruminant and nonruminant herbivores. *The Am. Nat.* 125:641–672.

Denevan, W. M. 1996. A bluff model of riverine settlement in prehistoric Amazonia. *Annals of the Association of American Geographers* 86:654–681.

Diaz, C. G. A. 1978. Social behavior of the collared peccary in captivity. *CEIBA* 22:73–126.

Dinerstein, E., D. Olson, D. J. Graham, A. L. Webster, S. A. Primm, M. P. Bookbinders, and G. Ledec. 1995. *A Conservation Assessment of the Terrestrial Ecoregions of Latin America and the Caribbean.* Washington, D.C.: World Bank.

Dixon, K. R. and J. A. Chapman. 1980. Harmonic mean measure of animal activity areas. *Ecology* 61:1040–1044.

Dobson, A. 1997. Some simple models for the dynamics of Cracid populations. In S. D. Strahl, S. Beaujon, D. M. Brooks, A. J. Begazo, G. Sedaghatkish, and F. Olmos, eds., *The Cracidae: Their Biology and Conservation*, pp. 107–115. Washington D.C.: Hancock House.

Dourojeanni, M. J. 1972. Impacto de la producción de la fauna silvestre en la economía de la Amazonía peruana. *Revista Forestal del Perú* 5:15–27.

Dourojeanni, M. J. 1985. Overexploited and underused animals in the Amazon region. In G. T. Prance and T. E. Lovejoy, eds., *Key Environments: Amazonia*, pp. 419–433. New York: Pergamon Press.

Dourojeanni, M. J. 1990. *Amazonia? Que Hacer?* Iquitos, Peru: CETA.

Dunbar, R. I. M. 1978. Competition and niche separation in a high altitude herbivore community in Ethiopia. *E.Afr. Wildl. J.* 16:183–199.

Durand, E. and D. McCaffrey. 1999. The Pacaya-Samiria Project: Enhancing conservation and improving livelihoods in Amazonian Peru. In C. Padoch, J. M. Ayres, M. Pinedo-Vasquez and A. Henderson, eds., *Várzea: Diversity, Conservation, and Development of Amazonia's Whitewater Floodplains*, pp. 233–246. New York: New York Botanical Garden Press.

Eaves, H. E. and R. C. Ruggiero. 2000. Socioeconomics and the sustainability of hunting in the forests of northern Congo (Brazzaville). In J. G. Robinson and E. L. Bennet, eds., *Hunting for Sustainability in Tropical Forests*, pp. 427–454. New York: Columbia University Press.

Eberhardt, L. L. and M. A. Simmons. 1992. Assessing rates of increase from trend data. *Journal of Wildlife Management* 56:603–610.

Eisenberg, J. F. 1979. Habitat, economy, and society: Some correlations and hypothesis for the Neotropical primates. In I. S. Bernstein and E. O. Smith, eds., *Ecological Influences on Social Organization: Evolution, and Adaptation*. New York: Garland S.T.P.M. Press.

Eisenberg, J. F. 1981. *The Mammalian Radiations: An Analysis of Trends in Evolution, Adaptation and Behavior*. Chicago: University of Chicago Press.

Eisenberg, J. F. 1989. *Mammals of the Neotropics. Vol. I. The Northern Neotropics*. Chicago.: University of Chicago Press.

Eisenberg, J. F. and R. W. Thorington, Jr. 1973. A preliminary analysis of a Neotropical mammal fauna. *Biotropica* 5:150–161.

Eiten, G. 1974. An outline of the vegetation of South America. *Symp. Congr. Int. Primatol. Soc. 5th*: 529–545.

Eletronorte. 1985. *Usina hidroelétrica de Tucurui, operação Curupira. Atividades Técnicas (ATE-015/85)*. Brasília.

Eletronorte. 1988. *Estudos de impacto ambiental (EIA). Vol. 2. Metodologia e diagnóstico ambiental*. Brasília.

Eletronorte. 1989. *Relatório da operação Jamari*. Brasília.

Ellis, R. 2002. *Diálogo entre conocimientos: Lecciones aprendidas de investigación participativa con pueblos indígenas.* Santa Cruz, Bolivia: CIDOB-DFID.

Ellison, L. N. and L. Cleary. 1978. Effects of human disturbance on breeding of Double-Crested Cormorants. *Auk* 95:510–517.

El Peruano. Normas Legales. 1995. *Régimen de la propiedad, comercialización, y sanciones por la caza de las especies de vicuña, Guanaco, y sus Híbridos No.5480.* Lima: Peruvian Federal Government.

Eltringham, S. K. 1984. *Wildlife Resources and Economic Development.* Chichester, U.K.: Wiley.

Emmons, L. 1999. *Neotropical Rain Forest Mammals: A Field Guide.* 2d ed. Chicago: University of Chicago Press.

Emmons, L. H. 1984. Geographical variation in densities and diversities of nonflying mammals in Amazonia. *Biotropica* 16:210–222.

Emmons, L. H. 1987. Comparative feeding ecology of felids in a neotropical rain forest. *Behav. Ecol. Sociobiol.* 20:271–283.

Engel, R. L. 1990. *A utilização de interações sociais do collared peccary (Tayassu tajacu) para a determinação da densidade de criação em cativeiro.* Undergraduate thesis. São Paulo: Biologia, Universidade Estadual Paulista.

Escamilla, A., M. Sanvicente, M. Sosa, and C. Galindo. 2000. Habitat mosaic, wildlife availability, and hunting in the tropical forest of Calakmul, Mexico. *Conservation Biology* 14:1592–1601.

Ewer, R. F. 1973. *The Carnivores.* Ithaca, N.Y.: Cornell University Press.

Fachín-Terán, A. 1992. Desove y uso de playas para nidificación de Taricaya (*Podocnemis unifilis*) en el río Samiria. Loreto-Perú. *Boletín de Lima* 79:65–75.

Fachín-Terán, A. 1994. Depredación de la taricaya *Podocnemis unifilis* en la Reserva Nacional Pacaya-Samiria, Loreto. *Boletín de Lima* 16 (91–96):417–423.

Fachín-Terán, A., M. Chumbe, and G. Taleixo. 1996. Consumo de tortugas de la Reserva Nacional Pacaya-Samiria. *Vida Silvestre Neotropical* 5:147–150.

Fachín-Terán, A., R. C. Vogt, and J. B. Thorbjarnarson. 2000. Padrões de caça e uso de quelônios na Reserva de Desenvolvimento Sustentável Mamirauá, Amazonas, Brasil. In: E. Cabrera, C. Mercolli, and R. Resquin, eds., *Manejo de fauna silvestre em Amazônia y Latinoamerica*, pp. 323–337. Asunción, Paraguay: CITES Paraguay, Fundación Moises Bertoni, University of Florida.

Falabella, P. G. R., ed. 1994. *A pesca no Amazonas: problemas e soluções.* Manaus, Brazil: Universidade do Amazonas.

Fang, T. F., R. E. Bodmer, R. Aquino, and M. H. Valqui, eds. 1997. *Manejo de fauna silvestre en la Amazonía.* La Paz: Editorial Instituto de Ecología.

Fang., T. F., O. L. Montenegro, and R. E. Bodmer, eds. 1999. *Manejo y conservación de fauna silvestre en América Latina.* La Paz: Instituto de Ecología.

Feare, C. J. 1976. The exploitation of Sooty Tern eggs in the Seychelles. *Biological Conservation* 10:169–181.

Fearnside, P. M. 1989. Brazil's Balbina dam: Environment versus the legacy of the pharaohs in Amazonia. *Environmental Management* 13:401–423.

Feinsinger, P., L. Margutti, and R. D. Oviedo. 1997. School yards and nature trails: Ecology education outside the university. *Trends in Ecology and Evolution* 12:115–20.

Fernandes, E. C. M. and Nair, P. K. R. 1986. An evaluation of the structure and function of tropical homegardens. *Agricultural Systems* 21:279–310.

Ferrari, S. F. 1993. Ecological differentiation in the Callitrichidae. In A. B. Rylands, ed., *Marmosets and Tamarins: Systematics, Behaviour, and Ecology.* Oxford: Oxford University Press.

Ferrari, S. F. and M. A. Lopes Ferrari. 1989. A reevaluation of the social organization of the Callitrichidae, with reference to the ecological differences between genera. *Folia Primatologica* 52:132–147.

Ferrari, S. F. and A. B. Rylands. 1994. Activity budgets and differential visibility in field studies of three marmosets (*Callithrix* spp.). *Folia Primatologica* 63:78–83.

Ferreira, A. R. 1971. *Viagem filosófica pelas capitanias do Grão Pará, Rio Negro, Mato Grosso e Cuiabá 1783–1792. Iconografia. Vol. 2. Zoologia.* Rio de Janeiro: Conselho Federal de Cultura.

Ferreira, A. R. 1972. *Viagem filosófica pelas capitanias do Grão Pará, Rio Negro, Mato Grosso e Cuiabá 1783–1792. Memórias: Zoologia, Botânica.* Rio de Janeiro: Conselho Federal de Cultura.

Fimbel, C., B. Curran, and L. Usongo. 2000. Enhancing the sustainability of duiker hunting through community participation and controlled access in the Lobéké region of southeastern Cameroon. In

J. G. Robinson and E. L. Bennett, eds., *Hunting for Sustainability in Tropical Forests*, pp. 356–374. New York: Columbia University Press.

Fischer, C. F. A. 1995. *Administração dos recursos pesqueiros na região do Médio Amazonas (Pará e Amazonas).* Brasilia: IBAMA.

Fischer, C. F. A., A. L. Chagas, and L. D. C. Dornelles. 1992. *Pesca de águas interiores.* Coleção Meio Ambiente: Série estudos: Pesca. 2. Brasilia: IBAMA.

Fittkau, E. J., U. Irmler, W. Junk, F. Reiss, and G. W. Schmidt. 1975. Productivity, biomass and population dynamics in Amazonian water bodies. In F.B Golley and E. Medina, eds., *Tropical Ecological Systems: Trends in Terrestrial and Aquatic Research*, pp. 284–311. New York: Springer-Verlag

Fitzgerald, L. A., J. M. Chani, and O. E. Donadío.1991. Tupinambis lizards in Argentina: Implementing management of a traditionally exploited resource. In J. Robinson and K. Redford, eds, *Neotropical Wildlife Use and Conservation*, pp. 303–316. Chicago: University of Chicago Press.

Flesher, K. 1999. Preliminary notes on the conservation status of Baird's tapir in northeastern Honduras. *Oryx* 33:294–300.

Flowers, N. M. 1983a. *Forager-Farmers: The Xavante Indians of Central Brazil.* Ph.D. dissertation. New York: City University of New York.

Flowers, N. M. 1983b. Seasonal factors in subsistence, nutrition, and child growth in a central Brazilian Indian community. In R. B. Hames and W. T. Vickers, eds., *Adaptive Responses of Native Amazonians*, pp. 357–392. New York: Academic Press.

Foerster, C. R. 1998. *Ambito de hogar, patrón de movimientos y dieta de la danta Centroamericana (Tapirus bairdii) en el Parque Nacional Corcovado, Costa Rica.* M.S. thesis. Heredia, Costa Rica: Universidad Nacional.

Fontenele, O. 1948. *Revisão bibliográfica sobre o pirarucu* (Arapaima gigas Cuvier 1815). Fortaleza, Brazil: DNOCS.

Fontes, G. M. 1966. *Alexandre Rodrigues Ferreira: Aspectos de sua vida e obras.* Manaus, Brazil: INPA.

Forman, R. T. T. 1995. *The Ecology of Landscapes and Regions.* New York: Cambridge University Press.

Forys, E. A. 1995. *Metapopulations of Marsh Rabbits: A Population Viability Analysis of the Lower Keys Marsh Rabbit* (Sylvilagus palustris hefneri). Ph.D. dissertation. Gainesville: University of Florida.

Fox, B. J. 1989. Community ecology of macropodoids. In G. Grigg, P. Jarman, and I. Hume, eds., *Kangaroos, Wallabies, and Rat-kangoroos*, pp 89–104. Chipping Norton, Australia: Beatty.

Fragoso, J. M. V. 1991. The effect of selective logging on Baird's tapir. In M. A. Mares and D. J. Schmidly, eds., *Latin American Mammalogy: History, Biodiversity, and Conservation*, pp. 295–304. Norman: University of Oklahoma Press.

Fragoso, J. M. V. 1994. *Large Mammals and the Community Dynamics of an Amazonian Rain Forest.* Ph.D. dissertation. Gainesville: University of Florida.

Fragoso J. M. V. 1998a. Home range and movement patterns of white-lipped peccary (*Tayassu pecari*) herds in the northern Brazilian Amazon. *Biotropica* 30:458–469.

Fragoso, J. M. V. 1998b. White-lipped peccaries and palms on the Ilha de Maracá. In W. Milliken and J.A. Ratter, eds., *Maracá: The Biodiversity and Environment of an Amazonian Rain Forest*, pp. 151–163. Chichester, U.K.: Wiley.

Fragoso, J. M. V. 1999. Perception of scale and resource partitioning by peccaries: Behavioral causes and ecological implications. *Journal of Mammalogy* 80:993–1003.

Fragoso, J. M. V., K. M. Silvius, and M. Prada Villa-Lobos. 2000. *Wildlife Management at the Rio das Mortes Xavante Reserve, MT, Brazil: Integrating Indigenous Culture and Scientific Methods for Conservation.* Brasilia: World Wildlife Fund-Brazil.

Frankham, R. 1995. Effective population size/adult population size ratios in wildlife: A review. *Genet Res. Camb.* 66:95–107.

Frankham, R. and I. R. Franklin. 1998. Response to Lynch and Lande. *Anim. Conserv.* 1:73.

Franklin, I. R. 1980. Evolutionary change in small populations. In M. E. Soulé and B. A. Wilcox, eds., *Conservation Biology: An Evolutionary-Ecological -Perspective*, pp. 135–149. Sunderland, Mass.: Sinauer.

Frasson, C. and Salgado, J. M. 1990. Capivara-Uma opção contra a fome e a deficiência de proteína animal. In *Anais interface nutrição X agricultura* (2° Simpósio), pp. 175–200. Piracicaba, Brazil: Fundação de Estudos Agrários "Luiz de Queiroz" (FEALQ).

Frederick, P. C. and M. W. Collopy. 1989. Researcher disturbance in colonies of wading birds: Effects of frequency of visits and egg-marking on reproductive parameters. *Colonial Waterbirds* 12:52–157.

Freese, C. H. 1997a. *Harvesting Wild Species: Implications for Biodiversity Conservation*. Baltimore: Johns Hopkins University Press.

Freese, C. H. 1997b. The "use it or lose it" debate: Issues of a consevation paradox. In C. H. Freese, ed., *Harvesting Wild Species: Implications for Biodiversity Conservation*, pp. 1–48. Baltimore: Johns Hopkins University Press.

Freese, C. H. 1998. *Wild Species as Commodities*. Washington, D.C.: Island Press.

Freese, C. H., P. G. Heltne, N. Castro, and G. Whitesides. 1982. Patterns and determinants of monkey densities in Peru and Bolivia, with notes on distribution. *International Journal of Primatology* 3:53–90.

Fritz, S. 1922. *Journal of the Travels and Labours of Father Samuel Fritz in the River of the Amazons between 1686 and 1723*. London: Hakluyt Society.

Fukushima, M., Takayama, Y., Habaguchi, T., and Nakano, M. 1997. Comparative hypocholesterolemic effects of capybara (*Hydrochoerus hydrochaeris*) oil, horse oil, and sardine oil in cholesterol-fed rats. *Lipids* 32:391–395.

Fyfe, D. A. and R. D. Routledge. 1991. *User's Manual for the Program TransAn. Version 1.00.* M.S. thesis. Burnaby, Canada: Simon Fraser University.

Gallagher, J. F., L. W. Varner, and W. E. Grant. 1984. Nutrition of the collared peccary in South Texas. *Journal of Wildlife Management* 48:749–761.

García Canclini, N. 1982. *Las culturas populares en el capitalismo*. México, D.F.: Editorial Nueva Imagen.

Garcia-Pelayo and R. Gross. 1979. *Nuevo Larousse básico*. Mexico, D.F.: Ediciones Larousse.

Gary, F. J. and B. C. Brown. 1984. Protein requirement of adult collared peccaries. *Journal of Wildlife Management* 49:351–355.

Gentry, A. 1988. Tree species richness of upper Amazonian forests. *Proceedings of the National Academy of Sciences* 85:156–159.

Gerrits, R. A. and H. Baas. 1997. *A Flood of Change: Appraisal on Development Impacts on Fish Stocks in the Middle Amazon Region and its Consequences for the Rural Population*. Unpublished report. Faro (PA), Brazil: Fundação Olho d'água.

Gibbons, J. W., J. L. Greene, and J. D. Congdon. 1990. Temporal and spatial movement patterns of slider and other turtles. In J. Whitfield Gibbons, ed., *Life History and Ecology of the Slider Turtle*, pp. 201–215. Washington, D.C.: Smithsonian Institution Press.

Gibbs, J. P., H. L. Snell, and C. E. Causton. 1999. Effective monitoring for adaptive wildlife management: Lessons from the Galápagos Islands. *Journal of Wildlife Management* 63:1055–1065.

Gilpin, M. 1997. Metapopulation and wildlife conservation: Approaches to modeling spatial structure. In D. R. McCullough, ed., *Metapopulation and Wildlife Conservation*, pp. 11–27. Washington, D.C.: Island Press.

Gilpin, M. E. 1987. Spatial structure and population vulnerability. In M. E. Soulé, ed., *Viable Populations for Conservation*, pp. 125–139. Cambridge: Cambridge University Press.

Glanz, W. E. 1982. Fauna de mamíferos terrestres de la Isla Barro Colorado: Censos y cambios a largo plazo. In E. G. Leigh, A. S. Rand, and D. M. Windsor, eds., *Ecología de un bosque tropical*, pp. 523–536. Balboa, Panamá.: Smithsonian Tropical Research Institute.

Godoy, R., N. Brokaw, and D. Wilkie. 1995. The effect of income on the extraction of forest products: Model, hypotheses, and preliminary findings from the Sumu Indians of Nicaragua. *Human Ecology* 23:29–52.

Godoy, R, K. Kirby, and D Wilkie. 2001. Tenure security, private time preference, and use of natural resources among lowland Bolivian Amerindians. *Ecological Economics* 38:105–118.

Godoy, R., D. Wilkie, and J. Franks. 1997. The effects of markets on neotropical deforestation. A comparative study of four Amerindian societies. *Current Anthropology* 38:875–878.

Godshalk, R. 1978. *El caimán del Orinoco, Crocodylus intermedius, en los llanos occidentales de Venezuela con observaciones sobre su distribución en Venezuela y recomendaciones para su conservación*. Caracas: FUDENA.

Gonzáles, L. 1998. La herpetofauna del Izozog. *Ecología en Bolivia* 31:45–52.

González, J. A. 1999a. Effects of harvesting of waterbirds and their eggs by native people in the northeastern Peruvian Amazon. *Waterbirds* 22 (2):217–224.

González, J. A. 1999b. *Propuesta de manejo de loros y guacamayos en la comunidad de Victoria*. Iquitos, Perú: Junglevagt for Amazonas, WWF/DK-AIF.

González, J. A. 1999c. Aves silvestres de importancia económica en el sector meridional de la Reserva Nacional Pacaya-Samiria (Loreto, Perú). In T. G. Fang, O. L. Montenegro, and R. E. Bodmer eds., *Manejo y conservación de fauna silvestre en América Latina*, pp. 315–328. La Paz: Editorial Instituto de Ecología.

González-Jiménez, E. 1977. Digestive physiology and feeding of capybaras. In: Rechcige, M., ed., *Nutrition and Food. Section G: Diets, Culture Media, Food Supplements*. Vol. 1, pp. 167–177. Cleveland: C.R.C. Press.

González-Jiménez, E., A. Escobar, and O. Caires. 1976. Un método para detectar coprofágia en chigüires: Resultados. In *Programa y resúmenes. II seminário sobre Chigüires (*Hydrochoerus hydrochaeris*) y Babas (*Caiman crododilus*)*, 2. Maracay, Venezuela: Facultad de Agronomia-UCV, CONICIT, IPA.

González-Jiménez, E. and R. Parra. 1972: Estudios sobre el chigüire (*Hydrochoerus hydrochaeris*). 1. Peso de los diferentes órganos y partes del cuerpo. A. *Cient. Ven*. 23:30.

González-Jiménez, E. G. 1995. *El capibara-estado actual de su producción*. FAO study. Rome: Animal Health Production, FAO.

Good, K. 1989. *Yanomami Hunting Patterns: Trekking and Garden Relocation as an Adaptation to Game Availability in Amazonia, Venezuela*. Ph.D. dissertation. Gainesville: University of Florida.

Gordon, I. J. 1989. Vegetation community selection by ungulates on the Isle of Rhum, III: Determinants of vegetation community selection. *J. Appl. Ecol*. 26:65–79.

Gottdenker, N. and R. E. Bodmer. 1998. Reproduction and productivity of white-lipped and collared peccaries in the Peruvian Amazon. *Journal of Zoology* 245:423–430.

Goulding, M. 1979. *Ecologia da pesca do Rio Madeira*. Manaus, Brazil: INPA.

Goulding, M. 1980. *The Fishes and the Forest*. Berkeley: University of California Press.

Goulding, M. 1983. The role of fishes in seed dispersal and plant distribution in Amazonian floodplain ecosystems. *Sonderbd. Naturwiss. Ver. Hamburg* 7:271–283.

Goulding, M. 1993. Flooded forests of the Amazon. *Scientific American* 268:114–120.

Goulding, M. 1999. Introduction: Fish and fisheries. In C. Padoch, J. M. Ayres, M. Pinedo-Vasquez, and A. Henderson, eds., *Várzea: Diversity, Conservation, and Development of Amazonia's Whitewater Floodplains*, pp. 3–6. New York: New York Botanical Garden Press.

Goulding, M., M. L. Carvalho, and E. G. Ferreira. 1988. *Rio Negro: Rich Life in Poor Water*. The Hague: SPB Academic Publishing.

Goulding, M., N. J. H. Smith, and D. J. Mahar. 1996. *Floods of Fortune: Ecology and Economy Along the Amazon*. New York: Columbia University Press.

Graham, L. R. 1995. *Performing Dreams: Discourses of Immortality Among the Xavante of Central Brazil*. Austin: University of Texas Press.

Graham, L. R. 2000. Lessons in collaboration: The Xavante.WWF wildlife management project in central Brazil. In R. Weber, J. Butler, and P. Larson, eds., *Indigenous Peoples and Conservation Organizations*, pp. 47–72. Washington, D. C.: World Wildlife Fund-U.S.

Grenand, P. 1996. Des fruits, des animaux, et des hommes: Stratégies de chasse et de pêche chez les Wayãpi d'Amazonie. In C. M. Hladik, A. Hladik, H. Pagezy, O. F. Linares, G. J. A. Koppert, and A. Froment, eds., *L'alimentation en forêt tropicale. Interactions bioculturelles et perspectives de développement*, pp. 671–684. Paris: UNESCO.

Gribel, R. 1990. The Balbina disaster: The need to ask why? *The Ecologist* 20:133–135.

Grimwood, I. R. 1969. *Notes on the Distribution and Status of Some Peruvian Mammals 1968*. Special publication No. 21. New York: American Committee for International Wild Life Protection and New York Zoological Society.

Gross, D., G. Eiten, N. M. Flowers, F. M. Leoi, M. L. Ritter, and D. Werner. 1979. Ecology and acculturation among native peoples of central Brazil. *Science* 206:1043–1050.

Gross, J. E., P. U. Alkon, and M. W. Demment. 1996. Nutrition ecology of dimorphic herbivores: Digestion and passage rates in Nubian ibex. *Oecologia* 107:170–178.

Guerrant, Jr., E. O. 1992. Genetic and demographic considerations in the sampling and reintroduction of rare plants. In P. L. Fiedler and S. K. Jain, eds., *Conservation Biology: The Theory and Practice of Nature Conservation, Preservation, and Management*, pp. 321–344. New York: Chapman and Hall.

Guerrero, J. and A. Arambiza. 2001. *Lista preliminar de las aves del P.N. Kaa-Iya del Gran Chaco e Izozog*. Santa Cruz, Bolivia: Kaa-Iya Project.

Guerrero, J., A. Arambiza, L. González, and E. Ity. 2000. Estudio de los Psittacidae sometidos a cacería por comunidades Guaraníes-Izozeñas. In W.R. Townsend, K. Rivero, C. Peña, and K. Linzer, eds., *Memorias del primer encuentro nacional de manejo de fauna en los Territorios Indígenas de Bolivia*, pp. 25–30. Santa Cruz, Bolivia: CIDOB.

Guinart, S. D. 1997. *Los mamíferos del bosque semideciduo neotropical de Lomerío (Bolivia): Interacción Indígena*. Ph.D. dissertation. Barcelona: Universidad de Barcelona.

Gumilla, J. 1963 [1745]. *El Orinoco illustrado y defendido*. Biblioteca Nacional de la Historia. No. 68. Caracas.

Hall, A. 1997. *Sustaining Amazonia: Grassroots Action for Productive Conservation*. Manchester: Manchester University Press.

Hames, R. B. 1980. Game depletion and hunting zone rotation among the Ye'kwanas and Yanomamo of Amazonas, Venezuela. *Working Papers on South American Indians* 2:31–66.

Hanek, G. 1982. *La pesquería en la Amazonía peruana: Presente y futuro*. FAO Technical Fisheries Document No 81. Rome: FAO.

Hansen, E. and C. Richard-Hansen. 2000. *Faune de Guyane. Guide des principales espèces soumises à réglementation*. Garies, France: Le Guen Editions.

Hansen, R. M., M. M. Mugambi, and S. M. Baun. 1985. Diets and trophic ranking of ungulates of the northern Serengeti. *Journal of Wildlife Management* 49:823–829.

Hanski, I. and M. Gilpin. 1991. Metapopulation dynamics: Brief history and conceptual domain. In M. Gilpin and I. Hanski, eds., *Metapopulation Dynamics: Empirical and Theoretical Investigations*, pp. 3–16. London: Academic Press.

Hanski, I. and M. E. Gilpin. 1997. *Metapopulation Biology: Ecology, Genetics, and Evolution*. San Diego: Academic Press.

Hanski, I. and D. Simberloff. 1997. The metapopulation approach, its history, conceptual domain, and application to conservation. In I. A. Hanski and M. E. Gilpin, eds., *Metapopulation Biology: Ecology, Genetics, and Evolution*, pp. 5–26. San Diego: Academic Press.

Harris, S. W. J. Cresswell, P. G. Forde, W. J. Trewhella, T. Whoollard, and S. Wray. 1990. Home range analysis using radio-tracking data: A review of problems and techniques particularly as applied to the study of mammals. *Mammal Review* 20:97–123.

Harrison, S. 1991. Local extinction in a metapopulation context: An empirical evaluation. In M. E. Gilpin and I. Hanski, eds., *Metapopulation Dynamics: Empirical and Theoretical Investigations*, pp. 73–88. London: Academic Press.

Hart, J. A. 2000. Impact and sustainability of indigenous hunting in the Ituri forest, Congo-Zaire: A comparison of unhunted and hunted duiker populations. In J. G. Robinson and E. L. Bennett, eds., *Hunting for Sustainability in Tropical Forests*, pp. 106–153. New York: Columbia University Press.

Hasler, R. 1996. *Agriculture, Foraging, and Wildlife Resource Use in Africa. Cultural and Political Dynamics in the Zambezi Valley*. New York: Kegan Paul International.

Haynes, A. M. 1987. Human exploitation of seabirds in Jamaica. *Biological Conservation* 41:99–124.

Hemming, J. 1978. *Red Gold: The Conquest of the Brazilian Indians, 1500–1760*. Cambridge, Mass.: Harvard University Press.

Henderson, P. A. and W. G. R. Crampton. 1997. A comparison of fish diversity and density between nutrient rich and poor lakes in the Upper Amazon. *J. Tropical Ecology* 13:175–198.

Henderson, P. A., W. D. Hamilton, and W. G. R. Crampton. 1998. Evolution and diversity in Amazonian floodplain communities. In D. M. Newbery, H. H. T. Prins, and N. D. Brown, eds., *Dynamics of Tropical Communities*. pp. 385–419. Oxford: Blackwell Science.

Herrera, E. A. 1985. Coprophagy in capybara, *Hydrochoerus hydrochaeris*. *J. Zool. Ser. A* 207:616–9.

Hill, K. and K. Hawkes. 1983. Neotropical hunting among the Aché of eastern Paraguay. In R. B. Hames and W. T. Vickers, eds., *Adaptive Responses of Native Amazonians*, pp. 139–188. New York: Academic Press.

Hill, K., and J. Padwe. 2000. Sustainability of Aché hunting in the Mbaracayú Reserve, Paraguay. In J. G. Robinson and E. L. Bennett, eds., *Hunting for Sustainability in Tropical Forests*, pp. 79–105. New York: Columbia University Press.

Hill, K., J. Padwe, C. Bejyvagi, A. Bepurangi, F. Jakugi, R. Tykuarangi, and T. Tykuarangi 1997. Impact of hunting on large vertebrates in the Mbaracayú Reserve, Paraguay. *Conservation Biology* 11:1339–1353.

Hirakawa, H. 2001. Coprophagy in leporids and other mammalian herbivores. *Mammal Review*, 31:61–80.

Hiraoka, M. 1995. Aquatic and land fauna management among the floodplain ribereños of the Peruvian Amazon. In T. Nishiza Wa and J. I. Uitto, eds., *The Fragile Tropics of Latin America: Sustainable Management of Changing Environments*, pp. 201–225. Tokyo: United Nations University Press.

Hitchcock, R. K. 2000. Traditional African wildlife utilization: Subsistence hunting, poaching, and sustainable use. In H. H. T. Prins, J. G. Grotenhuis, and T. T. Dolan, eds., *Wildlife Conservation by Sustainable Use*, pp. 389–416. Dordrecht, Netherlands: Kluwer Academic.

Hoces, D. 2000. Estado actual y perspectivas del mercado de fibra de vicuña. In B. Gonzalez, F. Bas, C. Tala, and A. Iriarte, eds., *Manejo sustentable de la vicuña*, pp. 223–231. Santiago: Pontificia Universidad Católica de Chile.

Hofmann, R. R. 1973. *The Ruminant Stomach: Stomach Structure and Feeding Habits of East African Game Ruminants.* Nairobi: Kenya Literature Bureau.

Hofmann, R. R. 1988. Anatomy of the gastrointestinal tract. In D. C. Church, ed., *The Ruminant Animal: Digestive Physiology and Nutrition*, pp. 14–43.Englewoods Cliffs, N.J.: Prentice Hall.

Hoogesteijn, R., A. Hoogesteijn, and E. Mondolfi. 1993. Jaguar predation and conservation: Cattle mortality caused by felines in three ranches in the Venezuelan Llanos. In N. Dunstone and M. L. Gorman, eds., *Mammals as Predators*, pp. 391–407. Symposia 65. London: Zoological Society.

Horn, H. S. 1966. Measurement of "overlap" in comparative ecological studies. *Am. Nat.* 100:419–424.

Howard, W. J., J. M. Ayres, D. Lima Ayres, and G. Armstrong. 1995. Mamirauá: A case study of biodiversity conservation involving local people. *Commonwealth Forestry Review* 74:76–79.

Howe, R. W., G. J. Davis, and V. Mosca. 1991. The demographic significance of 'sink' populations. *Biological Conservation* 57:239–255.

Hueck, K. 1972. *As florestas da América do Sul.* São Paulo: Editora Polígono e Editora Universidade de Brasilia.

Humboldt, A. 1975 [1859–1869]. *Del Orinoco al Amazonas. Viaje a las regiones equinocciales del Nuevo Continente.* Barcelona: Edt. Labor.

Hunter, M. L. 1996. *Fundamentals of Conservation Biology.* Cambridge, Mass.: Blackwell Science.

Hurt, R. and P. Ravn. 2000. Hunting and its benefits: An overview of hunting in Africa. In H. H. T. Prins, J. G. Grotenhuis, and T. T. Dolan, eds., *Wildlife Conservation by Sustainable Use*, pp. 295–314. Dordrecht, Netherlands: Kluwer Academic.

IBAMA. 1996. *Legislação pesqueira.* Brasilia: IBAMA.

IBAMA. 1997. *Administração participativa: Um desafio à gestão ambiental.* Brasilia: IBAMA.

IBAMA. 2000. *Projeto Iara.* www.ibama.gov.br.

IBGE. 1991. *Censo demográfico No. 4: Amazonas (Am): População residente por grupos de idade, seguindo as meso-regiões, os municípios, os destritos e o sexo.* Rio de Janeiro: IBGE.

IBGE. 2001. *Censo 2000: Sinopse preliminar.* www.ibge.gov.br.

IDRISIS. 1997. IDRISIS. Worcester, Mass.: Clark University Graduate School of Geography.

Illius, A. W. and I. J. Gordon. 1991. Prediction of intake and digestion in ruminants by a model of rumen kinetics integrating animal size and plant characteristics. *J. Agric. Scienc.* 116:145–157.

INEI. 1972. *Encuesta Nacional de Consumo de alimentos.* Lima: Instituto Nacional de Estadística-Perú.

INEI. 1993. *Censo Nacional de población y vivienda.* Lima: Instituto Nacional de Estadística e Informática-Perú.

INEI. 1995. *Compendo estadístico 1994–1995, Departamento Loreto.* Lima: Instituto Nacional de Estadística e Informática.

Iñigo-Elías, E. E. and M. A. Ramos. 1991. The psittacine trade in Mexico. In J. G. Robinson and K. H. Redford, eds., *Neotropical Wildlife use and Conservation*, pp. 380–392. Chicago: University of Chicago Press.

Ino, C., J. Kudrenecky, and W. R. Townsend. 2001. *Manejo de fauna en la TCO Sirionó: El recuento de taitetú (Pecari tajacu).* Publicaciones proyecto de investigación CIDOB-DFID No. 17. Santa Cruz, Bolivia: CIDOB.

INRENA 1994. *Evaluación poblacional de vicuñas a nivel nacional.* Lima: Ministerio de Agricultura.

INRENA. 1995 *Mapa ecológico del Perú. Guia explicativa.* Lima: Ministerio de Agricultura.

Instituto Nacional de Ecología (INE). 2000. *Programa de manejo, Reserva de la Biósfera Montes Azules, México.* México, D.F.: Secretaría de Medio Ambiente, Recursos Naturales y Pesca.

Instituto Nacional de Estadística, Geografía e Informática (INEGI). 2002. *Censo general de población y vivienda 2000.* Aguascalientes, México. (INEGI. *www.inegi.gob.mx/difusion/fpobla.html.*)

International Union for Conservation of Nature and Natural Resources (IUCN). 2000. 2000 IUCN Red list of threatened animals. Gland, Switzerland. (IUCN: Species Survival Commission. .)

IPAM 2000a. *Reservas de lago e o manejo comunitário da pesca. Meta 1: Desenvolver sistémas de manejo para os lagos de várzea.* www.ipam.org.br.

IPAM 2000b. *Instituto de Proteção do Meio Ambiente.* www.ipam.org.br.

Isaac, V. J., V. L. C. Rocha, and S. Mota. 1993. Considerações sobre a legislação da "piracema" e outras restrições da pesca da região do Médio Amazonas. In L. G. Furtado, W. Leitão, and A. F. d. Melo, eds., *Povos das Águas, Realidade e Perspectivas na Amazônia,* pp. 292–301. Belém, Brazil: PR/MCT/CNPQ-MPEG.

Isaac, V. J. and M. L. Ruffino. 1996. Population dynamics of tambaqui (*Colossoma macropomum* Cuvier 1818) in the lower Amazon, Brazil. *Fisheries Management and Ecology* 3:315–333.

Isaac, V. J., M. Ruffino, and D. McGrath. 1998. In search of a new approach to fisheries management in the middle Amazon region. In F. Funk, J. Heifetz, and J. Lanelle, eds., *Proceedings of the Symposium on Fishery Stock Assessment Methods for the 21st Century.* Anchorage: Alaska Sea Grant College Program.

IUCN. 1996. *Red List of Threatened Animals.* J. Baillie and B, Groombridge, comps. and eds. Gland, Switzerland: IUCN.

Jackson, J. E. and J. D. Giuletti. 1988. The food habit of pampas deer (*Ozotoceros bezoarticus celer*) in relation to its conservation in a relict natural grassland in Argentine. *Biol. Conserv.* 45:1–10.

Janzen, D. H. 1988. Tropical dry forests: The most endangered major tropical ecosystem. In E. O. Wilson, ed., *Biodiversity,* pp. 130–137. Washington, D.C.: National Academy Press.

Jarman, P. J. 1971. Diets of large mammals in the woodlands around Lake Kariba. *Oecologia* 8:157–178.

Jarman, P. J. and A. R. E. Sinclair. 1979. Feeding strategy and the pattern of resource partitioning in ungulates. In A. R. E. Sinclair and M. Norton-Griffiths, eds., *Serengeti: Dynamics of an Ecosystem,* pp 130–163. Chicago: University of Chicago Press.

Johns, A. D. 1987. Continuing problems for Amazon river turtles. *Oryx* 21:25–28.

Johns, A. D., and J. P. Skoruppa. 1987. Responses of rain forest primates to habitat disturbances: A review. *International Journal of Primatology* 8:157–191.

Johnson, E. G. and R. D. Routledge. 1985. The line transect method: A nonparametric estimator based on shape restrictions. *Biometrics* 41:669–679.

Johnson, M. K. 1982. Frequency sampling for microscopic analysis of botanical compositions. *J. Range. Manage.* 35:541–542.

Jorgenson, J. 2000. Wildlife conservation and game harvest by Maya hunters in Quintana Roo, Mexico. In J. Robinson and E. Bennett, eds., *Hunting for Sustainability in Tropical Forests,* pp. 251–266. New York: Columbia University Press.

Jorgenson, J. P. 1995. Maya subsistence hunters in Quintana Roo, Mexico. *Oryx* 29:49–57.

Jorgenson, J. P. and K. H. Redford. 1993. Humans and big cats as predators in the Neotropics. In N. Dunstone and M. L. Gorman, eds., *Mammals as Predators,* pp. 367–390. Symposia 65. London: Zoological Society.

Joshi, N. V. and M. Gadgil. 1991. On the role of refugia in promoting the prudent use of biological resources. *Theoretical Population Biology* 40:211–229.

Junglevagt for Amazonas. 1995. *Programa integrado para el desarrollo y la conservación Pacaya-Samiria.* Iquitos: AIF-WWF/DK.

Junk, W. J. 1973. Investigations on the ecology and production-biology of the "floating meadows" (Paspalo-Echinochloetum) on the middle Amazon. Part II. The aquatic fauna in the root zone of the floating vegetation. *Amazoniana* 4:9–102.

Junk, W. J. 1983. Aquatic habitats in Amazonia. *The Environmentalist* 3:24–34.

Junk, W. J. 1984. Ecology of the várzea floodplain of Amazonian white water rivers. In H. E. Sioli, ed., *The Amazon: Limnology and Landscape Ecology of a Mighty Tropical River and Its Basin.* Dordrecht, Netherlands: Junk.

Junk, W. J. 1997. General aspects of floodplain ecology. In W. Junk, ed., *The Central Amazon Flood-plain: Ecology of a Pulsing System*, pp. 3–20. Berlin: Springer.

Junk, W. J., P. B. Bayley, and R. E. Sparks. 1989. The flood pulse concept in river-floodplain systems. In D. P. Dodge, ed., *Proceedings of the International Large River Symposium*, pp. 110–127. Canadian Special Publications in Fisheries and Aquatic Sciences 106. Ottawa: National Research Council (NRC) Research Press.

Junk, W. J. and J. A. S. Nunes de Mello. 1987. Impactos ecológicos das represas hidrelétricas na bacia Amazônica Brasileira. *Tubinger Geographische Studien* 95:367–385.

Juste, J., J. E. Fa, J. Perez del Val, and J. Castroviejo. 1995. Market dynamics of bushmeat species in Equatorial Guinea. *Journal of Applied Ecology* 32:454–467.

Kaa-Iya Project. 2000. *Ore, mbaembimba kaa ipo reta. Nosotros, los animales del monte*. Santa Cruz, Bolivia: Kaa-Iya Project.

Kareiva, P. 1990. Population dynamics in spatially complex environments: Theory and data. *Phil. Trans. R. Soc. Lond. B*. 330:175–190.

Kellert, S. R., M. Black, C. Reid Rush, and A. J. Bath. 1996. Human culture and large carnivore conservation in North America. *Conservation Biology* 10:977–990.

Kettlewell, H. B. D. 1955. Selection experiments on industrial melanism in the lepidoptera. *Heredity* 9:323–342.

Kiltie, R. A. 1980. *Seed Predation and Group Size in Rain Forest Peccaries*. Ph.D. dissertation. Princeton. N.J.: Princeton University.

Kiltie, R. A. 1981. Stomach contents of rain forest peccaries (*Tayassu tajacu* and *T. pecari*). *Biotropica* 13(3):234–236.

Kiltie, R. A. and J. Terborgh. 1983. Observations on the behavior of rain forest peccaries in Peru: Why do white-lipped peccaries form herds? *Z Tierpsychologia* 62:241–255.

King, F. W. 1989. Conservation and Management. In C. H. Ross, ed., *Crocodiles and Alligators*, pp. 216–229. Silverwater, Australia: Golden Press.

Kinsey, W. G. 1981. The titi monkeys, genus *Callicebus*. In A. F. Coimbra-Filho and R. A. Mittermeier, eds., *Ecology and Behavior of Neotropical Primates. Vol. 1*.Rio de Janeiro: Academia Brasileira de Ciências.

Kirkwood, G. P., J. R. Beddington, and J. A. Rossouw. 1994. Harvesting species of different lifespans. In P. J. Edwards, R. M. May, and N. R. Webb, eds., *Large-Scale Ecology and Conservation Biology*, pp. 199–227. Oxford: Blackwell Science.

Kleymeyer, C. D. 1994. Cultural traditions and community-based conservation. In D. Western and R.M. Wright, eds., *Natural Connections: Perspectives in Community-Based Conservation*, pp. 323–336. Washington, D.C.: Island Press.

Knick, S. T. 1990. Ecology of bobcats relative to exploitation and a prey decline in southeastern Idaho. *Wildlife Monographs* 108:1–42.

Kohlhepp, G. 1984. Development planning and practices of economic exploitation in Amazonia: Recent trends in spatial organisation of a tropical frontier region in Brazil. In H. Sioli, ed., *The Amazon: Limnology and Landscape of a Mighty Tropical River and its Basin*, pp. 649–674. Dordrecht, Netherlands: Junk.

Konecny, M. J. 1989. Movement patterns and food habits of four sympatric carnivore species in Belize, Central America. In K. H. Redford and J. F. Eisenberg, eds., *Advances in Neotropical Mammalogy*, pp. 243–264. Gainesville, Fla.: Sandhill Crane Press.

Krebs, C. 1989. *Ecological Methodology*. New York: Harper Collins.

Krebs, C. J. and J. H. Myers. 1974. Population cycles in small mammals. *Adv. Ecol. Res.* 8:267–399.

Kuchling, G. 1988. Population structure, reproductive potential and increasing exploitation of the freshwater turtle *Erymnochelys madagascariensis*. *Biological Conservation* 43:107–113.

Laake, J. L., S. T. Buckland, D. R. Anderson, and K. P. Burnham. 1993. *DISTANCE User's Guide*. Fort Colins, Colo.: Colorado Cooperative Fish and Wildlife Service.

Lacy, R. C. 1992. The effects of inbreeding on isolated populations: Are minimum viable population sizes predictable? In P. L. Fiedler and S. K. Jain, eds., *Conservation Biology: The Theory and Practice of Nature Conservation, Preservation, and Management*, pp. 277–296. New York: Chapman and Hall.

Lama, M. 2000. Análisis de contenidos estomacales de *Tayassu pecari* y *Pecari tajacu* que habitan las tierras del Alto y Bajo Izozog. In E. Cabrera, C. Mercolli, and R. Resquin, eds., *Manejo de fauna sil-*

vestre en Amazonía y Latinoamérica, pp. 393–400. Asunción, Paraguay: CITES Paraguay, Fundación Moises Bertoni, University of Florida.

Lande, R. 1995. Mutation and Conservation. *Conservation Biology* 9:782–791.

Landeo, C. 1997. Usuarios del recurso taricaya (*Podocnemis unifilis*) en el río Manú. In T. F. Fang, R. E. Bodmer, R. Aquino, and M. H. Valqui, eds., *Manejo de fauna silvestre en la Amazonía*, pp.182–183. La Paz: Editorial Instituto de Ecología.

Langer, P. 1979. Adaptational significance of the forestomach of the collared peccary, *Dicotyles tajacu* (Mammalia: Artiodactyla). *Mammalia* 43:235–245.

Larghero, M. C. 2001. *Dieta y solapamiento de la población de venado de campo "Los Ajos"* (Ozotoceros bezoarticus L. 1758) *(Artiodactyla: Cervidae)*. M. S. thesis, Montevideo: University of Montevideo Republic.

Larkin, P. A. 1977. An epitaph to the concept of maximum sustained yield. *Transcripts of the American Fisheries Society* 106:1–11.

Lathrap, D. 1970. *The Upper Amazon.* New York: Praeger.

Laurance, W. F. 2000. Megadevelopment trends in the Amazon: Implications for global change. *Environmental Monitoring and Assessment* 61:113–122.

Laurance, W. F. 1991. Edge effects in tropical forests fragments: Application of a model for design of nature reserves. *Biological Conservation* 57:205–219.

Lavorenti, A. 1989. Domestication and potential for genetic improvement of capybara. *Revista Brasileira de Genética.* 12(2 supl.):137–144.

Lavorenti, A., P. B. Silva Neto, C. Meirelles, R. Machado Neto, I. U. Packer, W. R. S. Mattos, R. D. D'Arce, L. M. Verdade, and S. L. G. Nogueira-Filho. 1990. *Estudos dos aspectos biológicos, sociais e alimentares da capivara em cativeiro.* Unpublished scientific report. FINEP. Brasilia: Financiadora de Estudos e Projetos (FINEP).

Leaños, F. and R. L. Cuéllar. 2000. Monitoreo de cacería en comunidades Izoceñas: datos del automonitoreo. In E. Cabrera, C. Mercolli, and R. Resquin, eds., *Manejo de fauna silvestre en Amazonía y Latinoamérica*, pp. 83–88. Asunción, Paraguay: CITES Paraguay, Fundación Moises Bertoni, University of Florida.

Leeuwenberg, F. 1994. *Analise entno-zoológica e manejo de fauna cinegética na Reserva Indigena Xavante Rio das Mortes, aldeia Tenitipa, Mato Grosso, Brasil.* Unpublished final project report. Brasilia: WWF-Brazil.

Leeuwenberg, F. 1997a. Manejo de fauna cinegética na Reserva Indigena Xavante de Pimentel Barbosa, Estado de Mato Grosso, Brasil. In C. V. Padua and R. E. Bodmer, eds., *Manejo e conservação de vida silvestre no Brasil*, pp. 233–238. Brasília: CNPq/Belém, Brazil: Sociedade Civil Mamirauá.

Leeuwenberg, F. 1997b. Manejo adaptado para fauna cinégetica en reservas comunales indígenas: El ejemplo Xavante. In T. G. Fang, R. E. Bodmer, R. Aquino, and M. H. Valqui, eds., *Manejo de fauna silvestre en la Amazonia*, pp. 119–128 Lima: UNAP/University of Florida/UNDP-GEF/Instituto de Ecología.

Leeuwenberg, F. and J. Robinson 2000. Traditional management of hunting in a Xavante community in central Brazil: The search for sustainability. In J. Robinson and E. Bennett, eds., *Hunting for Sustainability in Tropical Forests*, pp. 375–394. New York: Columbia University Press.

Leite, P. A.G., A. Lima-Neto, R Fontana, and S. L.G. Nogueira Filho. 2001. Uso de uréia na alimentação de caititus (*Tayassu tajacu*) em cativeiro. In *Livro de resumos do XXVIII congresso Brasileiro de medicina veterinária*, p. 24. Salvador, Bahia, Brazil.

Leite, R. A. N. de. 1991. The effects of the Tucurui hydroelectric dam on the fish fauna of the lower Tocantins River (Pará, Brazilian Amazon). Abstract. *ASIH 71st Annual Meeting.* New York.

Lemos de Sá, R. 1995. *Effects of the Samuel Hydroelectric Dam on Mammal and Bird Communities in a Heterogeneous Amazonian Lowland Forest.* Ph.D. dissertation. Gainesville: University of Florida.

Levy, E. B. and E. A. Madden. 1933. The point method of pasture analysis. *N.Z.J. Agric.* 46:267–279.

Liao, W., D. S. Bhargava, and J. Das. 1988. Some effects of dams on wildlife. *Environmental Conservation* 15:68–70.

Lichtenstein, G., E. F. Oribe, M.Grieg-Gran, S. Mazzuchelli. 2001. Community Management of vicuñas in Peru. In B. Field, R. J. Warren, H. Okarma and P. R. Sievert, eds., *Wildlife, Land, and People: Priorities for the 21st Century. Proceedings of the 2d International Wildlife Management Congress* pp. 213–216. Bethesda, MD: The Wildlife Society.

Lima, D. M. 1992. *The Social Category Caboclo: History, Social Organization, Identity, and Outsider's*

Social Classification of the Rural Population of an Amazonian Region (the Middle Solimões). Ph.D. dissertation. Cambridge: University of Cambridge.

Lima, D. M. 1999. Equity, sustainable development, and biodiversity protection: Some questions about ecological partnerships in the Brazilian Amazon. In C. Padoch, J. M. Ayres, M. Pinedo-Vasquez, and A. Henderson, eds., *Várzea: Diversity, Conservation, and Development of Amazonia's Whitewater Floodplains,* pp. 247–263. New York: New York Botanical Garden Press.

Lima-Neto, A., S. S. C. Nogueira, and S. L. G. Nogueira-Filho. 2001. Aspectos comportamentais aplicados na determinação de técnicas de manejo alimentar para pecaris (Mammalia, Tayassuidae) em cativeiro. In *Livro de resumos do XXVIII congresso Brasileiro de medicina veterinária,* p. 22. Salvador, Bahia, Brazil.

Liva, H., L. F. D. Moraes, S. L. G. Nogueira-Filho, and A. Lavorenti. 1989. Aspectos da alimentação do cateto (*Tayassu tajacu*) em cativeiro. In *Anais do I congresso paulista de inicicação científica,* Piracicaba, Brazil.

Lochmiller, R. and Grant, W. 1982. Intraespecific agression results in death of a collared peccary. *Zoo Biology,* 1:161–162.

Lochmiller, R.L., E. C. Hellgren, J. F. Galagher, L. W. Varner, and W. E. Grant. 1989. Volatile fatty acids in the gastrointestinal tract of the collared peccary (*Tayassu tajacu*). *Journal of Mammalogy* 70:189–191.

López-Barbela, S. 1982. Determinación del ciclo estral en Chigüires (*Hydrochoerus hydrochaeris*). *Acta Científica Venezolana* 33:497–501.

López-Barbela, S. 1984. Una contribuición al conocimento de la reproducción del chigüire (*Hydrochoerus hydrochaeris*). *Informe Anual do IPA:*109–117.

López-Barbela, S. 1993. Gestación y reproducción post-parto en el báquiro de collar (*Tayassu tajacu*). *Rev. Fac. Agron.* 19:175–184.

Lowe McConnell, R. H. 1987. *Ecological Studies in Tropical Fish Communities.* Cambridge: Cambridge University Press.

Ludlow, M. E. and Sunquist, M. E. 1987. Ecology and behavior of ocelots in Venezuela. *Natl. Geogr. Res. Rep.* 3:447–461.

Lundberg, P., and N. Jonzén. 1999. Optimal population harvesting in a source-sink environment. *Evolutionary Ecology Research* 1:719–729.

Luthin, C. S. 1987. Status and conservation priorities for the world's stork species. *Colonial Waterbirds* 10:181–202.

Lynch, M. and Lande, R. 1998. The critically effective size for a genetically secure population. *Animal Conservation* 1:70–72.

MacArthur, R. H. 1972. *Geographical Ecology: Patterns in the Distribution of Species.* New York: Harper and Row.

Macdonald, D. W. 1981. Dwindling resources and the social behavior of capybara (*Hydrochoerus hydrochaeris*) (Mammalia). *Journal of Zoology, Lond.* 194:371–391.

MacMillan G. 1995. *At the End of the Rainbow? Gold, Land, and People in the Brazilian Amazon.* London: Earthscan.

Maffei, L. 2001. Estructura de edades de la urina (*Mazama gouazoubira*) en el Chaco boliviano. *Mastozoología Neotropical* 8:149–155.

Maffei, L. 2004. Age structure of two species of peccaries under hunting pressure in the Bolivian Chaco. *Mammalia* 67:575–578.

Maffei, L. and M. N. Becerra. 2000. Técnica básica para determinar la edad en ungulados silvestres en base al análisis de dientes. *Ecología en Bolivia* 34:39–44.

Magnat, C. 2000. *Le commerce de gibier en Guyane: Essai de caractérisation de la filière, et impact des personnes de passage dans la région.* DESS Report. CIRAD/EMVT.

Magnusson, W. E. 1993. Manejo da vida silvestre na Amazônia. In E. J. G. Ferreira, G. M. Santos, E. L M. Leão, and L. A. Oliveira, eds., *Bases científicas para estratégias de preservação e desenvolvimento da Amazônia,* Vol. 2, pp. 313–318. Manaus, Brazil: Instituto Nacional de Pesquisas da Amazônia.

Main, M. B., F. W. Weckerly, and V. Bleich. 1996. Sexual segregation in ungulates: New directions for research. *J. Mammal.* 77:449–461.

Malcolm, J. R. and J. C. Ray. 2000. Influence of timber extraction routes on central African small-mammal communities, forest structure, and tree diversity. *Conservation Biology* 14:1623–1638.

Mamani, A. M. 2000. *Historia natural y uso de* Ortalis canicollis *(charata) en Izozog, Provincia Cordillera, Departamento de Santa Cruz.* Undergraduate thesis. Santa Cruz, Bolivia: UAGRM.

Mamani, A. M. 2001. Aspectos biológicos y evaluación de la densidad poblacional de la charata *Ortalis canicollis* en Izozog, Prov. Cordillera del Dpto. Santa Cruz, Bolivia. In D. M. Brooks and F. González-García, eds., *Cracid Ecology and Conservation in the New Millenium*, pp. 87–100. Miscellaneous Publications No. 2. Houston: Houston Museum of Natural Science.

Mandujano, S., and V. Rico-Gray. 1991. Hunting, use, and knowledge of the biology of the white-tailed dee (*Odocoileus virginianus* Hays) by the Maya of central Yucatan, Mexico. *Journal of Ethnobiology* 11:175–183.

Mantovani, W and F. R. Martins. 1990. O método de pontos. *Acta Bot. Bras.* 4:95–122.

March, I. 1993. The white-lipped peccary. In W. L. R. Oliver, ed., *Pigs, Peccaries, and Hippos—Status Survey and Conservation Plan*, pp. 13–21. Gland, Switzerland: IUCN.

March, I. J. 1987. Los Lacandones de México y su relación con los mamíferos silvestres: Un estudio etnozoológico. *Biótica* 12:43–56.

March, I. J., E. J. Naranjo, R. Rodiles, D. A. Navarrete, M. P. Alba, P. J. Hernández, S. E. Domínguez, D. A. López, O. Jiménez, and V. H. Loaiza. 1996. *Diagnóstico para la conservación y manejo de la fauna silvestre en la selva Lacandona, Chiapas.* Final report. San Cristóbal de Las Casas, Chiapas, México: Secretaría de Medio Ambiente, Recursos Naturales y Pesca (SEMARNAP).

Marcum, C. L. and D. O. Loftsgaarden. 1980. A nonmapping technique for studying habitat preferences. *Journal of Wildlife Management* 44:963–968.

Martin, C. B. and E. B. Martin. 1991. Profligate spending exploits wildlife in Taiwan. *Oryx* 25:18–20.

Martin, C. H. G. 1983. Bushmeat in Nigeria as a natural resource with environmental implications. *Environmental Conservation* 10:125–132.

Martínez, T. and R. L. Cuéllar. 2003. Fenología de plantas frutales importantes para la fauna silvestre en la comunidad Kopere Brecha, Izozog, Santa Cruz, Bolivia. In R. Polanco-Ochoa, ed., *Manejo de Fauna silvestre en Amazonía y Latinoamérica: Selección de trabajos V Congreso Internacional*, pp. 174–177. Bogotá: Fundación Natura.

Matola, S., A. D. Cuarón, and H. Rubio-Torgler. 1997. Status and action plan of Baird's tapir (*Tapirus bairdii*). In D. M. Brooks, R. E. Bodmer, and S. Matola. eds.. *Tapirs: Status Survey and Conservation Action Plan*, pp. 29–45. IUCN/SSC Tapir Specialist Group. Gland, Switzerland: IUCN.

Maybury-Lewis, D. 1967. *Akwe-Shavante Society.* Oxford: Clarendon Press.

Mayer, J. J. and P. N. Brandt. 1982. Identity, distribution, and natural history of the peccaries, Tayassuidae. In M. A. Mares and H. H. Genoways, eds., *Mammalian Biology in South America*, pp. 433–455. Pittsburgh: University of Pittsburgh Press.

Mayer, J. J. and R. M. Wetzel. 1987. *Tayassu pecari. Mammalian Species* 293:1–7.

McCullough, D. 1987. The theory and management of *Odocoileus* populations. In C. Wemmer, ed., *Biology and Management of the Cervidae*, pp. 535–549. Washington, D.C.: Smithsonian Institution Press.

McCullough, D. R. 1996. Spatially Structured populations and harvest theory. *Journal of Wildlife Management* 60:1–9.

McDonald, D. R. 1977. Food taboos: A primitive environmental protection agency (South America). *Anthropos* 72:734–748.

McGoodwin, R. 1990. *Crisis in the World's Fisheries.* Palo Alto, Calif.: Stanford University Press.

McGrath, D., F. de Castro, E. Câmara, and C. Futemma. 1999. Community management of floodplain lakes and the sustainable development of Amazonian fisheries. In C. Padoch, J. M. Ayres, M. Pinedo-Vasquez, and A. Henderson, eds., *Várzea: Diversity, Conservation, and Development of Amazonia's Whitewater Floodplains*, pp. 59–82. New York: New York Botanical Garden Press.

McGrath, D., F. de Castro, and C. Futemma. 1994. Reservas de lago e o manejo comunitário de pesca no baixo Amazonas: Uma avaliação preliminar. In M. A. Dínção and I. M. Silveira, eds., *Amazônia e a crise da modernização.* Belém, Brazil: Museu Paraense Emilio Goeldi.

McGrath, D., F. de. Castro, C. Futemma, B. Amaral, and J. Calabria. 1994. Manejo comunitário de pesca nos lagos de várzea do Baixo Amazonas. In L. Furtado, A. Mello, and W. Leitão, eds., *Povo das águas: Realidade e perspectiva na Amazônia.* pp. 213–229. Belém, Brazil: Museu Paraense Emilio Goeldi.

McGrath, D. G., U. L. de Silva, and N. M. M. Crossa. 1998. A traditional floodplain fishery of the lower Amazon river, Brazil. *NAGA* (ICLARM quarterly) January-March 1998:4–11.

McNeely, J. A. 1988. *Economics and Biological Diversity: Developing and Using Economic Incentives to Conserve Biological Resources.* Gland, Switzerland: IUCN.

Mech, L. D. 1983. *Handbook of Animal Radio-Tracking.* Minneapolis: University of Minnesota Press.

Medellín, R. A. 1994. Mammal diversity and conservation in the Selva Lacandona, Chiapas, México. *Conservation Biology* 83:780–799.

Medem, F. 1981. *Los crocodylia de Sur America.* Vol. I. *Los crocodylia de Colombia.* Bogotá: Editorial Carrera 7a.

Medem, F. 1983. *Los crocodylia de Sur América.* Vol. II. Bogotá: Editorial Carrera 7a.

Medem, M. F. 1976. Recomendaciones respecto a contar el escamado y tomar las dimensiones de nidos, huevos y ejemplares de los Crocodylia y Testudines. *Lozania* 20:1–17.

Meggers, B. J. and C. Evans. 1983. Lowland South America and the Antilles. In. J. E. Jennings, ed., *Ancient South Americans.* pp. 286–335. San Francisco: W. H. Freeman.

Mena, P., J. Stallings, J. Regalado, and R. Cueva. 2000. The sustainability of current hunting practices by the Huaorani. In J. Robinson and E. Bennett, eds., *Hunting for Sustainability in Tropical Forests,* pp. 57–78. New York: Columbia University Press.

Mendes, A. 1999. *Determinação da ocorrência de cecotrofia em capivaras* (Hydrochoerus hydrochaeris hydrochaeris L. 1766). M. S. thesis. Piracicaba, São Paulo, Brazil: Escola Superior de Agricultura "Luiz de Queiroz."

Mendes, A., S. S. C. Nogueira, A Lavorenti, and S. L. G. Nogueira-Filho. 2000. A note on cecotrophy behaviour in capybara (Hydrochaeris hydrochaeris). *Applied Animal Behavior Science.* 66:161–167.

Mendoza, F. and A. Noss. 2003. Radiotelemetría de peni *Tupinambis rufescens* en el Chaco seco, Santa Cruz, Bolivia. In R. Polanco-Ochoa, ed., *Manejo de Fauna silvestre en Amazonía y Latinoamérica: Selección de trabajos V Congreso Internacional,* pp. 178–180. Bogotá: Fundación Natura.

Menezes, R. S. 1951. *Notas biológicas e econômicas sobre o pirarucu.* Serie de Estudos tecnicos No. 3. Rio de Janeiro: Ministerio da Agricultura.

Merona, B. and M. M. Bittencourt. 1988. A pesca na Amazônia através dos desembarques no mercado de Manaus: Resultados preliminares. *Mem. Soc. Cienc. Nat. La Salle* 48(suplemento):433–453.

Merona, B. de. 1990. Amazon fisheries: General characteristics based on two case studies. *Interciencia* 15:461–468.

Meschkat, A. 1961. *Reports to the Government of Brazil on the Fisheries of the Amazon Region.* Rome: BRA/TE/Fi.

Milliken, W. and J. A. Ratter. 1989. *The Vegetation of the Ilha de Maracá: First Report of the Vegetation Survey of the Maracá Rain Forest Project.* Edinburgh: Royal Botanical Garden.

Ministerio de Ecologia y Recursos Naturales Renovables. 1993. *Áreas protegidas de la selva paranaense en la provincia de Missiones, Argentina.* Missiones: MERNR.

Ministerio de Agricultura. 1997. Programa de fortalecimiento de la competitivida comunal en la crianza de ucuñas. Lima: Ministerio.

Miserendino, R. S. 2002. *Uso de hábitat y área de acción del taitetú* (Pecari tajacu) *en la zona de Cerro Cortado, Izozog (Gran Chaco), Provincia Cordillera, Santa Cruz, Bolivia.* M.S. thesis. La Paz: Instituto de Ecología, UMSA.

Mittermeier, R. A. 1975. A turtle in every pot. A valuable South American resource going to waste. *Animal Kingdom* April-May:9–14.

Mittermeier, R. A., N. Myers, J. B. Thomsen, G. A. B. da Fonseca, and S. Olivieri. 1998. Biodiversity hotspots and major tropical wilderness area: Approaches to setting conservation priorities. *Conservation Biology* 12:516–520.

Mittermeier, R. A., P. Robles-Gil, and C. Goettsch Mittermeier. 1997. *Megadiversity: Earth's Biologically Wealthiest Nations.* Mexico City: CEMEX.

Mittermeier, R. A., and M. G. M. van Roosmalen. 1981. Preliminary observations on habitat utilization and diet in eight Surinam monkeys. *Folia Primatologica* 36:1–39.

Moll, D. 1986. The distribution, status, and level of exploitation of the freshwater turtle *Dermatemys mawei* in Belize, Central America. *Biological Conservation* 35:87–96.

Mones, A. and J. Ojasti. 1986: Hydrochoerus hydrochaeris. *Mammalian Species.* 264:1–7.

Montaño, R. R. 2000. Monitoreo de cacería de *Tupinambis* spp. (Sauria: Teiidae) en el Izozog: Provincia Cordillera, Santa Cruz, Bolivia. In E. Cabrera, C. Mercolli, and R. Resquin, eds., *Manejo de fauna silvestre en Amazonía y Latinoamérica,* pp. 103–106. Asunción, Paraguay: CITES Paraguay, Fundación Moises Bertoni, University of Florida.

Montaño, R. R. 2001. *Monitoreo de cacería y aspectos de la biología de Tupinambis* spp. (Sauria: Teiidae) en el Izozog, Provincia Cordillera, Santa Cruz, Bolivia. Undergraduate thesis. Santa Cruz, Bolivia: UAGRM.

Montreuil, V., A. García, R. Rodriguez, and R. Del Aguila. 1997. Rendimiento maximo sostenible de la pesquería comercial de *Prochilodus nigricans* en la Amazonía peruana. Symposium internacional de fauna silvestre. Iquitos, Perú.

Montreuil, V. and S. Tello. 1990. *Rendimiento máximo sostenible de la pesquería comercial de Loreto.* Fishbyte 8. Manila, Philippines: ICLARM.

Moreira, G. and R. C. Vogt. 1990. Movements of *Podocnemis expansa* before and after nesting in the Trombetas river, Brazil. In *Abstracts of the 38th Annual Meeting Herpetologist' League and the 33d Annual Meeting of the Society for the Study of Amphibians and Reptiles,* p.79. New Orleans:Tulane University.

Moreira, J. R. and D. W. Macdonald.1997. Técnicas de manejo de capivaras e outros grandes roedores na Amazônia. In C. Valladares-Padua, R. E. Bodmer; and L. Cullen, Jr., eds., *Manejo e conservação de vida silvestre no Brasil,* pp. 186–213. Brasília: CNPq/Belém, Brazil: Sociedade Civil Mamirauá.

Moskovits, D. K. 1985. *The Behavior and Ecology of the Two Amazonian Tortoises,* Geochelone carbonaria *and* Geochelone denticulata, *in Northwestern Brazil.* Ph.D. dissertation. Chicago: University of Chicago.

Mourão, G. M. 1999. Uso comercial da fauna silvestre no Pantanal: Lições do passado. In *Anais do II simpósio sobre recursos naturais e sócio-econômicos do Pantanal,* pp. 39–45. Corumbá, Brazil: Ed Embrapa-Pantanal.

Mozans, H. J. 1910. *Up the Orinoco and Down the Magdalena.* New York: Appleton.

Mukerjee, M. 1994. What's in a name? *Scientific American* 271:26.

Munn, C. A. 1988. Macaw biology in Manu National Park, Peru. *Parrotletter* 1:18–21.

Munn, C. A., D. A. Blanco, E. Nycander, and D. Ricalde. 1991. Prospects for sustainable use of large macaws in southeastern Peru. In J. Clinton-Eitniear, ed., *Proceedings of the First Mesoamerican Workshop on Conservation and Management of Macaws,* pp. 42–47. Misc. Publication No. 1. San Antonio, Tex.: Center for the Study of Tropical Birds, Inc.

Murray, M. G. and D. Brown. 1993. Niche separation of grazing ungulates in the Serengeti: An experimental test. *J. Anim. Ecol.* 62:380–389.

Murray, M. G. and A. W. Illius. 2000. Vegetation modification and resource competition in grazing ungulates. *Oikos* 89:501–508.

Naranjo, E. J. 1995. Abundancia y uso de hábitat del tapir (*Tapirus bairdii*) en un bosque tropical húmedo de Costa Rica. *Vida Silvestre Neotropical* 4:20–31.

Naranjo, E. J. 2002. *Population Ecology and Conservation of Ungulates in the Lacandon Forest, Mexico.* Ph.D. dissertation. Gainesville: University of Florida.

Naranjo, E. J. and E. Cruz. 1998. Ecología del tapir en la Reserva de la Biósfera La Sepultura. *Acta Zoológica Mexicana* 73:111–125.

National Research Council. 1981. *Techniques for the Study of Primate Populations Ecology.* Washington, D.C.: National Academy Press.

Navarro, G. 1999. *Ecología biogeográfica y uso de especies forestales y leñas en el Parque Nacional y ANMI Kaa-Iya del Gran Chaco.* Technical Report No. 50. Santa Cruz, Bolivia: Kaa-Iya Project.

Neu, C. W., C. R. Byers, and J. M. Peek. 1974. A technique for analysis of utilization-availability data. *Journal of Wildlife Management* 38:541–545.

Nogueira, S. S. C. 1997. *Manejo reprodutivo da capivara* (Hidrochoerus hydrochaeris hydrochaeris) *em sistema intensivo de criação.* Ph.D. Dissertation. Sao Paulo: Instituto de Psicologia, Departamento de Psicologia Experimental, Universidade de São Paulo.

Nogueira, S. S. C., S. L .G. Nogueira-Filho, E. Otta, C. T. S. Dias, and A. Carvalho. 1999. Determination of the causes of infanticide in capybara (*Hydrochaeris hydrochaeris*) groups in captivity. *Applied Animal Behavior Science.* 62:351–357.

Nogueira, S. S. C., E. Otta, C. T. S. Dias, and S. L. G. Nogueira-Filho. 2000. Alloparental behavior in the capybara (*Hydrochoerus hydrochaeris*). *Revista de Etologia* 2:17–21.

Nogueira-Filho, S. L. G.1990. *Estudos da digestibilidade de nutrientes em caitetus (*Tayassu tajacu) *adultos submetidos a dietas com níveis crescentes de alimentos volumosos.* M.S. thesis. Piracicaba, Brasil: ESALQ/USP.

Nogueira-Filho, S. L. G. 1996. *Manual de criação de capivaras.* Viçosa, Brazil: Centro de Produções Técnicas (CPT).

Nogueira-Filho, S. L. G. 1999. *Manual de criação de catetos e queixadas.* Viçosa, Brazil: Centro de Produções Técnicas (CPT).

Nogueira-Filho, S. L. G., A. Lavorenti, R. L. Engel, E. K. Tabuchi, and P. S. Camargo. 1991. *A criação em manejo semiextensivo do cateto (*Tayassu tajacu*) e do queixada (*Tayassu pecari*).* Unpublished scientific report. Brasilia: CNPq.

Nogueira-Filho, S. L. G. and S. S. C. Nogueira. 2000. Criação comercial de animais silvestres: Produção e comercialização da carne e subprodutos na região sudeste do Brasil. *Revista Econômica do Nordeste* 31:188–195

Nogueira-Filho, S. L. G., S. S. C. Nogueira, and T. Sato. 1999. A estrutura social de pecaris (Mammalia, Tayassuidae) em cativeiro. *Revista de Etologia* 1:89–98.

Noss, A. 2000. La sostenibilidad de la cacería de subsistencia Izoceña. In E. Cabrera, C. Mercolli, and R. Resquin, eds., *Manejo de fauna silvestre en Amazonía y Latinoamérica*, pp. 535–544. Asunción, Paraguay: CITES Paraguay, Fundación Moises Bertoni, University of Florida.

Noss, A. and E. Cuéllar. 2000. Índices de abundancia para fauna terrestre en el Chaco boliviano: Huellas en parcelas y en brechas barridas. In E. Cabrera, C. Mercolli, and R. Resquin, eds., *Manejo de fauna silvestre en Amazonía y Latinoamérica*, pp. 73–82. Asunción, Paraguay: CITES Paraguay, Fundación Moises Bertoni, University of Florida.

Noss, A. J. 1999. Manejo de fauna comunitario en el Gran Chaco, Bolivia. In T. G. Fang, O. L. Montenegro, and R. E. Bodmer, eds., *Manejo y conservación de fauna silvestre en América Latina*, pp. 109–116. La Paz: Instituto de Ecología.

Noss, A. J. and R. L. Cuéllar. 2001. Community attitudes towards wildlife management in the Bolivian Chaco. *Oryx* 35:292–300.

Noss, A. J., R. L. Cuéllar, and J. Ayala. In press. Drive counts for grey brocket deer in the Bolivian Chaco. *Mammalia.*

Novaro, A. J. 1997. *Source-Sink Dynamics Induced by Hunting: Case Study of Culpeo Foxes on Rangelands in Patagonia, Argentina.* Ph.D. dissertation. Gainesville: University of Florida.

Novaro, A. J., K. H. Redford, and R. Bodmer. 2000. Effect of hunting in source-sink systems in the Neotropics. *Conservation Biology* 14:713–721.

Novoa, D. 1989. The multispecies fisheries of the Orinoco River: Development, present status, and management strategies. In D. P. Dodge, ed., *Proceeding of the International Large River Symposium,* pp. 422–428. Canadian Special Publication of Fisheries and Aquatic Sciences No.106. Ottawa: Department of Fisheries and Oceans.

NRC (National Research Council). 1991. *Microlivestock: Little-Known Small Animals with a Promising Economic Future.* Washington, D.C.: National Academic Press.

Nugent, S. 1993. *Amazonian Caboclo Society: An Essay on Invisibility and Peasant Economy.* Providence: Berg.

Nunes, A. P. 1992. *Uso do habitat, comportamento alimentar e organização social de Ateles belzebuth belzebuth (Primates: Cebidae).* M.S. thesis. Belém, Brazil: Universidade Federal do Pará.

OCEI, 1993. *El censo 90 en Cojedes.* Caracas: Oficina Central de Estadística e Informática.

Odum, E. P. and E. J. Kuenzler. 1955. Measurement of territory and home range size in birds. *Auk* 72:128–137.

Ojasti, J. 1971. La tortuga arrau del Orinoco. *Defensa de la Naturaleza* 2:3–9

Ojasti, J. 1973. *Estudio biológico del chiguüre o capibara.* Caracas: Fondo Nacional de Investigaciones Agropecuarias.

Ojasti, J. 1991. Human exploitation of Capybara. In J. Robinson and K. Redford, eds., *Neotropical Wildlife Use and Conservation*, pp. 236–252. Chicago: University of Chicago Press.

Ojasti, J. 1993. *Utilización de la fauna silvestre en América Latina: Situación y perspectivas para un manejo sostenible.* Guía FAO Conservación 25. Rome: FAO.

Ojasti, J. and F. Dallmeier. 2000. *Manejo de fauna silvestre Neotropical.* Washington, D.C.: Smithsonian Institution/Man and Biosphere Program.

Ouhoud-Renoux, F. 1998. *De l'outil à la prédation. Technologie culturelle et ethno-écologie chez les Wayãpi du haut Oyapock (Guyane Française).* Ph.D. dissertation. Paris: Université Paris X.

Overton, S. 1971. Estimating the numbers of animals in wildlife populations. In R. H. Giles, ed., *Wildlife Management Techniques*, pp. 403–455. Washington, D.C.: Wildlife Society.

Owen-Smith, N. 1989. Morphological factors and their consequences for resource partitioning among African savanna ungulates: A simulation modelling approach. In D. W. Morris, Z. Abramsky, B. J. Fox, and M. R. Willig, eds., *Patterns in the Structure of Mammalian Communities*, pp 155–165. Lubbock, Tex.: Texas Tech University Press.

Owen-Smith, R. N. 1988. *Megaherbivores: The Influence of Very Large Body Size on Ecology*. Cambridge: Cambridge University Press.

Pacheco, T. 1983. *Efectos positivos y negativos de la veda de caza de 1973 en la Amazonia Peruana*. Lima: Universidad Nacional Agraria la Molina.

Packard, J. M, K. J. Babbitt, P. G. Hannon, and W. E. Grant. 1990. Infanticide in captive collared peccaries (*Tayassu tajacu*). *Zoo Biology* 9:49–53.

Padoch C. 1988. People of the floodplain and forest. In J. Denslow, and C. Padoch, eds., *People of the Tropical Forest*, pp. 127–141. Berkeley: University of California Press.

Pádua, C. V. 1993. *The Ecology, Behavior, and Conservation of the Black-Lion Tamarin (*Leontopithecus chrysopygus*, Mikan, 1823)*. Ph.D. dissertation. Gainesville: University of Florida.

Padua, S. M. 1991. *Conservation Awareness Through an Environmental Education School Program at Morro do Diabo State Park, São Paulo State, Brazil*. M.S. thesis. Gainesville.: University of Florida.

Padua, S. M. 1997. Uma pesquisa em educação ambiental: A conservação do mico-leão-preto *Leontopithecus chrysopygus*. In C. Valladares-Padua, R. E. Bodmer, and Cullen, Jr. L., eds., *Manejo e conservação de vida silvestre no Brasil*, pp. 34–51. Belém, Brazil: MCT-CNPq/Sociedade Civil Mamirauá.

Painter, M. and A. Noss. 2000. La conservación de fauna con organizaciones comunales: Experiencia con el pueblo Izoceño en Bolivia. In E. Cabrera, C. Mercolli, and R. Resquin, eds., *Manejo de fauna silvestre en Amazonía y Latinoamérica*, pp. 167–180. Asunción, Paraguay: CITES Paraguay, Fundación Moisés Bertoni, University of Florida.

Painter, M., A. Noss, R. Wallace, and L. Painter. 2003. El manejo comunitario de fauna en Bolivia: Criterios de sostenibilidad. In R. Polanco-Ochoa, ed., *Manejo de Fauna silvestre en Amazonía y Latinoamérica: Selección de trabajos V Congreso Internacional*, pp. 304–316. Bogotá: Fundación Natura.

Palmatary, H. C. 1939. *The Archaeology of the Lower Tapajós Valley, Brazil*. Philadelphia: American Philosophical Society.

Pantoja, E. R. 2000. *Operação Mamirauá*. Unpublished internal report. Tefé, Brazil: Instituto de Desenvolvimento Sustentável Mamirauá.

Parada, E. and J. Guerrero. 2000. Encuestas sobre el consumo de carne de monte y carne doméstica en las comunidades Izoceñas. In E. Cabrera, C. Mercolli, and R. Resquin, eds., *Manejo de fauna silvestre en Amazonía y Latinoamérica*, pp. 499–506. Asunción, Paraguay: CITES Paraguay, Fundación Moisés Bertoni, University of Florida.

Parada, E. and R. Villarroel. 2001. Estudio de salud de los animales silvestres. In W. R. Townsend, K. Rivero, C. Peña, and K. Linzer, eds. *Memorias del primer encuentro nacional de manejo de fauna en los Territorios Indígenas de Bolivia*, pp. 21–24. Santa Cruz, Bolivia: CIDOB.

Parnell, J. F., D. G. Ainley, H. Blokpoel, B. Cain, T. W. Custer, J. L. Dusi, S. Kress, J. A. Kushlan, W. E. Southern, L. E. Stenzel, and B. C. Thompson. 1988. Colonial waterbird management in North America. *Colonial Waterbirds* 11:129–169.

Parra, R. 1978. Comparison of foregut and hindgut fermentation in herbivores. In G. G. Montgomery, ed., *The Ecology of Arboreal Folivores*, pp 205–230. Washington, D.C.: Smithsonian Institution Press.

Parra, R., R. Escobar, and E. G. González-Jiménez. 1978. El chigüire, su potencial biológico y su cria en confinamiento. In *Informe anual IPA*, pp. 83–94. Maracay, Venezuela: Facultad de Agronomía, UCV.

Payne, J. C. 1992. *A Field Study of Techniques for Estimating Densities of Duikers in Korup National Park, Cameroon*. M.S. thesis. Gainesville: University of Florida.

Peres, C. 2000. Evaluating the impact and sustainability of subsistence hunting at multiple Amazonian forest sites. In J. Robinson and E. Bennett, eds., *Hunting for Sustainability in Tropical Forests*, pp 31–56. New York: Columbia University Press.

Peres, C. A. 1990. Effects of hunting on western Amazonian primate communities. *Biological Conservation* 54:47–59.

Peres, C. A. 1994. Indigenous reserves and nature conservation in Amazonian forests. *Conservation Biology* 8:586–588.

Peres, C. A. 1997. Evaluating the sustainability of subsistence hunting in tropical forests. Working paper GEC 97-2. Norwich, U.K.: Centre for Social and Economical Research on the Global Environment (CSERGE), University of East Anglia.

Peres, C. A. and J. W. Terborgh. 1995. Amazonian nature reserves: An analysis of the defensibility status of existing conservation units and design criteria for the future. *Conservation Biology* 9:34–46.

Perez, B. F. J. and I. J. Gordon. 1999. The functional relationship between feeding type and jaw and cranial morphology in ungulates. *Oecologia* 118:157–165.

Perez, B. F. J., I. J. Gordon, and C. Nores. 2001. Evolutionary transitions among feeding styles and habitats in ungulates. *Evolutionary Ecology Research* 3:221–230.

Petrere, Jr., M. 1986. Variation in the relative abundance of *tambaqui (Colossoma macropomum* Cuvier, 1818) based on catch and effort data of the gill net fisheries. *Amazoniana* 9:527–547.

Petrere, J. M. 1978a. Pesca e esforço de pesca no Estado do Amazonas. I. Esforço e captura por unidade de esforço. *Acta Amazonica* 8:439–454.

Petrere, J. M. 1978b. Pesca e esforço de pesca no Estado do Amazonas. II. Locais, aparelhos de captura, e estatística de desembarque. *Acta Amazonica* 8(2/3):1–54.

Petrere, M. J. 1983. Yield per recruitment of the tambaqui *Colossoma macropomum* Cuvier, in the Amazonas state. Brazil. *J. Fish. Biol.* 22:133–144.

Petrullo, V. 1939. The Yaruros of the Capanaparo river, Venezuela. *Smithsonian Institution Bulletin* (Bureau of American Ethnology) 123:167–289.

Pierret, P. V. and M. J. Dourojeanni. 1967. Importancia de la caza para la alimentación humana en el curso inferior del río Ucayali, Perú. *Revista Forestal del Perú* 1:10–21.

Pimm, S., M. Ayres, A. Balmford, G. Branch, K. Brandon, T. Brooks, R. Bustamante, R. Costanza, R. Cowling, L. M. Curran, A. Dobson, S. Farber, G. A. B. da Fonseca, C. Gascon, R. Kitching, J. McNeely, T. Lovejoy, R. A. Mittermeier, N. Myers, J. A. Patz, B. Raffle, D. Rapport, P. Raven, C. Roberts, J. P. Rodríguez, A. B. Rylands, C. Tucker, C. Safina, C. Samper, M. L. Stiassny, J. Supriatna, D. H. Wall, and D. Wilcove. 2001. Can we defy nature's end? *Science* 293:2207–2208.

Pinder, L. 1997. *Niche Overlap Among Brown Brocket Deer, Pampas Deer, and Cattle in the Pantanal of Brazil.* Ph.D. dissertation. Gainesville: University of Florida.

Pinder, L. and S. Rosso. (1998) Classification and ordination of plant formations in the Pantanal of Brazil. *Plant Ecology* 136(2):151–165.

Pinedo-Vasquez, M. 1988. The river people of Maynas. In J. S. Denslow and C. Padoch, eds., *People of the Tropical Forest*, pp. 141–142. Berkeley: University of California Press.

Pinto, M. N. 1994. *Cerrado: Caracterização, ocupação e perspectivas.* Brasília: Edunb.

Pires, A. 1996. *Disperção de sementes na várzea do Solimões: Amazonas.* M.S. thesis. Belém, Brazil: Universidade Federal do Pará.

Plotkin, M. and L. Famolare. 1992. *Sustainable Harvest and Marketing of Rain Forest Products.* Washington, D.C.: Island Press.

Polanco-Ochoa, R. 2003. *Manejo de Fauna silvestre en Amazonía y Latinoamérica: Selección de trabajos V Congreso Internacional.* Bogotá: Fundación Natura.

Polisar, J. 1995. River turtle reproductive demography and explotation patterns in Belize: Implications for management. *Vida Silvestre Neotropical* 4:10–19.

Polisar, J. 2000. *Jaguars, Pumas, Their Prey Base, and Cattle Ranching: Ecological Perspectives of a Management Issue.* Ph.D. dissertation. Gainesville: University of Florida.

Ponce, M. 1979. *Podocnemis unifilis* Troschel 1848 "taricaya" (Chelonia, Pleurodira, Pelomedusidae) en el Bosque Nacional "Alexander von Humboldt," Loreto-Perú. Undergraduate thesis. Lima: Universidad Nacional Agraria.

Por, F. D. 1995. *The Pantanal of Mato Grosso.* Dordrecht, Netherlands: Kluwer Academic.

Porro, A. 1981. Os Omagua do alto Amazonas. Demografia e padrões de povoamento no século XVII: Contribuições a antropologia em homenagem ao professor Egon Schaden. Museu Paulista. *Série Ensaios da Universidade de São Paulo.* 4:207–231.

Porro, A. 1983. Os Solimões ou Jurimaguas. Território, migrações e comércio intertribal. *Revista do Museu Paulista (new series)* 29(23–38).

Posada, L. 1991. *Levantamientos de tipos de bosques y uso Actual del suelo del parque Nacional natural Utría y cuenca del río Valle, Departamento del Chocó.* Bogotá: Fundación Natura.

Posey, D. A. 1985. Indigenous management of tropical forest ecosystems: The case of the Kayapó Indians of the Brazilian Amazon. *Agroforestry Systems* 3:139–158.

Posey, D.A. and W. Baleé. 1989. *Resource Management in Amazonia: Indigenous and Folk Strategies. Advances in Economic Botany Vol. 7*. New York: New York Botanical Garden Press.

Poupard, J. P., A. Simão, A. T. Quintão, E. R. Porto, and E. M. Comastri. 1981. *Plano de manejo do Parque Nacional do Iguaçu*. Brasilia: IBDF-DN.

Prance, G.T. 1979. Notes on the vegetation of Amazonia. III. The terminology of Amazonian forest types subject to inundation. *Brittonia* 31:26–38.

Pratt, M.L. 1992. *Imperial Eyes: Travel Writing and Transculturation*. London: Routledge.

Prescott-Allen, R. 1996. *Assessing the Sustainability of Uses of Wild Species: Case Studies and Initial Assessment Procedure*. Gland, Switzerland: IUCN.

Primack, R. 1993. *Essentials of Conservation Biology*. Sunderland, Mass.: Sinauer.

Pritchard, P. C. H. and P. Trebbau. 1984. *The Turtles of Venezuela*. Society for the Study of Amphibians and Reptiles, Contrib. Herpetol., No. 2. Ann Arbor, Mich.

Puertas, P. and R. E.Bodmer. 1993. Conservation of a High Diversity Primate Assemblage. *Biodiversity and Conservation* 2:586–593.

Puertas, P. E. 1998. *Hunting Effort Analysis in Northeastern Peru: The Case of the Reserva Comunal Tamshiyacu-Tahuayo*. M.S. thesis. Gainesville: University of Florida.

Puig, S., F. Videla, M. I. Cona, and S. A. Monge. 2001. Use of food availability by guanacos (*Lama guanicoe*) and livestock in Northern Patagonia (Mendoza, Argentina). *Journal of Arid Environments* 47:291–308.

Pulliam, H. R. 1988. Sources, sinks, and population regulation. *American Naturalist* 132:652–661.

Pulliam, H. R. 1996. Sources and sinks: Empirical evidence and population consequences. In O. E. Rhodes, R. K. Chesser, and M. H. Smith, eds,. *Population Dynamics in Ecological Space and Time*, pp. 45–69. Chicago: University of Chicago Press.

Pulliam, H. R. and B. J. Danielson. 1991. Sources, sinks, and habitat selection: A landscape perspective on population dynamics. *American Naturalist* 137:50–56.

Putman. R. J. 1996. *Competition and Resource Partitioning in Temperate Ungulate Assemblies*. London: Chapman and Hall.

Pyrah, D. 1984. Social distribution and population estimates of coyotes in north central Montana. *Journal of Wildlife Management* 48:679–690.

Queiroz, H. L. 1999. A pesca, as pescarias e os pescadores de Mamirauá. In H. Queiroz and W. G. R Crampton, eds., *Estratégias para manejo de recursos pesqueiros em Mamirauá*. Brasília: SCM-CNPq.

Queiroz, H. L. and W. G. R. Crampton, eds. 1999a. *Estratégias para manejo de recursos pesqueiros em Mamirauá: Estudos do Mamirauá*. Brasílial: Sociedade Civil Mamirauá/CNPq.

Queiroz, H. L. and W. G. R. Crampton. 1999b. O manejo integrado dos recursos pesqueiros em Mamirauá. In H. Queiroz and W.G. R Crampton, eds., *Estratégias para manejo de recursos pesqueiros em Mamirauá*, pp. 177–191. Brasília: SCM-CNPq.

Queiroz, H. L. and A. D. Sardinha. 1999. A preservação e o uso sustentado dos pirarucus (*Arapaima gigas*, Osteoglossidae) em Mamirauá. In H. L. Queiroz and W. G. R. Crampton, eds., *Estratégias de manejo para recursos pesqueiros na Reserva de Desenvolvimento Sustentável Mamirauá*, pp. 108–141. Brasília: MCT-CNPq/Sociedade Civil Mamirauá.

Quigley, H. B. and Crawshaw, P. G., Jr. 1992. A conservation plan for the jaguar *Panthera onca* in the Pantanal region of Brazil. *Biol. Conserv.* 61:149–157.

Quintela, C. E. 1990. An SOS for Brazil's beleaguered Atlantic Forest. *Nat. Cons. Mag.* 40:14–19.

Rabinowitz, A. 1993. *Wildlife Field Research and Conservation Training Manual*. New York: Paul-Art Press.

Rabinowitz, A. R. and B. G. Nottingham. 1986. Ecology and behaviour of the jaguar (*Panthera onca*) in Belize, Central America. *J. Zool. London* 210:149–159.

Ralls, K. and J. Ballou. 1983. Extinctions: Lessons from zoos. In C. M. Schonewald-Cox, S. M. Chambers, B. MacBryde, and L. Thomas, eds., *Genetics and Conservation: A Reference for Managing Wild Animal and Plant Populations*, pp. 164–184. Menlo Park, Calif.: Benjamin/Cummings.

Ralls, K., P. H. Harvey, and A. M. Lyles. 1986. Inbreeding in natural populations of birds and mammals. In M. E. Soulé, ed., *Conservation Biology: The Science of Scarcity and Diversity*, pp. 35–56. Sunderland, Mass.: Sinauer.

Ramo, C. and B. Busto. 1986. Censo aéreo de caimanes (*Crocodylus intermedius*) en el río Tucupido (Portuguesa, Venezuela) con observaciones sobre su actividad de soleamiento. In J. P. Ross, ed.,

Crocodiles: Proceedings of the Seventh Working Meeting of Crocodile Specialist Group, pp. 109–119. Gland, Switzerland: IUCN.

Ramos, A. R. 1998. *Indigenism: Ethnic Politics in Brazil*. Madison: University of Wisconsin Press.

Rebêlo, C. H. 1985. *A situação dos quelônios aquáticos do Amazonas, comercio e conservação*. Unpublished internal report. Amazonas: Instituto Brasileiro de Desenvolvimento Florestal (IBDF)/AM Polamazonia.

Rebêlo, G. H. and L. Lugli. 1996. The conservation of freshwater turtles and the dwellers of the Amazonian Jaú National Park (Brazil). In S. K. Jain, ed., *Ethnobiology in Human Welfare*, pp. 253–258. New Delhi: Deep.

Rebêlo, G. H., G. Moreira, L. Lugli, L. Marajó, J. C. Raposo, A. L. Queiroz, R. F. Cruz, and C. Reimann. 1996. *Os quelônios do Parque Nacional do Jaú (AM)*. Unpublished technical report. Manaus, Brazil: Fundação Vitória Amazônica.

Redford, K. H. 1983. *Mammalian Myrmecophagy: Feeding, Foraging and Food Preference*. Cambridge, Mass.: Harvard University.

Redford, K. H. and J. F. Eisenberg 1992. *Mammals of the Neotropics. Vol. 2*. Chicago: University of Chicago Press.

Redford, K. H. and J. A. Mansour. 1996. *Traditional Peoples and Biodiversity Conservation in Large Tropical Landscapes*. Arlington, Va.: America Verde.

Redford, K. H. and J. G. Robinson. 1991. Subsistence and commercial uses of wildlife in Latin America. In J. G. Robinson and K. H. Redford, eds., *Neotropical Wildlife Use and Conservation*, pp. 6–23. Chicago: University of Chicago Press.

Redford, K. H. and A. M. Stearman. 1993. Forest-dwelling native Amazonians and the conservation of biodiversity: Interests in common or in collision? *Conservation Biology* 7:248–255.

Redford, K. H., A. Taber, and J. A. Simonetti. 1990. There is more to biodiversity than the tropical rain forests. *Conservation Biology* 4:328–330.

Regier, H. and H. Henderson. 1973. Towards a broad ecological model of fish communities and fisheries. *Transactions of the American Fishery Society* 12:56–72.

Reid, F. A. 1997. *A Field Guide to the Mammals of Central America and Southeast Mexico*. New York: Oxford University Press.

Reis, M. and P. Souza. 2000. *Monitoramento da proteção da RDSM: Relatório preliminar*. Unpublished internal report. Tefé, Brazil: Instituto de Desenvolvimento Sustentável Mamirauá.

Renkonen, O. 1938. Statisch-okologische Untersuchungen uber die terrestiche kaferwelt der finnischen bruchmoore. *Ann. Zool. Soc. Bot. Fenn. Vanamo* 6:1–231.

Ribeiro, M. C. and M. Petrere. 1990. Fisheries ecology and management of the jaraqui (*Semaprochilodus taeniurus* and *Semaprochilodus insiginis*) in Central Amazonia. *Regulated Rivers: Research and Management* 5:195–215.

Ribeiro, M. C., M. Petrere, and A. F. Juras. 1995. Ecological integrity and fisheries ecology on the Araguaia-Tocantins river basin, Brazil. *Regulated Rivers: Research and Management* 10:31–45.

Ricardo C.A. 1996. *Povos indígenas no Brasil: 1991–1995*. Sao Paulo: Insituto Socioambiental.

Richard-Hansen, C. 1998. *La gestion de la faune en Amazonie. Synthèse bibliographique*. Report. Diren/Region Guyane/FEDER: Silvolab.

Richard-Hansen, C. and Hansen, E. 1998. Gestion de la chasse en forêt tropicale Amazonienne. *JATBA, Revue d'Ethnobiologie* 40:541–558.

Richard-Hansen, C., J., C. Vié, and B. de Thoisy. 2000. Translocation of red howler monkeys (*Alouatta seniculus*) in French Guiana. *Biological Conservation* 93:247–253.

Ricker, W. E. 1975. Computation and interpretation of biological statistics of fish populations. *Bulletin of Fishery Researchs Board Canada*. 191:382 .

Ricklefs, R. E. and M. Lau. 1980. Bias and dispersion of overlap indices: Results of some Monte Carlo simulations. *Ecology* 61:1019–1024.

Riester, J. 1984. *Textos sagrados de los Guaraníes en Bolivia: una cacería en el Izozog*. La Paz: Los Amigos del Libro.

Riney, T. 1982. *Study and Management of Large Mammals*. New York: Wiley.

Ríos, M. A., M. J. Dourojeanni, and A. Tovar. 1973. La fauna y su aprovechamiento en Jenaro Herrera (Requena, Perú). *Revista Forestal del Perú* 5:73–92.

Robinson, J. G. 1996. Hunting wildlife in forest patches: An ephemeral resource. In J. Schellas and R. Greenberg, eds., *Forest Patches in Tropical Landscapes*, pp. 111–130. Washington, D.C. Island Press.

Robinson, J. G. 2000. Calculating maximum sustainable harvests and percentage offtakes. In J. Robinson and E. Bennett, eds., *Hunting for Sustainability in Tropical Forests*, pp. 521–524. New York: Columbia University Press.

Robinson, J. G. and E. L. Bennet. 2000a. Hunting for sustainability: The start of a synthesis. In J. G. Robinson and E. L. Bennet, eds., *Hunting for Sustainability in Tropical Forests*, pp. 499–520. New York: Columbia University Press.

Robinson, J. G. and E. L. Bennet, eds. 2000b. *Hunting for Sustainability in Tropical Forests.* New York: Columbia University Press.

Robinson, J. G., and E. L. Bennett. 2000c. Carrying capacity limits to sustainable hunting in tropical forests. In J. G. Robinson and E. L. Bennett, eds., *Hunting for Sustainability in Tropical Forests*, pp. 13–30. New York: Columbia University Press.

Robinson, J. G. and R. Bodmer 1999. Towards wildlife management in tropical forests. *Journal of Wildlife Management* 63:1–13.

Robinson, J. G. and C. H. Janson. 1987. Capuchins, squirrel monkeys, and atelines: Socioecological convergence with old world primates. In B. B. Smuts, D. L. Cheney, S. M. Seyfarth, R. W. Wrangham, and T. T. Struhsaker, eds., *Primate Societies*. Chicago: University of Chicago Press.

Robinson, J. G. and J. C. Ramirez. 1982. Conservation biology of neotropical primates. In M. A. Mares and H. M. Genoways, eds., *Mammalian Biology in South America*. Pymatuning Laboratory of Ecology Special Publication No. 6. Linnesville, Pa.: University of Pittsburg Press.

Robinson, J. G. and K. Redford 1991. Sustainable harvest of Neotropical forest mammals. In J. Robinson and K. Redford, eds., *Neotropical Wildlife Use and Conservation*, pp. 415–429. Chicago: University of Chicago Press.

Robinson, J. G. and K. Redford 1994. Measuring the sustainability of hunting in tropical forests. *Oryx* 28:249–256.

Robinson, J. G. and K. H. Redford. 1986a. Body size, diet, and population density of Neotropical forest mammals. *American Naturalist* 128:665–680.

Robinson, J. G. and K. H. Redford. 1986b. Intrinsic rate of natural increase in Neotropical forest mammals: Relationship to phylogeny and diet. *Oecologia* 68:516–520.

Robinson, J. G. and K. H. Redford. 1989. Body size, diet, and population variation in Neotropical forest mammal species: Predictors of local extinctions? In K. H. Redford and J. F. Eisenberg, eds., *Advances in Neotropical Mammalogy*. Gainesville, Fla.: Sandhill Crane Press.

Robinson, J. G., P. C. Wright, and W. G. Kinzey. 1987. Monogamous cebids and their relatives: Intergroup calls and spacing. In B. B. Smuts, D. L. Cheney, S. M. Seyfarth, R. W. Wrangham, and T. T. Struhsaker, eds., *Primate Societies*. Chicago: University of Chicago Press.

Robinson, S. K. 1985. Coloniality in the yellow-rumped cacique as a defence against nest predators. *Auk* 102:506–519.

Rodríguez, F., M. Rodríguez, and P. G. Vásquez. 1995. *Realidad y perspectivas: La Reserva Nacional Pacaya-Samiria*. Lima: Proyecto Pacaya-Samiria. ProNaturaleza/TNC/USAID.

Rodríguez, J. P. and F. Rojas. 1995. *Libro rojo de la fauna Venezolana*. Caracas: PROVITA, Fundación Polar.

Rojas, C. R. 2001. *Reproducción de Dasypus novemcinctus en el Izozog, Santa Cruz, Bolivia*. Undergraduate thesis. Santa Cruz, Bolivia: UAGRM.

Roosevelt, A. C. 1987. Chiefdoms in the Amazon and Orinoco. In R. D. Drennan and C. A. Uribe, eds., *Chiefdoms in the Americas*, pp. 153–184 Lanham, Md.: University Press of America.

Roosevelt, A. C. 1991. *Moundbuilders of the Amazon: Geophysical archaeology on Marajó Island, Brazil*. San Diego: Academic Press.

Roosevelt, A. C. 1999. Twelve thousand years of human-environment interaction in the Amazon Floodplain. In C. Padoch, J. M. Ayres, M. Pinedo-Vasquez, and A. Henderson, eds., *Várzea: Diversity, Conservation, and Development of Amazonia's Whitewater Floodplains*, pp. 371–392. New York: New York Botanical Garden Press.

Roosevelt, A. C., M. L. Costa, L. Lopes, M. Michab, N. Mercier, H. Valladas, J. Feathers, W. Barnett, M. I. da Silveira, A. Henderson, J. Silva, B. Chernoff, D. S. Resse, J. A. Holman, A. N. Toth, and K. Schick. 1996. Paleoindian cave dwellers in the Amazon: The peopling of the Americas. *Science* 272:373–384.

Roosevelt, A. C., R. A. Housley, M. I. da Silveira, S. Maranoca, and R. Johnson. 1991. Eighth millenium pottery from a prehistoric shell midden in the Brazilian Amazon. *Science* 254:1621–1624.

Ross, E. B. 1978. Food taboos, diet, and hunting strategy: The adaptation to animals in Amazon cultural ecology. *Current Anthropology* 19:1–36.

Ross, J. P., ed. 1998. *Crocodiles. Status Survey and Conservation Action Plan.* 2d ed. IUCN/SSC Crocodile Specialist Group. Gland, Switzerland: IUCN.

Routledge, R. D. and D. A. Fyfe. 1992. Confidence limits for line transect estimates based on shape restrictions. *Journal of Wildlife Management* 56:402–407.

Rubio-Torgler, H. 1992. *Evaluación de demanda de fauna con comunidades indígenas Embera en el P.N.N. Utría (Chocó, Colombia).* Report. Bogotá: Fundación Natura/Wildlife Conservation Society/Conservation Food and Health.

Rubio-Torgler, H. 1997. Estrategias para el manejo de especies de caza en el área de influenciea del parque nacional natural Utria. In T. G. Fang, R. E. Bodmer, R. Aquino, and M. H. Valqui, eds., *Manejo de fauna silvestre en la Amazonía*, pp. 135–144. La Paz: Editorial Instituto de Ecología.

Rubio-Torgler, H., A. Ulloa, and C. Campos-Rozo. 2000. *Manejo de fauna: Una construcción con lo local: Métodos y herramientas.* Bogotá: Fundación Natura, Ministerio del Medio Ambiente, Organización de Estados Iberoamericanos, OREWA, Instituto Colombiano de Antropología.

Rubio-Torgler, H., A. Ulloa; M, Rubio, and indigenous Embera communities. 1998. *Tras las huellas de los animales.* Bogotá: Fundación Natura, OREWA, MMA, Instituto Colombiano de Antropología (ICAN).

Ruffino, M. L. 1999. Fisheries development in the lower Amazon river. In C. Padoch, J. M. Ayres, M. Pinedo-Vasquez, and A. Henderson, eds., *Várzea: Diversity, Conservation, and Development of Amazonia's Whitewater Floodplains*, pp. 101–111. New York: New York Botanical Garden Press.

Ruffino, M. L. and V. J. Isaac. 1994. The fisheries of the lower Amazon: Questions of management and development. *Acta.Biol. Venez.* 15:37–46.

Ruffino, M. L. and V. J. Isaac. 2000. A pesca artesanal no Médio Solimões. In *Recursos pesqueiros do médio Soliões: Biologia e estatística pesqueira.* Série Estudos de Pesca No. 22. Brasília: IBAMA.

Ruffino, M. L., V. J. Isaac, and A. Milstein. 1998. Fisheries ecology in the lower Amazon: A typical artisanal practice in the tropics. *Ecotropica* 4:99–114.

Ruffino, M. L., S. Q. da Silva, and A. C. Castro. 1998. *Relatório: Desembarque pesqueiro em Santarém: Informe Estatístico (1994–1997).* Unpublished internal report. Santarém, Brazil: IBAMA Projeto Iara/Gopa/GTZ.

Rylands, A. B. and A. Keuroghlian. 1988. Primate populations in continuous forest and forest fragments in Central Amazonia. *Acta Amazonica* 18:291–307.

Rylands, A. B., and D. S. Faria. 1993. Habitats, feeding ecology, and home range size in the genus *Callithrix.* In A. B. Rylands, ed., *Marmosets and Tamarins: Systematics, Behaviour, and Ecology.* Oxford: Oxford University Press.

Saavedra, A. M. 2000. *Uso de aves por comunidades indígenas del Izozog.* Undergraduate thesis. Santa Cruz, Bolivia: UAGRM.

Sahley, C.T. 2000. Poblaciones de vicuñas en vías de recuperación: Un análisis de alternativas para su manejo. In B. Gonzalez, F. Bas, C. Tala, and A. Iriarte, eds., *Manejo sustentable de la vicuña y el guanaco*, pp.103–108. Santiago, Chile: Pontificia Universidad Católica de Chile.

Saint-Paul, U., J. Zuanon, M. A. V. Correa, M. Garcia, N. N. Fabre, U. Berger, and W. J. Junk. 2000. Fish communities in central Amazonian white- and blackwater floodplains. *Environmental Biology of Fishes* 57:235–250.

Sall, J. and A. Lehman. 1996. *JMP Start Statistics: A Guide to Statistical and Data Analysis Using JMP and JMP in Software.* Belmont, Calif.: Duxbury Press.

Salvatierra C., A., E. Barba P., J. D. Salvatierra M., and W. R. Townsend. 2001. *Manejo de lagarto: Recuento de Caiman yacare en la TCO Itonama.* Publicaciones Proyecto de Investigación CIDOB-DFID No. 18. Santa Cruz, Bolivia: CIDOB.

Sanchez, E. 1984. Sobrepoblación y necesidad de extracción de vicuñas en Pampa Galeras. In F. Villiger, ed., *La vicuña.* Lima: Editorial Los Pinos.

Sanderson, E. W., K. H. Redford, C. B. Chetkiewicz, R. A. Medellín, A. R. Rabinowitz, J. G. Robinson, and A. B. Taber. 2002. Planning to save a species: The jaguar as a model. *Conservation Biology* 16:58–72.

Santos, P. 1996. *Uso e plano de gestão da fauna silvestre numa área de Várzea Amazônica: A Estação Ecológica Mamirauá (Amazonas, Brasil).* M.S. thesis. Lisbon: University of Lisbon Faculty of Sciences.

Santos, R. V., N. M. Flowers, C. E. A. Coimbra, Jr., and S. A. Gugelmin. 1997. Tapirs, tractors, and tapes: The changing economy and ecology of the Xavante Indians of central Brazil. *Human Ecology* 25:545–566.

Schaller, G. B. and P. Crawshaw. 1981: Social organization in a capybara population. *Säugetierk Mitt* 29:3–16.

Schemitz E.D. 1980. *Wildlife Management Techniques Manual*. 4th ed. Bethesda, Md.: Wildlife Society.

Schmidt, G. W. 1973a. Primary production of phytoplankton in the three types of Amazonian waters. II. The limnology of a tropical floodplain lake in Central Amazonia (Lago do Castanho). *Amazoniana* 4:139–203.

Schmidt, G. W. 1973b. Primary production of phytoplankton in the three types of Amazonian waters. III. Primary productivity of phytoplankton in a tropical floodplain lake of Central Amazonia, Lago do Castanho, Amazonas, Brazil. *Amazoniana* 4:379–404.

Schmidt, G. W. 1976. Primary production of phytoplankton in the three types of Amazonian waters. IV. On the primary productivity of phytoplankton in a bay of the lower Rio Negro (Amazonas, Brazil). *Amazoniana* 5:517–528.

Schroder, G. D. and M. L. Rosenzweig. 1975. Perturbation analysis of competition and overlap in habitat utilization between *Dipodomys ordii* and *Dypodomys merriami*. *Oecologia* 19:9–28.

Schwartz, C. 2000. Até onde a Amazônia pode resistir. *Revista Veja* 33(47):66–72.

SCM. 1996. *Mamirauá Management Plan*. Brasilia: MCT-CNPq/ Sociedade Civil Mamirauá.

Seijas, A. E. 1998. *The Orinoco Crocodile* (Crocodylus intermedius) *in the Cojedes River System, Venezuela: Population Status and Ecological Characteristics*. Ph.D. dissertation. Gainesville: University of Florida.

Seijas, A. E. and C. Chávez. 2000. Population status of the Orinoco crocodile (*Crocodylus intermedius*) in the Cojedes River System, Venezuela. *Biological Conservation* 94:353–361.

Setzer, J., 1949. *Os solos do Estado de São Paulo*. 1ˢᵗ ed. Rio de Janeiro: Conselho Nacional de Geografia.

Shaw, J. H. 1991. The outlook for sustainable harvests in Latin America. In J. G. Robinson and K. H. Redford, eds., *Neotropical Wildlife use and Conservation*, pp. 24–34. Chicago: University of Chicago Press.

Shields, W. M. 1987. Dispersal and mating systems: Investigating their causal connections. In B. D. Chepko-Sade and Z. T. Halpin, eds., *Mammalian Dispersal Patterns*, pp. 3–24. Chicago: University of Chicago Press.

Shively, C. L.; Whiting, F. M.; Swingle, R. S.; Brown, W. H., and Solws, L. K. 1984. Some aspects of the nutritional biology of the CP. *Journal of Wildlife Management*, 49:729–732.

Shrimpton, R. and R. Giugliano. 1979. Consumo de alimentos e alguns nutrientes em Manaus, Amazonas 1973–1974. *Acta Amazonica* 9:117–141.

Silva, J. L. and S. D. Strahl. 1991. Human impact on populations of chachalacas, guans, and curassows (Galliformes: Cracidae) in Venezuela. In J. G. Robinson and K. H. Redford, eds., *Neotropical Wildlife Use and Conservation*, pp. 37–52. Chicago: University of Chicago Press.

Silva-Neto, P. B. 1989. *Alimentação e manejo de capivaras (Hydrochoerus hydrochaeris hydrochaeris L. 1766) em cativeiro*. M.S. thesis. São Paulo: Escola Superior de Agricultura "Luiz de Queiroz."

Slade, N. A., R. Gomulkiewicz, and H. M. Alexander. 1998. Alternatives to Robinson and Redford's method of assessing overharvest from incomplete demographic data. *Conservation Biology* 12:148–155.

Slough, B. G. and G. Mowat. 1996. Lynx population dynamics in an untrapped refugium. *Journal of Wildlife Management* 60:946–961.

Smith, J. N. M., P. R. Grant, B. R. Grant, I. Abbott, and L. K. Abbott. 1978. Seasonal variation in feeding habits of Darwin's finches. *Ecology* 59:1137–1150.

Smith, N. J. H. 1979a. Aquatic turtles of Amazonia: An endangered resource. *Biological Conservation* 16:165–176.

Smith, N. J. H. 1979b. *A pesca no Rio Amazonas*. Manaus, Brazil: INPA.

Smith, N. J. H. 1980. Antrosols and the human carrying capacity in Amazonia. *Annals of the American Association of Geographers* 70:533–566.

Smith, N. J. H. 1981. *Man, Fishes and the Amazon*. New York: Columbia University Press.

Smith, N. J. H. 1996 *The Enchanted Amazon Rain forest*. Gainesville: University Press of Florida.

Smith, N. J. H. 1999. *The Amazon River Forest: A Natural History of Plants, Animals, and People*. New York: Oxford University Press.

Snyder, N. F. R., F. C. James, and S. R. Beissinger. 1992. Toward a conservation strategy for neotropical psittacines. In S. R. Beissinger and N. F. R. Snyder, eds., *New World Parrots in Crisis: Solutions from Conservation Biology*, pp. 257–276. Washington, D.C.: Smithsonian Institution Press.

Sociedad Nacional de la Vicuña, 2000. Unpublished conference proceedings.

Soini, P. 1995. Características climáticas: Un resumen de cinco años de registros de la temperatura, pluviosidad y fluviometría en Cahuana, Río Pacaya. Informe de Pacaya No. 1. In P. Soini, A. Tovar, and U. Valdez, eds., *Reporte Pacaya-Samiria: Investigaciones en Cahuana: 1980–1994*, pp. 207–209. Lima: Centro de Datos para la Conservación, Universidad Nacional Agraria La Molina.

Soini, P. 1997. *Biología y manejo de la tortuga* Podocnemis expansa *(Testudines, Pelomedusidae)*. Caracas: Tratado de Cooperación Amazónica, SPT-TCA.

Soini, P. and M. Coppula. 1995. Estudio, reproducción y manejo de los quelonios del género *Podocnemis* (charapa, cupiso y taricaya) en la cuenca del Pacaya, río Pacaya, Loreto-Perú. Reporte No. 2. In Soini, P., A. Tovar, and U. Valdez, eds., *Reporte Pacaya-Samiria: Investigaciones en Cahuana: 1980–1994*, pp. 3–30. Lima: Centro de Datos para la Conservación, Universidad Nacional Agraria La Molina.

Soini, P., L. A. Sicchar, G. Gil, A. Fachín, R. Pezo, and M. Chumbe. 1996. *Una evaluación de la fauna silvestre y su aprovechamiento de la Reserva Nacional Pacaya-Samiria, Perú*. Documento Técnico no. 24. Iquitos, Perú: Instituto de Investigaciones de la Amazonia Peruana.

Soini, P. and M. Soini. 1995. Un resumen comparativo de la ecología reproductiva de los quelonios acuáticos. Informe No. 19. In P. Soini, A. Tovar, and U. Valdez, eds., *Reporte Pacaya-Samiria: Investigaciones en Cahuana: 1980–1994*, pp. 215–226. Lima: Centro de Datos para la Conservación, Universidad Nacional Agraria La Molina.

Sokal, R. R. and F. J. Rohlf. 1969. *Biometry*. San Francisco: Freeman Press.

Sokal, R. R. and F. J. Rohlf. 1995. *Biometry*. 3^d ed. New York: W.H. Freeman.

Soria, F. 2003. Radiotelemetría de la tortuga negra *Chelonoidis carbonaria* en el Chaco seco, Santa Cruz, Bolivia. In R. Polanco-Ochoa, ed., *Manejo de Fauna silvestre en Amazonía y Latinoamérica: Selección de trabajos V Congreso Internacional*, pp. 194–196. Bogotá: Fundación Natura.

Soria, F., F. Mendoza, and J. Ayala. 2001. Radio-telemetría de petas (*Chelonoidis carbonaria*) y peni (*Tupinambis rufescens*) en el campamento Cerro Cortado, Izozog, Santa Cruz, Bolivia. In W. R. Townsend, K. Rivero, C. Peña, and K. Linzer, eds., *Memorias del primer encuentro nacional de manejo de fauna en los Territorios Indígenas de Bolivia*, pp. 47–50. Santa Cruz, Bolivia: CIDOB.

SOS Mata Atlântica and INPE. 1993. *Evolução dos remanescentes Florestais e ecossistemas associados do domínio da Mata Atlântica*. São Paulo: SOS Mata Atlântica and Instituto de Pesquisas Espaciais.

Soulé, M. E., 1987. *Viable Population for Conservation*. Cambridge: Cambridge University Press.

Soulé, M. E. and B. A. Wilcox. 1980. *Conservation Biology: An Evolutionary-Ecological Perspective*. Sunderland, Mass.: Sinauer.

Souza-Mazurek, R., T. Pedrinho, X. Feliciano, W. Hilario, S. Geroncio, and E. Marcelo. 2000. Subsistence hunting among the Waimiri-Atroari Indians in central Amazonia, Brazil. *Biodiversity and Conservation* 9:579–596.

Sowls, L. K. 1984. *The Peccaries*. Tucson: University of Arizona Press.

Sowls, L. K. 1997. *Javelinas and Other Peccaries: Their Biology, Management and Use*. College Station, Tex.: Texas A and M University Press.

Sparks, D. R. and J. C. Malechek. 1968. Estimating percentage dry weight in diets using a microscope technique. *J. Range. Manage.* 21:264–265.

Spix, J. B. von and C. F. B. von Martius. 1822–1831. *Reise in Brasilien auf Befehl Sr. Majestat Maximillian Joseph I. Konigs von Baiern in den Jahren 1817 bis 1820*. Munich: Lindauer.

Srikosamatara, S., B. Siripholdej, and V. Suteethorn. 1992. Wildlife trade in Lao PDR and between Lao PDR and Thailand. *Natural History Bulletin of the Siam Society* 40:1–47.

Stearman, A. M. 1990. The effects of settler incursion on fish and game resources of the Yuquí, a native Amazonian society of eastern Bolivia. *Human Organization* 49:371–385.

Stearman, A. M. 1995. Neotropical foraging adaptations and the effects of acculturation on sustainable resource use: The Yuquí of lowland Bolivia. In L. E. Sponsel, ed., *Indigenous Peoples and the Future of Amazonia*, pp. 207–224. Tucson: University of Arizona Press.

Stearman, A. M. 1999. Cambio social, cacería y conservación en pueblos indígenas: puntos de conflicto y caminos hacia la resolución. In T.G. Fang, O. L. Montenegro, and R. E. Bodmer, eds., *Manejo y conservación de fauna silvestre en América Latina*, pp. 41–50. La Paz: Instituto de Ecología.

Stearman, A. M. 2000. A pound of flesh: Social change and modernization as factors in hunting sustainability among neotropical indigenous societies. In J. G. Robinson and E. L. Bennet, eds., *Hunting for Sustainability in Tropical Forests*, pp. 233–250. New York: Columbia University Press.

Stearman, A. M. and K. H. Redford. 1992. Commercial hunting by subsistence hunters: Sirionó Indians and Paraguayan caiman in lowland Bolivia. *Human Organization* 51:235–244.

Stehli, F. G. and S. D. Webb. 1985. *The Great American Biotic Interchange*. New York: Plenum Press.

Sternberg, H. O. R. 1995. Waters and wetlands of Brazilian Amazonia: An uncertain future. In T. Nishizawa and J. I. Uitto, eds., *The Fragile Tropics of Latin America: Sustainable Management of Changing Environments*. Tokyo: United Nations University Press.

Steward, J. H. and L. C. Faron. 1959. *Native Peoples of South America*. New York: McGraw-Hill.

Strahl, S. D. and J. L. Silva. 1997. Census methods for cracid populations. In S. D. Strahl, S. Beaujon, D. M. Brooks, A. J. Begazo, G. Sedaghatkish, and F. Olmos, eds., *The Cracidae: Their Biology and Conservation*, pp. 26–33. Blaine, Wash.: Hancock House.

Strauss, R. E. 1979. Reliability estimates for Ivlev's selectivity index, the forage ratio, and proposed linear index of food selection. *Trans. Am. Fish. Soc.* 108:344–352.

Strey, O. F. and Brown, R. D. 1989. In vivo and in vitro digestibilities for collared peccaries. *Journal of Wildlife Management* 53:607–612.

Stüwe, M. 1985. Manual for the Micro-Computer Program for Animal Locations: MCPALL. Washington, D.C.: National Zoo.

Sunquist, M. E. 1992. The ecology of the ocelot: The importance of incorporating life history traits into conservation plans. In *Felinos de Venezuela: Biologia, ecologia y conservacion*. Caracas: Fudeci.

Sunquist, M. E., F. Sunquist, and D. E. Daneke. 1989. Ecological separation in a Venezuelan llanos carnivore community. In K. H. Redford and J. F. Eisenberg, eds., *Advances in Neotropical Mammalogy*, pp. 197–232. Gainesville, Fla.: Sandhill Crane Press.

Swank, W. and J. Teer 1989. Status of the jaguar—1987. *Oryx* 23:14–21.

Swanson, T. M. 1992. The role of wildlife utilization and other policies for diversity conservation. In T. M. Swanson and E. B. Barbier, eds., *Economics for the Wilds: Wildlife, Diversity, and Development*, pp. 65–102. Washington, D.C.: Island Press.

Swanson, T. M. and E. B. Barbier, eds. 1992. *Economics for the Wilds: Wildlife, Diversity, and Development*. Washington, D.C.: Island Press.

Sweitzer, R A, B. J. Gonzalez, I. A. Gardner, D. Van Vuren, J. D. Waithman, and W. M. Boyce. 1997. A modified panel trap and immobilization technique for capturing multiple wild pig. *Wildlife Society Bulletin* 25:699–705.

Swihart, R. K. and N. A. Slade. 1985. Testing for independence of observations in animal movements. *Ecology* 66:1176–1184.

Taber, A. B. 1991. The status and conservation of the Chacoan peccary in Paraguay. *Oryx* 25:147–155.

Taber, A. B., C. P. Doncaster, N. N. Neris, and F. H. Colman. 1993. Ranging behavior and population dynamics of the Chacoan peccary, *Catagonus wagneri*. *Journal of Mammalogy* 74:443–454.

Taber, A. B., C. P. Doncaster, N. N. Neris, and F. H. Colman. 1994. Ranging behaviour and activity patterns of two sympatric peccaries, *Catagonus wagneri* and *Tayassu tajacu*, in the Paraguayan Chaco. *Mammalia* 58:61–71.

Taber, A. B., A. J. Novaro, N. N. Neris, and F. H. Colman. 1997. The food habits of sympatric jaguar and puma in the Paraguayan Chaco. *Biotropica* 29:204–213.

Taber, A. B., G. Navarro, and M. A. Arribas. 1997. A new park in the Bolivian Gran Chaco: An advance in tropical dry forest conservation and community-based management. *Oryx* 31:189–198.

Taber, A. B. and W. L. R. Oliver. 1993. Review of priorities for the conservation action and future research on Neotropical peccaries. In W. L. R. Oliver, ed., *Peccaries and Hippos: Status, Survey, and Conservation Action Plan*, pp. 37–40. Gland, Switzerland: IUCN.

Tablante-Garrido, P. N. 1961[1831]. *Provincia de Apure. Monografía del Gobernador General José Cornelio Muñoz*. Cuadernos Geográficos No. 1. Merida, Venezuela: Instituto de Geografía y de Conservación de Recursos Naturales, Universidad de Los Andes.

Tello, S. 1993. Unpublished Trip Report. Iquitos, Peru: IIAP.

Terborgh, J. 1983. *Five New World Primates: A Study in Comparative Ecology*. Princeton, N.J.: Princeton University Press.

Terborgh, J. 1988. The big things that run the world: A sequel to E. O. Wilson. *Conservation Biology* 2:402–403.

Terborgh, J., L. H. Emmons, and C. Freese. 1986. La fauna silvestre de la Amazonía: El despilfarro de un recurso renovable. *Boletín de Lima* 46:77–85.

Thomas, B. T. 1987. Philopatry of banded Maguari Storks and their decline in Venezuela. *Boletín de la Sociedad Venezolana de Ciencias Naturales* 41:137–157.

Thomas, L., J. L. Laake, J. F. Derry, S. T. Buckland, D. L. Borchers, D. R. Anderson, K. P. Burnham, S. Strindberg, S. L. Hedley, M. L. Burt, F. Marques, J. H. Pollard, and R. M. Fewster. 1998. *DISTANCE 3.5: Research Unit for Wildlife Population Assessment.* St. Andrews, Fife, Scotland: University of Saint Andrews.

Thomsen, J. B. and A. Brautigam. 1991. Sustainable use of neotropical parrots. In *Neotropical Wildlife Use and Conservation*, pp. 359–379. J. G. Robinson and K. H. Redford, eds., Chicago: University of Chicago Press.

Thomsen, J. B. and T. A. Mulliken. 1992. Trade in Neotropical psittacines and its conservation implications. In S. R. Beissinger and N. F. R. Snyder, eds., *New World Parrots in Crisis: Solutions from Conservation Biology*, pp. 221–239. Washington, D.C.: Smithsonian Institution Press.

Thorbjarnarson, J., comp. 1992. *Crocodiles: An Action Plan for their Conservation.* H. Messel, F. W. King. and J. P. Ross, eds. Gland, Switzerland: IUCN/SSC Crocodile Specialist Group.

Thorbjarnarson, J. and G. Hernández. 1992. Recent investigation on the status and distribution of the Orinoco crocodile, *Crocodylus intermedius*, in Venezuela. *Biological Conservation* 62:179–188.

Thorbjarnarson, J. B., N. Perez, and T. Escalona. 1993. Nesting of *Podocnemis unifilis*. *Journal of Herpetology* 27:344–347.

Thorbjarnarson, J. B., N. Perez, and T. Escalona. 1997. Biology and conservation of aquatic turtles in the Cinarucu-Capanaparo National Park, Venezuela. In *Proceedings of the International Conference on Conservation, Restoration, and Management of Tortoises and Turtles*, pp. 109–112. New York: New York Turtle and Tortoise Society.

Tilman, D. and P. Kareiva. 1997. *Spatial Ecology: The Role of Space in Population Dynamics and Interspecific Interactions.* Princeton, N.J.: Princeton University Press.

Torres, J. 1998. *Estudio comparativo de las principales características de la fibra de la alpaca y del cruce inter-especifico (Alpaca y Llama) en el anexo de Sumbay, Arequipa, Perú.* Undergraduate thesis. Arequipa, Perú: Universidad Nacional de San Agustín.

Townsend, W. R. 1995a. *Living on the Edge: Sirionó Hunting and Fishing in Lowland Bolivia.* Ph.D. dissertation. Gainesville: University of Florida.

Townsend, W. R. 1995b. Cultural teachings as an ecological database: Murui (Witoto) knowledge about primates. *Latinamericanist*, 31(1) 1–7.

Townsend, W. R. 1996a. *Participatory Wildlife Management Workshop in Lomerio.* Tech Doc. 45/1966. Santa Cruz, Bolivia: BOLFOR.

Townsend, W. R. 1996b. *Nyao Itõ: Caza y pesca de los Sirionó.* La Paz: Instituto de Ecología. Universidad Mayor de San Andrés.

Townsend, W. R. 1996c. La utilidad del monitoreo del uso de la cacería para la defensa del territorio. In C. Campos, A. Ulloa, and H. Rubio, eds., *Manejo de Fauna con Comunidades Rurales*, pp. 177–189. Bogotá: Organización Regional Indígena Embera Wounaan (Orewa), Fundación Natura.

Townsend, W. R. 1997. La participación comunal en el manejo de la vida silvestre en el oriente de Bolivia. In T.G. Fang, R.E. Bodmer, R. Aquino, and M. H. Valqui, eds., *Manejo de fauna silvestre en la Amazonia*, pp 105–109. La Paz: Instituto de Ecologiá.

Townsend, W. R. 2000a. *El Monitoreo de la cacería (registro): una herramienta para el manejo de la fauna silvestre.* Santa Cruz, Bolivia: CIDOB-DFID.

Townsend, W. R. 2000b. The sustainability of subsistence hunting by the Sirionó Indians of Bolivia. In J. Robinson and E. Bennett, eds., *Hunting for Sustainability in Tropical Forests*, pp. 267–281. New York: Columbia University Press.

Townsend, W. R. 2002. Investigación participativa en el manejo comunitario de la fauna silvestre: Experiencia de la Reserva de la Biosfera-Tierra Comunitaria de Origen Pilón Lajas. In L. O. Rodríguez, ed., *El manu y otras experiencias de investigación y manejo de Bosques Neotropicales*, pp 206–212. Cusco, Peru: Pro Manu.

Townsend, W. R., R. Nuñez, and V. Macuritofe. 1984. Contribuciones a la etnozoologia de la Amazonia colombiana: El conocimiento zoológico entre los Huitotos. *Colombia Amazónica* 1:37–74.

Townsend, W. R., K. Rivero, C. Peña, and K. Linzer. 2001. *Memorias del primer encuentro nacional de manejo de fauna en Territorios Indígenas de Bolivia.* Publicaciones Proyecto de Investigación CIDOB-DFID No. 25. Santa Cruz, Bolivia: CIDOB.

Tremblay, J. and L. N. Ellison. 1979. Effects of human disturbance on breeding of Black-crowned Night Herons. *Auk* 96:364–369.

Tronc, E. and S. Vuillemin. 1973. Contribution a l'etude ostéologique de *Erymnochelys madagascariensis* Grandidier, 1867. *Bull. Acad. Malg.* 51:189–224.

Turner, J. L. 1985. Changes in multispecies fisheries when many species are caught at the same time. *FAO Fisheries Report.* 338:201–211.

Tyburn, J. J. 1994. *De la chasse et de la consommation de gibier sur la bande côtière guyanaise.* DESS report. Montpellier, France: CIRAD/EMVT.

Ulloa, A., H. Rubio-Torgler, and C. Campos-Rozo. 1996. *Trua wuandra: Estrategias para el manejo de fauna de caza con comunidades indígenas Embera en el Parque Nacional Natural Utría, Chocó, Colombia.* Bogotá: Fundación Natura, Ministerio del Medio Ambiente, OEI, OREWA.

Vaillant, L. and G. Grandidier. 1910. Histoire naturelle des reptiles, 1er partie: Crocodiles et tortues. In A. and G. Grandidier, eds., *Histoire physique, naturelle et politique de Madagascar*, Vol. 17, pp. 26–86. Paris, Hachette.

Val, A. L. and V. M. F. Almeida-Val. 1995. *Fishes of the Amazon and Their Environment. Physiological and Biochemical Aspects.* Berlin: Springer-Verlag.

Valladares-Padua, C. 1987. *Black Lion Tamarin* Leontopithecus chrysopygus: *Status and Conservation.* M.S. thesis. Gainesville: University of Florida.

Valladares-Padua, C. 1993. *The Ecology, Behavior, and Conservation of the Black Lion Tamarins, Leontopithecus chrysopygus*, MIKAN, 1823. Ph.D. dissertation. Gainesville: University of Florida.

Valladares-Padua, C. and R. E. Bodmer, eds., 1997. *Manejo e conservação de vida silvestre no Brasil*, pp. 52–69. Brasilia: CNPq and Sociedade Civil Mamirauá.

van Horne, B. 1981. Demography of *Peromyscus maniculatus* populations in serial stages of coastal coniferous forest in southeast Alaska. *Canadian Journal of Zoology* 59:1045–1061.

van Roosmallen, M. G. M. 1980. *Habitat Preferences, Diet, Feeding Strategy, and Social Organization of the Black Spider Monkey* (Ateles p. paniscus Linnaeus 1758) *in Surinam.* Arnhem, Netherlands: Rijksinstituut voor Natuurbeheer.

van Roosmalen, M. G. M. and L. L. Klein. 1988. The spider monkeys, genus *Ateles.* In R. A. Mittermeier, A. B. Rylands, A. Coimbra-Filho, and G. A. B. Fonseca, eds., *Ecology and Behavior of Neotropical Primates.*, Vol. 2, pp. 455–537. Washington, D.C.: World Wildlife Fund.

Van Soest, P. J. 1982. *Nutritional Ecology of the Ruminant.* Corvallis, Ore.: O and B Books.

Vanzolini, P. E. 1977. A brief biometrical note on the reproductive biology of some South American *Podocnemis* (Testudines, Pelomedusidae). *Papeis Avulsos de Zoologia* 31:79–102.

Vásquez, M. A. and M. A. Ramos. 1992. *Reserva de la biósfera montes Azules, selva Lacandona: Investigación para su conservación.* Ecosfera No.1. San Cristóbal de Las Casas, Chiapas, México: Publ. Ocas.

Ventocilla, J. 1992. *Caceria y subsistencia en Cangandi (una comunidad de los indígenas Kunas).* Hombre y Ambiente. Vol. 23. Quito: Ediciones ABYA-YALA.

Vickers, W. 1991. Hunting yields and game composition over ten years in an Amazonian village. In J. G. Robinson and K. H. Redford, eds., *Neotropical Wildlife Use and Conservation*, pp. 53–81. Chicago: University of Chicago Press.

Vickers, W. T. 1994. From opportunism to nascent conservation: The case of the Siona-Secoya. *Human Nature* 5:307–337.

Vieira, A. 1925–1928. Carta ao Rei Afonso VI. Maranhão 28 November 1659. In J. L. Azevedo, ed., *Cartas do Padre Antonio Vieira*, Vol. 1, pp. 549–571. Coimbra, Portugal: Imprensa da Universidade de Coimbra.

Villarroel, R. 2000. Parásitos de contenidos estomacales de *Tupinambis rufescens.* In E. Cabrera, C. Mercolli, and R. Resquin, eds., *Manejo de fauna silvestre en Amazonía y Latinoamérica*, pp. 355–360. Asunción, Paraguay: CITES Paraguay, Fundación Moises Bertoni, University of Florida.

Vogt, R. C. and P. Soini. In Press. *Podocnemis unifilis* Troschel, 1848: Tracajá, Terecay, Yellow-spotted Amazon River Turtle. In A. Rhodin and P. Pritchard, eds., *Conservation Biology of Freshwater Turtles*, Vol. II. Gland, Switzerland: IUCN/SSC.

Wallace, A. R. 1853. *A Narrative of Travels on the Amazon and Rio Negro, with an Account of the Native Tribes, and Observations on the Climate, Geology, and Natural History of the Amazon Valley.* London: Reeve.

Walters, C. J. 1986. *Adaptive Management of Renewable Resources.* New York: Macmillan.

Watkinson, A. R. and W. J. Sutherland. 1995. Sources, sinks, and pseudosinks. J. *Animal Ecology* 64:126–130.

WCI. 1992. *The Wild Bird Trade: When a Bird in the Hand Means None in the Bush*. Wildlife Conservation International Policy Report No. 2. New York: New York Zoological Society.

Weber, W. and A. Rabinowitz. 1996. A global perspective on large carnivore conservation. *Conservation Biology* 10:1046–1054.

Welcomme, R. L. 1979. *Fisheries Ecology of Floodplain Rivers*. New York: Longman.

Welcomme, R. L. 1985. River fisheries. *FAO Fisheries Technical Paper* 262:327.

Western, D., R. M. Wright, and S. Strum. 1994. *Natural Connections: Perspectives in Community-Based Conservation*. Washington, D.C.: Island Press.

Wheeler, J., M. Fernandez, R. Rosadio, D. Hoces, M. Kadwell, and M.W. Bruford. 2000. *Diversidad genética y manejo de poblaciones de vicuñas en el Perú: Final Grant Report*. British Council, U.K.: Darwin Initiative on Species Survival.

Wheeler, J. and D. Hoces. 1997. Community participation, sustainable use, and vicuña conservation in Peru. *Journal of Mountain Research and Development*. 17:283–287.

Wiens, J. A. 1989. Spatial scaling in ecology. *Functional Ecology* 3:385–397.

Wilbur, H. M., P. J. Morin, and R. N. Harris. 1983. Salamander predation and the structure of experimental communities: Anuran responses. *Ecology* 64:1423–1429.

Wilson, E. O. 1988. *Biodiversity*. Washington, D.C.: National Academy Press.

Wilson, E. O. 1988. The current state of biological diversity. In E. O. Wilson, ed., *Biodiversity*. Washington, D.C., National Academy Press.

Wray, S. 1994. Competition between muntjac and other herbivores in a commercial coniferous forest. *Deer* 9:237–242.

Wright, T. F., C. A. Toft, E. E. Enkerlin-Hoeflich, J. González-Elizondo, M. Albornoz, A. Rodríguez-Ferraro, F. Rojas-Suárez, V. Sanz, A. Trujillo, S. R. Beissinger, V. Berovides, X. Gálvez, A. T. Brice, K. Joyner, J. Eberhard, J. Gilardi, S. E. Koenig, S. Stoleson, P. Martuscelli, J. M. Meyers, K. Renton, A. M. Rodríguez, A. C. Sosa-Asanza, F. Vilella, and J. F. Wiley. 2001. Nest poaching in Neotropical parrots. *Conservation Biology* 15:710–720.

Wright, V. L. 1978. Causes and effects of biases on waterfowl harvest estimates. *Journal of Wldlife Management* 42:251–262.

Yost, J. A. and P. Kelley. 1983. Shotguns, blowguns, and spears: The analysis of technological efficiency. In R. B. Hames and W. T. Vickers, eds., *Adaptive Responses of Native Amazonians*, pp. 198–224. New York: Academic Press.

Young, T. P. 1994. Natural die-offs of large mammals: Implications for conservation. *Cons. Biol.* 8:410–418.

Zent, S. 1997. Piaroa and the Cracidae: Game management under shifting cultivation. In S. D. Strahl, S. Beaujon, D. M. Brooks, A. J. Begazo, G. Sedaghatkish, and F. Olmos, eds., *The Cracidae: Their Biology and Conservation*, pp. 177–194. Washington, D.C.: Hancock House.

Zúñiga, M. 1997. Módulo de uso sustentable de la vicuña. Talk presented at the 5th Extraordinary reunion on the international convention on the vicuña. La Paz.

Index